AF441710

Springer Series in
OPTICAL SCIENCES

Herbert Venghaus (Ed.)

Wavelength Filters in Fibre Optics

With 210 Figures

 Springer

Dr. Herbert Venghaus
Fraunhofer Institute for Telecommunications
Heinrich-Hertz-Institut
Einsteinufer 37
10587 Berlin
Germany
E-mail: venghaus@hhi.fraunhofer.de

ISSN 0342-4111

ISBN-10 3-540-31769-4 Springer Berlin Heidelberg New York

ISBN-13 978-3-540-31769-2 Springer Berlin Heidelberg New York

Library of Congress Control Number: 2006923694

Springer is a part of Springer Science+Business Media

springer.com

© Springer-Verlag Berlin Heidelberg 2006
Printed in Germany

Typesetting: LE-TeX Jelonek, Schmidt & Vöckler GbR, Leipzig
Cover concept by eStudio Calamar Steinen using a background picture from The Optics Project. Courtesy of John T. Foley, Professor, Department of Physics and Astronomy, Mississippi State University, USA.
Cover production: *design & production* GmbH, Heidelberg

Printed on acid-free paper   57/3100/YL   5 4 3 2 1 0

Preface

Information and communication technologies have been growing and developing steadily for as long as any of us can remember. Growth was particularly strong in the last few decades, and fibre optic communication systems have become predominant whenever information is to be transmitted over medium or long distances. Even when the 'telecommunication bubble' burst at the beginning of the new millennium, the only thing which vanished was the expectation of making a fortune by buying and reselling telecom device and equipment manufacturing companies; the upgrading of existing fibre optic links and the deployment of new ones continued unabated.

The reason for the predominance of communication via optical fibres is the vast amount of information a single fibre can carry. However, in order to take advantage of this potential, it is mandatory to transmit different wavelength channels simultaneously over a single optical fibre, and the handling of these wavelength channels requires wavelength selective devices or wavelength filters. Among the functionalities optical filters have to accomplish are the selection of single or several channels out of a larger number of channels, the separation of one channel wavelength from unwanted spurious noise at different wavelengths, filters have to support routing, enable the lossless (or low loss) combination of wavelengths, and they have to compensate for wavelength dispersion effects.

Filters can be fabricated by many different techniques, and the optimum solution for a specific task depends on many parameters such as the number of wavelength channels involved and their separation, the temperature and polarisation dependence of the device, physical size and last but not least cost. There are a number of high-level books on the market, which focus on one type of optical filter and provide a corresponding in-depth treatment of all aspects from comprehensive theory to practical implementation and detailed description of experimental results. On the other hand, filters used in fibre optic communication systems are treated briefly in many books covering fibre optic systems in a more generic way, and as a consequence the corresponding filters get an only rather short treatment.

The present book is intended to bridge this gap: It focuses on filters only, but it covers all relevant wavelength filters used in fibre optic communication systems. The individual chapters have been written by international experts in the respective field, and the presentations always comprise a general description of the physics behind the filter under consideration, technical implementations, typical characteristics of devices commercially available, and performance of devices still in the research stage as well. The book is intended for graduate students, engineers working in fibre optics or related fields, and for all those wishing to gain a better understanding of a group of key devices enabling the exploitation of fibre optic communication systems.

The continuing demand for ever greater bandwidth by professional and private users will also induce a permanent pressure to upgrade existing fibre optic communication systems. This will require on the one hand the use of more advanced electronic circuits (in particular ultra-high speed electronics), and it will also bring more optical functionalities into fibre optic networks, for example optical routing, but eventually also all-optical signal processing such as all-optical signal regeneration, all-optical wavelength conversion, or all-optical header processing. In any case, wavelength as a parameter is very likely to provide more functionality, and as a consequence wavelength filters will become even more important than they are already today.

Berlin, April 2006   *Herbert Venghaus*

Contents

Abbreviations

ADM	add-drop multiplexer
AM	amplitude modulation, -modulated
AR	auto-regressive
AR	anti-reflection
ARMA	autoregressive moving average (filter)
ASE	amplified spontaneous emission
AWG	arrayed waveguide grating
BBO	beta barium borate
BER	bit error rate, bit error ratio
BPF	band pass filter
CAD	computer aided design
CATV	cable television
CD	chromatic dispersion
CFBG	chirped fibre Bragg grating
CPA	chirped pulse amplification
CROP	cationic ring-opening polymerisation
CTE	coefficient of thermal expansion
CVD	chemical vapour deposition
CWDM	coarse wavelength division multiplexing
CWL	centre wavelength
DC	directional coupler
DCF	dispersion-compensating fibre
DCM	dispersion-compensating module
DCG	dichromated gelatine
DFB	distributed feedback
DGEF	dynamic gain equalising filter
DH	double hetero (-structure)
DIBS	dual ion beam sputtering
DMX	demultiplexer
DOE	diffractive optical element
DSF	dispersion-shifted fibre

DTFT	discrete-time Fourier transform
DUT	device under test
DWDM	dense wavelength division multiplexing
EAM	electro-absorption modulator
EDFA	Erbium-doped fibre amplifier
EELED	edge-emitting LED
EMI	electromagnetic interference
EOP	eye-opening penalty
EPF	edge-pass filter
FBG	fibre Bragg grating
FEA	finite element analysis
FEC	forward error correction
FFP	fibre Fabry–Perot (filter)
FFT	fast Fourier transform
FHD	flame hydrolysis deposition
FIR	finite impulse response
FP	Fabry–Perot (filter)
FPR	free propagation region
FSAN	full services access network(s)
FSDG	free space diffraction grating
FSR	free spectral range
FT	Fourier transform
FTTC	Fibre-to-the-curb, -cabinet
FTTH	Fibre-to-the-home
FTTX	Fibre-to-the-x
FWHM	full-width at half-maximum
GFF	gain-flattening filter
GD	group delay
GDR	group delay ripple
GRIN	gradient-index
GTR	Gires–Tournois resonator
HBT	heterojunction bipolar transistor
HNA	high numerical aperture (~fibre)
HPF	high-pass filter
HR	high reflector, high reflectance
HWP	half wave plate

IAD   ion-assisted deposition
IBS   ion beam sputtering
IIR   infinite impulse response
IL   insertion loss
ISI   intersymbol interference
ITU   International Telecommunication Union

LAN   local area network
LED   light emitting diode
LPCVD   low pressure chemical vapour deposition
LPF   low-pass filter
LPG   long period grating

MA   moving average (filter)
MI   Michelson interferometer
MMF   multi-mode fibre
MMI   multi-mode interference (coupler)
MPS   modulation phase-shift (method)
MR   microring
MUX   multiplexer
MW   multi-wavelength
MZI   Mach–Zehnder interferometer
MZM   Mach–Zehnder modulator

NLO   non-linear optic(al)
NRZ   non-return-to-zero
NZDSF   non-zero dispersion shifted fibre

OADM   optical add-drop multiplexer
OLCR   optical low coherence reflectometry
OPL   optical path length
OSA   optical spectrum analyser
OXC   optical cross-connect

PBS   polarisation beam splitter
PDL   polarisation-dependent loss
PECVD   plasma enhanced chemical vapour deposition
PICVD   plasma impulse chemical vapour deposition
PLC   planar lightwave circuit
PMD   polarisation mode dispersion
PON   passive optical network
PRBS   pseudo-random bit sequence

PVA	polyvinyl alcohol
PVD	physical vapour deposition
PWR	passive wavelength router
RC	resonant coupler
RF	radio frequency
RIE	reactive ion etching
ROADM	reconfigurable OADM
RW	ridge waveguide
RZ	return-to-zero
SC	semiconductor
SCG	slanted chirped grating
SiON	silicon oxinitride (- oxynitride)
SMF	single-mode fibre
SNR	signal-to-noise ratio
SOI	silicon-on-insulator
SoS	silica-on-silicon (SiO_2/Si)
SSB	single side-band
SSE	source spontaneous emission
SWP	short wave pass
TCF	twin-core fibre
TDCM	tuneable dispersion compensating module
TEC	thermoelectric cooler
TIR	total internal reflection
VLSI	very large scale integration
VOA	variable optical attenuator
VPG	volume-phase grating
VPHG	volume-phase holographic grating
VSB	vestigial side band (filtering)
WAN	wide area network
WDM	wavelength division multiplexing
WG	waveguide
WLL	wireless local loop
WWDM	wide WDM

Definition of symbols

Symbol (and simple relations)	Name	Primary unit	Numerical value (or practical unit)
c	speed of light in vacuum	m/s	$2.997\ 924\ 58 \cdot 10^8$
$D = \mathrm{d}\tau_g / \mathrm{d}\lambda$	dispersion of a component	s/m	(ps/nm)
E	electric-field amplitude	V/m	
e	elementary charge	C	$-1.602 \cdot 10^{-19}$
f	frequency	Hz	
f_0	anchor frequency for ITU-T grid	THz	193.1
H, $\mathbf{H}$	transfer function, transfer matrix	1	
h $(\hbar = h/2\pi)$	Planck's constant	Js	$6.626\ 069\ 3 \cdot 10^{-34}$
k_0, $k = 2\pi/\lambda = \omega/c$	propagation constant of light in vacuum	rad/m	
k	extinction coefficient	m^{-1}	
L_π	coupling length	m	
n	refractive index	1	
$n_{eff} = \beta/k$	effective refractive index	1	
$n_g = c/v_g$	group index	1	
n_o, n_e	ordinary and extraordinary refractive index of birefringent crystals	1	
P	signal power	W	
R	reflectivity (power)	1	
r	reflectivity (field)	1	
$\Re(s)$	real part of complex variable s	-	
T	transmission	1	
T	temperature	K	
$v_{ph} = c/n_{eff}$	phase velocity of light in matter	m/s	

Ton Koonen
COBRA Institute
Dept. Electrical Engineering
Eindhoven University of Technology
E-Hoog Building, room 11.33,
Den Dolech 2, P.O. Box 513,
5600 MB Eindhoven, The Netherlands
email: a.m.j.koonen@tue.n

Berndt Kuhlow
Fraunhofer Institute for Telecommunications,
Heinrich-Hertz-Institute
Einsteinufer 37
D 10587 Berlin, Germany
email: Kuhlow@hhi.fhg.de

Jean-Pierre Laude
Consultant
13 bis rue de Chilly
Longjumeau, 91160, France
http://monsite.wanadoo.fr/jplaude
email: jplaude@wanadoo.fr

Xaveer J.M. Leijtens
COBRA Research Institute
Technische Universiteit Eindhoven
Den Dolech 2
EH 10.10
P.O. Box 513
5600 MB Eindhoven, The Netherlands
email: X.J.M.Leijtens@tue.nl

Alain Mugnier
Quantel group
4, rue Louis de Broglie – Building D
22304 Lannion cedex - France
email: alain.mugnier@quantel.fr

Andreas Othonos
University of Cyprus
Department of Physics
P.O. Box 20537
Nicosia
Cyprus
email: othonos@ucy.ac.cy

Christophe Peucheret
COM•DTU
Department of Communications, Optics & Materials
Technical University of Denmark
DK-2800 Kgs. Lyngby, Denmark
email: cp@com.dtu.dk

David Pureur
Quantel group
4, rue Louis de Broglie – Building D
22304 Lannion cedex - France
email: david.pureur@quantel.fr

René M. de Ridder
Integrated Optical MicroSystems (IOMS)
MESA+ Institute for Nanotechnology, Faculty of EWI
University of Twente
P.O. Box 217
7500 AE Enschede, The Netherlands
email: R.M.deRidder@utwente.nl

Chris G. H. Roeloffzen
Telecommunication Engineering Group
CTIT, Faculty of EWI
University of Twente
P.O. Box 217
7500 AE Enschede, The Netherelands
email: C.G.H.Roeloffzen@utwente.nl

Robert B. Sargent
Custom Optics
JDSU
2789 Northpoint Parkway
Santa Rosa, CA 95407, USA
email: robert.sargent@jdsu.com

Meint K. Smit
COBRA Research Institute
Technische Universiteit Eindhoven
Den Dolech 2
EH 10.33
P.O. Box 513
5600 MB Eindhoven, The Netherlands
email: M.K.Smit@tue.nl

Markus K. Tilsch
Science & Technology
JDSU
2789 Northpoint Parkway
Santa Rosa, CA 95407, USA
email: markus.tilsch@jdsu.com

Carl-Michael Weinert
Fraunhofer Institute for Telecommunications,
Heinrich-Hertz-Institute
Einsteinufer 37
10587 Berlin, Germany
email: Weinert@hhi.fhg.de

0 Introduction

Herbert Venghaus

Fibre optic networks are the backbone of today's communication systems. Due to their superior data transmission characteristics optical fibres are used whenever bit rates above a few tens of Mbit/s are to be transmitted over distances beyond about one km. Consequently long distance terrestrial and submarine links as well as metropolitan area (METRO) networks rely completely on optical fibres, and only access networks (the "last mile") are based on a variety of techniques with optical fibres being of limited relevance so far.

The optical fibres under consideration are made of oxide glasses, and silicate glass fibres are the most widely used ones (halide and chalcogenide glasses being other variants). Fibre optical communication started with the classical paper by Kao and Hockham in 1966 [1]: Optical communication was expected to become competitive with existing coaxial cables if 20 dB/km attenuation was achieved, and that target was met in 1970 – the same year in which the first CW semiconductor laser operation at room temperature was reported [2, 3]. Early fibre optic systems were based on multimode fibres and lasers operating in the 800...900 nm wavelength range. By 1980 a new generation of optical fibres and corresponding lasers enabled the set-up of transmission links at 1.3 µm with characteristics superior to those of the short-wave systems, and subsequently transmission at 1.55 µm emerged as well, while these two transmission windows were separated by a peak of higher absorption around 1390 nm due to OH^--vibrations [4], see upper half of Fig. 0.1.

Lasers emitting in the 1.3 µm and 1.55 µm wavelength ranges soon became commercial commodities and since then they constitute the basis for long and intermediate range transmission [5]. The 1.39 µm absorption could significantly be reduced or even suppressed in more recently developed fibres (see Fig. 0.1, lower part), and as a consequence the whole range from about 1270 nm and extending beyond 1630 nm is available for optical transmission. However, since the vast majority of deployed fibre still has the 1.39 µm absorption peak, transmission is still predominantly in the 1.3 µm and the 1.55 µm windows.

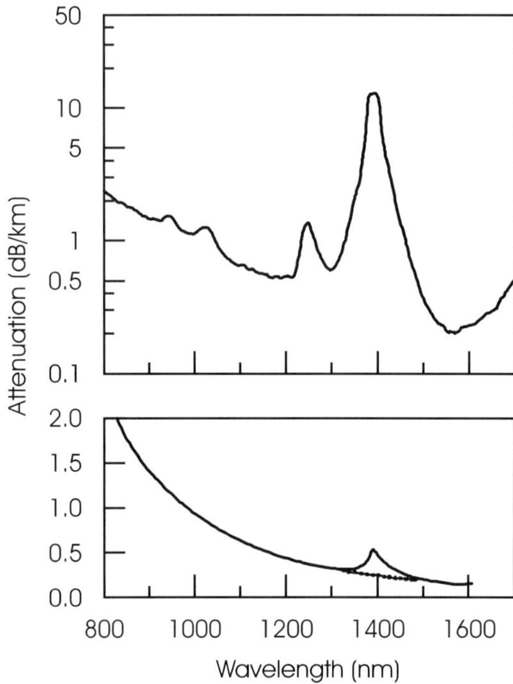

Fig. 0.1. Transmission characteristics single-mode optical fibres. Upper part: data given in Ref. 4, lower part: Corning SMF-28^{TM} optical fibre (*full line*), SMF-28e^{TM} (*dotted line*)

Laying optical fibres is rather expensive in most cases, and thus right from the beginning of fibre optic communication there was a strong strive to make optimum use of the huge transmission capacity of an optical fibre, which amounts to about 50 THz or even more in the wavelength range between 1.2 and 1.6 µm [6].

Raising the transmission capacity of a single fibre can be achieved in essentially two different ways: the first approach is to multiplex lower bit rate data streams to higher aggregate data rates. The technique is called 'Time Division Multiplexing' (TDM). Maximum single channel bit rates in deployed systems are essentially limited to about 10 Gbit/s, while 40 Gbit/s systems are beginning to be installed in selected links.

Single channel bit rates up to 2.56 Tbit/s have been achieved (by combining TDM with polarisation multiplexing and differential quadrature phase shift keying (DQPSK) modulation) in laboratory and field demonstrations [7], but such high bit rates are far from being employed in real systems in the foreseeable future. The cost of generating and detecting ultra-high bit rate channels in combination with mitigating the degradation of high-speed data streams due to chromatic and polarisation mode dispersion and nonlinear effects as well, constitutes a considerable impediment for the large scale installation of long-distance optical systems operating at 40 Gbit/s and beyond. However, these obstacles will gradually become less important as high-speed electronics will become less expensive, so that electronic forward error correction (FEC) will relax the requirements on acceptable bit error ratios by orders or magnitude similar to the case of 10 Gbit/s transmission. Signal degradation is less of an issue for shorter distances (several 100 m up to a few km), and consequently single channel bit rates may well be raised to 100 Gbit/s in the next few years (100 Gigabit Ethernet for example).

In any case, however, TDM alone enables the use of a very small fraction of the total fibre transmission capacity only, even if cheaper electronics and the further development of advanced modulation formats will enhance the robustness, performance, and cost of high bit-rate systems.

Besides TDM a second approach to exploit the fibre transmission capacity is using different transmission wavelengths at the same time ('Wavelength Division Multiplexing', WDM). To first order simultaneously transmitted wavelengths do not interfere with each other [8] and each wavelength can be modulated independently from any other wavelength (which includes using different bit rates and different modulation schemes for different wavelengths). In addition, different wavelengths can either be launched into the same direction or counter propagating as well.

WDM received attention in the seventies already (see for example [9–11]) and has made tremendous progress since then with record values of several hundred channels transmitted over a single fibre (see below).

Wavelength channel separations have initially been a few 10 nm as a consequence of technical restrictions, while such large separations are chosen today as a matter of convenience since no temperature stabilisation of such systems is needed (see Appendix).

The development of WDM systems with narrow channel spacing ('Dense' WDM, DWDM) has been strongly spurred by the advent of the Erbium-doped fibre amplifier (EDFA) in the mid-nineties. Due to the characteristics of the Er^{3+} ion these WDM systems operate primarily in the 1530–1565 nm range (C-band) or up to 1625 nm (L-band).

The TDM and WDM techniques are normally combined, and (undersea) systems with up to about 100 wavelength channels and 10 Gbit/s per channel are currently operated [12], while record laboratory values of transmission over a single fibre such as 6.4 Tbit/s (160×40 Gbit/s channels with 50 GHz separation [13]), 3.73 Tbit/s (373×10 Gbit/s RZ-DPSK channels with 25 GHz spacing [14]), or 1022 ultra-dense WDM channels with

1 Optical Filters in Wavelength-Division Multiplex Systems

Carl M. Weinert

1.1 Network Aspects

Wavelength filters in optical transmission systems are a special subgroup of physical components defined in such a way that they select or modify parts of the spectrum of the signal. In fact, optical wavelength filters are defined with respect to the modifications which they induce on the frequency spectrum.

We will restrict our considerations to the application of optical filters in optical networks. Figure 1.1 schematically shows a global optical network combining local and regional networks via the long haul network. Most of today's network concepts are based on Wavelength Division Multiplexing (WDM), which means that WDM filters are mainly needed in order to route and select specific wavelength channels.

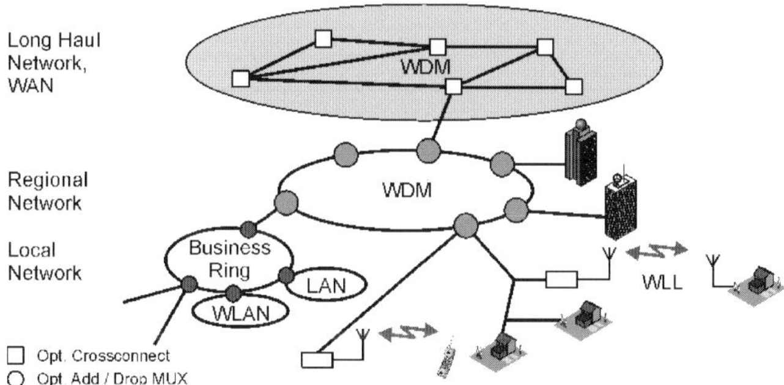

Fig. 1.1. Global optical network. WDM: wavelength division multiplexing, LAN: local area network, WLAN: Wireless LAN, WAN: wide area network, WLL: wireless local loop

The filters in optical WDM-systems are classified as bandpass filters, low pass filters, high pass filters, and notch filters.

Bandpass filters (BPFs) transmit optical power within a certain wavelength window only and reflect the rest. In the case of single channel transmission the role of an optical bandpass filter is to separate the channel information from the noise which has been added for example by optical amplifiers. This noise is in general broadband and can often be described as white noise, i.e it has a constant level in the power spectrum. By applying a bandpass filter to select the wavelength channel, the useful information is retained and most of the noise is rejected resulting in an improvement of signal-noise ratio (SNR).

In the case of many wavelength channels, in addition to rejecting the noise the bandpass filter rejects all the undesired WDM channels of the multitude of transmitted wavelength channels (see Fig. 1.2a). Furthermore, BPFs are essential components used for multiplexing and demultiplexing wavelengths in a WDM system. As shown in Fig. 1.2b, a multiplexer combines different sources with different wavelengths into a single fibre.

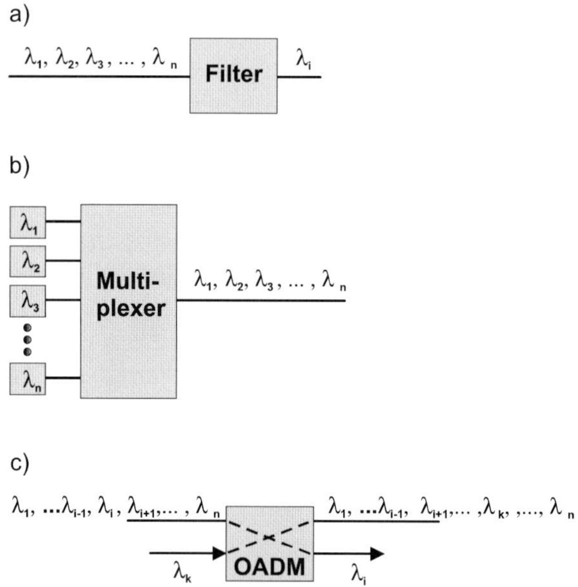

Fig. 1.2. Filter devices for WDM: (**a**) bandpass filter which selects the desired channel, (**b**) multiplexer which combines sources with different wavelengths into a single output. In reverse direction, the structure is used as demultiplexer (**c**) optical add-drop multiplexer where channel λ_k is added to and channel λ_i is dropped from the WDM spectrum [1]

In the reverse direction, the same device acts as a demultiplexer to separate different wavelengths to different outputs. Another important device for building WDM networks is the optical add-drop multiplexer (OADM) shown in Fig. 1.2c where a particular wavelength channel is added to and another wavelength channel is extracted from the WDM spectrum. Bandpass filters which are periodic in frequency can be used as so-called interleavers, which allow multistage multiplexing of channels. For example, with a periodic filter every second wavelength channel could be demultiplexed from a multitude of equally spaced channels (cf. Chap. 9).

The optical crossconnect (OXC) is used for routing different WDM channels. This means that the OXC can separate wavelength channels from incoming fibre bundles and redistribute them appropriately to outgoing fibres. In general, an OXC consists of many WDM components. An example of a wavelength crossconnect is depicted in Fig. 1.3. The wavelength channels of an input port are spatially separated by demultiplexers and can then be connected via multiplexers to an output port. The OXC is called a static wavelength crossconnect if the combinations of input ports and output ports are fixed as shown in Fig. 1.3. In this crossconnect architecture a wavelength (denoted by the lower index) of an input port (denoted by the upper index) is connected to an output port. In such a way, the incoming wavelength channels can be routed according to their wavelength. OXCs with more complex functionality also have OADMs to add and drop channels. Combining optical switches with multiplexers and demultiplexers, dynamic OXCs can be built. In dynamic OXCs the connections between input ports and output ports can be changed. This enables the dynamic reconfiguration of optical networks.

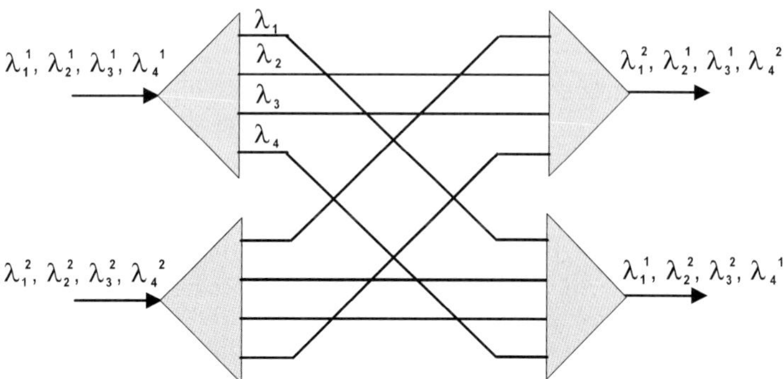

Fig. 1.3. Schematic of a static optical crossconnect. According to the wavelength a channel from an input port is routed to an output port

Notch filters reflect a specified wavelength or a narrow wavelength region with high transmission outside that region.

Low-pass filters (LPF) and **high-pass filters** (HPF) are filters which provide a sharp cut-off either above or below a particular wavelength. They are useful for isolating specific regions of the spectrum. Often referred to jointly as edge-pass filters (EPFs), low-pass filters and high-pass filters are used to pass (or transmit) a range of wavelengths and to block (or reflect) other wavelengths on one side of the passband. In the case of low-pass filters, the transmitted wavelength is long wavelength radiation, while short wavelength radiation is reflected. Conversely, high-pass filters transmit a wide spectral band of short wavelength radiation and block long wavelength radiation.

In addition, other types of optical filters exist in WDM systems. One of them are **power equalization filters**. Wavelength channels should have equal power levels. However, there are a number of reasons why different wavelength channels acquire different power levels. Adding and dropping of channels, nonuniform gain of the amplifiers, power inequalities of source lasers are some of the causes. Even if the differences in power are small in one span, they may accumulate over the transmission spans to yield large inequalities in power. Therefore, gain equalizing filters are needed in WDM systems. Furthermore, since the network configuration changes with time, this equalizing has to be done dynamically. Thus, Dynamic Gain Equalizing Filters (DGEF) are required for WDM networks.

Filters which transmit the complete frequency band but induce phase changes are called **allpass filters**. Since the group delay is the first derivative of the phase with respect to angular frequency, such filters can be designed to compensate for group velocity dispersion which accumulates during fibre transmission.

In order to illustrate generic applications of optical filters we depict an optical connection in Fig. 1.4. At the beginning and the end of the connection BPF-based multiplexers or demultiplexers are needed to combine or separate the different wavelength channels. After transmission through the fibre span dispersion compensating filters have to be applied to reduce signal degradation due to residual dispersion which has accumulated along the fibre. BPFs are needed in the OADM for adding or dropping channels. After the Erbium-doped fibre amplifier (EDFA) has amplified the WDM channels, a power equalizing filter is needed to ensure equal power on all channels. In addition, an edge filter is used to avoid perturbations of the data channels by the amplifier pump power at lower wavelengths.

Although different kinds of filters are necessary in an optical WDM transmission system, bandpass filters are by far the most important since

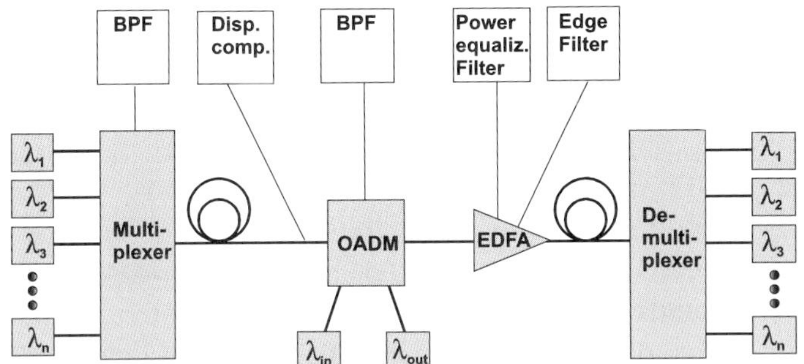

Fig. 1.4. Schematic of an optical transmission path showing examples of different filters needed in multiplexers, OADMs, or amplifiers

they are prerequisite for the add and drop, multiplex, and interleave and routing functionalities which are essentials for a WDM network.

1.2 Mathematical Description

Signals and physical components can be expressed mathematically by complex functions describing amplitude (real part) and phase (imaginary part). In the time domain the change or response of the input signal due to a component is evaluated by the convolution of the function representing the component with the complex expression for the incoming signal. According to the Fourier Transform theorem the response can equivalently be evaluated in frequency space by the product of the Fourier transform (FT) of the input signal and the FT of the function describing the physical component. Since the components we are dealing with are filters in the wavelength (or frequency) domain, we will adopt a description in the frequency domain.

For the ideal bandpass filter centred at ω_0 with filter width $\Delta\omega$ the transfer function is defined by $H(i\omega)$, ($\omega = 2\pi f$ where ω is the angular frequency and f is the frequency):

$$H(i\omega') = e^{-i\omega'} \quad 0 \le |\omega'| \le \Delta\omega \tag{1.1}$$

with

$$\omega' = \omega - \omega_0 = 2\pi(f - f_0) \tag{1.2}$$

Here f_0 is the centre frequency of the filter. In general, f_0 is chosen to match the carrier frequency of the wavelength channel which has to be selected by the BPF. For the ideal bandpass the squared amplitude function

$|H(\mathrm{i}\omega')|^2$, which determines the filter transmission, has rectangular shape. The ideal filter therefore perfectly separates wanted from unwanted parts of the spectrum. However, such filters are physically not realizable. Therefore filter functions have to be found which describe the properties of available physical filters.

The mathematical description of filters is useful under several aspects. Filter functions are needed as input data for numerical simulation of transmission systems, and the mathematical expressions can be used to determine the properties and requirements of filters and their physical realization. There exists a vast literature on the design of filters treating digital and analogue filters, finite and infinite impulse response filters, recursive and non-recursive filter design, which has been especially developed for the design of electronic filter circuits. Optical WDM-filters, on the other hand, are realized mainly as analogue, recursive, infinite impulse response filters (see also Chap. 2), the properties of which can be described by a complex transfer function $H(s)$ with complex variables [2]

$$H(s) = \frac{\sum_{i=0}^{m} b_i s^i}{1 + \sum_{i=1}^{n} a_i s^i} \tag{1.3}$$

m and n denote the number of zeros and poles which describe the special type of filter (see also Chap. 9, Sect. 9.3). In many cases, WDM-filters are all-pole filters, i. e. all b_i except b_0 are zero. The number of poles n is called the filter order. The distribution of the poles in the complex plane completely determines the filter properties. Tables and expressions for many filter functions can be found e.g. in [2, 3].

In order to evaluate the filter properties for the angular frequency, the transfer function is expressed as

$$H(s)|_{s=\mathrm{i}\omega'} = |H(\mathrm{i}\omega')| e^{\mathrm{i}\Theta(\omega')} \tag{1.4}$$

The transmission properties of the filter are given by the squared amplitude function and the phase behaviour is described by $\Theta(\omega')$ which can be evaluated from (1.4)

$$\Theta(\omega') = \tan^{-1}(\mathrm{Im}(H(\mathrm{i}\omega')/\mathrm{Re}(H(\mathrm{i}\omega')) \tag{1.5}$$

The group delay $\tau(\omega')$ is defined as

$$\tau(\omega') = -\frac{d\Theta(\omega')}{d\omega'} \tag{1.6}$$

The filter curves given above describe bandpass filters and are easily converted into high-pass or low-pass filters by the transformations given in [3]. In the case of low-pass filters it can be shown that the transformation from the BPF to the LPF can be done by simply choosing $\omega_0 = 0$ (cf. (1.2)).

We exemplify the above formulas with a normalized Bessel filter of order 10 as a bandpass filter at centre frequency f_0. We choose a Bessel filter since it shows regions both of constant and varying group delay. With increasing order of the filter the transfer function becomes more and more Gaussian and group delay becomes constant for all frequencies. For commercially available BPFs the measured transmission and phase behaviour can often be simulated by a Bessel filter of appropriate filter order. The form of $H(s)$ is found in [2]. In Fig. 1.5 we show the squared amplitude function both in linear and logarithmic scale.

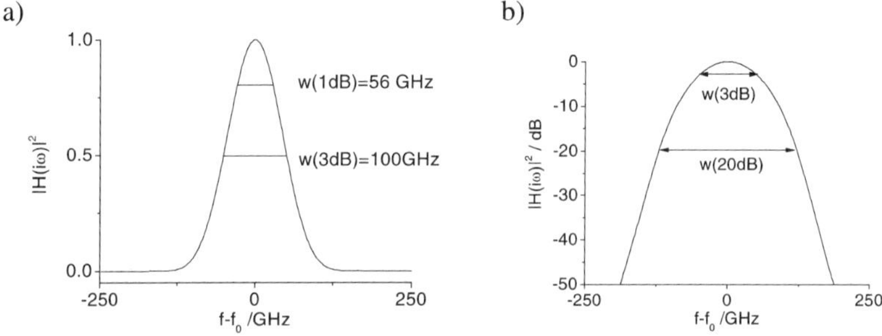

Fig. 1.5. Squared amplitude function $|H(i\omega)|^2$ for a Bessel filter of order $n = 10$ and FWHM = 100 GHz: linear scale (**a**) and logarithmic scale (**b**)

For filters with non ideally steep bandpass skirts as shown in Fig. 1.5 we can define filter bandwidths w given for different attenuations in dB. Common choices found in data tables of physical filters are $w(1\,\mathrm{dB})$, $w(3\,\mathrm{dB})$, and $w(20\,\mathrm{dB})$, which are 56 GHz, 100 GHz, and 240 GHz for the example given in Fig. 1.5. $w(3\,\mathrm{dB})$ is also called the full-width at half-maximum (FWHM) of the filter. The flat top of the filter is called the pass-band. A common definition for the passband width is $w(1\,\mathrm{dB})$.

In Fig. 1.6 we depict the phase function and group delay of the 10th order Bessel filter.

The phase function consists of several segments which are almost parallel. Accordingly, the group delay is constant in a frequency range of about 240 GHz which corresponds to $w(20\,\mathrm{dB})$ as shown in Fig. 1.5. Thus, such a filter will add negligible dispersion only to the signal.

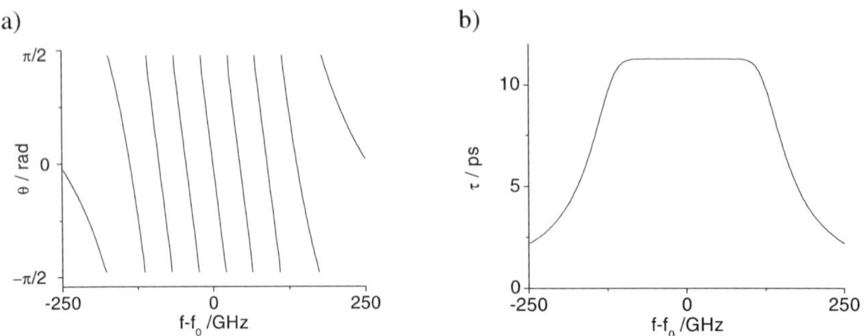

Fig. 1.6. Phase function $\Theta(\omega)$ (**a**) and group delay $\tau(\omega)$ (**b**) for a 10th order Bessel filter (FWHM = 100 GHz)

The transmission curves of most physical BPFs have a similar bell shape as depicted in Fig. 1.5, but can largely differ by their phase and group delay behaviour. These special types of filter functions have been investigated widely in the literature. They differ only in the type of polynomial in the denominator of (1.3), but have specific properties. For example, **Butterworth** filters describe an optimised flat behaviour of $|H(i\omega)|^2$ near f_0 (the passband width increases with the filter order). Butterworth filters are used to describe filters with flat tops. **Bessel** filters describe filters which have optimum flat group delay as has been shown in the filter example in Fig. 1.6. For increasing filter order the amplitude transfer function of Bessel filters becomes Gaussian. **Chebychev** filters have specific ripple properties of the transmission curves in the passband and stopband. They could therefore be used to simulate ripples occuring in the transmission function of physical filters. Further filter types as well as tables for the filter polynomials are given in [2, 3]. For a good approximation of physical filters by mathematical filter functions, measurements of the transfer function as well as phase and group delay measurements have to be made. The appropriate mathematical function can then be found by fitting the filter curves to the measured data.

1.3 Physical Realization of Filters

The physical realisation of WDM filters based on different principles of operation and fabrication on different materials will be discussed in the following chapters of the book. Here, we will therefore only define the main specifications of WDM filters. Further filter parameter definitions are given in the glossary.

In addition to the filter transfer function we have to consider optical loss, polarisation, temperature behaviour, and crosstalk.

1. *Insertion loss.* Insertion loss is the input to output loss of the filter for the passband and is, of course, required to be low.
2. *Polarisation-dependent loss (PDL).* Loss should be independent of the state of polarisation of the incoming signals.
3. *Temperature shift.* The temperature shift is the amount of wavelength shift per unit degree change in temperature. Temperature shift is required to be low.
4. *Passband width.* The passband width is defined by the 1 dB width $w(1\,\mathrm{dB})$. It is a measure for the flatness of the filter top. Flat passbands are required to allow for small shifts in laser wavelength over time. As more and more filters are cascaded the overall passband becomes narrower. Thus the individual filter passband width does also determine the number of filters which can be cascaded.
5. *Crosstalk.* Crosstalk energy is defined as the relative amount of energy passed through from adjacent channels. Different kinds of crosstalk are defined in the glossary.
6. *Group delay.* Group delay is defined in (1.6). For bandpass filters it should be constant to avoid the build-up of chirp in the signals when passing through the filter. For filters used for dispersion compensation, a suitable change in group delay over the wavelength is required.
7. *Free spectral range (FSR).* The FSR describes the spectral periodicity of the filter transmission function, i. e. the frequency difference between two filter maxima.
8. *Finesse.* The finesse of a filter is defined by the FSR divided by the 3 dB width of the filter.
9. *Tuneability.* Tuneability is defined as the ability to change properties of the filter, mainly its centre wavelength.

In addition we summarize the main physical principles of operation for WDM filters:

1. *Coupler* type filters make use of the change of coupling length with wavelength to separate different wavelengths. Normal couplers operate as broadband devices, and consequently they can separate only wavelengths which are far apart. A typical example is combining and separating the pump wavelength of a fibre amplifier at $0.98\,\mu\mathrm{m}$ wavelength from the data channel at $1.55\,\mu\mathrm{m}$ by couplers. Standard dielectric waveguide couplers will not be treated in detail in the present book, but the reader is referred for example to the classical paper by Kogelnik [4].

2. As discussed before, signals passing optical filters are in general affected with respect to amplitude and phase. In many cases the amplitude-wavelength dependence is of higher interest, but phase effects become particularly relevant for high bit rates and narrow channel widths. As a consequence, a discussion of phase characteristics is included in most of the individual chapters, but a separate chapter (Chap. 2) provides a more coherent treatment of phase-related phenomena.

3. *(Bulk) Diffraction Gratings* are based on the interference of multiple optical signals originating from the same source but with different phase shifts. Gratings can be operated in reflection/transmission or as diffraction gratings. In the case of diffraction gratings, signals of different wavelength can be resolved spatially. They can therefore be used for routing. Bulk diffraction gratings for WDM applications are described in Chap. 3.

4. *Fibre Bragg gratings* are gratings which consist of one dimensional periodic perturbations along the fibre. They can be used for add-drop functions and for dispersion compensation and are treated in detail in Chap. 5.

5. *Fabry–Perot* type filters are based on resonance in a cavity formed by two highly reflective mirrors placed parallel to each other, also called etalon (Chap. 6). Cascading of many cavities leads to very flat passbands and to sharp skirts. Examples of filters based on this principle are the thin-film resonant multicavity filters, discussed in Chap. 7. Related filter types are the ring filters where the resonator is a waveguide ring. Ring filters are treated in Chap. 8. Fabry–Perot type filters are also suited for constructing allpass filters which only induce a desired phase change without attenuation in the amplitude transfer function.

6. *Mach–Zehnder Interferometer (MZI)* type filters make use of two interfering paths of different lengths to separate different wavelengths. (Cascaded) MZIs constitute the fundamental building block of wavelength interleavers (Chap. 9), and the *Arrayed Waveguide Grating* filter can be considered as a generalisation of the MZI filter (Chap. 4), while its characteristics look like the 2-dimensional equivalent of a diffraction grating.

References

1. R. Ramaswami and K. N. Sivarajan: *Optical Networks* (Morgan Kaufmann, San Francisco, CA, USA, 2002)
2. L. R. Rabiner and B. Gold: *Theory and Application of Digital Signal Processing* (Prentice-Hall Inc., Englewood Cliffs, NJ, 1975)
3. A. Poularikas: *The Handbook of Formulas and Tables for Signal Processing* (CRC-Press, Boca Raton, FL, USA, 1999)
4. H. Kogelnik: "Theory of optical waveguides" in: *Guided-Wave Optoelectronics* (T. Tamir, ed.), Chap. 2 (Springer, Heidelberg, Berlin, 1988)

2 Phase Characteristics of Optical Filters

Christophe Peucheret

2.1 Introduction

Group velocity dispersion is a well known effect in optical fibres where the frequency dependence of the group index is responsible for pulse spreading, leading to inter-symbol interference (ISI) and power penalty [1]. However, it has only been realised recently that dispersive effects arising from optical filters might be detrimental to propagation of high bit-rate signals in wavelength division multiplexing (WDM) systems and networks [2, 3]. The need to take the dispersion of wavelength selective elements into account is being made more acute due to a number of technology trends. First, the increase of individual channel bit rates to 10 Gbit/s and the expected migration towards 40 Gbit/s per channel mean that even small dispersion values can no longer be ignored. Second, the quest for increased capacity in a single fibre has led to intense research in order to maximise the spectral efficiency and therefore reduce the channel spacing in WDM systems. Consequently, the relative bandwidth available to each channel is reduced, meaning that the channel experiences the effect of the edge of the passband of the filter transfer function, where the filter dispersion is expected to be most significant. Adjacent channel crosstalk reduction also triggers the need for WDM filters with steeper passband edges that, depending on the technology, are likely to result in increased dispersion [4, 5]. Finally, the evolution of WDM systems from point-to-point links to more complex network structures including optical add-drop multiplexers and optical cross-connects means that a number of filtering elements will be cascaded over the path of a specific channel. As the effect of dispersion is accumulative, more severe signal degradation is to be expected in future all-optical transparent networks [3, 6].

It has therefore become essential to be able to control the dispersion of optical WDM filters, either to limit potential signal degradation as in multiplexers and demultiplexers to be used in terminal equipment or within add-drop and cross-connect nodes, or to provide tailorable dispersion as for example in chirped fibre Bragg gratings for dispersion compensation. The need for dispersion characterisation of optical filters has appeared in the mid-nineties, and suitable measurement methods have essentially been inspired from the experience gained in the characterisation of single mode optical fibres [7, 8]. However, specificities due the wavelength selective nature of the components often mean that known optical fibre characterisation solutions might not be directly transferred to the case of WDM filters. First, the desired bandpass or bandstop filter characteristics of most of the filters of interest will induce new requirements in terms of dynamic range of the measurement method, especially since the wavelength region where dispersion is expected to be significant corresponds to the edges of the transfer functions, where the attenuation of the device might be large. Second, WDM filters exhibit spectral features that are strongly wavelength dependent, as opposed to optical fibres where both attenuation and dispersion vary relatively slowly with wavelength. Finally, the dispersion values exhibited by WDM components are usually relatively small, whereas in the case of optical fibres the dispersion values to be measured can very often be made arbitrarily large by increasing the length of fibre over which the measurement is performed, assuming uniform distribution of the dispersion over the fibre length. The consequences of those new requirements on the choice of a suitable measurement method are discussed in more depth in this chapter. It should be noted that another branch of optics where similar challenges are met is the characterisation of components for femtosecond laser design, for which proper dispersion engineering is essential, owing to the short pulse widths involved [9].

This chapter gives a general introduction to the topic of phase-related characteristics of wavelength filters and further presents a number of techniques suitable for the characterisation of phase-related quantities (including group delay and dispersion) complemented by typical experimental results measured on relevant wavelength filters. Section 2.2 starts with a compilation of useful definitions and then focuses on causality arguments, which may be used to infer the phase response of some types of optical filters from their wavelength-dependent attenuation. A discussion of the applicability of the method is illustrated by a practical example. Two categories of techniques enabling the determination of dispersion-related quantities are presented in Sect. 2.3, namely interferometric methods and radio-frequency (RF) amplitude modulation techniques. Emphasis is given to low coherence interferometry and the modulation phase-shift methods as they are very often considered as the methods of choice for WDM filter characterisation. General dispersion properties of WDM filters are presented in Sect. 2.4 together with their consequences on optical communication system design.

Methods for evaluating the impact of WDM filter dispersion on the limitation of their usable bandwidth and cascadability are introduced. The effect of group delay ripples is also discussed. Finally, some examples of passband and dispersion engineering for the design of advanced WDM filters are provided. Properties specific to a given filter technology are described in more detail in the relevant chapters elsewhere in this book.

2.2 Theoretical Considerations

2.2.1 Definitions

Throughout this chapter the transfer function of an optical linear element, such as a wavelength filter, will be written

$$H(\omega) = |H(\omega)| e^{-i\phi(\omega)} . \tag{2.1}$$

Note that the definition of the sign of the phase in (2.1) might differ from the one used in some other chapters of this book. According to the theory of linear systems, the transfer function is the Fourier transform of the impulse response of the filter

$$H(\omega) = \int_{-\infty}^{+\infty} h(t)\, e^{-i\omega t} dt . \tag{2.2}$$

The group delay of the filter can be calculated from the phase of its transfer function according to

$$\tau = \frac{d\phi}{d\omega} . \tag{2.3}$$

The definition of τ is sometimes also found with negative sign (cf. e. g. Chaps. 8 and 9). This is a consequence of the choices of the sign for the phase in the definition of transfer functions such as (2.1), as well as of the time dependence in the complex representation of the electric field.

In practice, only the relative group delay is of practical interest as it is the variations of τ with wavelength that result in pulse distortion, therefore making it unnecessary to characterise the absolute group delay. The dispersion, usually expressed in ps/nm, is the derivative of the group delay with respect to wavelength

$$D = \frac{d\tau}{d\lambda} . \tag{2.4}$$

It can be checked that the set of definitions above is fully consistent with that customarily used for optical fibres, where D would represent the total dispersion accumulated over a length L of fibre.

2.2.2 Minimum-phase Filters and Amplitude-phase Relations

It is known from the theory of linear systems that the phase response of a filter can be inferred from its amplitude response provided the so-called "minimum-phase" condition is satisfied [10, 11]. It is therefore natural to consider such a numerical approach to retrieve the dispersion from the amplitude transfer function of optical filters. Provided the minimum-phase condition holds, the mathematical relations linking the amplitude and the phase of an optical filter are analogous to the Kramers–Kronig relations between the absorption and refractive index (or real and imaginary parts of the dielectric constant) of a material [12]. Mathematically, quantities satisfying such relations are known as Hilbert transform pairs. As a consequence, "amplitude-phase", "Kramers–Kronig", or "Hilbert transform" relations are often used indifferently in the context of wavelength filter characterisation. In this section, the conditions for the existence of such relations are presented, followed by a discussion of their practical use for the determination of the dispersive properties of WDM filters.

The Minimum-phase Condition

Let us consider a passive linear optical filter with impulse response $h(t)$. Such a physical system is causal and stable, therefore its impulse response is real valued and satisfies the conditions $h(t) = 0$ for $t < 0$ and $|h(t)| < \infty$. Let $H(\omega)$ be the transfer function of the optical filter (i. e. the Fourier transform of its impulse response) and $H_L(s)$, where s is a complex variable, its Laplace transform. The Fourier transform can be evaluated from the Laplace transform according to $H_L(i\omega) = H(\omega)$. The fact that $h(t)$ is causal translates into $H_L(s)$ being analytic in the right-hand plane. Under those assumptions it can be shown that

$$H(\omega) = \frac{1}{i\pi} P \int_{-\infty}^{+\infty} \frac{H(\Omega)}{\Omega - \omega} \, d\Omega \qquad (2.5)$$

where P denotes the Cauchy principal value. Equation (2.5) can be derived either from causality considerations for the impulse response (see e. g. [13]) or, equivalently, from contour integration along a path where $H(\Omega)$ is analytic and avoiding the singularity at $\Omega = \omega$.

One important consequence of (2.5) is that the real (respectively imaginary) part of the transfer function $H(\omega)$ can be determined from the knowledge of its imaginary (respectively real) part. The real and imaginary parts of $H(\omega)$ are said to constitute a Hilbert transform pair.

The logarithm of the transfer function of a linear device as defined in (2.1) can be expressed as

$$\ln H(\omega) = \ln|H(\omega)| - i\phi(\omega).\qquad(2.6)$$

Therefore, by applying the results of (2.5) to the function $\ln H(\omega)$, it is tempting to derive Hilbert transform relations between the logarithm of the amplitude transfer function and the phase of an optical filter. However, this would require the function $\ln H_L(s)$ to fulfil the initial assumption of being analytic in the right-hand plane. One important issue is that $H_L(s)$ might have zeros in the right-hand plane where the logarithm is not defined. If we moreover assume that $H_L(s)$ has no zeros for $\Re(s) \geq 0$, where $\Re$ denotes the real part, then its phase will be uniquely determined by its amplitude response according to

$$\phi(\omega) = \frac{1}{\pi} P \int_{-\infty}^{+\infty} \frac{\ln|H(\Omega)|}{\Omega - \omega}\, d\Omega.\qquad(2.7)$$

Filters for which the logarithm of the Laplace transform of their impulse response is analytic in the right-hand plane are said to be of the "minimum-phase" type.

Amplitude-phase expressions that are equivalent to (2.7) are often found in the literature. For instance, starting from (2.7) and performing the change of variable $u = \ln \Omega/\omega$, a new expression for the phase can be written which highlights the relation between the variations of the amplitude response with frequency and the phase response

$$\phi(\omega) = \frac{1}{\pi} \int_{-\infty}^{+\infty} \frac{d}{du}\left[\ln|H(\omega\, e^{u})|\right]\ \ln \coth \frac{|u|}{2}\, du.\qquad(2.8)$$

The factor $\ln \coth|u|/2$ peaks around $u = 0$ corresponding to $\Omega = \omega$ and exhibits a fast decrease when $|u|$ increases. Consequently, the main contribution to the integral (2.8) arises from values in the vicinity of $u = 0$, and the phase of the transfer function at ω depends mostly on the slope of the amplitude transfer function around the same frequency ω. An immediate consequence is that any attempt to realise sharp edges in the transfer function of a minimum-phase optical band-pass filter will result in increased dispersion at those edges. Recalling our discussion of Sect. 2.1, it can be concluded that, for a minimum-phase optical filter, the two goals of achieving low crosstalk and low dispersion cannot be reached simultaneously. Filters which are not minimum-phase will offer more degrees of freedom in order to achieve these two desired features. Note, however, that the fact that a filter is non minimum-phase does not prevent it from exhibiting a high dispersion at the edges of its passband, but simply means that its dispersion cannot be calculated from the attenuation spectrum.

Practical Applicability of Amplitude-phase Relations

Some early work on the applicability of the Kramers–Kronig relations to the determination of the group delay of optical filters was first presented in [14] in the context of sub-picosecond laser design at 800 nm, where it was clearly demonstrated that such relations do not necessarily exist for some types of filters. Their applicability to etalon filters was further discussed in [15] where the need for a careful consideration of the zeros of the transfer function as well as of the frequency dependence of all optical parameters was highlighted. At the same time, the determination of the dispersion of components to be used in WDM systems around 1550 nm became the object of increased attention. Kramers–Kronig relations were successfully applied to the reflectivity of fibre Bragg gratings (FBG) [16, 17], showing good agreement between recovered and theoretical group delay. A comparison between measured group delay and delay recovered from measured amplitude responses was later presented in [18] for uniform fibre gratings, showing good agreement for some devices. It was later on nevertheless suggested that, for real imperfect gratings, the modelling of the group delay of the corresponding perfect grating will often provide a better estimate than the recovery of the group delay by applying the Kramers–Kronig relations to the measured reflectivity [19].

The first step in determining the dispersive properties of a particular WDM filter based on Kramers–Kronig analysis is therefore to analyse whether this filter is of the minimum-phase type. An in-depth discussion of the applicability of the minimum-phase condition for different optical filter technologies has been presented in [5]. It was shown that generalised Mach–Zehnder filters (including arrayed waveguide gratings) are in general not minimum-phase. Interference filters such as Fabry–Perot and thin-film filters are inherently of the minimum-phase type when used in transmission. It has also been shown that grating filters are minimum-phase in transmission but that it is not always the case in reflection [20]. Nevertheless, in case the grating is symmetric, its group delay is identical in reflection and transmission, therefore enabling the reflection group delay to be recovered from the grating transmittivity.

Whether a given filter is of the minimum-phase type can in theory be determined from checking the analyticity of its transfer function in the right-hand plane provided the filter response can be accurately modelled. However, this does not necessarily imply that the real, imperfect, implementation of the filter belongs to the same category (minimum or non-minimum phase). Indeed, it has been reported that the effect of loss in arrayed waveguide grating filters can move the zeros of their complex transfer functions from the imaginary axis to the left-hand plane [5], making the real filters satisfy the minimum-phase condition. Note that in the case when its zeros in the right-hand plane are known, a transfer function can be decomposed into the product of a minimum-phase function and an all-pass transfer function [15, 19] from which the phase response can be calculated. However, such an approach is not usable when the only information available about the filter is a measured amplitude response over a small part of the real frequency axis.

Once it has been ensured that the filter is minimum phase, practical considerations such as the frequency range over which the integration (2.7) needs to be performed as well as the implementation of the phase retrieval algorithm and its robustness to noisy measurement data and close to the zeros of the transfer function need to be taken into account. Numerous techniques have been proposed in the literature in order to compute Kramers–Kronig relations (see e. g. [21, 22] and the references in [23]). In the context of WDM filters, a calculation algorithm based on a non-linear frequency transformation known as the Wiener–Lee transform [10] has been applied to the case of fibre Bragg gratings [17] as well as a method using digital signal processing techniques [24]. More recently, an iterative approach [25] has shown its effectiveness at reconstructing the group delay of fibre gratings based on their transmission, even in the presence of significant noise.

As an illustration the amplitude-phase algorithm described in [17] has been applied to the case of a uniform fibre Bragg grating used in reflection. Such a symmetrical grating is known to be of the minimum-phase type [5, 20]. The measured reflectivity of a uniform fibre Bragg grating with

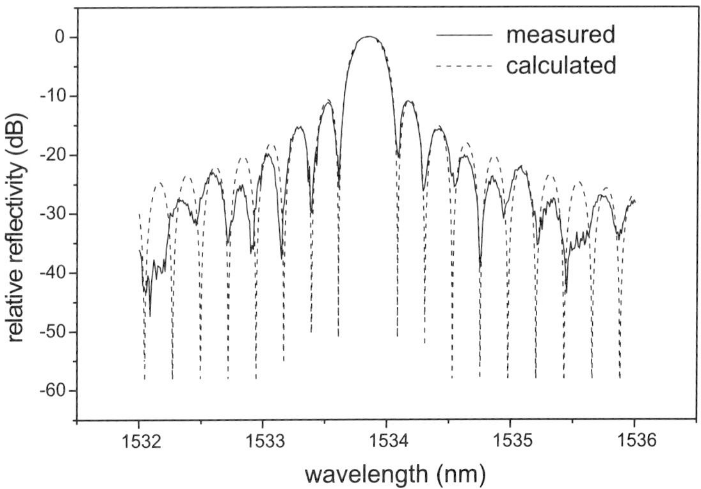

Fig. 2.1. Comparison of the measured and calculated power transfer functions in reflection for a 33 GHz uniform fibre Bragg grating

full-width half-maximum (FWHM) bandwidth equal to 33 GHz is plotted in Fig. 2.1 together with the corresponding theoretical transfer function calculated using the coupled mode equations formalism [26], showing good agreement for the main lobe and the first two sidelobes on both sides.

The phase responses calculated from the measured and theoretical reflectivities using the Wiener–Lee transform algorithm are shown in Fig. 2.2 (top). Good agreement is observed at the centre of the reflectivity main lobe. The phase discontinuities occur at the minima of the reflectivity where, due to imperfections in the real grating and the limited resolution of the optical spectrum analyser used for amplitude transfer function measurements, the measured and calculated reflectivities deviate significantly, hence the discrepancies observed for the phase.

The group delay calculated by differentiating the phase retrieved using the Wiener–Lee transform algorithm applied to the theoretical reflectivity and the theoretical group delay obtained using coupled mode equations are compared in Fig. 2.2 (bottom). Apart from the spikes in the recovered group delay occurring close to the zeros of the reflectivity, a good agreement is obtained, especially in the main lobe of the transfer function. Note that the theoretical group delay curve has been slightly up-shifted in Fig. 2.2 in order to allow for an easier comparison.

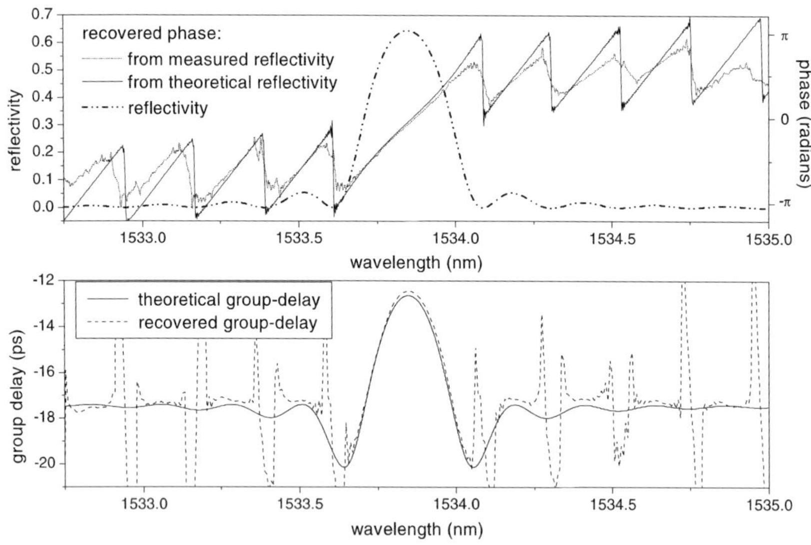

Fig. 2.2. (*Top*) phase responses calculated by applying the Wiener–Lee transform algorithm to the measured and calculated reflectivities of the uniform fibre grating whose transfer function is represented in Fig. 2.1. The calculated reflectivity is also plotted on a linear scale. (*Bottom*) corresponding theoretical group delay and group delay recovered from the theoretical reflectivity

The example above nevertheless confirms the concerns raised in [19] regarding the practical applicability of the method. Even if the main features of the phase or group delay of a filter that is known to be of the minimum-phase type can be retrieved by Kramers–Kronig analysis, the accuracy of the method might be questioned when it deals with the characterisation of the phase of imperfect real devices.

2.3 Measurement of the Dispersion of WDM Components

Due to the aforementioned difficulties associated with the use of amplitude-phase relations, direct measurement of the phase, group delay or dispersion of WDM filters is often preferred. In this section, two broad categories of phase-related quantities measurement techniques, based on either interferometry or small signal amplitude modulation of a continuous lightwave, are presented, and their advantages and limitations discussed.

2.3.1 Interferometric Techniques

A broad class of techniques enabling the characterisation of phase-related quantities makes use of interferometers where phase shifts can be converted into intensity variations that can be detected using a conventional photodiode. Several approaches have been followed in order to measure the phase properties of optical filters. A common difficulty to most implementations is the need for stabilisation schemes aimed at suppressing phase drifts as well as a precise calibration of the measurement set-up in the absence of the device under test.

Some early measurements have made use of coherent sources for the characterisation of the phase of optical filters. For instance in [27], measuring the power oscillations at the output of a Michelson interferometer while a laser was tuned over the device passband enabled to characterise the dispersion of a bulk grating pair as well as that of a linearly chirped waveguide grating. An all-fibre Michelson interferometer using phase modulation in the reference arm and lock-in detection was proposed in [28] and was used for some of the first characterisations of the group delay of a variety of fibre gratings [29]. The technique directly measures the phase change induced by the device under test while a laser is tuned over its passband. Phase-locked interferometry, where the delay of the reference arm of a Michelson interferometer is continuously adjusted while the wavelength is scanned, has been shown to enable group delay measurements with high temporal and spectral resolution, however at the price of a complex experimental set-up where the delay is measured using the interference pattern of an auxiliary coherent source [30].

A second approach consists of a wavelength domain analysis of the interference fringes obtained at the output of a Michelson or Mach–Zehnder interferometer under broadband illumination. Initially applied to the characterisation of short lengths of optical fibres [31], the technique has also been used for the characterisation of wavelength-selective elements such as a grating pair [32] and, more recently, short lengths of photonic bandgap fibres [33]. Dispersion affects the wavelength periodicity of the interference fringes measured using an optical spectrum analyser. The local maxima or minima of the phase versus wavelength correspond to slow modulation of the interference fringes. Once the phase turning points have been identified, the phase can be fully reconstructed by keeping in mind that consecutive local maxima of the fringes on either side are obtained for phase jumps of $\pm 2\pi$. The interferometer needs to be balanced within the coherence length of the source, requiring a tuneable path length in the reference arm and possibly stabilisation of its operating point. Both temporal and spectral resolution depend on the number of interference fringes visible in the bandwidth of the device under test, making the method unsuitable for narrow filters with low dispersion. On the other hand, wide bandwidth and highly dispersive devices could be characterised by this technique as long as the knowledge of small group delay ripples is not required.

Low Coherence Interferometry

A powerful approach to the characterisation of the group delay of WDM filters is low coherence interferometry [34–36], also sometimes known as Fourier transform spectroscopy. A typical low coherence interferometry set-up is represented in Fig. 2.3 in a Mach–Zehnder configuration suitable for the characterisation of transmission filters such as arrayed waveguide gratings [37]. Reflective devices such as fibre Bragg gratings can be characterised in the same set-up using a circulator or in an equivalent Michelson interferometer configuration [38]. A low coherence source generates a broadband signal that is input to a 3 dB coupler. The device under test is inserted into one of the arms of the Mach–Zehnder interferometer while the second arm contains a variable length reference path. When the optical path length difference between the light contributions propagating in the two arms of the interferometer is within the coherence length of the source, interference fringes are obtained at the output 3 dB coupler and can be recorded via a photodetector followed by an analogue-to-digital converter for further processing.

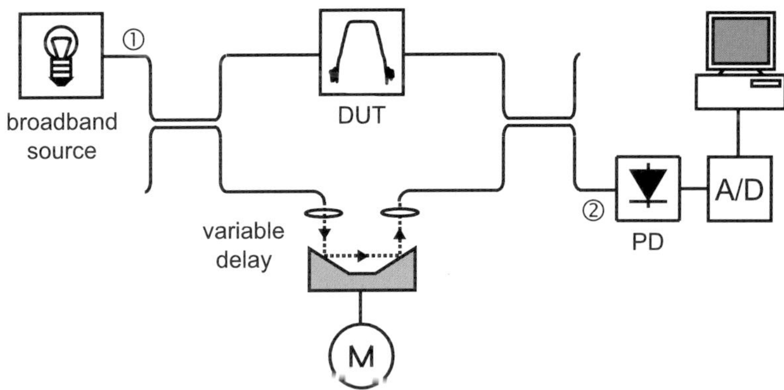

Fig. 2.3. Low coherence Mach–Zehnder interferometer for the measurement of the complex transfer function of optical filters used in transmission. An equivalent Michelson structure can be used for reflective devices. DUT: device under test; PD: photodiode; A/D: analogue-to-digital converter; M: motor driving the translation stage setting the variable delay

The method can be understood as follows. Let $E_1(t)$ and $E_2(t)$ be the contributions to the electric field at the output 3 dB coupler that have propagated from the source through the upper and lower arm of the interferometer, respectively. Let t_1 and t_2 be the delays from port ① to port ② corresponding to propagation through the device under test (DUT) and the reference arm, respectively. Let $E(t)$ be the field emitted by the source and $S(\omega)$ its spectral density. The intensity detected at port ② is proportional to the square of the modulus of the field averaged over a large number of optical cycles

$$I(t) = \left\langle \left| E_1(t) + E_2(t) \right|^2 \right\rangle, \tag{2.9}$$

where the angled brackets denote time averaging. Assuming identical polarisations for E_1 and E_2, the intensity becomes

$$I(t) = \left\langle E_1(t)E_1^*(t) \right\rangle + \left\langle E_2(t)E_2^*(t) \right\rangle + 2\Re\left\langle E_1(t)E_2^*(t) \right\rangle. \tag{2.10}$$

The first two terms are constant (dc) offsets, while the interference term can be shown to contain information about the transfer functions of the two arms of the interferometer. If the transfer functions of the device under test and of the reference path are denoted $H(\omega)$ and $H_{ref}(\omega)$, respectively, the two contributions to the total field at port ② can be written, assuming an ideal 3 dB coupler and neglecting the equal constant phase shifts experienced in the couplers along the two paths,

$$E_1(t) = \frac{1}{4\pi} \int_{-\infty}^{+\infty} \tilde{E}(\omega)\, H(\omega)\, e^{i\omega(t-t_1)}\, d\omega, \tag{2.11}$$

where $\tilde{E}(\omega)$ is the Fourier transform of the input field $E(t)$. In a similar way

$$E_2(t) = \frac{1}{4\pi} \int_{-\infty}^{+\infty} \tilde{E}(\omega)\, H_{ref}(\omega)\, e^{i\omega(t-t_2)}\, d\omega . \tag{2.12}$$

Hence the interference term

$$\langle E_1(t)E_2^*(t)\rangle = \left(\frac{1}{4\pi}\right)^2 \iint \langle \tilde{E}(\omega)\tilde{E}^*(\omega')\rangle H(\omega)H_{ref}^*(\omega')e^{i(\omega-\omega')t}e^{i(\omega't_2-\omega t_1)}d\omega d\omega' \tag{2.13}$$

With the usual assumptions of ergodicity and stationarity for the field $E(t)$, the time average can be replaced by ensemble average and it can be shown that

$$\langle \tilde{E}(\omega)\tilde{E}^*(\omega')\rangle = 2\pi\, \delta(\omega'-\omega)\, S(\omega) \tag{2.14}$$

where δ is the Dirac delta function. The interference term in (2.10) can therefore be written

$$I_{ac}(\tau_{21}) = 2\Re\langle E_1(t)E_2^*(t)\rangle = \frac{1}{2}\Re F^{-1}\left[S(\omega)H(\omega)H_{ref}^*(\omega)\right] \tag{2.15}$$

where F^{-1} denotes inverse Fourier transform. The argument of the interference term is the difference in time delay $\tau_{21} = t_2 - t_1$ that is linked to the optical path difference Δx between the light propagating in each of the arms of the interferometer. Consequently, if the interference fringes are recorded while Δx is changed by continuously increasing the length of the reference path along a translation stage, the interferogram $I(\Delta x)$ contains information about the complex transfer function (therefore including the desired phase information) of the device under test. In practice the Fourier transform of the fringes also contains information about other devices in the test path (such as the optical fibre patch cords, imperfections in the 3 dB coupler, etc.). However, performing a measurement scan without the device under test enables a total calibration of the set-up from which the parasitic contributions to the measured $H(\omega)$ as well as $H_{ref}^*(\omega)S(\omega)$ can be determined.

A variant of the technique makes use of a tuneable narrow band source and detects the shift of the centre of the fringes' envelope when the wavelength of the source is changed, enabling a direct determination of the variations of the group delay with wavelength [9, 34]. The time resolution depends on the visibility of the fringes' envelope which improves with increasing source bandwidth, while the spectral resolution is also obviously limited by the bandwidth of the source, making the method only suitable for the characterisation of broad bandwidth components. An all-fibre implementation of a low coherence reflectometry technique suitable for the characterisation of a broadband grating was also presented in [39].

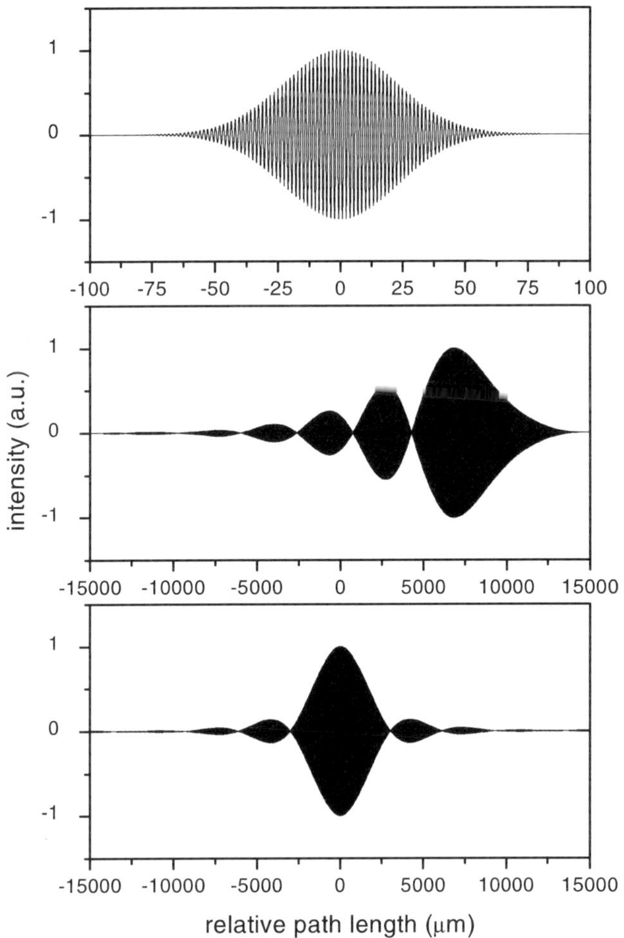

Fig. 2.4. Calculated low coherence interference fringes obtained from a broadband Gaussian source with 40 nm FWHM (*top*) and after reflection by a fibre Bragg grating filter designed for 50 GHz channel spacing (*centre*). Fourier transforming the fringes enables the determination of the spectral density of the source and the complex transfer function of the filter. Interference fringes that would be obtained from a linear phase grating with identical amplitude transfer function are also represented for comparison (*bottom*)

Figure 2.4 shows calculated low coherence interferometry fringes obtained by using a broadband source modelled as having a Gaussian spectrum with 40 nm full-width half-maximum (FWHM) bandwidth. The interferograms have been calculated with and without a Gaussian apodised grating designed for WDM systems with 50 GHz channel spacing in the test arm of the Michelson interferometer. Fourier transforming the interference fringes would enable to recover the grating complex transfer function. In order to illustrate that the phase transfer function is indeed influencing the recorded interferogram, an additional calculation has been performed where a hypothetical device having the same amplitude response as the 50 GHz grating, but linear phase, was considered in the test arm of the interferometer. Note that the spatial spread of the interferograms is larger in the presence of the optical filter as a consequence of band limitation. Consequently the larger scale used in the centre and bottom graphs of Fig. 2.4 does not enable to resolve the interference fringes but only their envelope.

The method is fast since both, amplitude and phase characteristics of the filter, can be obtained with a single scan of the tuneable delay line. However, it requires a high linearity of the translation stage in the reference arm. In practice a second interferometer making use of e. g. an He-Ne laser is used to monitor the change in path length while the variable delay is scanned. The wavelength resolution depends on the scanning range and can be increased by zero padding the interferogram before applying a fast Fourier transform (FFT) algorithm. The technique has also been shown to exhibit a large dynamic range [38], making it suitable for the characterisation of the dispersion at the edges of the passband of WDM filters. However, the visibility of the fringes decreases when a bandpass element such as a WDM filter is included in the test arm. As the group delay and dispersion are obtained by differentiation of the phase, a sufficient signal-to-noise ratio should be ensured to avoid numerical errors. Smoothing of the measured data, averaging over multiple scans, or enhanced balanced detection can be applied.

The use of low coherence interferometry has been reported for the characterisation of fibre Bragg gratings [38, 40], showing clear benefits in terms of accuracy and acquisition speed compared to the widely used modulation phase-shift method that will be described in Sect. 2.3.2, however at the price of increased complexity of the measurement set-up. One unique feature of the method is the possibility to retrieve the group delay characteristics of individual gratings in a cascade, provided the contribution of each component to the interferogram can be isolated, or by processing of the entire interferogram if the reflection bands of the gratings do not overlap [41]. The method has also been successfully applied to the characterisation of the dispersion of arrayed waveguide gratings (AWGs). In one approach the entire interferogram is Fourier transformed, directly leading to the device dispersion [37]. Resolving the respective contributions of each waveguide in the array, from which the phase and amplitude error distribution can be determined, has also been shown to enable full accurate modelling of the AWG including its dispersion [37, 42, 43].

2.3.2 RF Modulation Methods

The Modulation Phase-shift Method

Due to its relative simplicity of implementation, the modulation phase-shift (MPS) technique [44, 45] has become the method of choice for characterisation of the dispersion of optical fibres as well as of optical components. Early reports of the use of the MPS method for WDM filters focused on devices such as Mach–Zehnder planar dispersion equalisers [46] or arrayed waveguide grating multiplexers [3]. A typical phase-shift set-up is shown in Fig. 2.5. Light from a continuous wave (CW) tuneable laser is externally modulated with a sinusoidal signal at frequency f_m using a Mach–Zehnder modulator (MZM). The choice of the modulation frequency, typically from a few tens of megahertz up to a few gigahertz, will be discussed in detail later. The modulated light is then input to the device under test before being detected by a photodiode. The photocurrent is bandpass filtered around f_m before being input to a vector voltmeter (VM). In practice, a network analyser can conveniently be used to provide the radio frequency signal used to drive the modulator and perform relative phase measurements. Comparison of the phase of the detected photocurrent at f_m with the reference phase of the modulating signal enables the determination of the group delay of the device under test according to

$$\Delta\varphi = 2\pi f_m \tau(\lambda_0)\tag{2.16}$$

where λ_0 is the wavelength of the tuneable laser source that is precisely monitored using a wavelength meter (WM). Repeating the measurement while the wavelength is tuned over the wavelength range of interest enables

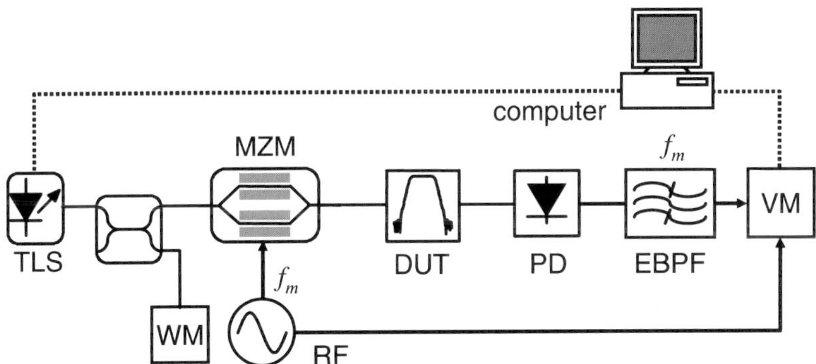

Fig. 2.5. Modulation phase-shift measurement set-up. TLS: tuneable laser source; WM: wavelength meter; MZM: Mach–Zehnder modulator; RF: radio frequency signal generator; DUT: device under test; PD: photodiode; EBPF: electrical bandpass filter; VM: vector voltmeter

the determination of the relative group delay as a function of wavelength from which the dispersion can theoretically be calculated according to (2.4).

In order to understand the limitations of the MPS method, it is essential to clarify the assumptions under which (2.16) has been derived. The complex representation of the electric field at the output of the MZM can be written

$$E_{in}(t) = E_0[1 + m\cos(\omega_m t - \varphi)]e^{i\omega_0 t} \tag{2.17}$$

where ω_0 is the angular frequency of the continuous lightwave generated in the tuneable laser, ω_m is the modulation angular frequency, φ is the phase of the modulating signal, and m is the modulation index. The spectrum of the modulated signal consists of a carrier at ω_0 and of two side-bands at $\omega_0 - \omega_m$ and $\omega_0 + \omega_m$. The electric field at the output of the DUT, whose complex transfer function has been defined according to (2.1), can easily be calculated by considering the attenuation and phase shifts experienced by the carrier and the two side-bands. If it is assumed that the carrier and the two side-bands experience the same amount of attenuation from the DUT (i. e. the magnitude of its transfer function can be considered constant over the bandwidth of the amplitude-modulated signal), therefore

$$\left| H\left(\omega_0 + \omega_m\right)\right| = \left| H\left(\omega_0 - \omega_m\right)\right| = \left| H\left(\omega_0\right)\right|, \tag{2.18}$$

then the component of the photocurrent that is retained after electrical bandpass filtering around ω_m can be written

$$i_{\omega_m}(t) = i_0 m \left| H\left(\omega_0\right)\right|^2 \cos\left(\phi(\omega_0) - \frac{\phi_+ + \phi_-}{2}\right)\cos\left(\omega_m t + \frac{\phi_- - \phi_+}{2} - \varphi\right) \tag{2.19}$$

where ϕ_+ and ϕ_- are shorthand notations for the phase shifts experienced by the upper and lower side-bands, respectively. Hence the phase difference between the detected signal and the reference signal used to drive the modulator is

$$\Delta\varphi = \frac{\phi_+ - \phi_-}{2}. \tag{2.20}$$

Equation (2.20) shows that the amplitude-modulated (AM) signal effectively probes the DUT at two frequencies corresponding to its two side-bands, and that the method effectively returns an average of the phase of the DUT at the side-bands' frequencies. If, furthermore, the phase can be assumed to vary linearly around the laser carrier frequency,

$$\phi(\omega) = \phi(\omega_0) + \frac{\partial\phi}{\partial\omega}(\omega_0)(\omega - \omega_0), \tag{2.21}$$

then the phase difference $\Delta\varphi$ can be directly related to the group delay at ω_0 according to (2.16). This last assumption is equivalent to considering the group delay constant over the bandwidth of the amplitude-modulated signal.

Equation (2.16) shows that, for a given phase resolution of the vector voltmeter or network analyser, typically of the order of 0.1°, the group delay resolution of the method can be improved by increasing the modulation frequency. However, this will result in a larger frequency separation between the side-bands of the AM signal. Consequently, the assumption that the phase of the DUT can be considered linear over the bandwidth of the amplitude modulated signal might no longer be satisfied.

The modulation phase-shift technique is widely used for the characterisation of the dispersion of optical fibres [47] where both, dispersion and attenuation, are expected to be slowly varying functions of wavelength, meaning that the assumptions of (2.18) and (2.21) are fulfilled over a frequency range up to a few gigahertz corresponding to the separation between the AM signal side-bands. However, the situation is radically different in the case of the characterisation of WDM filters where both, the dispersion and attenuation, are expected to vary significantly with wavelength, especially at the edges of the pass-band or stop-band of the device. Furthermore, as the dispersion values associated with optical filters are expected to be relatively small, the use of relatively high modulation frequencies might be required for their characterisation, hence increasing the side-bands' separation. Consequently, a compromise has to be found between group delay resolution and distortion induced by the averaging effect.

This point is illustrated in Fig. 2.6 where results of phase-shift measurements performed on a fibre Fabry–Perot filter with a FWHM bandwidth of 40 GHz are represented together with the calculated group delay response of the filter. While reasonably good agreement is obtained for a modulation

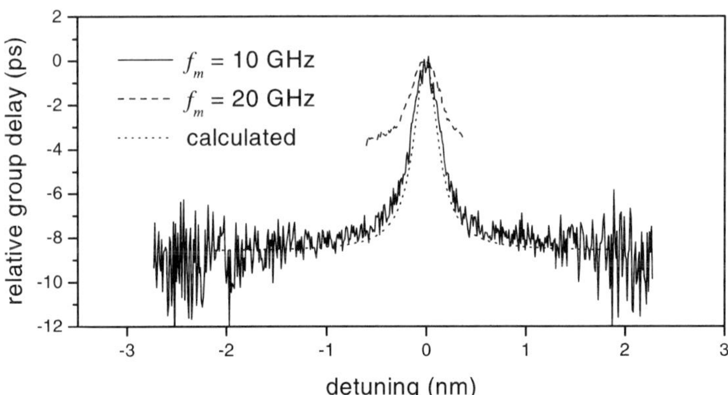

Fig. 2.6. Relative group delay of a 40 GHz fibre Fabry–Perot filter measured by the phase-shift technique using modulation frequencies of 10 and 20 GHz. The calculated group delay curve is shown for comparison. Note that raw measurement data without curve fitting nor averaging are presented

frequency of 10 GHz, this is no longer the case at 20 GHz where the measured group delay departs significantly from the theoretical value.

The relative error induced by the use of a too high modulation frequency can be assessed numerically if the theoretical group delay or dispersion responses are known. Not only the relative phase experienced by the two sidebands affects the accuracy of MPS measurements, but also their relative attenuation. Figure 2.7 shows the theoretical group delay response of a Gaussian apodised fibre Bragg grating designed for 50 GHz channel spacing (FWHM bandwidth: 44 GHz) together with simulated phase-shift measurement results that would be obtained when using modulation frequencies of 1, 5, 10, and 20 GHz. Relatively good agreement is observed between the theoretical group delay and the simulated measurement performed with a modulation frequency of 1 GHz. However, the group delay features at the edges of the passband are no longer properly characterised when the modulation frequency is increased to 5 or 10 GHz. An inversion of the variations of the measured group delay close to the passband centre frequency even appears when a modulation frequency of 20 GHz is used. In contrast to the case of a real measurement, only the choice of a too high modulation frequency affects the retrieved group delay in those simulations. In practice, the resolution of the group delay at low modulation frequencies will be limited by the phase accuracy of the vector voltmeter and the signal-to-noise ratio of the detected signal. The latter will limit the bandwidth over which accurate measurements can be performed due to the filter attenuation.

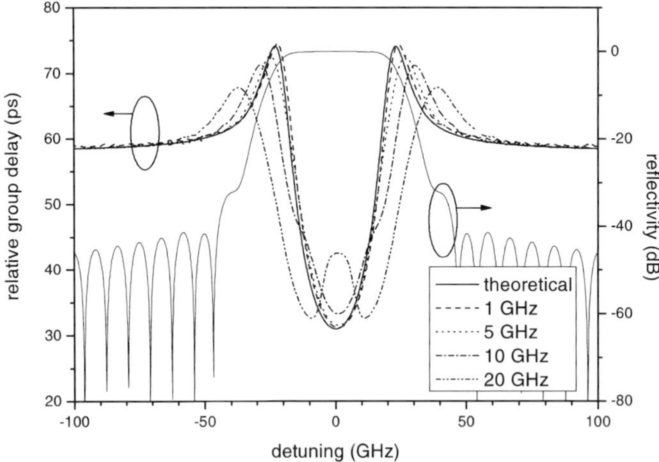

Fig. 2.7. Theoretical transfer function of a Gaussian apodised fibre Bragg grating designed for 50 GHz channel spacing and simulated group delay curves obtained with the phase-shift method using modulation frequencies of 1, 5, 10, and 20 GHz

In the case of the characterisation of the group delay ripples typically associated with fibre Bragg gratings (those will be discussed in more details in Sect. 2.4.3 as well as in Chap. 5), increasing the modulation frequency will decrease the amplitude of the measured ripples [48]. Depending on their wavelength periodicity, some modulation frequencies might even result in an inversion of the polarity of the measured group delay ripples [49].

A drawback of the modulation phase-shift method is that it enables the determination of the group delay of the device according to (2.16), and not of its dispersion. The dispersion can be obtained by numerical differentiation based on (2.16). However, the numerical differentiation of noisy measurement data turns out to be difficult unless a fitting or smoothing procedure is used [50]. Consequently, it has become customary to use a representation of the group delay as a function of wavelength to describe the measured dispersive properties of WDM filters in scientific publications or product data sheets. An adaptation of the MPS method, known as the "differential phase-shift technique" [51], uses low speed modulation of the frequency of the laser source operated in an otherwise conventional phase-shift set-up to directly detect the change in group delay occurring over the wavelength excursion of the laser $\delta\lambda$, hence enabling a direct determination of the dispersion according to

$$D(\lambda) = \frac{\Delta\varphi(\lambda + \delta\lambda/2) - \Delta\varphi(\lambda - \delta\lambda/2)}{2\pi f_m \delta\lambda}. \qquad (2.22)$$

The method suffers from the same limitations as the conventional phase-shift technique regarding accuracy of the measurement of devices having strong wavelength-dependent features such as WDM filters, but it removes the need for curve fitting in order to obtain the dispersion from the measured group delay. A detailed comparison of the MPS and differential phase-shift techniques in a metrology environment can be found in [52, 53].

Enhancement of the Accuracy of the Phase-shift Technique

Since the shortcomings of the modulation phase-shift method for the characterisation of WDM filters have been realised, a number of approaches have been suggested in order to improve its accuracy.

If a single side-band (SSB) signal is substituted to the double side-band amplitude-modulated signal in a conventional phase-shift set-up, the difference between the phase of the detected signal at ω_m and the phase of the sinusoidal signal driving the modulator becomes

$$\Delta\varphi = \pm(\phi_\pm - \phi_0), \qquad (2.23)$$

where the sign depends on the selected side-band, and ϕ_0 is the phase shift induced by the filter at the carrier frequency ω_0. Such a single side-band signal can easily be generated by creating a 90 degree phase-shift between the sinusoidal signals applied to the two arms of a dual-drive Mach–Zehnder modulator [54]. By keeping the CW laser frequency ω_0 constant and sweeping the modulating frequency, it becomes possible to directly map the phase transfer function of the device under test with high accuracy [55], including group delay ripples that would not be resolved using the conventional MPS method.

Another approach consists in locking one of the side-bands of the AM signal to a precise absolute frequency and increasing the modulation frequency simultaneously with the CW laser carrier frequency, so that the second side-band scans over the wavelength range of interest [56]. As for the SSB method, the wavelength range over which the measurement is performed is then limited by the maximum frequency at which amplitude modulation and detection are possible, typically up to a few tens of gigahertz. The characterisation of typical WDM filters will require the measurement to be repeated at different carrier frequencies or locked side-band absolute frequencies in order to scan over the full bandwidth of the device.

Post-processing of measurement data obtained using the conventional phase-shift technique has also been proposed as a way to circumvent the averaging effect of the method at high modulation frequencies. In [57] the measured group delay spectrum $\tau_{meas}(\omega)$ has been shown to be equal to the convolution of the real group delay with a rectangular function Π equal to $1/2\omega_m$ over the interval $[-\omega_m, \omega_m]$,

$$\tau_{meas}(\omega) = \tau(\omega) * \Pi(\omega/2\omega_m). \tag{2.24}$$

Fourier transforming (2.24) leads to

$$\mathrm{T}_{meas}(u) = \mathrm{T}(u)\, \mathrm{sinc}(\omega_m u), \tag{2.25}$$

where $\mathrm{T}(u)$ and $\mathrm{T}_{meas}(u)$ are the Fourier transforms of the real and measured group delay, respectively, and where the Fourier variable u is related to the inverse of the frequency period of the ripples. This approach has been successfully applied to explain the reduction of the amplitude of measured group delay ripples when the modulation frequency is increased, as well as the inversion of polarity observed for some modulation frequencies, depending on the spectral content of the ripples. However, deconvolution of measured phase-shift data is made difficult due to the zeros of the sinc function. Performing phase-shift measurements at two different modulation frequencies and using a weighted average of their Fourier transforms has been shown to overcome this difficulty [58].

Beyond the influence of the modulation frequency, other limitations of the accuracy of the MPS method such as the residual chirp of the Mach–Zehnder modulator, the phase linearity of the electrical devices when the power of the detected optical signal is varying, for instance due to the edge of the transfer function of a filter [59], or the influence of the amplified spontaneous emission noise of the laser source [60] have been pointed out. Phase drifts of the measurement system might also constitute a limiting factor, especially when either high spectral resolution characterisation is performed by using small tuning steps for the CW laser, or when high accuracy is sought at lower modulation frequencies by averaging over a large number of measurements of the electrical phase difference $\Delta\varphi$. Fast wavelength scanning has been demonstrated using continuously swept lasers [61], however, at the price of reduced wavelength accuracy due to the impossibility to use conventional wavelength meters.

In spite of its known limitations, the modulation phase-shift method has established itself as the standard for group delay measurements of WDM filters due to is relative simplicity. Careful optimisation of the measurement set-up and procedure has been shown to enable high accuracy characterisation [53, 59, 62]. It is nevertheless essential to be aware of its limitations in order to allow for a correct interpretation of MPS measurement results, as provided for instance in WDM filter manufacturers' data sheets.

The Dispersion Offset Method

The analysis of the modulation phase-shift method shows that the amplitude of the photocurrent at the modulation frequency also depends on the phase properties of the device under test through the even orders of its Taylor expansion, as can be seen in (2.19). If a second order Taylor expansion is sufficient to describe the phase around the carrier wavelength λ_0, the cosine term describing the amplitude of the photocurrent in (2.19) cancels for modulation frequencies satisfying

$$f_k^2 = \frac{(k-1/2)\,c}{D\lambda_0^2},\qquad (2.26)$$

where k is a strictly positive integer. It is therefore possible to relate the frequencies at which the magnitude of the photocurrent cancels to the value of the dispersion at the carrier frequency. Such an RF modulation method, also known as "fibre transfer function" method, has been proposed for the measurement of dispersion in optical fibres [63, 64] and can be extended to the direct characterisation of the dispersion of filters [65]. The method relies on the fact that, due to the dispersive nature of optical components, the propagation constants of the two side-bands of the amplitude-modulated signal are different. For a given dispersion value, modulation frequencies can be found where the components of the beat signal between the carrier and the side-bands are in counterphase, resulting in cancellation of the photocurrent seen as dips in the small-signal frequency response. The experimental approach consists of sweeping the frequency of the modulating signal and detecting the cancellations of the photocurrent on a network analyser in order to measure the frequencies f_k from which the dispersion at the CW laser frequency can be calculated according to (2.26). Such a measurement can be performed by using a set-up similar to the one represented in Fig. 2.8.

From (2.26) it can be seen that the minimum dispersion that can be measured by this method depends on the maximum frequency at which the amplitude modulated signal can be generated and detected. This limitation arises from the fact that, for a given amount of dispersion, the phase mismatch between the side-bands required for cancellation of the photocurrent is only achieved when the side-band separation is large enough. For instance, a minimum dispersion of 156 ps/nm can be measured at 1550 nm if the maximum modulation frequency is 20 GHz. By inserting a constant dispersion offset, such as a length of standard single mode fibre, small values of both positive and negative dispersions can be measured, making the method suitable for the characterisation of optical filters. Such an offset will increase the total amount of dispersion to a value large enough to be measured by the set-up. The dispersion of the device under test is then equal to the change in total dispersion after it has been inserted. Stability of the dispersion offset over time should therefore be ensured, for instance by keeping it at a constant temperature.

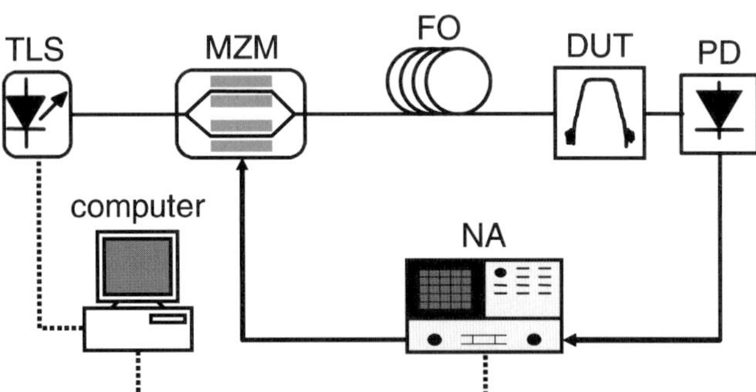

Fig. 2.8. Dispersion offset measurement set-up. TLS: tuneable laser source; MZM: Mach–Zehnder modulator; FO: fibre offset; DUT: device under test; PD: photodiode; NA: network analyser

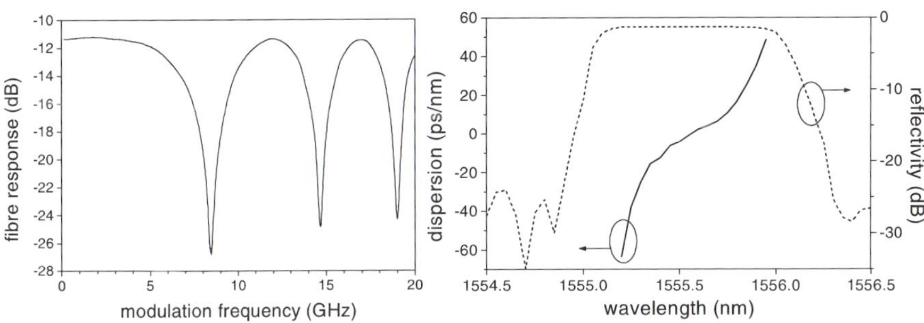

Fig. 2.9. (*Left*) measured frequency response of 50 km of standard single mode fibre. The shift of the frequency dips enables to determine small dispersion values introduced by optical components such as WDM filters. (*Right*) dispersion of a fibre grating Mach–Zehnder optical add-drop multiplexer measured with the dispersion offset technique

When characterising optical filters, the two side-bands of the AM signal might experience different attenuation. In this case, it can easily be shown [65] that imperfect cancellation of the detected signal occurs, resulting in shallower dips in the modulation transfer function, but that the frequencies of the dips are not affected by the relative amplitude of the side-bands. However, this effect limits the accuracy of the determination of the dips' frequencies. Consequently, the applicability of the method depends on the steepness of the slope of the amplitude transfer function of the filter under test.

As an example, the dispersion offset method has been applied to the characterisation of a commercial fibre grating Mach–Zehnder optical add-drop multiplexer (OADM) based on the design initially proposed in [66], showing values in excess of ±40 ps/nm in the considered wavelength range, as illustrated in Fig. 2.9.

In the remaining part of this chapter, it will be examined how the knowledge of the dispersion of optical filters can be related to their impact on transmission systems and networks, thus allowing proper selection of a filter type for a given application.

2.4 Dispersion of WDM Filters and System Implications

The dispersive properties of WDM filters have been the object of increased attention over the past few years. Consequently, those are now systematically studied, either based on measurements performed on real devices using one of the methods described in Sect. 2.3, or based on filter modelling. The phase behaviour of various WDM filter technologies has been discussed in depth in previous work [4, 5, 67], and the phase characteristics of some of the filter types covered in the present book are presented individually in the relevant chapters. In this section, some general issues about the dispersion of optical filters and its system implications are discussed. First, the group delay behaviour of three of the most widely used WDM filter technologies (fibre Bragg gratings, multilayer interference filters, and arrayed waveguide gratings) is compared based on measurements performed on commercially available components. The system impact of filter dispersion and methods for its numerical and experimental evaluation are then reviewed. The special case of group delay ripples presented by some types of filters such as chirped Bragg gratings is treated separately. Finally, novel filter designs where additional degrees of freedom are introduced in order to simultaneously tailor their amplitude and phase responses are briefly presented.

2.4.1 Dispersive versus Linear–phase Filters

Filter technologies such as fibre Bragg gratings (Chap. 5), thin-film interference filters (Chap. 7) and arrayed waveguide gratings (Chap. 4) have a strong potential for WDM systems applications due to their versatility and design degrees of freedom. They are typically used in subsystems such as optical add-drop multiplexers and optical cross-connects (OXCs), as well as in terminal multiplexers and demultiplexers. In what follows, typical group delay curves measured on commercial devices are compared in order to highlight the characteristic features of each type of filter. All group delay curves presented here have been measured using the standard modulation phase-shift method.

One of the main advantages of the fibre Bragg grating technology is that the transfer function of the filter can be tailored by the proper choice of the distribution of the coupling coefficient along the fibre length, which is itself determined by the longitudinal effective refractive index profile. In this way it becomes possible to "square" the passband and to reduce the crosstalk level of fibre grating filters used in reflection, a process known as apodisation. Whether this process necessarily results in increased group delay at the edges of the transfer function depends on whether the filter is minimum-phase, which is not systematically the case when a fibre grating is used in reflection, as will be described in Sect. 2.4.4. For instance, a Gaussian apodisation profile can be used to simultaneously square the pass-band while maintaining an acceptable crosstalk level [26]. As an illustration, two commercial-grade devices designed for dense wavelength division multiplexing (DWDM) systems with 100 and 50 GHz channel spacing have been characterised by the phase-shift technique. The results

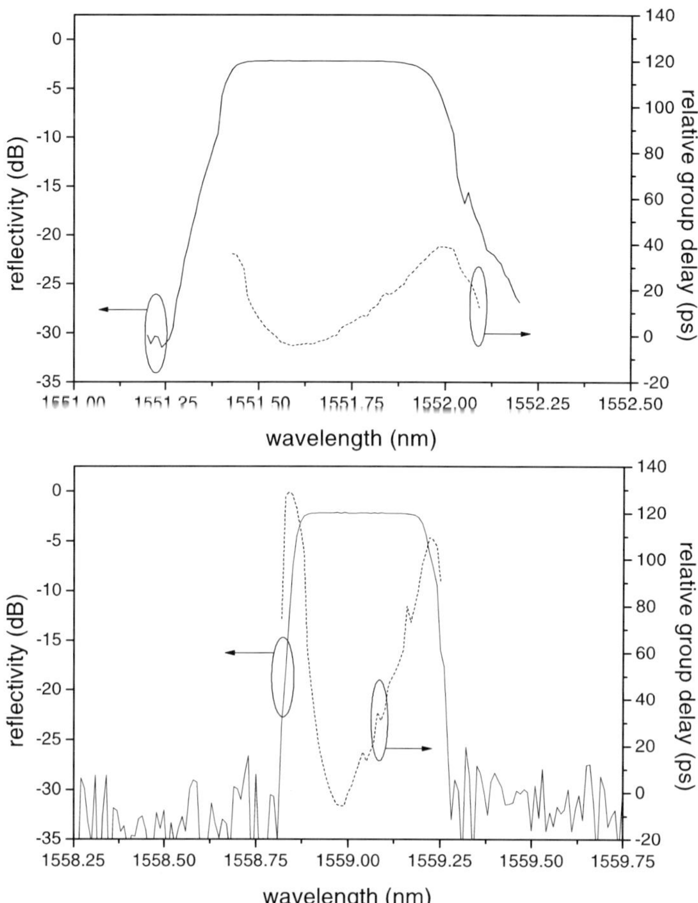

Fig. 2.10. Measured reflectivity and group delay of Gaussian apodised fibre Bragg gratings designed for 100 GHz (*top*) and 50 GHz channel spacing (*bottom*). The modulation phase-shift frequency used for the group delay measurements is $f_m = 2$ GHz

obtained with a modulation frequency equal to 2 GHz are shown in Fig. 2.10. It can be seen that the shape of the group delay curve is similar for both bandwidths. However, the extent of the variations of the group delay with wavelength is much higher in the 50 GHz grating case, indicating that this device exhibits higher values of dispersion in the passband than its 100 GHz counterpart.

Dielectric multilayer interference filters, also known as thin-film filters, consist of several Fabry–Perot like cavities separated by reflectors made of stacks of alternating low and high refractive index quarter-wave layers. The possibility to tailor their bandwidth and the steepness of the slope of

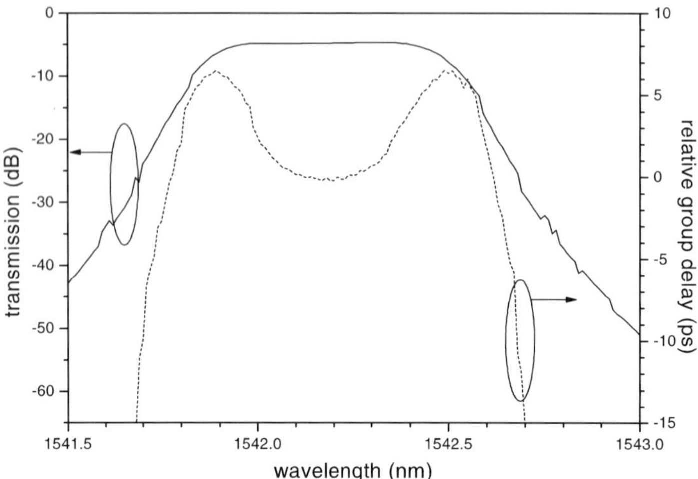

Fig. 2.11. Measured transmittivity and group delay (modulation frequency $f_m = 2.5$ GHz) of a thin-film filter-based demultiplexer

their transfer function by engineering the number of cavities and $\lambda/4$ dielectric layers makes them particularly attractive for DWDM applications [68, 69]. A typical transfer function of a thin-film (de)multiplexer is represented in Fig. 2.11, showing a characteristic flat-top and relatively steep-slope response. The 3 dB and 20 dB bandwidths of the transfer function are 80 and 120 GHz, respectively. The dispersion is equal to 0 ps/nm at the centre wavelength and is found to vary between −60 and +20 ps/nm within the 3 dB bandwidth. The dispersion behaviour of thin-film filters depends on the device structure and can also be influenced by imperfections in layer deposition resulting in non-uniformities and surface roughness [70]. As such devices can be shown to be minimum-phase in transmission [5], any attempt to square their transfer function, for instance in order to accommodate smaller channel spacing in DWDM systems, will result in increased dispersion at the edges of the passband. However, exploiting their design degrees of freedom can be used to mitigate dispersion effects while preserving an acceptable amplitude response [71, 72]. Additionally, a second stage consisting of a reflective all-pass filter can be added in order to partially compensate for the dispersion of conventional transmission bandpass filters [73] (cf. Fig. 7.7).

As pointed out in [5], the mechanism of resonant coupling in grating devices means they could be considered to the limit as being equivalent to thin-film filters with a large number of cavities. The measured group delay responses represented in Figs. 2.10 and 2.11 confirm the similar behaviour of these two types of filters, although the characterised thin-film filter

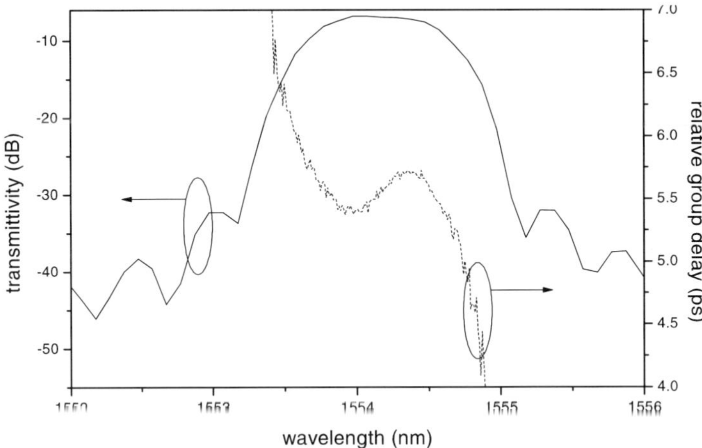

Fig. 2.12. Measured transmittivity and group delay (modulation frequency $f_m = 10\,\mathrm{GHz}$) of an arrayed waveguide grating demultiplexer designed for 200 GHz channel spacing

appears to be less dispersive than a Gaussian apodised grating of equivalent bandwidth.

Arrayed waveguide gratings consist of two free-propagation regions to which the input and output fibres are connected, linked by an array of waveguides designed in such a way that the optical length difference between two consecutive waveguides is constant [74, 75, see also Chap. 4]. Those devices can be used as (de)multiplexers as well as wavelength routers. It can easily be shown that, if the distribution of the excitation of the arrayed waveguides is symmetric, then an AWG is a linear-phase device, consequently dispersion-free [5, 67].

Figure 2.12 shows the measured power transfer function and group delay of a conventional (i. e. whose passband has not been flattened by some of the techniques discussed briefly in Sect. 2.4.4 as well as in Chap. 4) AWG designed for 200 GHz channel spacing and having a 3 dB bandwidth of 125 GHz. A modulation frequency as high as 10 GHz was necessary in order to be able to measure the group delay in the passband, indicating that the device exhibits low dispersion values. At such a high modulation frequency, the limitations of the measurement method described in Sect. 2.3.2 should be kept in mind. The maximum dispersion value in the passband is estimated to be ±2 ps/nm, confirming the nearly linear-phase nature of conventional non passband flattened AWGs.

The origin of residual dispersion has been investigated in AWG multiplexers with a Gaussian spectral response, as well as for passband flattened devices [42]. Phase and amplitude errors in the transmission of the arrayed waveguides have been identified as the main source of the dispersion. Fourier transform spectroscopy measurements performed on InP and silica-on-silicon devices showed that slowly-varying phase errors (i. e. non-random contributions to the phase distribution of each path in the arrayed waveguides) were responsible for dispersion imperfections [43, 76], as confirmed by device modelling [77].

The three examples shown above highlight the importance of the choice of a proper technology if the dispersive properties of DWDM filters are relevant. The degrees of freedom offered by apodisation of fibre Bragg gratings enable to realise nearly ideal flat-top transfer functions offering low adjacent channel crosstalk and reduced passband narrowing when cascaded, however, at the expense of dispersion at the edges of the passband. This has been shown to be even more critical for narrow bandwidth devices necessary to accommodate reduced channel spacing. On the other hand, concepts such as AWGs do theoretically offer dispersion-free devices (in practice limited by manufacturing imperfections resulting in amplitude and phase errors between the arrayed waveguides) at the expense of a less ideal amplitude transfer function. It will be shown in Sect. 2.4.4 how extra degrees of freedom can be introduced for advanced components design in order to mitigate those usual trade-offs. Even though their properties have not been detailed in this section, it is essential to keep dispersion in mind for the other filter technologies presented elsewhere in this book, including bulk grating based devices [78, 79], ring resonators [80], and wavelength interleavers [81].

2.4.2 System Impact of WDM Filter Dispersion

Since it has been realised that the dispersion of optical filters might affect the performance of WDM systems [2, 3], a number of theoretical, numerical, and experimental studies attempting to assess the impact of filter dispersion have been reported. Such investigations aim at either getting a better understanding of the impact of a given complex transfer function on the quality of the filtered signal, hence potentially enabling to optimise the design trade-offs for the amplitude and phase responses, or at assessing the tolerances or scalability limitations of optical links or networks making use of such filters.

From a component design point of view, the focus is on evaluating signal distortion induced by the combined effects of the amplitude and phase responses, in order to assess whether it is possible to reach a desired amplitude response target while keeping the phase induced distortion under control. In the case of wavelength (de)multiplexers, a so-called flat-top transfer function, presenting low adjacent channel crosstalk level, reduced bandwidth narrowing when cascaded, as well as small dispersion values at the passband edges will be sought. In other types of filtering elements such as dispersion compensating devices, the target will be to achieve a desired dispersion profile over a given bandwidth while minimising the effects of undesired group delay variations known as group delay ripples (cf. Sect. 2.4.3).

Optical network designers will have interest in determining the maximum number of devices that can be cascaded over a link, hence the number of nodes over which a signal can be routed transparently without resorting to optical or electro-optic regeneration [82]. Clearly, this number will depend on the rate at which the effective bandwidth of a cascade of filters decreases, justifying the need for flat-top transfer functions that are more resilient to bandwidth narrowing, but also on the dispersion accumulated over the path, including filter-induced dispersion. Additionally, the knowledge of system margins is of paramount importance for the design of all-optical WDM networks. These margins include the tolerance to misalignment between the signal centre frequency and the centre frequencies of the filters present along the link, which may be limited by the dispersion at the edges of the filters' passbands.

It is therefore essential to be able to evaluate dispersion-induced limitations arising because of WDM filters. Together with other key filter parameters such as:

- insertion loss,
- passband shape,
- adjacent channel crosstalk,
- in-band crosstalk for devices such as wavelength routers,
- polarisation-dependent loss and polarisation mode dispersion,
- thermal and mechanical stability,

the knowledge of the dispersive properties of a given filter and its system implications will help assessing the suitability of this particular filter technology to the system under design.

It is beyond the scope of this chapter to present results on dispersion-induced signal degradation for all the filter technologies presented elsewhere in this book. Those will depend not only on the filter type, but also on its practical realisation, as well as on numerous system parameters, including bit-rate, modulation format, signal extinction ratio, and chirp. Consequently, such results will not be general enough to be of significant value, and general trends of the dispersive behaviour of the most widely used filter technologies have already been presented in Sect. 2.4.1. Instead, general techniques used to evaluate the influence of WDM filter dispersion on signal degradation are introduced and illustrated by specific examples.

Many of the early demonstrations of the impact of filter dispersion refer to fibre gratings. Some of the first evaluations of the dispersion-induced limitation of the usable bandwidth of grating filters have been performed by assuming a maximum tolerable value of dispersion, for instance 1000 ps/nm corresponding to about 1 dB power penalty for a 10 Gbit/s non-return-to-zero (NRZ) system [83]. By calculating the grating complex transfer function, typically by using a transfer matrix approach based on coupled-modes equations [84], it becomes possible to evaluate the frequency range over which the filter dispersion remains within this limit [85]. While such an approach enables to easily link grating design parameters (such as the apodisation function, grating length, and refractive index modulation value) to a system related quantity (the grating usable bandwidth), it ignores the interaction between amplitude and phase filtering as well as the fact that the dispersion may vary significantly over the modulated spectrum width, especially at the edges of the filter passband. Consequently, more accurate system modelling is required in practice. Calculating the amount of broadening experienced by a single Gaussian pulse has been used to evaluate the dispersive behaviour of a chirped moiré grating exhibiting almost constant in-band group delay [86]. However, in digital optical communication systems, a linear device such as an optical filter may introduce intersymbol interference, and it is therefore desirable to evaluate its effect on a whole data pattern, usually modelled as a pseudo-random bit sequence (PRBS). It is customary to resort to numerical simulation in order to calculate the complex envelope of a modulated signal after it has been filtered by one or a cascade or WDM filters. In the absence of suitable receiver models able to properly evaluate the bit-error-rate (BER) of signals strongly deteriorated by intersymbol interference, qualitative comparisons based on calculations of the eye-opening penalty (EOP) are often used. However, such a figure of merit is difficult to relate to the BER-based quantities, such as power penalty, that are used in practice to define the system margins.

It is therefore preferable, whenever possible, to perform an experimental evaluation of the filtering-induced signal degradation. Such early investigations have been reported for instance in [87, 88] for a single apodised fibre Bragg grating used in reflection or in transmission. The influence of the chirp of a 10 Gbit/s NRZ modulated signal was further investigated. Asymmetries in the penalty against wavelength detuning curves for chirped signals enabled to conclude on the influence of the grating dispersion that exhibits opposite signs on the short and long wavelength sides of the passband. A first experimental comparison of the power penalty induced by a fibre Bragg grating and an arrayed waveguide grating filter was presented in [89], clearly outlining the inherent dispersive limitations of fibre gratings as opposed to linear phase devices [90]. Other early reports on the evaluation of the system impact of filter dispersion include penalty measurements in transmission and reflection for a single tuneable grating [91]. As an illustration, Fig. 2.13 shows the measured penalty as a function of wavelength for a 10 Gbit/s NRZ signal reflected by a Gaussian apodised FBG designed for 50 GHz channel spacing. The usable bandwidth for a maximum tolerable power penalty can be determined from such a measurement. However, distinguishing the relative contribution of phase and amplitude filtering is difficult. Numerical simulations, where it is possible to independently calculate the effect of amplitude, phase, and conjugated filtering, might provide qualitative insights into the main source of signal degradation and can be used to complement such penalty measurements.

When a fibre grating is used in transmission, as is typically the case in optical add-drop multiplexer structures, either using optical circulators [92] or in a Mach–Zehnder interferometer configuration [66] (cf. Figs. 5.28 and 5.29), its out-of-band dispersion might induce degradation to the channels that are immediately adjacent to the dropped one. This might ultimately limit the channel spacing in DWDM networks making use of such devices. This potential limitation was recognised in [93] where a single Gaussian pulse analysis was performed based on an analytical expression for the out-of-band dispersion of a grating used in transmission. An evaluation of grating dispersion in transmission and its consequence on pulse degradation was also performed in [94, 95] where the use of the grating in

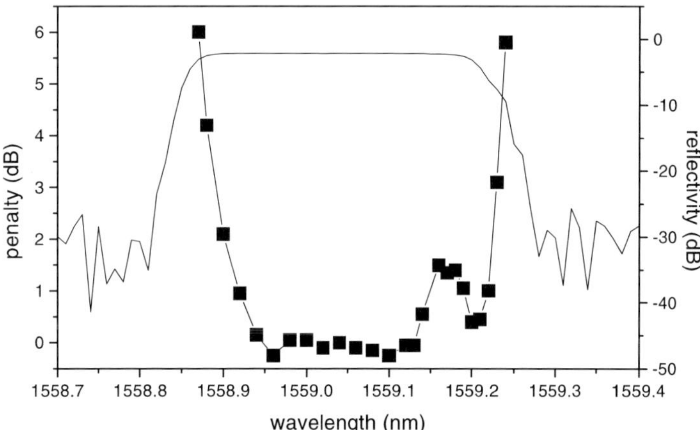

Fig. 2.13. Power penalty (at a BER of 10^{-9} – non-preamplified receiver) for a 10 Gbit/s NRZ signal reflected by the fibre grating designed for 50 GHz channel spacing whose complex transfer function is represented in Fig. 2.10. The conjugate effects of amplitude and phase filtering are accounted for in the penalty curve. Black squares: power penalty, full line: reflectivity

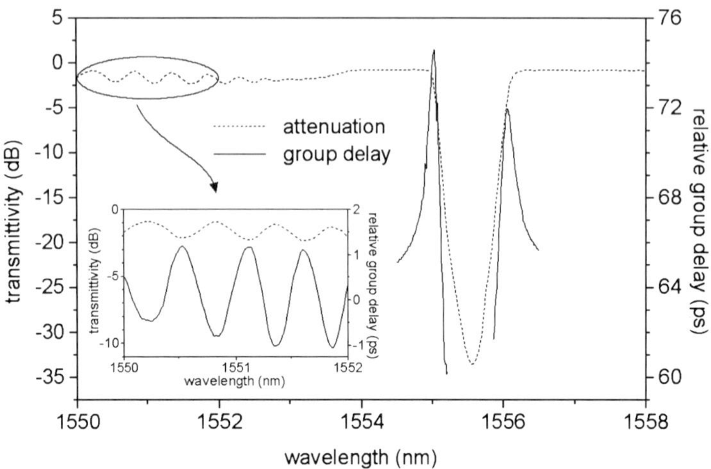

Fig. 2.14. Measured transmittivity and relative group delay in transmission of a fibre grating Mach–Zehnder OADM. The group delay around the stop-band was measured at a modulation frequency of $f_m = 5$ GHz, while $f_m = 2.5$ GHz was used to characterise the group delay associated to the amplitude ripples on the short wavelength side (inset)

an OADM was considered. The cascadability of Gaussian apodised gratings was also investigated in [96] based on an analytical expression for the out-of-band dispersion and a maximum tolerable value of dispersion. The fact that fibre gratings are minimum-phase filters in transmission has also been exploited to provide bounds on their dispersion and to examine the consequences for grating design [97]. Figure 2.14 shows the group delay of a fibre grating-based Mach–Zehnder OADM close to its stop-band. No dispersion impairments are expected if this device is used in a WDM system with 200 GHz (1.6 nm at 1550 nm) channel spacing. However, operation with 100 GHz channel spacing would result in dispersion of the order of ±5 to ±15 ps/nm for the closest channels on both sides of the stopband. The attenuation and group delay ripples observed on the short wavelength side of the stop-band will be discussed in Sect. 2.4.3.

As already mentioned, for network and system design, it is often desirable to know not only the impact of a single WDM filter on signal distortion, but also that of a full cascade of filters, as encountered by the signal propagating along a given path in the network. In this case the full range of analytical and numerical techniques described previously remains available [95]. For instance, in [6] it was found that dispersion accumulation is the main limitation to the cascadability of flat-top FBGs and thin-film filters. Approximating the dispersion at the centre of the passband with a linear function of wavelength enabled to numerically assess the detuning tolerance of a cascade of filters for a given value of power penalty.

From the experimental side, transmission over a filter cascade can be emulated by re-circulating loop experiments where the signal is propagated a number of times through a single component. One obvious limitation to the scheme is that it emulates propagation through perfectly aligned filters having strictly identical transfer functions which is certainly not the case in a real network. However, unless one is able to do experiments on a specific link or path through a network, taking into account the relative detuning of the filters as well as manufacturing-induced differences between their transfer functions would require a statistical treatment in order to obtain results of a sufficiently general value, especially when the system is made nonlinear due to propagation over optical fibre spans. Other known limitations of re-circulating loop experiments are the extra loss induced by the loop switch, as well as the fact that some length of fibre is necessary to store data into the loop. Consequently, if one is interested in evaluating the degradation induced by the filter alone, it should be ensured that this fibre span is perfectly dispersion-compensated and operated in the linear regime. In spite of these limitations, re-circulating loops are convenient tools to emulate filtering-induced degradation in large scale WDM networks when only a few filter samples are available, as would be the case for device prototyping.

Such an approach has been used in [3] to successfully demonstrate that the cascadability of WDM filters is also influenced by their dispersive characteristics. The detuning tolerance of dispersive multilayer interference filters was found to be less than half that of dispersion-free AWG designs. In Fig. 2.15 it is shown how the usable bandwidth of the fibre Bragg grating designed for 50 GHz channel spacing, whose transfer function is represented in Fig. 2.10, narrows when the device is cascaded up to 5 times in a re-circulating loop. Furthermore, its optimum operation wavelength is shifted towards the short wavelength side of its passband (1558.98 nm) corresponding to an extremum of the group delay curve, hence zero-dispersion [98].

Another reported approach has consisted in measuring the power penalty induced by a single grating, and after ensuring good agreement with simulations, numerically extrapolating the results to a cascade of filters [99]. Straight line experiments aiming at evaluating the power penalty induced by a cascade of FBGs have also been reported [100]. In [101], an experimental and numerical comparison of the power penalty induced by cascading fibre Bragg gratings and thin-film filters was further conducted. It was confirmed in this study that the dispersion of the gratings was largely responsible for the measured penalty. More recently, a detailed comparison of the cascadability behaviour of conventional Blackman apodised gratings with dispersion-optimised designs has confirmed the

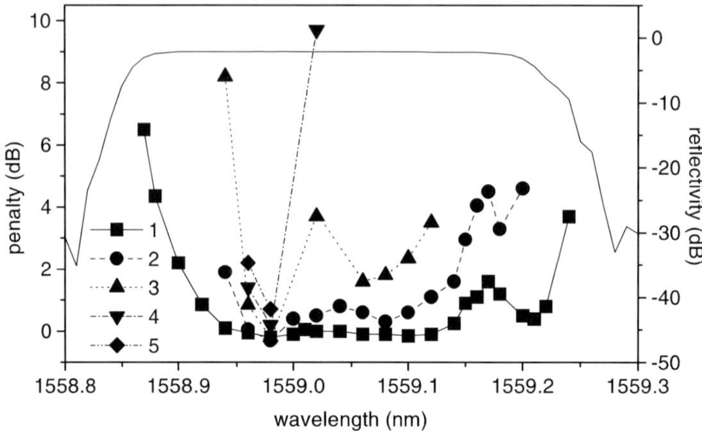

Fig. 2.15. Power penalty as a function of wavelength for 1 to 5 round trips in a recirculating loop set-up where a 10 Gbit/s NRZ signal is reflected by a Gaussian apodised FBG designed for 50 GHz channel spacing

importance of proper dispersion engineering for high bandwidth utilisation in FBG-based devices [102].

The use of vestigial side-band (VSB) filtering has enabled the demonstration of DWDM systems with high spectral efficiency and record capacity [103]. In such schemes the redundancy of the information contained in the two side-bands of e. g. an NRZ signal is exploited in order to reduce the channel spacing after partial suppression of one of the side-bands by optical filtering. In contrast to the conventional use of filtering elements as (de)multiplexers, the edge of the passband of optical filters is now used to effectively suppress the undesired side-band. Alternatively, a notch filter can be used for the same purpose [104]. In both cases, the need for a sharp filter cut-off necessary in order to obtain good side-band suppression may cause the preserved side-band to experience significant dispersion, depending on the filter technology. Consequently, a trade-off between the side-band suppression ratio and the dispersion-induced pre-chirping of the signal needs to be considered since both result from the steepness of the VSB filter transfer function [105]. The negative dispersion associated with the short wavelength edge of the passband or stopband of a fibre grating might actually contribute to the observed enhanced dispersion tolerance of VSB signals whose corresponding side-band has been suppressed, as observed in [104]. To illustrate this point, the eye-opening penalty of a filtered 10 Gbit/s NRZ signal has been calculated for various values of the detuning of its centre frequency with respect to the centre of the passband of a Gaussian apodised FBG designed for 50 GHz channel spacing. The results are shown

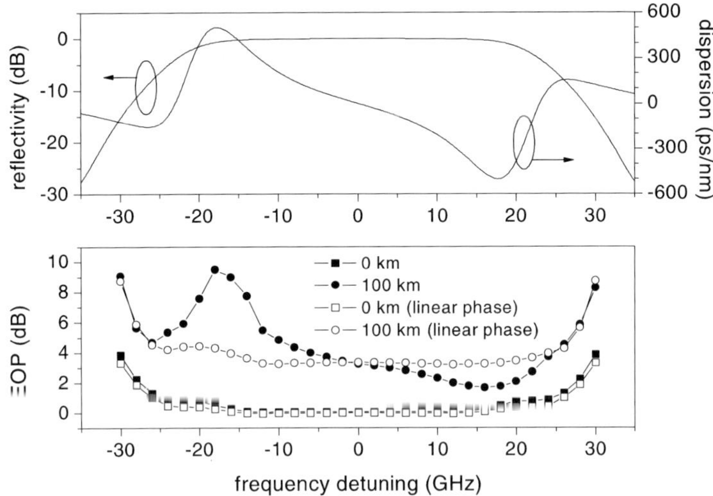

Fig. 2.16. Top: calculated transfer function (reflectivity and dispersion) of a Gaussian apodised fiber Bragg grating designed for 50 GHz channel spacing. Bottom: eye-opening penalty as a function of NRZ signal detuning with respect to the grating centre frequency. The EOP was calculated using a $2^{10}-1$ pseudo-random sequence directly after the VSB filter (0 km) and after transmission over 100 km SMF. For comparison, EOPs have also been calculated assuming the VSB filter is linear-phase

in Fig. 2.16 for transmission over 100 km standard single mode fibre, as well as directly at the output of the VSB filter. For comparison, the corresponding eye opening penalties have also been calculated with a hypothetical VSB filter whose amplitude transfer function is the same as that of the FBG, but exhibiting no dispersion (linear-phase filter). It is clearly seen that, beyond reduction of the signal spectral width potentially allowing for increased spectral efficiency, the negative filter dispersion is mostly responsible for the enhanced dispersion tolerance observed for positive frequency detunings. Consequently, the dispersion of the VSB filter needs to be taken into account for the proper design of such systems, as outlined in [104], as well as in [106] where a low dispersion bulk diffraction grating multiplexer was used for side-band suppression.

2.4.3 Group Delay Ripples

Fast oscillations of the group delay of an optical filter with respect to wavelength are usually referred to as group delay ripples (GDR). Here, fast oscillations means that their frequency period is small compared to the filter bandwidth and to the average trend of the group delay change within the filter passband. Such ripples can be found in a number of resonant WDM filter types including chirped fibre Bragg gratings used for dispersion compensation [107] or for gain equalisation [108], long period fibre gratings [109], and various structures of all-pass filters designed for dispersion compensation [110, 111]. The amplitude and periodicity of the GDRs depend on the filter design as well as on imperfections in the fabrication process. In the following, GDRs are illustrated in more detail in the context of chirped fibre Bragg gratings. However, the subsequent discussion on their impact on optical fibre communication systems holds independently of the filter technology.

In the case of FBGs, group delay ripples occur due to interference induced by reflections at the grating ends [112]. Those ripples can therefore be reduced by a proper apodisation of the grating [107, 113], cf. also Chap. 5. Furthermore, imperfections in the fabrication process are responsible for random variations of the period and amplitude of the grating refractive index modulation that also induce GDRs through residual multiple reflections [114, 115]. Consequently, beyond the optimisation of the apodisation profile, improvements in the fabrication process are also necessary in order to allow better control of GDRs in chirped fibre gratings [116].

Figure 2.17 shows the attenuation and group delay (measured using the phase shift technique with a modulation frequency of 130 MHz) of an early chirped FBG. The average dispersion of the device (obtained from the slope of a linear fit to the measured group delay as a function of wavelength over the 1555.0 to 1555.4 nm range) is of the order of −800 ps/nm,

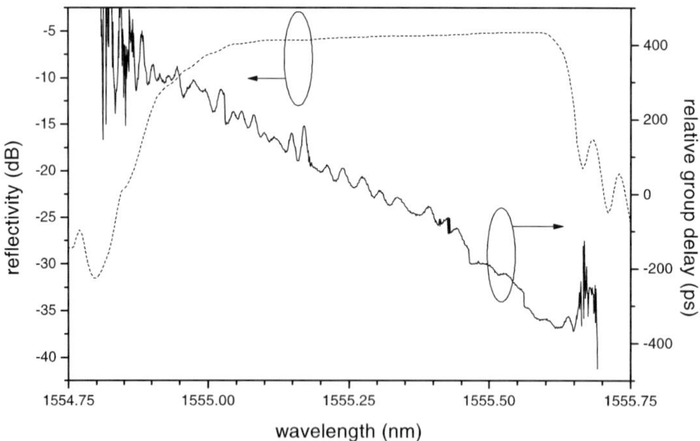

Fig. 2.17. Measured reflectivity and relative group delay of an early chirped fibre Bragg grating for dispersion compensation, showing typical group delay ripples associated with this type of component

enabling compensation of the dispersion accumulated over about 50 km of standard single mode fibre. GDRs with a period of about 4 GHz and peak-to-peak amplitude of up to several tens of picoseconds are clearly visible, resulting in large variations of the local dispersion. This example is provided for illustrative purpose only as contemporary chirped FBGs exhibit much lower GDRs thanks to optimised apodisation and improved fabrication techniques.

In order to specify the amount of GDR that can be tolerated for a given system, it is essential to be able to assess its influence on the filtered signals. The impact of GDRs on system performance is difficult to predict since it depends on their amplitude, period (relative to the signal spectral width), and phase (with respect to the channel centre frequency). The purpose of a typical chirped FBG is to compensate for the dispersion accumulated by the signal over propagation in optical fibre. Consequently, provided higher order dispersion effects can be ignored in the fibre, a linear dependence of the group delay of the FBG with respect to wavelength is expected. Any departure from this situation is likely to result in unwanted signal distortion. For instance, a signal tuned close to the quadrature point of a sinusoidal group delay ripple having a frequency period sufficiently large compared to the bit rate will experience nearly constant dispersion, while the effect will be equivalent to that of higher order dispersion in case the signal is tuned close to an extremum of the ripple [117, 118]. Consequently, the ripple will be responsible for the creation of echo pulses that may interact with neighbouring information bits, hence leading to inter-symbol interference. This point is illustrated in Fig. 2.18, where the effect of a sinusoidal GDR on an isolated raised cosine pulse has been calculated for three different relative detunings between the ripple and the centre frequency of the pulse spectrum. In the case of a sinusoidal GDR, the delay of the echo pulses can be shown to be equal to the inverse of the frequency period of the ripple. The worst case signal degradation is then induced by GDRs whose period is close to the bit rate [118–120]. Beyond this qualitative understanding some figure of merit needs to be found in order to be able to quantify the effects of GDRs [121]. The problem is complicated by the fact that the ripples of real fabricated components are not sinusoidal and that, as pointed out earlier, the induced system penalty depends also on the phase and amplitude of the ripple.

Numerous theoretical and numerical studies have contributed to a better understanding of the effects of the different relevant GDR parameters such as peak-to-peak amplitude, frequency, and phase (on top of some of the previous references, some relevant discussions can be found in e. g. [122–124]). It was shown in [125] that the peak-to-peak value of the phase ripple over the signal bandwidth (as opposed to that of the group delay

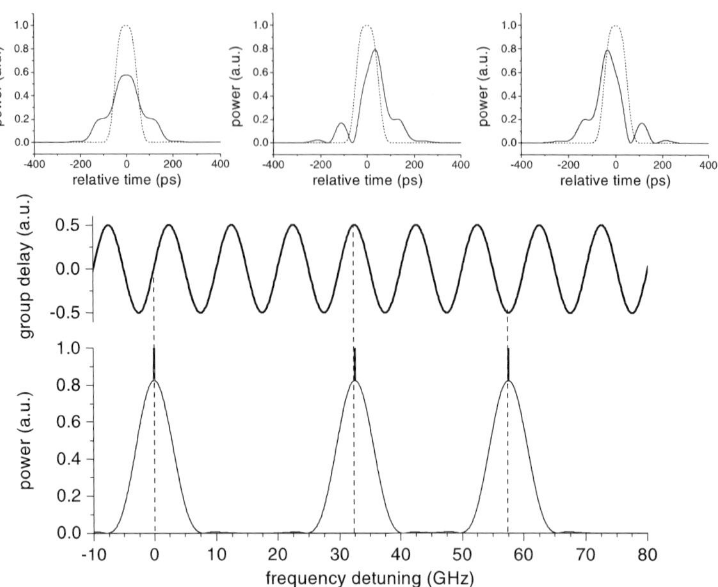

Fig. 2.18. Calculated illustration of the effect of a sinusoidal group delay ripple on an isolated pulse depending on the phase of the ripple with respect to the modulated spectrum centre frequency. The dotted line represents the pulse shape at the filter input. Here, the pulse spectral width compared to the GDR period means that the pulse will experience higher order dispersion even when the signal is tuned to the quadrature point of the ripple. In this example, an arbitrarily high value of the peak-to-peak GDR (100 ps) has been used in order to enhance its effect

ripple) could be used to accurately predict the worst case penalty experienced by the filtered signal, even in the presence of realistic GDR profiles. A method enabling the direct determination of the phase ripple with high resolution was subsequently proposed [126]. Good correlation between system performance and two figures of merit that can be extracted from GDR measurements (the residual dispersion and the variance of the residual phase ripple in the signal bandwidth) has also been reported in [127]. Ultimately, as in the general case for dispersive devices discussed in Sect. 2.4.2, experimental investigations can be performed to assess the real impact of GDRs on the performance of a fabricated filter [128, 129]. Furthermore, as the GDRs induced by imperfections in the fabrication process are not reproducible from one device to another, investigations of cascaded filtering, e. g. when chirped FBGs for dispersion compensation are used in multi-span systems, require some form of statistical analysis [130].

Coupling to cladding modes is responsible for amplitude ripples on the short wavelength side of the transfer function of fibre gratings used in transmission [26, 131], as can be observed on the measured transfer function of an early FBG Mach–Zehnder OADM depicted in Fig. 2.14. Such amplitude ripples are associated with group delay ripples having the same periodicity, as shown in the inset of Fig. 2.14. In this particular case, the period of the ripples is estimated to be about 0.5 nm around 1551 nm, and the maximum dispersion values are ±15 ps/nm. The evaluation of system impairments related to those dispersion and amplitude ripples would require statistical considerations on their amplitude, periodicity, and phase, as well as on the topology and wavelength assignment of the WDM network where such OADMs would be used. Fortunately, techniques exist that can successfully suppress these short-wavelength ripples [132, 133].

2.4.4 Advanced Filter Design

Optical filters for WDM systems have to satisfy a number of system requirements. As discussed previously, those include low out-of-band crosstalk, large detuning tolerance in order to allow for possible system drifts, as well as low dispersion, especially when multiple filtering elements are encountered by a signal over its path in a transparent network domain. Depending on the filter technology, it might be contradictory to attempt to achieve several of those goals simultaneously. For instance, approaching a flat-top power transfer function is a desirable feature for a wavelength (de)multiplexer as it results in low adjacent channel crosstalk as well as reduced passband narrowing due to cascading. However, if the filter is minimum-phase, increased dispersion at the edges of its passband will result from the action of squaring its amplitude response. Inversely, conventional arrayed waveguide grating (de)multiplexers have been shown to exhibit very low dispersion in their passband. Unfortunately, their amplitude response can be well approximated by a Gaussian transfer function [75], and consequently exhibits smooth roll-off, making them prone to severe passband narrowing when concatenated over a given path in a network.

In this section, the implications of those design trade-offs on filter dispersion are illustrated based on two examples. First, the effect of passband flattening of AWGs on their dispersive properties is examined. Second, it is shown how advanced apodisation profiles can be used in order to reduce the dispersion of fibre Bragg gratings while maintaining their ideal square amplitude response.

Passband-flattened AWGs

Several methods have been proposed in order to approximate the desired flat-top transfer function for AWGs [134–138]. One promising approach

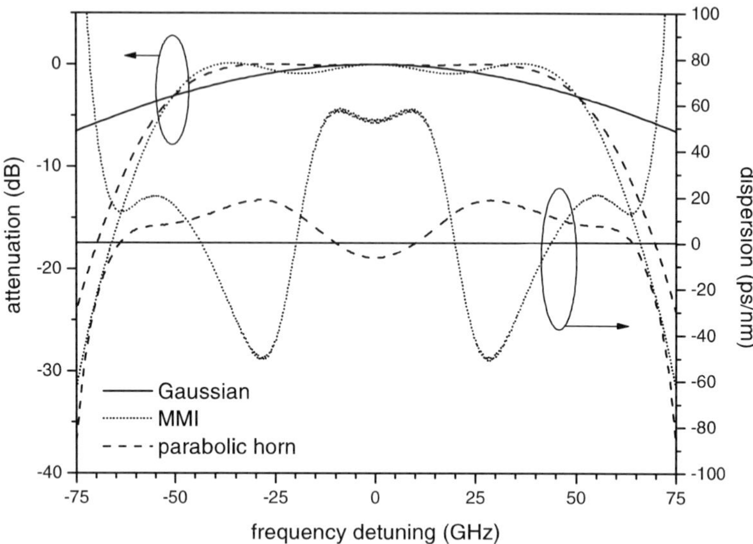

Fig. 2.19. Calculated power transfer function and group delay of a conventional Gaussian and of passband-flattened AWGs using an MMI coupler or a parabolic horn. The three devices are designed for the same 3 dB bandwidth of 100 GHz (device modelling by Chrétien Herben, METEOR project)

consists in approximating a flat-top response by the summation of two Gaussian distributions. This principle can be realised in practice by using a multi-mode interference (MMI) coupler [139] before the input slab, as proposed in [140, 141]. Alternatively, using a parabolic horn before the input slab has also been shown to result in passband flattening, as proposed in [142]. These last two approaches are the ones considered here. Issues to take into account when introducing a passband flattening process in the design of an AWG (de)multiplexer are the possible presence of ripples in the amplitude response, the introduction of excess loss, as well as the introduction of dispersion [143].

Numerical simulations have been performed to compare the dispersion properties of devices having the same 3 dB bandwidth. The results of accurate device modelling are shown in Fig. 2.19 for a conventional Gaussian, as well as for AWGs whose passbands have been flattened by using either an MMI coupler or a parabolic horn at the input of the free propagation region.

As expected, the standard Gaussian design is dispersion-free, whereas relatively large dispersion values are observed in conjunction with ~1 dB amplitude ripples when an MMI coupler is employed to square the field distribution before the input slab. Clearly, such high dispersion values would be unacceptable for high bit rate applications and the large amplitude ripple would also prevent cascading the device. Using a parabolic horn, both amplitude ripple and dispersion are reduced. Further, a proper design of the parabolic horn has been shown to result in flat-top transmission with reduced chromatic dispersion [144]. Low dispersion passband flattened designs based on engineering the relative attenuation and phase shifts of the arrayed waveguides have also been demonstrated [145].

Low Dispersion Fibre Bragg Gratings

Since fibre Bragg gratings are compact, low-loss devices that are inherently compatible with optical fibres, considerable interest exists for realising advanced transfer functions that would present a rectangular passband in conjunction with linear phase.

It can be shown that, in the weak coupling regime, the coupling potential due to the refractive index corrugation $\kappa(z)$, where z is the longitudinal coordinate of the grating, and the reflectivity are related by a Fourier transform-like expression [146]. Although this relation no longer holds for strong gratings, it can nevertheless provide some rough design guidelines for the optimisation of the shape of the reflectivity of grating filters. The coupling potential $\kappa(z)$ depends directly on the amplitude and phase of the refractive index perturbation [147]. Therefore, in order to achieve a flat-top transfer function, the refractive index profile should follow a $\sin z / z$ dependence. However, this requires the ability to synthesise π phase-shifts in the refractive index profile. Gratings realised following this approach have been shown in [148] to exhibit far less dispersion than gratings with conventional Gaussian apodisation profiles. The use of other symmetric refractive index profiles with phase-shifts has enabled the demonstration of filters with square amplitude response and reduced dispersion [149, 150].

It has been mentioned in Sect. 2.2.2 that grating filters are generally not of the minimum-phase type when used in reflection. This opens the possibility to achieve the double goal of synthesising devices with an ideal rectangular transfer function and low dispersion at the edges of the passband. However, it has also been demonstrated in [20] that, if the grating is symmetric, its group delay is identical in transmission and in reflection. A grating being minimum-phase in transmission [20], the group delay in reflection of a symmetric device is then uniquely determined by its reflectivity. Therefore, using an asymmetric refractive index profile offers additional degrees of freedom for the design of low dispersion filters with nearly square amplitude responses. In [151–154] asymmetric refractive index profiles with multiple phase-shifts have been demonstrated, allowing

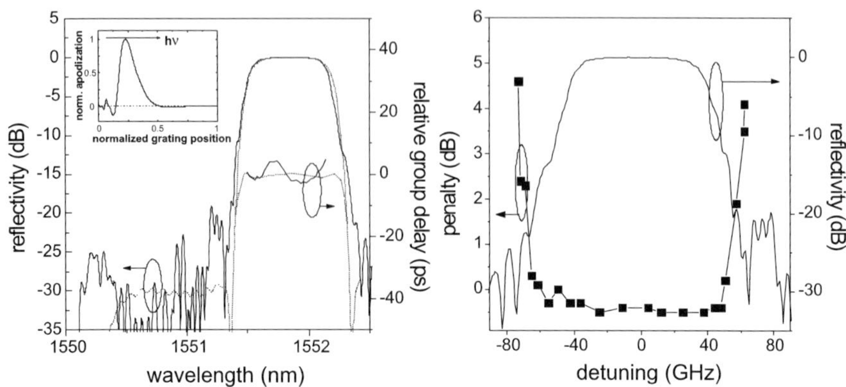

Fig. 2.20. (*Left*) measured (*solid line*) and calculated (*dotted line*) reflectivity and group delay of a low dispersion fibre Bragg grating whose normalized apodisation profile is shown as inset. (*Right*) measured penalty as a function of detuning for 10 Gbit/s return-to-zero (RZ) modulation [157]. Device modelling and characterisation by Hans-Jürgen Deyerl

the fabrication of nearly dispersion-free, high reflectivity, rectangular gratings. However, unlike the symmetric designs presented above, the dispersion-free operation of those filters is dependent on the direction of light propagation. Consequently, they may not be used in conventional OADM structures. The development of advanced grating writing techniques, together with efficient algorithms enabling to calculate the required refractive index profile to achieve a given target filter response [155], have allowed the fabrication of grating structures where both amplitude and phase responses can be engineered [156].

Figure 2.20 shows the asymmetric apodisation profile of a low dispersion fibre Bragg grating [157]. The grating was fabricated using the polarisation control method [158], and its measured reflectivity and group delay are represented together with simulation results. A reflectivity of 99.7% for a 20 mm long device as well as a group delay fluctuation of less than 10 ps were obtained within the 3 dB bandwidth of a device designed for 100 GHz channel spacing. The good dispersion properties of the grating were confirmed by penalty measurements at 10 Gbit/s, showing that up to 99% of the filter's 20 dB bandwidth could be used with less than 1 dB power penalty.

Consequently, recent results have demonstrated that advanced asymmetric refractive index profiles with multiple phase shifts can be used in order to simultaneously realise nearly ideal square amplitude responses with low dispersion.

2.5 Conclusion

In this chapter, the need for accurate determination of the dispersive properties of WDM filters and estimate of their impact on systems has been highlighted. Over the past few years, the awareness of both component and system designers towards this sensitive issue has grown, and it is now customary to present group delay curves and specifications together with the amplitude response of optical filters.

From the characterisation side, two key experimental methods, namely low coherence interferometry and the modulation phase-shift technique, have been presented in detail and their benefits and limitations discussed. These limitations should be kept in mind when analysing group delay curves found in the literature and product descriptions.

Beyond component characterisation, it is also essential to be able to quantify the effect of filters' complex transfer functions on the signal quality in an optical link or network. It has been reviewed how the in-band dispersion of WDM filters may limit the usable bandwidth of the devices as well as the number of devices that can be cascaded along a given path in a network. For some types of components, group delay ripples might induce some further signal degradation. The out-of-band dispersion in OADMs might also limit the channel spacing in DWDM systems.

Improved component manufacturing techniques are effective at reducing spurious dispersive effects due to e. g. group delay ripples or coupling to cladding modes which were critical in early generations of devices. However, basic physical limitations will remain which result from the fact that various filters are of minimum-phase type, and hence have dispersive properties fully determined by their amplitude response. It has also been illustrated how the introduction of extra degrees of freedom can be used to circumvent some of the limitations of conventional filter designs, allowing the realisation of advanced WDM filters whose amplitude and phase responses can be engineered to meet specific targets inferred from system requirements.

There is no reason to doubt that the increase in capacity needs observed since the dawn of optical communications will stop in the near future, meaning that the limits of high bit-rate and high spectral efficiency systems will need to be pushed further. Visions of complex transparent optical networks may also come closer to reality. These trends will make the requirements for understanding and control of dispersion in WDM filters even more important.

References

1. G. P. Agrawal: *Fiber-optic communication systems*, Chap. 2 (Wiley, New York, 1997)

2. N. N. Khrais, F. Shehadeh, J.-C. Chiao, R. S. Vodhanel, and R. E. Wagner: "Multiplexer eye-closure penalties for 10 Gb/s signals in WDM networks," *Techn. Digest Opt. Fiber Commun. Conf. (OFC'96)*, San Jose, California, USA, PD33 (1996)

3. C. Caspar, H.-M. Foisel, C. v. Helmolt, B. Strebel, and Y. Sugaya: "Comparison of cascadability performance of different types of commercially available wavelength (de)multiplexers," Electron. Lett. **33**, 1624–1626 (1997)

4. G. Lenz, B. J. Eggleton, C. K. Madsen, C. R. Giles, and G. Nykolak: "Optimal dispersion of optical filters for WDM systems," IEEE Photon. Technol. Lett. **10**, 567–569 (1998)

5. G. Lenz, B. J. Eggleton, C. R. Giles, C. K. Madsen, and R. E. Slusher: "Dispersive properties of optical filters for WDM systems," IEEE J. Quantum Electron. **34**, 1390–1402 (1998)

6. M. Kuznetsov, N. M. Froberg, S. R. Henion, and K. A. Rauschenbach: "Power penalty for optical signals due to dispersion slope in WDM filter cascades," IEEE Photon. Technol. Lett. **11**, 1411–1413 (1999)

7. L. G. Cohen: "Comparison of single-mode fiber dispersion measurement techniques," J. Lightwave Technol. **LT-3**, 958–966 (1985)

8. P. Hernday: "Dispersion measurements," in *Fiber Optic Test and Measurement* (D. Derickson, ed.), Chap. 12 (Prentice Hall, Upper Saddle River, NJ 1998)

9. W. H. Knox, N. M. Pearson, K. D. Li, and C. A. Hirlimann: "Interferometric measurements of femtosecond group delay in optical components," Opt. Lett. **13**, 574–576 (1988)

10. A. Papoulis: *The Fourier integral and its applications*, Chap. 10 (McGraw-Hill, New York, 1962)

11. H. W. Bode: *Network analysis and feedback amplifier design*, (Van Nostrand, New York, 1945)

12. L. D. Landau, E. M. Lifschitz, and L. P. Pitaevskii, *Electrodynamics of continuous media*, 2nd edition, pp. 279–283, (Butterworth-Heinemann, Oxford, 1984)

13. D. C. Hutchings, M. Sheik-Bahae, D. J. Hagan, and E. W. van Stryland: "Kramers–Krönig relations in nonlinear optics," Opt. Quant. Electron. **24**, 1–30 (1992)

14. M. Beck, I. A. Walmsley, and J. D. Kafka: "Group delay measurements of optical components near 800 nm," IEEE J. Quantum Electron. **27**, 2074–2081 (1991)

15. R. H. J. Kop, P. de Vries, R. Sprik, and A. Lagendijk: "Kramers–Kronig relations for an interferometer," Opt. Commun. **138**, 118–126 (1997)

16. M. A. Muriel and A. Carballar: "Phase reconstruction from reflectivity in uniform fiber Bragg gratings," Opt. Lett. **22**, 93–95 (1997)

17. A. Carballar and M. A. Muriel: "Phase reconstruction from reflectivity in fiber Bragg gratings," J. Lightwave Technol. **15**, 1314–1322 (1997)

18. D. Pastor and J. Capmany: "Experimental demonstration of phase reconstruction from reflectivity in uniform fibre Bragg gratings using the Wiener–Lee transform," Electron. Lett. **34**, 1344–1345 (1998)

19. J. Skaar and H. E. Engan: "Phase reconstruction from reflectivity in fibre Bragg gratings," Opt. Lett. **24**, 136–138 (1999)

20. L. Poladian: "Group-delay reconstruction for fiber Bragg gratings in reflection and transmission," Opt. Lett. **22**, 1571–1573 (1997)

21. F. W. King: "Analysis of optical data by the conjugate Fourier-series approach," J. Opt. Soc. Am. **68**, 994–997 (1978)

22. B. Harbecke: "Application of Fourier's allied integrals to the Kramers–Kronig transformation of reflectance data," Appl. Phys. A **40**, 151–158 (1986)

23. F. W. King: "Efficient numerical approach to the evaluation of Kramers–Kronig transforms," J. Opt. Soc. Am. B **19**, 2427–2436 (2002)

24. K. B. Rochford and S. D. Dyer: "Reconstruction of minimum-phase group delay from fibre Bragg grating transmittance/reflectance measurements," Electron. Lett. **35**, 838–839 (1999)

25. A. Ozcan, M. J. F. Digonnet, and G. S. Kino: "Group delay recovery using iterative processing of amplitude of transmission spectra of fibre Bragg gratings," Electron. Lett. **40**, 1104–1106 (2004)

26. T. Erdogan: "Fiber grating spectra," J. Lightwave Technol. **15**, 1277–1294 (1997)

27. C. J. Brooks, G. L. Vossler, and K. A. Winick: "Phase response measurement technique for waveguide grating filters," Appl. Phys. Lett. **66**, 2168–2170 (1995)

28. S. Barcelos, M. N. Zervas, R. I. Laming, and D. N. Payne; "Interferometric fibre grating characterization," *IEE Colloquium on Optical Fibre Gratings and their Applications*, pp. 5/1–5/7, (1995)

29. S. Barcelos, M. N. Zervas, R. I. Laming, D. N. Payne, L. Reekie, J. A. Tucknott, R. Kashyap, P. F. McKee, F. Sladen, and B. Wojciechowicz: "High accuracy dispersion measurements of chirped fibre gratings," Electron. Lett. **31**, 1280–1282 (1995)

30. M. Beck and I. A. Walmsley: "Measurement of group delay with high temporal and spectral resolution," Opt. Lett. **15**, 492–494 (1990)

31. P. Merritt, R. P. Tatam, and D. A. Jackson: "Interferometric chromatic dispersion measurements on short lengths of monomode optical fiber," J. Lightwave Technol. **7**, 703–716 (1989)

32. X. S. Yao and J. Feinberg: "Simple in-line method to measure the dispersion of an optical system," Appl. Phys. Lett. **62**, 811–813 (1993)

33. D. Müller, J. West, and K. Koch: "Interferometric chromatic dispersion measurement of a photonic bandgap fiber," Proc. SPIE **4870**, 395–403, (2002)

34. P.-L. Francois, M. Monerie, C. Vassallo, Y. Durteste, and F. R. Alard: "Three ways to implement interferencial techniques: application to measurements of chromatic dispersion, birefringence, and nonlinear susceptibilities," J. Lightwave Technol. **7**, 500–513 (1989)

35. K. Naganuma, K. Mogi, and H. Yamada: "Group-delay measurement using the Fourier transform of an interferometric cross correlation generated by white light," Opt. Lett. **15**, 393–395 (1990)

36. S. Diddams and J.-C. Diels: "Dispersion measurements with white-light interferometry," J. Opt. Soc. Am. B **13**, 1120–1129 (1996)

37. J. Gehler and W. Spahn: "Dispersion measurement of arrayed-waveguide gratings by Fourier transform spectroscopy," Electron. Lett. **36**, 338–339 (2000)

38. S. D. Dyer, K. B. Rochford, and A. H. Rose: "Fast and accurate low coherence interferometric measurements of fiber Bragg grating dispersion and reflectance," Opt. Express **5**, 262–266 (1999)

39. M. Volanthen, H. Geiger, M. J. Cole, R. I. Laming, and J. P. Dakin: "Low coherence technique to characterise reflectivity and time delay as a function of wavelength within a long fibre grating," Electron. Lett. **32**, 757–758 (1996)

40. S. D. Dyer and K. B. Rochford: "Low-coherence interferometric measurements of fibre Bragg grating dispersion," Electron. Lett. **35**, 1485–1486 (1999)

41. S. D. Dyer and K. B. Rochford: "Low-coherence interferometric measurements of the dispersion of multiple fiber Bragg gratings," IEEE Photon. Technol. Lett. **13**, 230–232 (2001)

42. H. Yamada, K. Okamoto, A. Kaneko, and A. Sugita: "Dispersion resulting from phase and amplitude errors in arrayed-waveguide grating multiplexers-demultiplexers," Opt. Lett. **25**, 569–571 (2000)

43. H. Yamada, H. Sanjoh, M. Kohtoku, K. Takada, and K. Okamoto: "Measurement of phase and amplitude error distributions in arrayed-waveguide grating multi/demultiplexers based on dispersive waveguide," J. Lightwave Technol. **18**, 1309–1320 (2000)

44. B. Costa, D. Mazzoni, M. Puleo, and E. Vezzoni: "Phase shift technique for the measurement of chromatic dispersion in optical fibers using LED's," IEEE J. Quantum Electron. **QE-18**, 1509–1514 (1982)

45. S. Ryu, Y. Horiuchi, and K. Mochizuki: "Novel chromatic dispersion measurement method over continuous Gigahertz tuning range," J. Lightwave Technol. **7**, 1177–1180 (1989)

46. K. Takiguchi, K. Okamoto, S. Suzuki, and Y. Ohmori: "Planar lightwave circuit optical dispersion equaliser," *Proc. European Conf. Opt. Commun. (ECOC'93)*, vol. 3, pp. 33–36, ThC 12.9 (1993)

47. ITU-T Recommendation G.650.1, *Definition and test methods for linear, deterministic attributes of single-mode fibre and cable* (International Telecommunications Union, Geneva, Switzerland, 2004)

48. T. Niemi, G. Genty, and H. Ludvigsen: "Group-delay measurements using the phase-shift method: improvement on the accuracy," *Proc. European Conf. Opt. Commun. (ECOC'01)*, Amsterdam, The Netherlands, Th.M.1.5 (2001)

49. T. Niemi, M. Uusimaa, and H. Ludvigsen: "Limitations of phase-shift method in measuring dense group delay ripple of fiber Bragg gratings," IEEE Photon. Technol. Lett. **13**, 1334–1336 (2001)

50. W. H. Press, S. A. Teukolsky, W. T. Vetterling, and B. P. Flannery: *Numerical recipes in C*, 2nd edition, Chap. 14.8, (Cambridge University Press, Cambridge, 1992)

51. A. J. Barlow, R. S. Jones, and K. W. Forsyth: "Technique for direct measurement of single-mode fiber chromatic dispersion," J. Lightwave Technol. **LT-5**, 1207–1213 (1987)

52. J. B. Schlager, S. E. Mechels, and D. L. Franzen: "Precise laser-based measurements of zero-dispersion wavelength in single-mode fibers," *Techn. Digest Opt. Fiber Commun. Conf. (OFC'96)*, vol. 2, pp. 293–294, FA2 (1996)

53. S. E. Mechels, J. B. Schlager, and D. L. Franzen: "Accurate measurements of the zero-dispersion wavelength in optical fibers," J. Res. Natl. Inst. Stand. Technol. **102**, 333–347 (1997)

54. G. H. Smith, D. Novak, and Z. Ahmed: "Technique for optical SSB generation to overcome dispersion penalties in fibre-radio systems," Electron. Lett. **33**, 74–75 (1997)

55. J. E. Román, M. Y. Frankel, and R. D. Esman: "Spectral characterization of fiber gratings with high resolution," Opt. Lett. **23**, 939–941 (1998)

56. R. M. Fortenberry: "Enhanced wavelength resolution chromatic dispersion measurements using fixed side-band technique," *Techn. Digest Opt. Fiber Commun. Conf. (OFC'00)*, Baltimore, Maryland, USA, vol 1, pp 107–109, TuG8 (2000)

57. G. Genty, T. Niemi, and H. Ludvigsen: "New method to improve the accuracy of group delay measurements using the phase-shift technique," Opt. Commun. **204**, 119–126 (2002)

58. R. Fortenberry, W. V. Sorin, and P. Hernday: "Improvement of group delay measurement accuracy using a two-frequency modulation phase-shift method," IEEE Photon. Technol. Lett. **15**, 736–738 (2003)

59. T. Dennis and P. A. Williams: "Relative group delay measurements with 0.3 ps resolution: toward 40 Gbit/s component metrology," *Techn. Digest Opt. Fiber Commun. Conf. (OFC'02)*, Anaheim, California, USA, pp 254–256, WK3 (2002)

60. T. Dennis and P. A. Williams: "Chromatic dispersion measurement error caused by source amplified spontaneous emission," IEEE Photon. Technol. Lett. **16**, 2532–2534 (2004)

61. S. Y. Set, M. K. Jablonski, K. Hsu, C. S. Goh, and K. Kikuchi: "Rapid amplitude and group-delay measurement system based on intra-cavity modulated swept-lasers," IEEE Trans. Instrum. Meas. **53**, 192–196 (2004)

62. A. H. Rose, C.-M. Wang, and S. D. Dyer: "Round robin for optical fiber Bragg grating metrology," J. Res. Natl. Inst. Stand. Technol. **105**, 839–866 (2000)

63. F. Devaux, Y. Sorel, and J. F. Kerdiles: "Simple measurement of fiber dispersion and of chirp parameter of intensity modulated light emitter," J. Lightwave Technol. **11**, 1937–1940 (1993)

64. B. Christensen, J. Mark, G. Jacobsen, and E. Bødtker: "Simple dispersion measurement technique with high resolution," Electron. Lett. **29**, 132–134 (1993)

65. C. Peucheret, F. Liu, and R. J. S. Pedersen: "Measurement of small dispersion values in optical components," Electron. Lett. **35**, 409–411 (1999)

66. D. C. Johnson, K. O. Hill, F. Bilodeau, and S. Faucher: "New design concept for a narrowband wavelength-selective optical tap and combiner," Electron. Lett. **23**, 668–669 (1987)

67. C. K. Madsen and J. H. Zhao: *Optical filter design and analysis, a signal processing approach* (Wiley, New York, 1999)

68. M. Scobey and R. Hallock: "Thin film filter based components for optical add/drop," in *OSA Trends in Optics and Photonics, WDM components,* (D. A. Nolan, ed.), vol 29, pp. 25–33, (Opt. Soc. America, Washington, DC, 1999)

69. B. Nyman, M. Farries, and C. Si: "Technology trends in dense WDM demultiplexers," Opt. Fib. Technol. **7**, 255–274 (2001)

70. K. Zhang, J. Wang, E. Schwendeman, D. Dawson-Elli, R. Faber, and R. Sharps: "Group delay and chromatic dispersion of thin-film-based, narrow bandpass filters used in dense wavelength-division-multiplexed systems," Appl. Opt **41**, 3172–3175 (2002)

71. R. B. Sargent: "Recent advances in thin film filters," *Techn. Digest Opt. Fiber Commun. Conf. (OFC'04)*, Los Angeles, California, USA, TuD6 (2004)

72. R. M. Fortenberry, M. E. Wescott, L. P. Ghislain, and M. A. Scobey: "Low chromatic dispersion thin film DWDM filters for 40 Gb/s transmission systems," *Techn. Digest Opt. Fiber Commun. Conf. (OFC'02)*, Anaheim, California, USA, pp. 319–320, WS2 (2002)

73. M. Tilsch, C. A. Hulse, F. K. Zernik, R. A. Modavis, C. J. Addiego, R. B. Sargent, N. A. O'Brien, H. Pinkney, and A. V. Turukhin: "Experimental demonstration of thin-film dispersion compensation for 50–GHz filters," IEEE Photon. Technol. Lett. **15**, 66–68 (2003)

74. M. K. Smit and C. van Dam: "PHASAR-based WDM-devices: principles, design and applications," IEEE J. Select. Topics Quantum Electron. **2**, 236–250 (1996)

75. H. Takahashi, K. Oda, H. Toba, and Y. Inoue: "Transmission characteristics of arrayed waveguide N×N wavelength multiplexer," J. Lightwave Technol. **13**, 447–455 (1995)

76. H. Yamada, H. Sanjoh, M. Kohtoku, and K. Takada: "Measurement of phase and amplitude error distributions in InP-based arrayed-waveguide grating multi / demultiplexers," Electron. Lett. **36**, 136–138 (2000)

77. P. Muñoz, D. Pastor, J. Capmany, and S. Sales: "Analytical and numerical analysis of phase and amplitude errors in the performance of arrayed waveguide gratings," IEEE J. Select. Topics Quantum Electron. **8**, 1130–1141 (2002)

78. C. X. Yu, D. T. Neilson, C. R. Doerr, and M. Zirngibl: "Dispersion free (de)mux with record figure-of-merit," IEEE Photon. Technol. Lett. **14**, 1300–1302 (2002)

79. R. Ryf, Y. Su, L. Möller, S. Chandrasekhar, X. Liu, D. T. Neilson, and C. Randy Giles: "Wavelength blocking filter with flexible data rates and channel spacing," J. Lightwave Technol. **23**, 54–61 (2005)

80. O. Schwelb: "Transmission, group delay, and dispersion in single-ring optical resonators and add/drop filters – a tutorial overview," J. Lightwave Technol. **22**, 1380–1394 (2004)

81. X. Liu: "Can 40–Gb/s duobinary signals be carried over transparent DWDM systems with 50–GHz channel spacing," IEEE Photon. Technol. Lett. **17**, 1328–1330, (2005)

82. N. Hanik, A. Ehrhardt, A. Gladisch, C. Peucheret, P. Jeppesen, L. Molle, R. Freund, and C. Caspar: "Extension of all-optical network-transparent domains based on normalized transmission sections," J. Lightwave Technol. **22**, 1439–1453 (2004)

83. A. H. Gnauck and R. M. Jopson: "Dispersion compensation for optical fiber systems," in *Optical Fiber Telecommunications IIIA* (I. P. Kaminow and T. L. Koch, eds.), Chap. 7 (Academic Press, San Diego, 1997)

84. M. Yamada and K. Sakuda: "Analysis of almost-periodic distributed feedback slab waveguides via a fundamental matrix approach," Appl. Opt. **26**, 3474–3478, (1987)

85. M. Ibsen, H. Geiger, and R. I. Laming: "In-band dispersion limitations of uniform apodised fibre gratings," *Proc. European Conf. Opt. Commun. (ECOC'98)*, Madrid, Spain, vol. 1, pp. 413–414 (1998)

86. L. R. Chen and P. W. E. Smith: "Fibre Bragg grating transmission filters with near-ideal filter response," Electron. Lett. **34**, 2048–2050 (1998)

87. G. Nykolak, G. Lenz, B. J. Eggleton, and T. A. Strasser: "Impact of fiber grating dispersion on WDM system performance," *Techn. Digest Opt. Fiber Commun. Conf. (OFC'98)*, San Jose, California, USA, vol. 2, pp. 4–5, TuA3 (1998)

88. G. Nykolak, B. J. Eggleton, G. Lenz, and T. A. Strasser: "Dispersion penalty measurements of narrow fiber Bragg gratings at 10 Gb/s," IEEE Photon. Technol. Lett. **10**, 1319–1321 (1998)

89. G. Lenz, G. Nykolak, and B. J. Eggleton: "Dispersion of optical filters in WDM systems: theory and experiment," *Proc. European Conf. Opt. Commun. (ECOC'98)*, Madrid, Spain, vol. 1, pp. 271–272 (1998)

90. G. Lenz, G. Nykolak, and B. J. Eggleton: "Waveguide grating routers for dispersionless filtering in WDM system at channel rate of 10 Gbit/s," Electron. Lett. **34**, 1683–1684 (1998)

91. P. Leisching, H. Bock, A. Richter, D. Stoll, and G. Fischer: "Optical add/drop multiplexer for dynamic channel routing," Electron. Lett. **35**, 591–592 (1999)

92. K. P. Jones, M. S. Chaudhry, D. Simeonidou, N. H. Taylor, and P. R. Morkel: "Optical wavelength add drop multiplexer in installed submarine WDM network," Electron. Lett. **31**, 2117–2118 (1995)

93. B. J. Eggleton, G. Lenz, N. Litchinitser, D. B. Patterson, and R. E. Slusher: "Implications of fiber grating dispersion for WDM communication systems," IEEE Photon. Technol. Lett. **9**, 1403–1405 (1997)

94. N. M. Litchinitser, B. J. Eggleton, G. Lenz, and G. P. Agrawal: "Dispersion in cascaded-grating-based add/drop filters," Techn. Dig. Conf. Lasers and Electro-Optics, (CLEO'98), CTh015 (1998)

95. N. M. Litchinitser, B. J. Eggleton, and G. P. Agrawal: "Dispersion of cascaded fiber gratings in WDM lightwave systems," J. Lightwave Technol. **16**, 1523–1529 (1998)

96. H. Geiger and M. Ibsen: "Complexity limitations of optical networks from out-of-band dispersion of grating filters," *Proc. European Conf. Opt. Commun. (ECOC'98)*, Madrid, Spain, vol. 1, pp. 405–406 (1998)

97. L. Poladian: "Design constraints for wavelength-division-multiplexed filters with minimal side-channel impairment," Opt. Lett. **26**, 7–9 (2001)

98. C. Peucheret, A. Buxens, T. Rasmussen, C. F. Pedersen, and P. Jeppesen: "Cascadability of fibre Bragg gratings for narrow channel spacing systems using NRZ and duobinary modulation," *Proc. OptoElectron. Commun. Conf. (OECC'01)*, Sydney, Australia, pp. 92–93, TUA3 (2001)

99. H. Bock, P. Leisching, A. Richter, D. Stoll, and G. Fischer: "System impact of cascaded optical add/drop multiplexers based on tunable fiber Bragg gratings," *Techn. Digest Opt. Fiber Commun. Conf. (OFC'00)*, Baltimore, Maryland, USA, vol. 2, pp. 296–298 (2000)

100. M. Kuznetsov, N. M. Froberg, S. R. Henion, C. Reinke, and C. Fennelly: "Dispersion-induced power penalty in fiber-Bragg-grating WDM filter cascades using optically preamplifed and nonpreamplified receivers," IEEE Photon. Technol. Lett. **12**, 1406–1408 (2000)

101. M. Kuznetsov, N. M. Froberg, S. R. Henion, C. Reinke, C. Fennelly, and K. A. Rauschenbach: "Dispersion-induced power penalty in fiber-Bragg-grating WDM filter cascades," *Techn. Digest Opt. Fiber Commun. Conf. (OFC'00)*, Baltimore, Maryland, vol. 2, pp. 311–313 (2000)

102. K. H. Ylä-Jarkko, M. N. Zervas, M. K. Durkin, I. Barry, and A. B. Grudinin: "Power penalties due to in-band and out-of-band dispersion in FBG cascades," J. Lightwave Technol. **21**, 506–510 (2003)

103. S. Bigo: "Multiterabit/s DWDM terrestrial transmission with bandwidth-limiting optical filtering," IEEE J. Select. Topics Quantum Electron. **10**, 329–340 (2004)

104. H. Kim and A. H. Gnauck: "10 Gbit/s 177 km transmission over conventional singlemode fibre using a vestigial side-band modulation format," Electron. Lett. **37**, 1533–1534 (2001)

105. Y. Kim, S. Kim, I. Lee, and J. Jeong: "Optimization of transmission performance of 10–Gb/s optical vestigial side-band signals using electrical dispersion compensation by numerical simulation," IEEE J. Select. Topics Quantum Electron. **10**, 371–375 (2004)

106. C. X. Yu, S. Chandrasekhar, T. Zhou, and D. T. Neilson: "0.8 bit/s/Hz spectral efficiency at 10 Gbit/s via vestigial-side-band filtering," Electron. Lett. **39**, 225–227 (2003)

107. K. Ennser, M. N. Zervas, and R. I. Laming: "Optimization of apodized linearly chirped fiber gratings for optical communications," IEEE J. Quantum Electron. **34**, 770–778 (1998)

108. H. Chotard, Y. Painchaud, A. Mailloux, M. Morin, F. Trépanier, and M. Guy: "Group delay ripple of cascaded Bragg grating gain flattening filters," IEEE Photon. Technol. Lett. **14**, 1130–1132 (2002)

109. D. B. Stegall and T. Erdogan: "Dispersion control with use of long-period fiber gratings," J. Opt. Soc. Am. A **17**, 304–312 (2000)

110. C. K. Madsen, G. Lenz, A. J. Bruce, M. A. Cappuzzo, L. T. Gomez, and R. E. Scotti: "Integrated all-pass filters for tunable dispersion and dispersion slope compensation," IEEE Photon. Technol. Lett. **11**, 1623–1625 (1999)

111. C. K. Madsen: "Subband all-pass filter architectures with applications to dispersion and dispersion-slope compensation and continuously variable delay lines," J. Lightwave Technol. **21**, 2412–2420 (2003)

112. L. Poladian: "Graphical and WKB analysis of nonuniform Bragg gratings," Phys. Rev. E **48**, 4758–4767 (1993)

113. L. Poladian: "Understanding profile-induced group-delay ripple in Bragg gratings," Appl. Opt. **39**, 1920–1923 (2000)

114. M. Sumetsky, B. J. Eggleton, and C. Martijn de Sterke: "Theory of group delay ripple generated by chirped fiber gratings," Opt. Express **10**, 332–340, 2002

115. R. L. Lachance, M. Morin, and Y. Painchaud: "Group delay ripple in fibre Bragg grating tunable dispersion compensators," Electron. Lett. **38**, 1505–1507 (2002)

116. M. Sumetsky, P. I. Reyes, P. S. Westbrook, N. M. Litchinitser, B. J. Eggleton, Y. Li, R. Deshmukh, and C. Soccolich: "Group-delay ripple correction in chirped fiber Bragg gratings," Opt. Lett. **28**, 777–779 (2003)

117. S. G. Evangelides Jr., N. S. Bergano, and C. R. Davidson: "Intersymbol interference induced by delay ripple in fiber Bragg gratings," *Techn. Digest Opt. Fiber Commun. Conf. (OFC'99)*, San Diego, California, USA, vol. 4, pp. 5–7, FA2 (1999)

118. T. N. Nielsen, B. J. Eggleton, and T. A. Strasser: "Penalties associated with group delay imperfections for NRZ, RZ and duo-binary encoded optical signals," *Proc. European Conf. Opt. Commun. (ECOC'99)*, Nice, France, vol. 1, pp. 388–389 (1999)

119. C. Scheerer, C. Glingener, G. Fischer, M. Bohn, and W. Rosenkranz: "Influence of filter group delay ripples on system performance," *Proc. European Conf. Opt. Commun. (ECOC'99)*, Nice, France, vol. 1, pp. 410–411 (1999)

120. C. Riziotis and M. N. Zervas: "Effect of in-band group delay ripple on WDM filter performance," *Proc. European Conf. Opt. Commun. (ECOC'01)*, Amsterdam, The Netherlands, vol. 4, pp. 492–493, Th.M.1.3 (2001)

121. K. Hinton: "Metrics for dispersion ripple in optical systems," Opt. Fib. Technol. **10**, 50–72 (2004)

122. K. Ennser, M. Ibsen, M. Durkin, M. N. Zervas, and R. I. Laming: "Influence of nonideal chirped fiber grating characteristics on dispersion cancellation," IEEE Photon. Technol. Lett. **10**, 1476–1478 (1998)

123. D. Penninckx, S. Khalfallah, and P. Brosson: "System impact of phase ripples in optical components," *Techn. Digest Opt. Fiber Commun. Conf. (OFC'01)*, Anaheim, California, USA, ThB4 (2001)

124. X. Liu, L. F. Mollenauer, and X. Wei: "Impact of group-delay ripple in transmission systems including phase-modulated formats," IEEE Photon. Technol. Lett. **16**, 305–307 (2004)

125. H. Yoshimi, Y. Takushima, and K. Kikuchi, "A simple method for estimating the eye-opening penalty caused by group-delay ripple of optical filters," *Proc. European Conf. Opt. Commun. (ECOC'02)*, Copenhagen, Denmark, vol. 4, 10.4.4 (2002)

126. N. Cheng, D. J. Krause, and J. C. Cartledge: "Measuring frequency response of dispersion compensating fibre Bragg grating using Fourier coefficients," Electron. Lett. **41**, 402–403 (2005)

127. M. H. Eiselt, C. B. Clausen, and R. W. Tkach: "Performance characterization of components with group delay fluctuations," IEEE Photon. Technol. Lett. **15**, 1076–1078 (2003)

128. B. J. Eggleton, A. Ahuja, P. S. Westbrook, J. A. Rogers, P. Kuo, T. N. Nielsen, and B. Mikkelsen: "Integrated tunable fiber gratings for dispersion management in high-bit rate systems," J. Lightwave Technol. **18**, 1418–1432 (2000)

129. B. J. Eggleton, A. Ahuja, P. S. Westbrook, J. A. Rogers, P. Kuo, T. N. Nielsen, and B. Mikkelsen: "Correction to : Integrated tunable fiber gratings for dispersion management in high-bit rate systems," J. Lightwave Technol. **18**, 1591 (2000)

130. S. Jamal and J. C. Cartledge: "Variation in the performance of multispan 10–Gb/s systems due to the group delay ripple of dispersion compensating fiber Bragg gratings," J. Lightwave Technol. **20**, 28–35 (2002)

131. V. Mizrahi and J. E. Sipe: "Optical properties of photosensitive fiber phase gratings," J. Lightwave Technol. **11**, 1513–1517 (1993)

132. E. Delevaque, S. Boj, J.-F. Bayon, H. Poignant, J. Le Mellot, M. Monerie, P. Niay, and P. Bernage: "Optical fiber design for strong gratings photoimprinting with radiation mode suppression," *Techn. Digest Opt. Fiber Commun. Conf. (OFC'95)* PD5 (1995)

133. F. Bakhti, F. Bruyère, X. Daxhelet, X. Shou, J. Da Loura, S. Lacroix, and P. Sansonetti: "Grating-assisted Mach–Zehnder OADM using photosensitive-cladding fibre for cladding mode coupling reduction," Electron. Lett. **35**, 1013–1014 (1999)

134. K. Okamoto and H. Yamada: "Arrayed-waveguide grating multiplexer with flat spectral response," Opt. Lett. **20**, 43–35 (1995)

135. Y. P. Ho, H. Li, and Y. J. Chen: "Flat channel-passband-wavelength multiplexing and demultiplexing devices by multiple-Rowland-circle design," IEEE Photon. Technol. Lett. **9**, 342–344 (1997)

136. C. Dragone, T. Strasser, G. A. Bogert, L. W. Stulz, and P. Chou: "Waveguide grating router with maximally flat passband produced by spatial filtering," Electron. Lett. **33**, 1312–1314 (1997)

137. A. Rigny, A. Bruno, and H. Sik: "Multigrating method for flattened spectral response wavelength multi/demultiplexer," Electron. Lett. **33**, 1701–1702 (1997)

138. C. Dragone: "Efficient techniques for widening the passband of a wavelength router," J. Lightwave Technol. **16**, 1895–1906 (1998)

139. L. B. Soldano and E. C. M. Pennings: "Optical multi-mode interference devices based on self-imaging: principles and applications," J. Lightwave Technol. **13**, 615–627 (1995)

140. J. B. D. Soole, M. R. Amersfoort, H. P. LeBlanc, N. C. Andreadakis, A. Rajhel, C. Caneau, R. Bhat, M. A. Koza, C. Youtsey, and I. Adesida: "Use of multimode interference couplers to broaden the passband of wavelength-dispersive integrated WDM filters," IEEE Photon. Technol. Lett. **8**, 1340–1342 (1996)

141. M. R. Amersfoort, J. B. D. Soole, H. P. LeBlanc, N. C. Andreadakis, A. Rajhel, and C. Caneau: "Passband broadening of integrated arrayed waveguide filters using multimode interference couplers," Electron. Lett. **32**, 449–451 (1996)

142. K. Okamoto and A. Sugita: "Flat spectral response arrayed-waveguide grating multiplexer with parabolic waveguide horns," Electron. Lett. **32**, 1661–1662 (1996)

143. M. E. Marhic and X. Yi: "Calculation of dispersion in arrayed waveguide grating demultiplexers by a shifting-image method," IEEE J. Select. Topics Quantum Electron. **8**, 1149–1157 (2002)

144. T. Kitoh, Y. Inoue, M. Itoh, M. Kotoku, and Y. Hibino: "Low chromatic-dispersion flat-top arrayed waveguide grating filter," Electron. Lett. **39**, 1116–1118 (2003)

145. B. Fondeur, A. L. Sala, H. Yamada, R. Brainard, E. Egan, S. Thedki, N. Gopinathan, D. Nakamoto, and A. Vaidyanathan: "Ultrawide AWG with hyper-Gaussian profile," IEEE Photon. Technol. Lett. **16**, 2628–2630, (2004)

146. H. Kogelnik: "Filter response of nonuniform almost-periodic structures," Bell Syst. Techn. J. **55**, 109–126 (1976)

147. G. Hugh Song: "Theory of symmetry in optical filter responses," J. Opt. Soc. Am. A. **11**, 2027–2037 (1994)

148. M. Ibsen, M. K. Durkin, M. J. Cole, and R. I. Laming: "Optimised square passband fibre Bragg grating filter with in-band flat group delay response," Electron. Lett. **34**, 800–802 (1998)

149. T. Shibata, M. Shiozaki, M. Ohmura, K. Murashima, A. Inoue, and H. Suganuma: "The dispersion-free filters for DWDM systems using 30 mm long symmetric fiber Bragg gratings," *Techn. Digest Opt. Fiber Commun. Conf. (OFC'01)*, Anaheim, California, USA, WDD84 (2001)

150. T. Shibata, K. Murashima, K. Hashimoto, M. Shiozaki, T. Iwashima, T. Okuno, A. Inoue, and H. Suganuma: "The novel dispersion reduced fiber Bragg grating suitable for 10 Gb/s DWDM systems," IEICE Trans. Electron. **E85-C**, 927–933, (2002)

151. M. Ibsen, R. Feced, P. Petropoulos, and M. N. Zervas: "99.9% reflectivity dispersion-less square-filter fibre Bragg gratings for high speed DWDM networks," *Techn. Digest Opt. Fiber Commun. Conf. (OFC'00)*, Baltimore, Maryland, USA, PD21 (2000)

152. M. Ibsen, R. Feced, P. Petropoulos, and M. N. Zervas: "High reflectivity linear-phase fibre Bragg gratings," *Proc. European Conf. Opt. Commun. (ECOC'00)*, Munich, Germany, vol. 1, pp. 53–54 (2000)

153. M. Ibsen, P. Petropoulos, M. N. Zervas, and R. Feced: "Dispersion-free fibre Bragg gratings," *Techn. Digest Opt. Fiber Commun. Conf. (OFC'01)*, San Diego, California, USA, MC1 (2001)

154. M. Ibsen, R. Feced, J. A. J. Fells, and W. S. Lee: "40 Gbit/s high performance filtering for DWDM networks employing dispersion-free fibre Bragg gratings," *Proc. European Conf. Opt. Commun. (ECOC'01)*, Amsterdam, The Netherlands, vol. 4, pp. 594–595, Th.B.2.1 (2001)

155. R. Feced, M. N. Zervas, and M. A. Muriel: "An efficient inverse scattering algorithm for the design of nonuniform fiber Bragg gratings," IEEE J. Quantum Electron. **35**, 1105–1115 (1999)

156. M. Ibsen and R. Feced: "Fiber Bragg gratings for pure dispersion-slope compensation," Opt. Lett. **28**, 980–982 (2003)

157. H.-J. Deyerl, C. Peucheret, B. Zsigri, F. Floreani, N. Plougmann, S. J. Hewlett, M. Kristensen, and P. Jeppesen: "A compact low dispersion fiber Bragg grating with high detuning tolerance for advanced modulation formats," Opt. Commun. **247**, 93–100 (2005)

158. H. J. Deyerl, N. Plougmann, J. B. Jensen, F. Floreani, H. R. Sørensen, and M. Kristensen: "Fabrication of advanced Bragg gratings with complex apodization profiles by use of the polarization control method," Appl. Opt. **43**, 3513–3522 (2004)

3 Diffraction Gratings WDM Components

Jean-Pierre Laude

3.1 Introduction

The key property of a grating is its ability to simultaneously diffract all incoming wavelengths, so that it is possible to construct simple wavelength division multiplexers, passive wavelength routers, wavelength cross-connects, and add-drops with a very large number of close channels or tuneable filters with a narrow bandwidth.

A diffraction grating [1–5] is an optical surface on which a large number of grooves, N_o, (several tens to several thousands per millimetre) are located. The grating has the property of diffracting light in a direction related to its wavelength (Fig. 3.1). Hence an incident beam with several wavelengths is angularly separated into different directions. Conversely, several wavelengths λ_1, λ_2, ... λ_n coming from different directions can be combined into the same direction. The diffraction angle depends on the groove spacing and on the incidence angle.

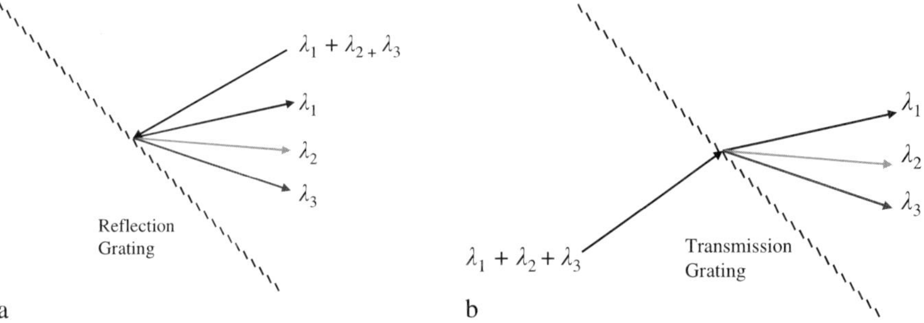

Fig. 3.1. Principle of demultiplexing by diffraction of an optical grating: wavelengths λ_1, λ_2, λ_3 coming from a single transmission fibre are diffracted into different directions. The diffraction grating works in reflection (**a**) or in transmission (**b**)

3.2 Grating Interference-diffraction Principles

3.2.1 Diffraction Orders

Let us consider (cf. Fig. 3.2) a transparent and equidistant slit array in vacuum, refractive index $n = 1$, and a plane wave incident at an angle θ (measured with respect to the normal of the grating). Each slit diffracts light in transmission. In the direction θ', measured from the perpendicular to the grating, the waves coming from different slits are in phase, if the optical path difference Δ_0 between successive optical paths $(AB) + (CD)$ is

$$\Delta_0 = \Lambda (\sin \theta + \sin\theta') = m\lambda_m , \qquad (3.1)$$

where m is an integer, λ is the wavelength, and Λ the distance between two successive slits. Direct transmission corresponds to $m = 0$, while $m = \pm 1$ correspond to the first diffraction orders on each side of the direct transmission.

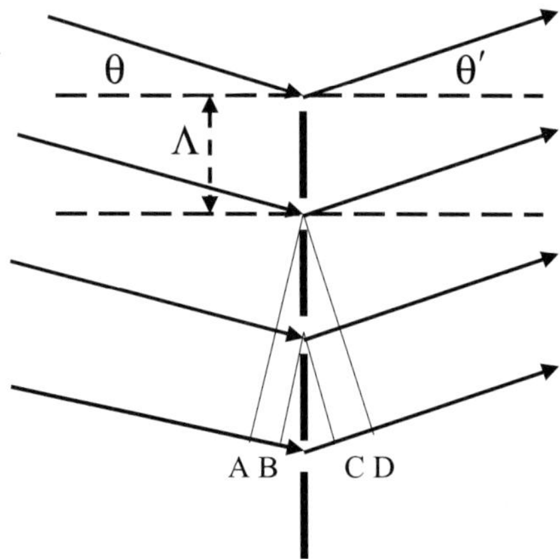

Fig. 3.2. Calculation of the diffraction order angles

3.2.2 Dispersion

It is easily demonstrated that the angular dispersion, corresponding to the wavelength variation is

$$\frac{d\theta'}{d\lambda} = \frac{m}{\Lambda \cos\theta'} \qquad (3.2)$$

or

$$\frac{d\theta'}{d\lambda} = \frac{\sin\theta + \sin\theta'}{\lambda\cos\theta'} \tag{3.3}$$

In the Littrow condition (corresponding to a reflection grating working with $\theta = \theta'$)

$$\frac{d\theta'}{d\lambda} = 2\frac{\tan\theta'}{\lambda} \tag{3.4}$$

Equation (3.4) shows that the angular dispersion versus the relative wavelength variation depends on $\tan\theta'$ only. The angular dispersion tends towards an infinite value when θ' tends to 90°. Consequently, wavelengths can be angularly separated.

3.2.3 Resolution

The resolution limit corresponds to the angular width of the diffraction of the whole surface of the grating projected into the direction θ'. One can show that the maximum resolution $R_{s,\,max}$ that can be obtained is

$$R_{s,\max} = \frac{\lambda}{d\lambda} = mN_0, \tag{3.5}$$

where N_0 is the total number of grooves. Very high resolution, typically $\lambda/d\lambda = 0.5\times10^6$, is currently obtained in spectrometric instrumentation. It is less in WDM components, in which the dimensions cannot be too high: for example, a 1800 lines/mm, 58 mm grating buried in silica corresponds to a resolution of 10^5 (or 16 picometres at a wavelength of 1.55 µm).

3.2.4 Free Spectral Range

The grating equation (3.1) is satisfied for a different wavelength for each integral diffraction order m. For a given set of incident and diffracted angles we get several wavelengths corresponding to different values of m. The free spectral range (FSR) of a grating is the largest wavelength interval in a given order, which does not overlap with the equivalent interval in an adjacent order. The free spectral range may be expressed by

$$FSR = \lambda_m - \lambda_{m+1} = \lambda_m / (m+1). \tag{3.6}$$

3.3 Grating Manufacturing

Two processes are currently used for the manufacturing of masters of diffraction gratings: diamond ruling on interferometrically controlled engines and, since its practical introduction in 1966 by Jobin Yvon, holographic recording.

3.3.1 Ruled Surface-relief Gratings

Since the first known diffraction grating was constructed by the American astronomer David Rittenhouse in 1785, who reported a half-inch grating with fifty-three apertures, several attempts have been made to produce gratings for scientific use. The first practical diffraction grating was realised in 1820 by the German scientist Joseph von Fraunhofer (1787–1826) with fine parallel wires stretched between two parallel threaded rods. With his grating he was already able to resolve the sodium D lines. In 1850 the Prussian F.A. Norbert was able to make gratings superior to Fraunhofer's. In the late 19th century Professor Henry Rowland (1848–1901) of Johns Hopkins University in Baltimore, with his mechanical engineer Schneider, began ruling plane and concave gratings on speculum metal blanks from John Brashear (1840–1920) on sophisticated ruling engines. Near 1880 Rowland and his co-workers were able to rule gratings with 1700 lines/mm.

A.A. Michelson of the University of Chicago originally designed his engine in the 1910s. This engine has later been rebuilt by B. Gale. Before the second world war the gratings of A.A. Michelson, typically 500 lines/mm, 30 cm long high quality gratings, made a tremendous impact on the field of spectroscopy. Around 1950 Richardson and R. Wiley of Bausch & Lomb rebuilt the engine of Milchelson, and in 1990 a new servo control system based on a laser interferometer was added.

G. Harrison of MIT, Bausch & Lomb (using the engine originally built by D. Mann of Lincoln in 1953), and Jobin Yvon (J.P. Laude and M. Gouley, 1965), following the technique of G. Harrison, made interferometric control ruling engines that are still today among the best operating ruling engines in the world. The needs of spectroscopy have given rise to the ruling of gratings used at high angle, comprising a higher and higher number of grooves per millimetre, or equivalently with a smaller number of grooves but used at high order. A further difficulty arises from the fact that the quality of the grating is closely connected with the degree of precision with which the straightness, the parallelism, and the equidistance of the grooves are controlled. Finally, the profile of the grooves, determined as will be seen later by the purpose for which the grating is intended, must be maintained constant from the first to the last groove, thus excluding any wear of the ruling tool in the course of operation. For example, in the Jobin Yvon engines the carriage of the grating rides on perfectly smooth ways under the precise control of a Michelson interferometer which controls the carriage displacement in order to maintain absolute parallelism and displacement accuracy in such a way that the mean error in the position of the grooves is less than 4 nm.

In the specific case of concave gratings, the profile remains uniform along the groove because the ruling diamond is not displaced parallel to itself, but rather its motion axis is adjusted on the centre of the grating concave blank.

The production of high quality gratings is linked to the solution of problems concerning temperature, vibration, the reliability of operating conditions during the ruling of the grating master that can last several days. The temperature of the machine is maintained constant to a few 1/100 of a degree. Particular attention is paid to the elimination of vibrations in the design of the room, and mechanical connections with the building are reduced to a minimum with dampers. The interferential control has the function of relating the advancement of the grating carriage between the ruling of two consecutive grooves to the wavelength of a monochromatic source. Such control eliminates residual periodic errors of the drive mechanism. Any short interruption in the operation of the machine would almost irremediably spoil the quality of the grating being produced.

3.3.2 Classical Holographic Gratings

These gratings, which are material recordings of interference fringes produced by laser beams, show characteristics of performance supplementing and extending the limitations of machine ruled gratings. At the very beginning of the laser era (1962), the theoretical possibility of obtaining such gratings was first mentioned. However, it was necessary to wait until 1968 and the innovations of the Jobin Yvon research team to get the first effective holographic plane and concave gratings [6–7]. Basically, this method consists of recording a system of interference fringes in the bulk of a photosensitive layer deposited on a blank of optical quality. An adequate chemical treatment develops a modulated profile which constitutes the grooves of the grating.

3.3.3 Holographic Gratings on Silicon

V-shape grooves can be fabricated by anisotropic etching of a silicon wafer under a mask of SiO_2 or Si_3N_4 manufactured using an intermediate photoresist holographic mask. The etchants are usually alkaline ions such as NaOH or preferably KOH at elevated temperatures. The etching rate is smaller in the [111] direction than in the other directions. If the surface of the wafer is (100), then symmetrical V grooves with an angle of 70.53° within two (111) side walls are obtained (Fig. 3.3). Asymmetric grooves are obtained on silicon substrates polished with a proper inclination γ of the crystal surface (becoming an hkk plane) with the (111) plane. The top angle will be 70.53° (k larger than h) or 180° − 70.53° = 109.47° (k smaller than h, Fig. 3.4). Excellent Si gratings for telecommunication applications have been proposed early [8]. Today such gratings give efficiencies exceeding 80% [9].

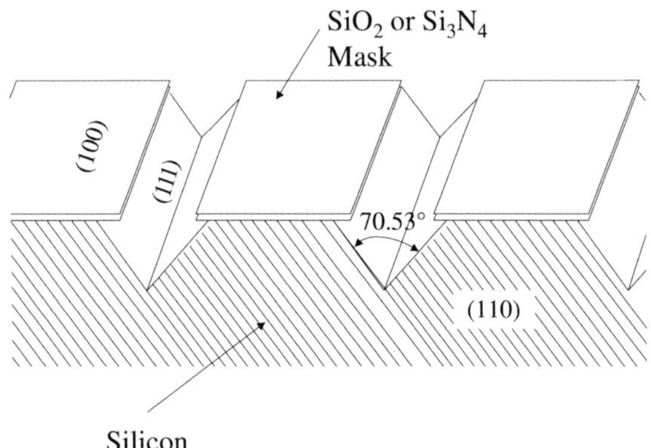

Fig. 3.3. V-shape grooves fabricated by anisotropic etching of a silicon wafer

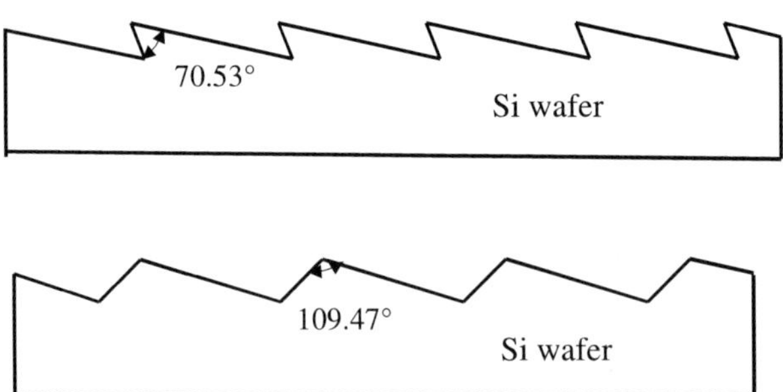

Fig. 3.4. Asymmetric grooves are obtained on silicon substrates polished with a proper inclination γ of the crystal surface. The top angles are 70.53° or 180° − 70.53° = 109.47°, respectively

3.3.4 Surface Relief on Multilayer Dielectric Gratings

The compression of high energy, high repetition rate, ultra-short (picosecond and sub-picosecond) laser pulses using a pair of gratings in chirped pulse amplification (CPA) chains stimulated the development of new high efficiency low damage threshold gratings. The initial short seed pulse coming from an oscillator is stretched by a factor ten thousand or more before passing through the optical amplifier chain where it grows in energy. The amplified pulses are then compressed back into short pulses (typically a few 100 femtoseconds in duration: $1\,\text{fs} = 10^{-15}\,\text{s}$) of high intensity (100 TW and more). The compression is done with a pair of large parallel gratings. In the compression stage the gratings operate with a variable pulse duration from typical 1 ns stretched pulses to typical 300 fs compressed pulses (case of LULI in France [10–11]). Normally in CPA the compression is done at wavelengths around $1.06\,\mu\text{m}$. In this wavelength range a properly designed and carefully ruled gold-coated grating has a diffraction that exceeds 95% in polarised light, however, the threshold for optical damage for short high energy pulses is typically $0.8\,\text{J/cm}^2$ for ns pulses. This threshold, related to the metal absorption, is not strongly dependent on the pulse duration, but is not high enough for this application. In order to overcome the unavoidable absorption loss due to the metallic reflection on the traditional surface-relief gratings it was shown that efficient and larger damage threshold gratings could be obtained by etching the grating surface into the top layer of a multilayer reflective dielectric stack [12–17].

So far a very high efficiency in polarised light has only been demonstrated theoretically and experimentally with a higher damage threshold on short pulses. For example, 96% to 98% diffraction efficiency at $1.06\,\mu\text{m}$ were obtained on Jobin Yvon gratings with 1740 lines/mm as early as 1998. In dielectric materials the damage threshold is notably dependent on the pulse duration: the threshold is lower with shorter pulses. At 0.5 ps duration these gratings had a damage threshold above $2\,\text{J/cm}^2$, twice the value observed on the best equivalent gold-coated gratings [10–11, 18–19].

The grating can be formed on top of the multilayer by ion etching or material deposition through a periodic mask, recorded using the traditional holographic means (interference pattern of laser light on the surface) in a photoresist layer deposited by spin-coating, followed by removal of the mask in a solvent or by other chemical, or plasma means. The grating can also simply be prepared in the photosensitive material deposited on the multilayer: in this case the "mask" will not be removed. The multilayer stack consists of dielectric layers with low and high index alternatively (typically SiO_2/HfO_2 with indexes 1.46 and 1.9 at $1.053\,\mu\text{m}$, respectively). The dispersion is generally high as the grating, with a high frequency, works at a high angle (typically 60° to 80°) in the first order. It can be designed to work in transmission or in reflection. Two typical structures of multilayer dielectric HfO_2/SiO_2 gratings with a periodic structure etched in SiO_2 or in HfO_2 are presented in Fig. 3.5. Such multilayer dielectric gratings made by the Diffractive Optics Group at Lawrence Livermore National Laboratory with 565-nm period grooves ion beam etched with 650 nm depth in a SiO_2 low-index top layer and 675-nm period grooves etched with 300 nm depth in a HfO_2 high-index top layer, both exhibit diffraction efficiency above 95% into the −1 reflected order at 1053 nm, 52°, TE-polarisation. This laboratory also announced gratings with 96.5% mean and 99.2% maximum efficiencies at 1053 nm, 66°, in first order [14, 19]. 96% and 98% were measured on Jobin Yvon gratings etched in a SiO_2 low-index top layer and a HfO_2 high-index top layer, respectively [16, 17].

The multilayer dielectric gratings initially developed for the stretching and compression of laser pulses can be used in WDM components. Of course, the results obtained at 1 μm for the pulse compressors can be extrapolated to 1.3 or 1.5 μm. In this application the most important factor is not the high damage threshold, but the possibility to obtain nearly 100% diffraction efficiency in TE mode polarised light in the 1st order. Another unique feature is that the optical bandwidth and the efficiency of the grating are adjustable with the angle of incidence and with the grating and the multilayer structures. For example, the grating can be designed to reflect one wavelength while transmitting all other wavelengths for add-dropping with very low losses. Of course, as is generally the case for WDM components using high-frequency gratings, the input polarisation has to be conditioned, independent on orientation in order to get low polarisation-dependent loss (PDL).

3.3.5 Stratified Volume Diffractive Optical Elements

A new class of high efficiency gratings was introduced in 1999 by Chambers and Nordin [20]. It uses multiple binary grating layers interleaved with homogeneous layers as shown in Fig. 3.6. The gratings are shifted laterally relative to one another. A plane wavefront, for example, from a parallel vertical incoming beam is diffracted by the layers when it passes through the component. Preferential constructive interference occurs for given wavelengths in given directions and orders of interference. Depending on the characteristics of the layers and on their relative positions, efficiencies comparable to those obtained in a volume holographic grating with slanted fringes can be obtained. In [21] an efficiency of 82% with 3 grating layers and 92–96% with 5 grating layers is predicted and confirmed in

Fig. 3.5. Typical multilayer dielectric gratings

experiments with TM-polarisation, in first order at 2.05 µm with a grating period of 4 µm. The deflection angle in the transmission of a normally incident beam is 32°.

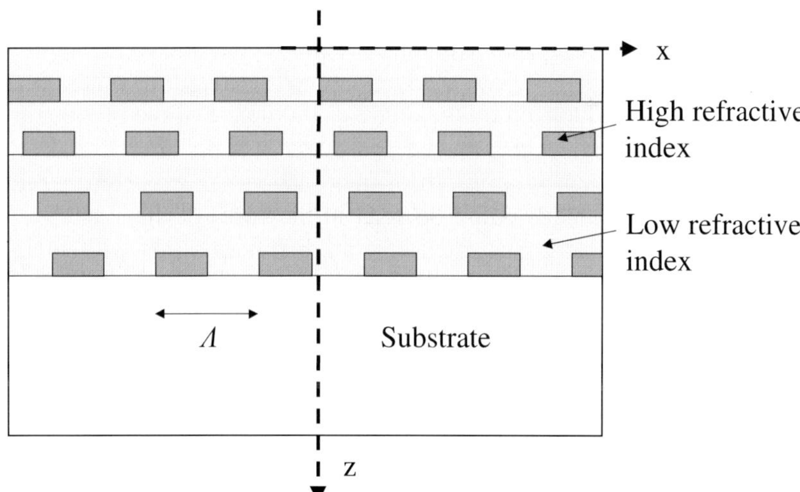

Fig. 3.6. Stratified volume diffraction grating structure

Up to now, to our knowledge, such gratings have not been applied to WDM. However, their excellent characteristics in efficiency and maybe their reliability, as they are made of stable transparent inorganic materials, could be interesting in particularly demanding applications.

3.3.6 Volume-phase Holographic Gratings

Introduction

Volume-phase holographic gratings (VPHG) have been developed over the past three decades in parallel with the most common holographic surface-relief gratings. VPHGs are based on recording interference stationary waves in the volume of the photosensitive coating in order to obtain efficient holograms. In his autobiography Dennis Gabor, the inventor of holography in 1947, underlines how he was "fascinated ... by Gabriel Lippmann's method of colour photography [22], which played such a great part in [his] work", and the early work of Y.N. Denisyuk (1958–1962) combines holography with Lippmann's work to produce white light reflection holograms.

In 1968 T.A. Shankoff [23] published the first paper on dichromated gelatine holograms followed quickly by L.W. Lin [24] and others. The process was improved later by M. Chang and others [25–29]. In particular, from 1974–1984 R. Rallison [30] pioneered the production of glass sandwich dichromate holograms. The principle can be applied to the manufacturing of masters and replicas as well.

It was noted early that the brightness of the reconstructed image of the hologram obtained in the finite thickness sensitive material of the photographic plate was influenced by the angle of incidence of the reconstructed beam. In 1972, Akagi et al. showed the cross section of a Kodak 469F holographic plate taken by an electron microscope. As expected, a microstructure of index changes could be seen in the form of the fins of a Venetian window-blind acting as small mirrors diffracting light with an angular distribution centred around the direction of reflection [31].

In 1983 Dickson patented a "method for making holographic optical elements with high efficiency" in large quantities by optically replicating a previously recorded interference pattern consisting of parallel Bragg surfaces on a silver halide master, in the dichromated gelatine film of a copy element [32]. The patent describes how to get the Bragg surfaces properly reoriented within the gel to maximize the efficiency of the final replica. This was proposed for multi-element optical scanners used to read barcoded labels (in supermarket operation for example). Today Dickson^{TM} gratings are proposed at wavelengths in the C-, S-, or L-bands (see for example [33]).

Physics of Volume-phase Holographic Gratings

Figure 3.7 schematically displays the recording of index modulation in the fringe planes located in the interfering volume of two parallel coherent laser beams of wavelength λ_0.

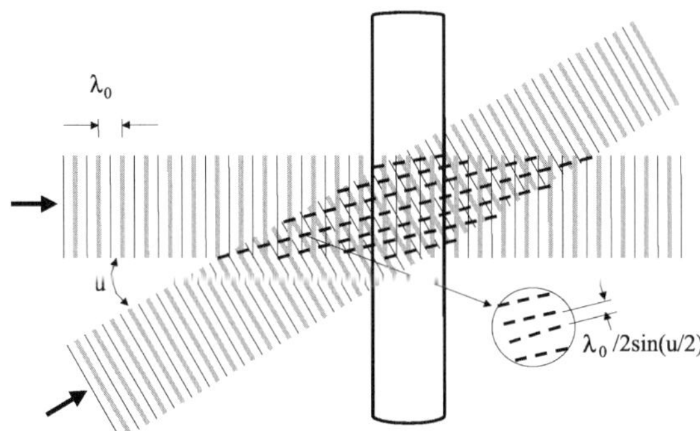

Fig. 3.7. Recording of index modulation in the fringe planes. The angle between the two incident beams being u the distance between the fringe planes is $\lambda_0/2\sin(u/2)$

With v being defined as the spatial frequency (number of lines per unit length) of the grating, the spacing in the plane of incidence on the grating surface is $1/v$ (Fig. 3.8). The angle θ' of the diffraction order m for an angle of incidence θ at wavelength λ in free space is given by the classical transmission grating equation:

$$(\sin\theta + \sin\theta') = m\,\lambda\,v. \tag{3.7}$$

Lets assume a grating material with mean refractive index n. When the optical path difference $\Delta = n \times (H_1OH_2)$ between two successive rays diffracted from adjacent planes of equal index, $n + \Delta n$, is equal to an even number of wavelengths, the Bragg phase condition is fulfilled and a maximum of diffraction in the order m is obtained. This can be calculated easily from the projections of the segment OO' on the incident and reflected beams:

$$H_1O = OH_2 = \Lambda\,\sin\theta_B, \tag{3.8}$$

where Λ is the periodic distance between planes of equal index spacing and θ_B the Bragg angle in the material (Fig. 3.9). The Bragg condition is given by:

$$\Delta = 2\,n\,\Lambda\,\sin\theta_B = m\lambda_B. \tag{3.9}$$

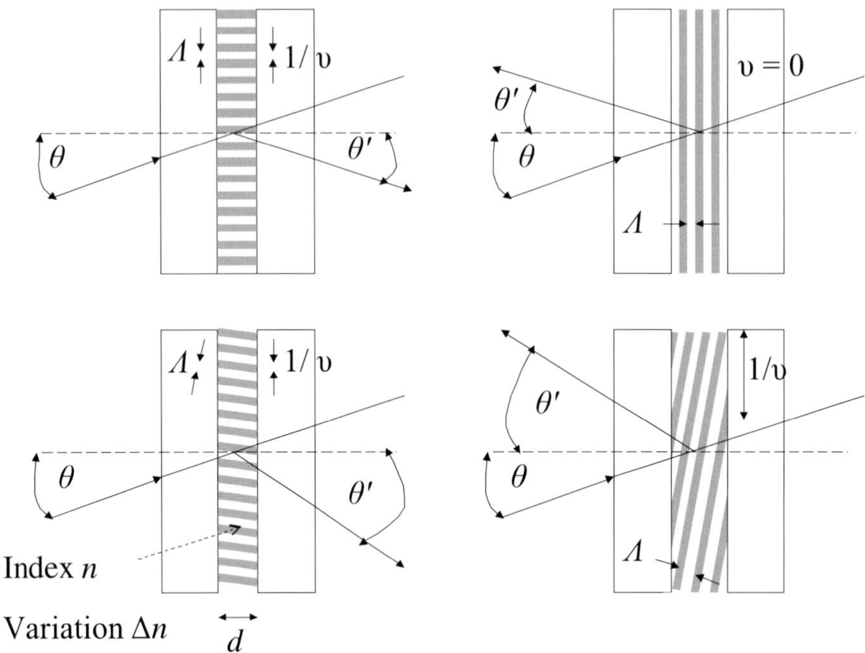

Fig. 3.8. Volume-phase holographic gratings (VPHG) in transmission or in reflection

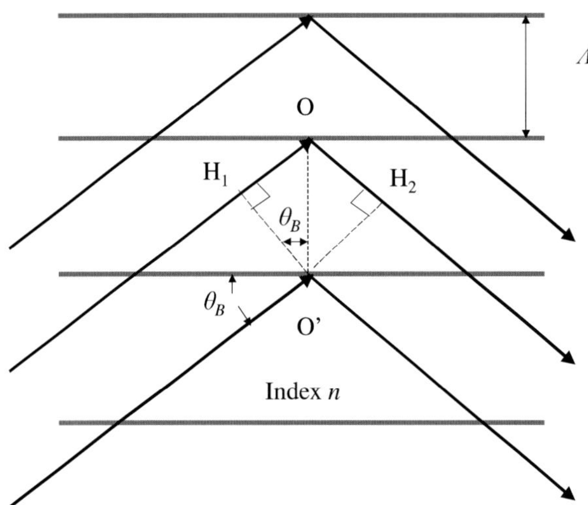

Fig. 3.9. Bragg condition: the condition of phase necessary to get a maximum of diffraction in the direction θ_B corresponds to $n \times (H_1OH_2)$ equal to an even number of wavelengths

Applications of Volume-phase Gratings

Today volume phase gratings and in particular "s-p phased gratings" become particularly important for high resolution spectroscopy, which requires large incidence angles. They are in use in very different applications such as astronomy [34–38], optical scanners, polarisation switches, optical memories, and WDM components. We will present hereafter some examples of results obtained with VPGs.

Dichromated gelatine has a good transparency from 300 nm to 2800 nm and high resolution allowing line densities up to 6000 lines/mm. The typical index, index modulation, and depth are 1.35, 0.02 to 0.1, and up to 20 μm, respectively. Excellent transmission gratings at 300, 600, 1200, and 2400 lines/mm have been obtained with peak efficiencies in the first order of 80% and more. For 600 to 2400 lines/mm transmission gratings this is generally larger than the efficiency obtained with triangular-groove profile purely dielectric surface-relief gratings. The reflective grating has the ability to "saturate" the diffraction efficiency. The theoretical efficiency curve of such a grating can be maintained at 100% over a given spectral bandwidth that is unfortunately relatively narrow. But this bandwidth can be enhanced by a change of index modulation as a function of the grating depth. In Ref. 36 Barden et al. present such a grating operating at 532 nm with a practical efficiency of 90% in a square bandwidth larger than 20 nm. Such "saturated" reflection gratings can be used for DWDM filters (50 GHz width at 1295.5 nm in [39]). VPG technology has the noticeable advantage to allow the stacking of multiple gratings into a single component. For example, in Ref. 36 again Barden et al. describe a transmission grating multiplexing 656 nm (1200 lines/mm) and 486 nm (1620 lines/mm) into the same direction 23° with peak efficiencies of 93% and 80%, respectively.

Applications of multiple VPGs include CWDM transmission filters (4 channels at 1275 nm, 1350 nm, 1490 nm, and 1550 nm recorded on a small photorefractive glass cube are presented in [39]). In Ref. 40 Yaqoob and Riza give the specifications of a free-space wavelength-multiplexed optical scanner using a 940 lines/mm VPG in dichromated gelatine. The Bragg wavelength is 1550 nm, with a Bragg angle outside the grating of 46.76° ($\approx 36°$ inside). The estimated thickness and index modulation are 8.39 μm and 72.4×10^{-3}, respectively. The absolute efficiency is 93% at the maximum of 1550 nm and 87% at 1520 nm and 1590 nm, while the PDL, being negligible from 1530 nm to 1570 nm, remains smaller than 0.12 dB from 1500 nm to 1600 nm.

A polarisation switch, using a thick, 1118 lines/mm VPG in a polymer-dispersed liquid crystal, is described in Ref. 49. With a high monomer concentration a high isotropy without external field is ensured. Designed for working in the 1550 nm wavelength window at a Bragg angle of 35.26°, the alignment of the peaks without external field is obtained with a product of index modulation and grating thickness $\Delta n \times d = 1.9\,\mu m$. The PDL remains lower than 0.2 dB across the whole C-band (1.53 to 1.57 µm). The s and p diffraction efficiencies can be shifted under an electrical field for polarisation switching.

High dispersion VPGs in prisms for operating wavelength ranges 1525 to 1565 nm with 1790.5 lines/mm, 1575 to 1608 nm with 1742.0 lines/mm, and 1555 to 1608 nm also with 1742.0 lines/mm with angular dispersion at the centre wavelength of 0.3°/nm, resolution 12 GHz, 6.5 dB loss, 0.3 dB PDL and other gratings with different characteristics for DWDM or polarisers are currently available [42–44]. VPGs for operating wavelength ranges 1530 to 1570 nm, with 940 lines/mm and efficiencies better than 93% for s- and p-polarisation are proposed as well [45]. Mux/demux devices with 100 GHz spacing based on transmission VPGs used in double pass are commercially available [46]. For 40 channel devices the insertion loss is typically 3 dB and 5 dB for Gaussian (pass band of 0.4 nm at −3 dB down maximum) and flat top (pass band of 0.51 nm at −3 dB down maximum) channel shape, respectively, with PDL < 0.3 dB.

Volume holography is also being proposed in holographic memory or optical information processing. Research performed on optical databases using VPGs demonstrated large storage capacity and fast parallel access (see for example [47–48]). It is possible to record information databases on multiple holograms in a small volume. For example, three 4-bit digital words have been recorded in a 2 cm^3 iron-doped LiNbO$_3$ crystal via angle multiplexing at 488 nm and retrieved in three directions at 1550 nm by a 200 GHz spacing WDM read-out beam at 1550 nm [49].

Recently it was shown by S. Han et al. that chirped VPGs could have interesting applications in compensation or management of chromatic dispersion and polarisation mode dispersion [50]. However, the practical dispersion of −10 ps/nm obtained in their experiment is still by far too small for application to standard chromatic dispersion compensators. In fact VPGs are or have been a solution for many different applications [51].

Photo-materials for VPGs

One of the major challenges to implementing volume holography has been the development of suitable materials. These materials need to offer high photosensitivity, optical clarity, dimensional stability, thickness, index modulation, manufacturability, non-destructive readout, and environmental and thermal stability. Many references on the phase materials for this application can be found in [52].

Silver-Halide in gelatine. Useful information on these recording materials is given in [53]. (The film section of this paper is out of date, but the chemistry part is outstanding and still very useful). The material can be exposed at wavelengths up to 750 nm. The sensitivity is high (up to $3 \mu J/cm^2$ at the optimum wavelength), and Δn can reach 0.1. The grain size, a significant source of scattering, can be as low as 10 nm.

Dichromated gelatine (DCG). The DCG material is recommended for VPGs for spectroscopic applications. The sensitivity is lower ($4 mJ/cm^2$ at 442 nm to $100 mJ/cm^2$ at 514 nm). Δn can reach 0.2 to 0.25. Thicknesses d up to $100 \mu m$ have been used, but R. Rallison recommends $d = 25$ to $30 \mu m$ as practical limits for DCG with maximum $d \times \Delta n$ of about $2.5 \mu m$, no matter what the thickness of the film is [52]. One drawback is its sensitivity to moisture. But in principle a baking at 150°C supplemented by a glass protective cap solves this problem.

Polymers. DMP-128 from Polaroid is generally used at thicknesses of 5 to $15 \mu m$, where it has about the same index modulation as DCG. So it has the ability to produce a high reflection efficiency in a thin layer. The material is sensitive in the red. Its sensitivity is about $25 mJ/cm^2$. It can be filled with liquid crystals. HRF Dupont photopolymers show modulations up to $\Delta n = 0.06$. This relatively low value is not a drawback in thick film applications. For example, an HRF 600×001–20 film with a thickness of $20 \mu m$ was used to record a guided wave holographic grating WDM with four channels at 1551, 1553, 1555, and 1557 nm, and another WDM structure operating at 800 nm with results consistent with the theoretical rigorous coupled wave analysis (RCWA) model [54]. The Shipley photo-resist is one of the materials of choice for relief holographic gratings. Its sensitivity is relatively low, $2000 mJ/cm^2$ at 488 nm, but it can be doped with sensitisers. Highly-efficient (96%) multiplexed VPGs designed for playback at 1550 nm have been fabricated recently using 200-, 300-, and 500-μm thick layers of ULSH-500 photopolymer from Aprilis Inc. based on cationic ring-opening polymerization (CROP) [55–57]. The material was originally developed by Polaroid and optimized to achieve low transverse shrinkage without sacrificing sensitivity [58–59]. Typical exposure irradiance of $1 mW/cm^2$, and $\Delta n = 1.3 \times 10^{-2}$ at the read wavelength of 514.5 nm are given. More information on optical data storage materials can be found in [60]. Polyvinyl alcohols (PVA) are interesting as photopolymer materials despite their low adherence to glass. They do not require wet processing and show good stability. For example, up to 1500 lines/mm transmission gratings have been recorded in PVA with

$6\,mW/cm^2$ at $514\,nm$ in [61] in order to obtain quantitative information on the higher harmonics in the Fourier expansion of the recorded refractive index. These gratings show a good efficiency in the first on-Bragg angular replay condition. Epoxy resin-photopolymer composites are also good candidates for volume holography. A large selection of epoxies, amines and photopolymers are available allowing compromises between the optical, mechanical and environmental properties [62]. Many years ago doped and partially polymerized poly(methylmethacrylate) had been used to record strip waveguides using a variation of index up to 5×10^{-2} [63]. Such a material has been used recently to record single or superimposed volume gratings for DWDM [64]. A modulation $\Delta n = 6.4 \times 10^{-4}$ is deduced from the experiments. With four Bragg gratings working at $100\,GHz$ spacing efficiencies from 20 to 83% have been obtained. A diffraction efficiency of more than 99% with a flat top bandwidth of $1\,nm$ was demonstrated on a single reflection grating with a typical length of $15\,mm$. This material seems extremely interesting. However, the high volume shrinkage during polymerization and the strong thermal expansion of the polymer need further attention before industrialization.

For specific applications photorefractive lithium-niobate may become very attractive. It was shown that polarisation-independent DWDM components with VPGs working at high diffraction angles can be achieved by superposition of two gratings superimposed in a photorefractive $LiNbO_3$ crystal working on ordinary and extraordinary waves in such a way that both diffracted waves are collinear outside the crystal [65].

Refractive index differences can be induced in ZnS or PbS doped glasses by femtosecond laser pulses [66]. Using this method in a Ti-sapphire laser focused beam, a refractive index difference up to 20% has been obtained on gratings with periods $3\,\mu m$ and $15\,\mu m$ with 90% diffraction efficiency in the near IR. This was recorded at $750\,mW$ average power, $250\,kHz$ repetition rate, $200\,fs$ pulse duration with a beam focused to a diameter of a few microns [67].

3.3.7 Ruled and Holographic Concave Gratings

Aberration Correction

The possibility of controlling the focal properties of concave surface-relief gratings by a proper distribution of the grooves has been known for a long time. Concave gratings are capable of stigmatic imaging without the need for auxiliary focusing optics. Concave surface-relief gratings can be manufactured on computerized ruling engines. However, it is generally easier to record these gratings through holographic techniques. Moreover, the holographic recording is also applicable to the manufacturing of the VPGs and of almost all other grating structures.

Blazing of Surface-relief Concave Gratings

The groove shape allows the concentration of the diffracted energy into a given spectral range: the grating is then said to be "blazed". The blaze angle of surface-relief concave gratings has to be changed continuously on the grating surface (Fig. 3.10). In principle, in the scalar theory it is necessary to keep (angle AMN) = (angle NMB) for each position M along the concave surface. On holographic gratings, a profile variation with surface location can be obtained by variable incidence ion etching. Following production of the holographic master, which has pseudo-sinusoidal groove profiles, the grating is then used as a mask subjected to an oriented ion beam to remove surface atoms until the groove structure presented by the surface hologram is brought into the substrate itself. To "shape" the grooves, the angle of incidence of the ions to the substrate can be adjusted to produce triangular blazed grating profiles. High efficiencies have been achieved with concave aberration-corrected holographic gratings [68]. Typically 60% to 80% efficiencies and sub-nanometre wavelength resolution from 1500 to 1560 nm on all single-mode concave holographic grating tuneable demultiplexers are obtainable.

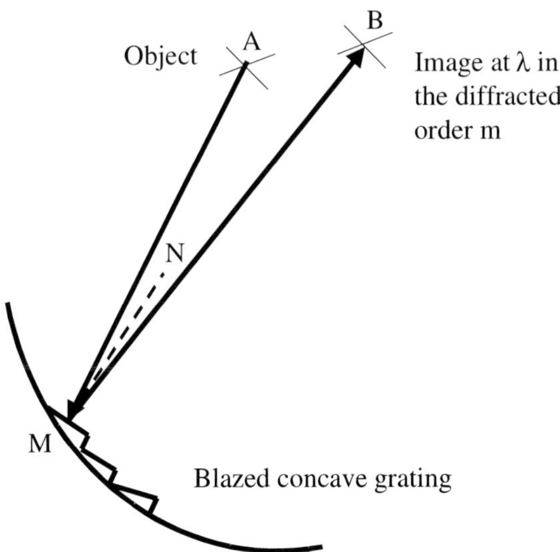

Fig. 3.10. Blazed concave grating

Another method has been proposed in which a blazed grating can be fabricated through grey-scale X-ray lithography. As the gratings can be generated over a considerable range of distances from the mask, the ability to write gratings maintaining a constant blaze on a substrate of arbitrary shape seems promising [69].

On classically ruled gratings, one can rule three or four zones with a constant but optimized angle and with a corresponding loss of resolution. The resolution of such a grating ruled in N parts will be at least N times less than that of the same grating ruled as a single part, but, in most cases, an efficiency approximately equal to the efficiency of the equivalent plane grating can be obtained with a very small number of zones [70].

3.4 Efficiency and Polarisation-dependent Loss versus Wavelength

3.4.1 Plane Reflection Surface-relief Metallic Gratings

We use Fig. 3.11 with the following notations:

N: normal to the mean grating surface
M: normal to the facet
θ: angle of incidence (with respect to N)
θ':   diffraction angle (with respect to N)
i: angle of incidence (with respect to M)
i':   diffraction angle (with respect to M)
Λ: groove spacing
γ: blaze angle

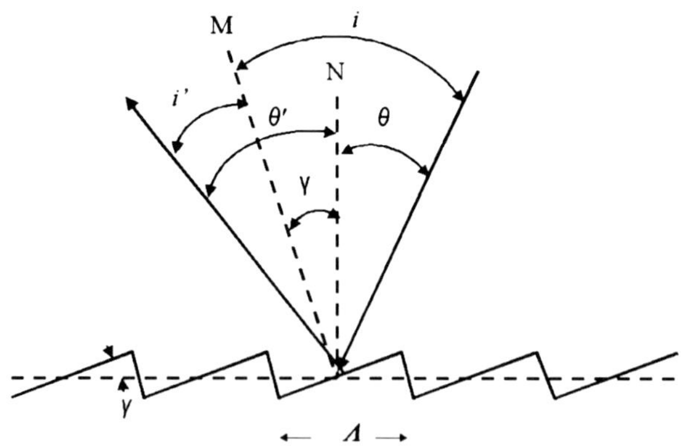

Fig. 3.11. Diffraction plane grating

If the index outside the grating is n, equation (3.1), which gives the position of the wavelength λ_m in order m, becomes

$$n \Lambda (\sin\theta + \sin\theta') = m\lambda_m. \tag{3.10}$$

Scalar Approximation

When the groove spacing is much larger than the wavelength and at small angles, the efficiency can be usefully approximated by a purely scalar theory. A small angle approximation for θ, θ', and γ is used, and it is assumed that the number of grooves is large enough to have an angular width of the diffraction by the total grating surface much smaller than the angular width of the diffraction by a facet. We get rough results, but this analysis is very useful to start with.

Let us remark that the diffracted energy is maximum in the direction corresponding to a reflection on each grating facet, i. e. when $i = i'$. From the relations $i = \theta - \gamma$ and $i' = \theta' - \gamma$ we get the blaze angle value necessary to get a maximum at given incidence and diffraction angles θ and θ':

$$\gamma = \frac{\theta + \theta'}{2}. \tag{3.11}$$

This angle γ determines the shape of the diamond to be used for the ruling of the grating master.

For reflection gratings the blaze angle is generally calculated in the Littrow condition, in which $\theta = \theta'$, corresponding to an incident and exit beam in the same direction. Under these conditions the blaze wavelength is:

$$\lambda_{1b} = 2\Lambda\sin\gamma \qquad \text{in the first order,} \tag{3.12}$$

$$\lambda_{2b} = \Lambda\sin\gamma \qquad \text{in the second order, and} \tag{3.13}$$

$$\lambda_{mb} = (2\Lambda/m) \sin\gamma \quad \text{in the } m^{\text{th}} \text{ order.} \tag{3.14}$$

When $\theta \neq \theta'$ the blaze wavelength in order m is:

$$\lambda_{mb\ \theta \neq \theta'} = (2\Lambda / m) \sin\gamma \cos\frac{(\theta - \theta')}{2} \tag{3.15}$$

The efficiency η_m in the order m can be approximated by:

$$\eta_m = R \frac{\cos\theta}{\cos\theta'} \operatorname{sinc}^2 \left[m\pi \cos\theta \frac{\sin((\theta + \theta')/2 - \gamma)}{\sin((\theta + \theta')/2)} \right] \tag{3.16}$$

where R is the reflectivity of the coating. The wavelength-dependent efficiency $\eta_m(\lambda)$ can be obtained from (3.16) combined with (3.1). A high efficiency can be obtained with the simple smooth triangular groove profile of Fig. 3.11. Gratings with such a profile are so-called echelette or echelle gratings. The border between these two classes of gratings is rather badly defined and there is no consensus between the different manufacturers for this. The term echelette is generally reserved for fine spacing gratings (i. e., typically 600 lines/mm and more) and used in orders lower than $|m| = 5$ or $|m| = 6$. Echelles are "coarse" gratings (i. e., typically 20 lines/mm to 600 lines/mm) and are used at high diffraction angles, generally in orders larger than $|m| = 5$ or $|m| = 6$ and sometimes in orders beyond 100. Echelles operate in high orders with correspondingly short free spectral ranges (cf. (3.6)). Echelles and echelettes are often specified by the "R number", which equals $\tan\theta'$ in the Littrow condition. The "R number" is an important characteristic, linearly related to the angular dispersion (see Sect. 3.2.2). The typical values of this "R number" 1, 2, 3, and 4 correspond to $\theta' = 45°$, $63.43°$, $71.56°$, and $75.96°$, respectively. Of course, the same high "R number" can be obtained for an echelle grating with a "large" groove spacing Λ_1 working in a high order m_1, or with a fine enough groove spacing Λ_2 echelette grating working in a low order m_2 when

$$m_1/\Lambda_1 = m_2/\Lambda_2 \tag{3.17}$$

The expression of the spectral distribution $E_m(\lambda)$ of the diffracted energy from an echelette grating in the approximation of small blaze angle and the curve giving its value in the first order are given in [Ref. 71, pp. 41–43].

Electromagnetic Theory

With gratings having a small groove spacing ($\Lambda = $ a few λ or less), and/or at high angles the scalar approximation is no longer valid. Anomalies appear in the efficiency curves, and the spectral distribution of the diffracted energy depends on the polarisation. The maximum efficiency with unpolarised incident light is lower than the efficiency at the blaze wavelength λ_{mb} given by the scalar equation. The efficiencies in polarised light were fully calculated from Maxwell's equations (M. Petit, thesis, Faculté d'Orsay, France, 1966 and [72]). Efficiencies of perfectly conducting sinusoidal gratings or perfectly conducting triangular-profile gratings typically used in WDM components calculated at the Laboratoire d'Optique Electromagnétique de Marseille are reproduced in [Ref. 71, pp. 44–45]. With respect to polarisation we will use the following conventions: An electromagnetic wave with electric field polarised perpendicular to the plane of incidence (i. e. parallel to the grooves of the grating) is called s-polarised or TE, and a wave with electric field vector parallel to the plane of incidence (i. e. perpendicular to the grating grooves) is called p-polarised or TM (see also Glossary).

Today electromagnetic theory software enabling analysis and optimization of the grating groove profiles is commercially available. The PC Grate software is based on rigorous modified integral methods [73]. These methods are applicable to the calculation of a broad range of Λ/λ values of one to several hundreds including the calculation of the efficiencies of the echelette and echelle gratings [74–75]. The GSOLVER software uses hybrid rigorous coupled wave analysis and modal theory based on algorithms similar to those of the coupled-wave analysis of Moharam and Gaylord [76–79]. GSOLVER can model low order classical surface-relief gratings in transmission or in reflection as well as VPGs with sinusoidal or non-sinusoidal profiles and VPHGs with fringes normal to the grating surface or with tilted fringes.

A very high efficiency can be obtained for TM-polarised light over a large spectral range at λ/Λ values almost up to 2 [71]. This means that a high dispersion with a high efficiency can be obtained in polarised light. (The dispersion tends to infinity when λ/Λ tends to 2). Devices operating with TM-polarisation only will exhibit very low losses. It is possible when dealing with unpolarised light to angularly separate TE and TM in the incident beam. Then the TM incident beam can go directly to the grating and the TE incident beam is rotated by 90° before reaching the grating. This increases the complexity and the cost of the WDM device, but without such compensations it is necessary to design gratings with the same efficiency for both s- and p-polarised electric fields. We tried to optimize the theoretical profiles of a few typical relief gratings for a maximum theoretical efficiency at low PDL from 1500 nm to 1600 nm. Theoretical efficiency curves are given in Fig. 3.12 for triangular profiles, and in Fig. 3.13 for sinusoidal profiles. In theory, with gold-coated gratings, operating in air in the first order from 1500 nm to 1600 nm, absolute efficiencies of 80 to 90% and low PDL can be obtained with up to 600 to 800 lines/mm and optimized groove profiles. However, above 600 lines/mm it becomes increasingly difficult to manufacture gratings with such profiles. Moreover, whatever the profile optimization for good efficiencies at low PDL, one gets a smaller and smaller efficiency as the grating frequency increases as can be seen from the curves (less than 50/70% above 950 lines/mm corresponding to a Littrow angle of 47.41°at 1550 nm). But of course, when necessary, it is always possible to increase the linear dispersion at a given Littrow angle using a longer focal length.

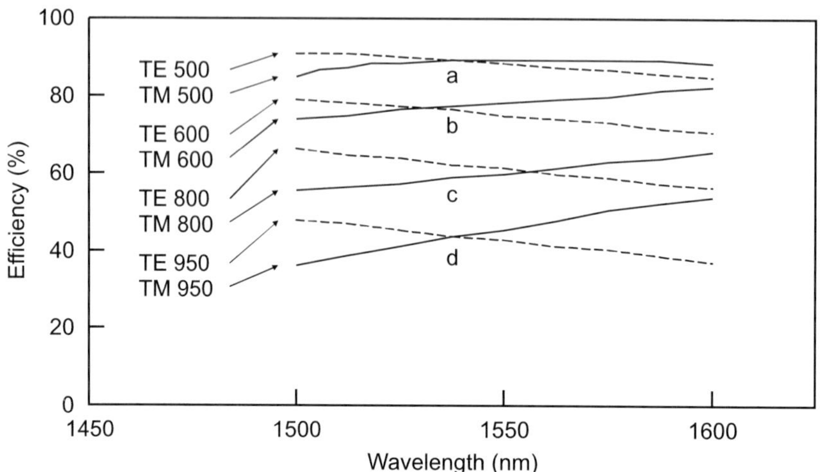

Fig. 3.12. Theoretical efficiency in the first order of four triangular profile gratings with gold coating optimized for low PDL and maximum efficiency in air from 1500 nm to 1600 nm at fixed incidence corresponding to Littrow conditions at 1550 nm (**a**) 500 lines/mm, incidence 22°, (**b**) 600 lines/mm, incidence 27.71°, (**c**) 800 lines/mm, incidence 38.32°, (**d**) 950 lines/mm, incidence 47.41°, TE and TM correspond to s- and p-polarisation, respectively

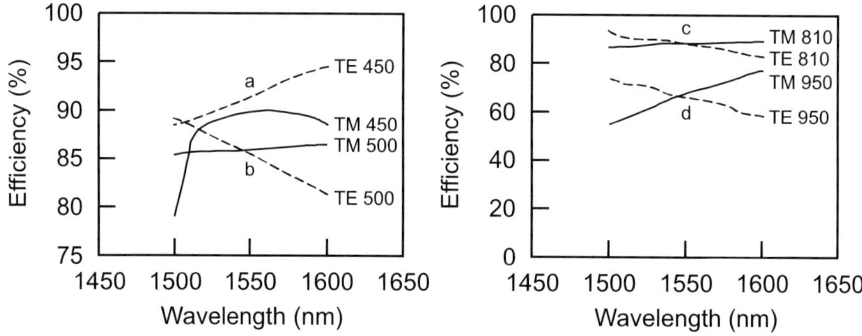

Fig. 3.13. Theoretical efficiency in the first order of four sinusoidal profile gratings with gold coating optimized for low PDL and maximum efficiency in air from 1500 nm to 1600 nm at fixed incidence corresponding to Littrow conditions at 1550 nm. On all curves, as for triangular profiles, the efficiency for TE-polarisation is larger than the efficiency for TM-polarisation at 1500 nm and the opposite at 1600 nm (except in case (a)). (**a**) 450 lines/mm, incidence 20.41°, (**b**) 500 lines/mm, incidence 22.80°, (**c**) 810 lines/mm, incidence 38.31°, (**d**) 950 lines/mm, incidence 47.41°

In order to use gratings with larger dispersion at limited PDL and acceptable efficiency, a solution is to work in a higher order and with a grating immersed in a transparent material. Then the wavelength inside the

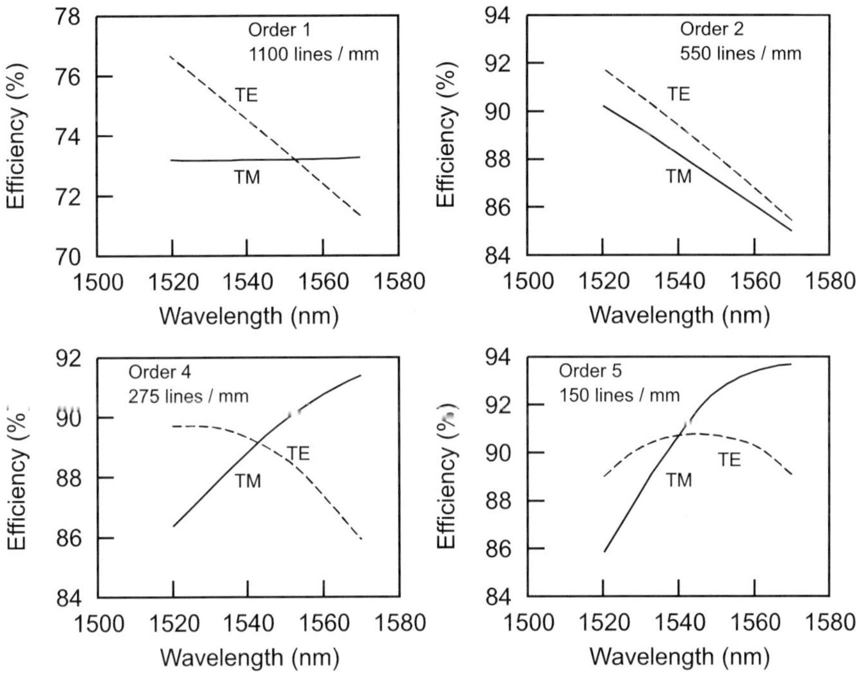

Fig. 3.14. Theoretical efficiency of four triangular gratings, coated with gold, and diffracting immersed in a material of index $n = 1.52$. The four gratings correspond to the same dispersion, TM-polarisation: full lines, TE-polarisation: dashed lines

material is divided by the index n of the material. With a grating on silicon diffracting inside silicon, this solution is very effective as the index is very high, but the interface with the device generally needs to be coated with a low PDL anti-reflection multilayer coating. On the other hand, if the grating is immersed into a low index polymer material this coating can be avoided.

In Fig. 3.14 we show how a 1100 lines/mm triangular grating coated with gold could be optimized in first order in a material of index 1.52. The efficiency is larger than 70% and the PDL very low from 1520 nm to 1570 nm. With equivalent dispersion gratings, 550 lines/mm in second order, 275 lines/mm fourth order, 150 lines/mm fifth order the efficiency is progressively increased up to 90 % at low PDL.

In Fig. 3.15 we give the theoretical efficiencies of higher dispersion gratings. With 1400 lines/mm it is still possible to optimize the profile to get a theoretical efficiency of about 70% in the first order and up to about 90% with the equivalent dispersion grating 466.66 lines/mm operating in the third order. However, with a 1800 lines/mm grating (Littrow angle of 66.6°

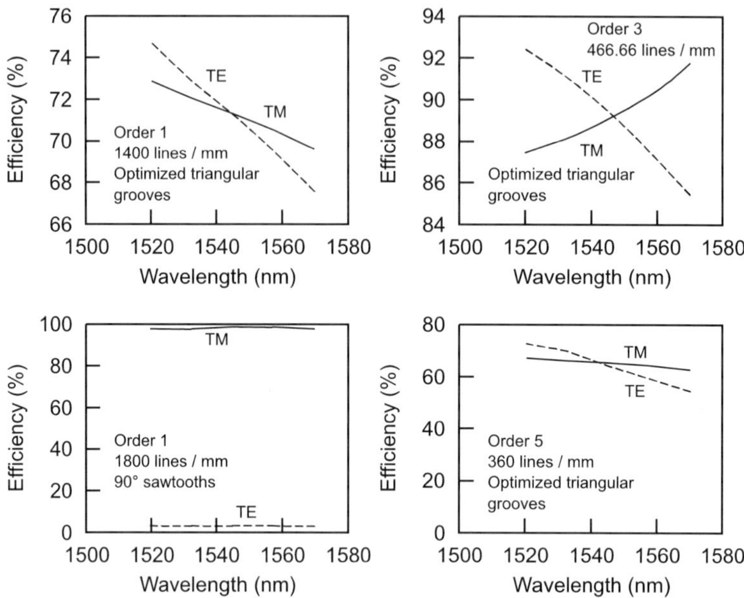

Fig. 3.15. Theoretical efficiencies of high dispersion gratings immersed in index $n = 1.52$, Littrow condition at 1550 nm. 1400 lines/mm and 466.66 lines/mm, incidence 45.5°, 1800 lines/mm and 360 lines/mm, incidence 66.6°

at 1550 nm in index 1.52) it was not possible to avoid the PDL in the first order. This grating with a saw-tooth profile can be used for TM-polarisation with a high efficiency (see [71] for more information). An equivalent dispersion grating with 360 lines/mm, of different profile, working in the fifth order, could be optimized for low PDL at about 65% efficiency. In theory, echelle gratings with special groove profiles operating in higher orders would give better results, but are still difficult to manufacture.

Characteristics of medium and high dispersion ruled diffraction gratings used by Confluent Photonics in their DWDM components are given in [80]. For medium-dispersion reflection "covered/immersed" gratings a measured efficiency of 90% with a very low PDL from 1520 nm to 1570 nm is reported for a 750 lines/mm equivalent dispersion grating. For high dispersion reflection gratings an efficiency of 68% and a PDL lower than 0.21 dB from 1525 nm to 1565 nm is given for a 1080 lines/mm equivalent dispersion grating. (Unfortunately in both cases the orders m, necessarily larger than 1 considering the predictions of the electromagnetic theory, are not reported).

For echelle gratings operated in the order m the free spectral range has been given in (3.6). However, in order to limit the noise caused by the scattering of adjacent orders it is preferable to use a free spectral range larger than the operating spectral range. For example, in [81–82] for a grating working in air in the C-band from 1528 nm to 1560 nm the maximum order m calculated from (3.6) would be 47, but a 52.63 lines/mm grating operating at $m = 22$ was preferred. The grating efficiency of 61%–75% of the DWDM component measured in [81] in the wavelength range (1549.32–1560.61 nm) and the polarisation dependence lower than 1 dB (including the PDL of the circulator) of the reconfigurable add/drop multiplexer (ROADM) measured in [82] perfectly match the prediction of the theory for a triangular profile blazed at an angle of 63°–64°.

Optimization of Groove Profile in Silicon Substrate for Low Polarisation-dependent Loss

Low PDL at high efficiency can also be obtained in gratings etched on silicon substrates. It was shown that gratings with symmetrical V-grooves (with the natural 70.5° apex angle obtained in oriented silicon crystal etching) and with a specific flat top between groove profiles permit high dispersion with high efficiency at low PDL. Popov et al. [83] report 80% efficiency and less than 5% efficiency difference between s- and p-polarisations on such a grating with a groove spacing of 0.4 μm working inside a Si prism at 33.7° of incidence. They show good agreement with their modelling [84].

Coating of the Grating Surface

It is worth noting that the efficiency of a standard reflection surface-relief grating can often be increased by a multi-dielectric coating of the surface [85]. However, in such a case it is necessary to take into account the high perturbations of the electric field in the grating structure. This perturbation is often difficult to model as the structure of the over-coating depends on many manufacturing parameters and is generally not known very precisely.

3.4.2 Transmission Surface-relief Gratings

The grating grooves are transferred onto a resin coating on a glass blank with both faces polished within a quarter of a fringe. The grooves can be considered as a set of small diffracting prisms, and n represents the resin index, γ the facet angle, and Λ the grating period (see Fig. 3.16).

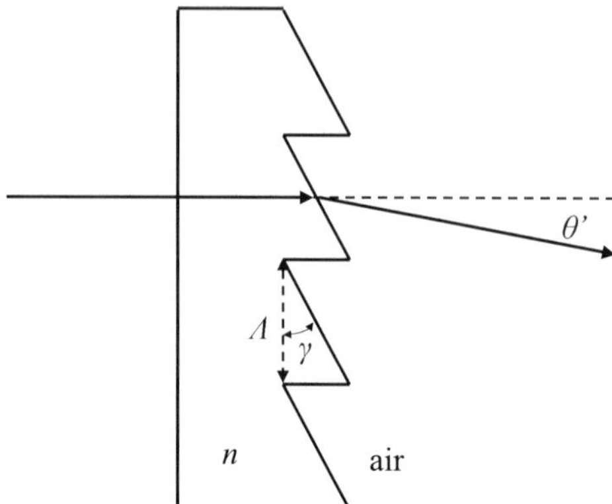

Fig. 3.16. Transmission grating

Scalar Approximation

This approximation is only valid for gratings with groove spacing much larger than one wavelength and small blaze angle. When the incident light is perpendicular to the blank (Fig. 3.16), the blaze wavelength λ_b is given by the formula:

$$\lambda_b = \Lambda\,(n-1)\,\sin\gamma \tag{3.18}$$

Electromagnetic Theory

Here again it is necessary to use the electromagnetic theory to obtain the correct efficiency value. With classical echelette transmission gratings and grisms, low PDL gratings with efficiencies of 60% to 70% in the C-band have been obtained [86]. Using (3.18) it can be demonstrated that one can obtain 100% efficiency with a metallic coating on the small facet of the grooves. This is obtained with an electromagnetic field perpendicular to the conductive facet [87–88]. A similar idea (single-side metal coating grooves in reflection) was applied recently in order to obtain efficient and low PDL gratings in reflection [89].

A very high efficiency can also be obtained for one polarisation with relatively deep sinusoidal profiles on transmission gratings with Λ small enough to present only $m = 0$ and $m = 1$ orders.

3.4.3 Surface-relief Concave Metallic Gratings

Concave gratings are generally used in reflection; therefore, their efficiencies are calculated like those of plane reflection gratings. However, the calculations are generally more complicated as the blaze angle, the angle of incidence, and the angle of diffraction of such gratings continuously change along the surface. More information and references to WDM configurations using these gratings can be found in [Ref. 71, pp. 46–47, p. 179].

3.4.4 Surface Relief on Multilayer Dielectric Gratings

These gratings have been proposed and developed for high power laser resonators [90] and pulse-compressors in chirped-pulse amplification [91]. The problem is to get highly dispersive gratings with low absorption and high efficiency on one given polarisation at least. High dispersion and near 100% efficiency in TM polarisation is routinely obtained with surface-relief holographic gratings coated with gold (gratings with λ/Λ from 0.7 to 1.9, see Fig. 3.17 in [71]). But the absorption of the metal (due to its finite conductivity) reduces the damage-power threshold of the grating. It was shown, that high diffraction efficiency could be obtained using a transmission grating interface on a reflective waveguiding dielectric multilayer stack (Fig. 3.17). Generally the grating is used in the Littrow condition.

The first condition is to get a grating with only two orders (0 and −1) in both, the incident medium (index n_0) and under the grating (index n). It was shown that this condition, necessary to trap the transmitted order 1 in the dielectric (proposed in the past as a condition to get 100% dielectric guide grating couplers [92]), corresponds to:

$$2/3 < \lambda/(n\Lambda) < 2 \text{ for } n_0 = 1. \tag{3.19}$$

The second condition is that the wave incident on the grating generates two transmitted diffracted waves (order 0 and −1) equal in amplitude and out of phase by 90°[93].

The physical interpretation is given in Fig. 3.17, showing the effect of the first diffraction through the grating structure, the reflection from the reflective multilayer stack, and the second diffraction through the grating.

Assume a wave A incident in the Littrow condition. In practice the 0th order in reflection in the incident medium on the grating interface can be reduced to a very low value. The 0th order in reflection can be ignored compared to the transmitted diffraction waves B_0 and B_{-1} in 0th and -1st order, respectively. According to the second condition the grating structure is such that B_0 and B_{-1} are equal in amplitude, but 90° out of phase. Let use suppose that $\Delta\varphi = 0$ in 0th order and $\Delta\varphi = 90°$ in 1st order. The waves B_0 and B_{-1} are reflected by the multilayer coating with the same reflectivity and phase change. They are transmitted again through the grating interface in two 0th and -1st order waves, each with the same

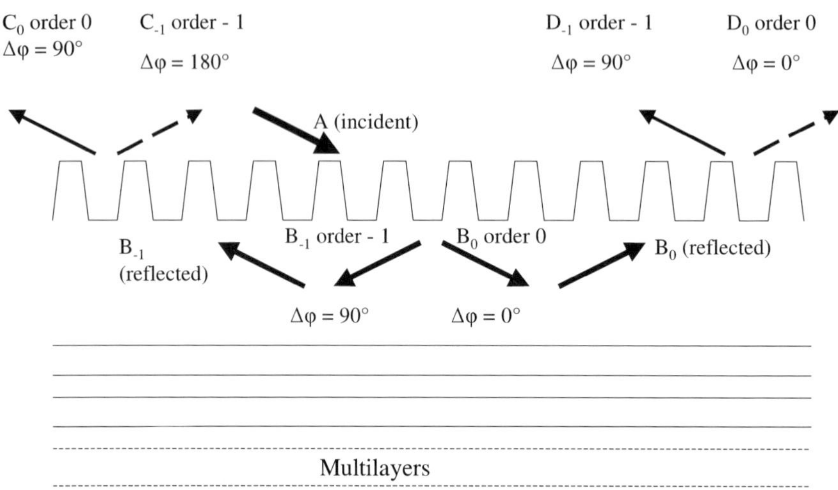

Fig. 3.17. Blazing a multilayer dielectric grating on a given polarisation at 100% in order −1. Physical interpretation: a wave A arrives on the grating in the Littrow condition. After transmission and diffraction by the grating the waves B_0 and B_{-1} are equal in amplitude but 90° out of phase. After reflection from the multilayer we get waves C_0 and D_{-1} travelling backward through the grating in the Littrow direction with C_0 and D_{-1} being equal and in phase. In addition, there are waves reflected into direction C_{-1} and D_0, which have opposite phases. As a consequence, C_0 and D_{-1} are added, while C_{-1} and D_0 mutually cancel each other

phase change. B_{-1} produces C_0 in the 0th order without additional phase change and C_{-1} in the first order with an additional phase change of 90°. B_0 produces D_0 in the 0th order without additional phase change and D_{-1} in the first order with a phase change of 90°. Finally we get waves C_0 and D_{-1} reflected backward into the first order Littrow direction. C_0 and D_{-1} are equal in amplitude and in phase with $\Delta\varphi = 90°$. In addition we get waves C_{-1} and D_0 reflected with a phase difference of $\Delta\varphi = 180°$. Thus C_0 and D_{-1} are added and C_{-1} and D_0 mutually cancel each other. These considerations can be extended to the other waves reflected several times and diffracted through the structure. As $\lambda/(n\Lambda)$ is small and as the grooves must be relatively deep compared to the grating period, the optimization of the structure is generally done using a rigorous method based on the full-vector Maxwell's equations. Up to now high efficiency and low absorption have been demonstrated only in TE-polarisation.

The efficiency of a lamellar grating single interface in the 0th and first order in air is given in Fig. 3.18. We used grating parameters not far from those given in [93]. The grating frequency is 1477 lines/mm. We assume a material of index 1.9, a groove depth of 400 nm, a ridge width of 203 nm,

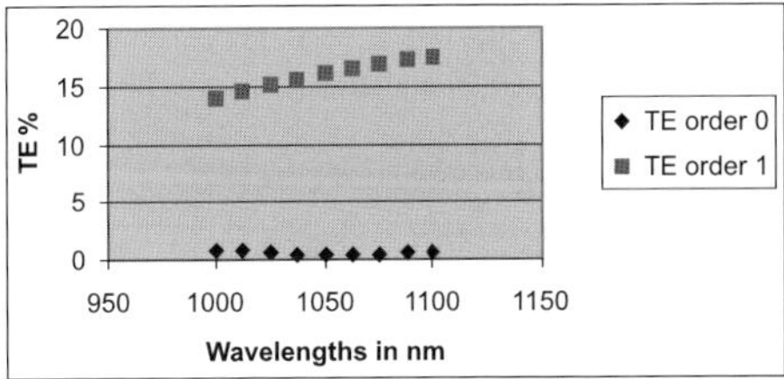

Fig. 3.18. Efficiency of a lamellar grating interface with 1477 lines/mm, on a material of index 1.9 calculated in order to get no reflectivity in the 0th order and as much as possible in the first order in TE-polarisation. With a high reflectivity multilayer under this grating the efficiency in the first order will reach more than 95%

and an incidence angle of $-51.2°$. We calculated the efficiency in a wavelength range from 1000 nm to 1100 nm. Such a grating used in front of a high reflectivity coating will give an efficiency close to 100%. An efficiency exceeding 95% for TE-polarisation is predicted in Ref. 85 with a similar grating interface added to a high reflectivity dielectric multilayer coating.

3.4.5 Stratified Volume Diffractive Optical Elements

Since this method is rather new, we have not much information on these components. As described in the paper by D. Chambers et al. [21] the efficiencies and manufacturing tolerance of these gratings were calculated using RCWA. At $2.05\,\mu m$ operating wavelength, on gratings of $4\,\mu m$ period, efficiencies of 82% with a 3 layer grating and up to 96% with a 5 layer grating were predicted. The polarisation state and PDL have not been given. A tolerance of layer thickness of 50 nm and a tolerance of lateral offset between grating layers of 100 nm have been reported.

3.4.6 Volume-phase Holographic Gratings

The efficiencies of the volume-phase gratings have been analyzed either through a modal theory [94] or a coupled-wave theory [95–97]. The efficiency η of the volume-phase transmission grating is a function of the index modulation Δn, of the thickness of the recorded layer d, of the Bragg angle θ_B in the material, and of the wavelength λ. In the first order, at the Bragg condition, for plane gratings with planes of equal isotropic index orthogonal to the grating surface (Fig. 3.8, first case and Fig. 3.9),

Kogelnik [95] gave the following approximate expressions for the spectral bandwidth $\Delta\lambda_{FWHM}$ and the efficiency η_i $(i = s,p)$

$$\Delta\lambda_{FWHM} = \lambda \ (\Lambda/d) \cot\theta_B \tag{3.20}$$

$$\eta_s = \sin^2 [\ \pi \ \Delta n \ d \ / \ (\lambda \ cos\theta_B)] \tag{3.21}$$

$$\eta_p = \sin^2 [\ \pi \ \Delta n \ d \ cos(2\theta_B) \ / \ (\lambda \ cos\theta_B)], \tag{3.22}$$

where η_s and η_p refer to s- (TE) and p- (TM) polarised light, respectively.

The factor $cos(2\theta_B)$, determining the difference of efficiency between the two polarisations, comes from unparallel electric fields between the incident and diffracted beams in p-polarisation. The Kogelnik approximation, retaining only the 0th and 1st transmitted orders in coupled wave equations, is valid only for

$$\lambda^2/(n\Delta n \ \Lambda^2) \geq 10. \tag{3.23}$$

Efficiencies η_s and η_p reach 1 (100% efficiency) when the argument of the sin functions is $\pi/2$, $3\pi/2$, $5\pi/2$ … In the case of isotropic index (meaning that Δn does not depend on the polarisation) it is possible in this approximation to get 100% for both polarisations at specific angles $cos(2\theta_B) = 1/3$, 1/5, 3/5 … (The corresponding Bragg angles inside the grating are: 35.264°, 39.231°, 26.565° …, respectively). At the Bragg wavelength λ_B, in a given order m, with a material index n, these angles correspond to specific values of the grating period Λ determined by the Bragg equation:

$$2 \ n \ \Lambda \ sin\theta_B = m \ \lambda_B. \tag{3.24}$$

Better theoretical predictions for the efficiency of VPGs can be performed with the GSOLVER program.

In the telecommunication range, for thin dichromated gelatine VPGs (Dickson gratings) with wavelengths around 1550 nm, it was demonstrated both theoretically and experimentally, that the optimal spatial frequency for a high efficiency and a low PDL was 940 lines/mm (Fig. 3.19, [42]).

Practically it is possible to align the peaks of the s- and p-polarisation by adjustment of the thickness of the grating layer. From (3.20) to (3.22) it can be seen that a broad passband requires a thin grating with a high index modulation.

In Fig. 3.20 the absolute diffraction efficiency and the PDL of a 940 lines/mm Dickson transmission grating is reproduced [40]. This grating corresponds to $\theta_B = 35.97°$, $\lambda_B = 1550$ nm, $\Delta n = 0.0724$, and to a thickness of 8.4 µm. It can be seen that the efficiency is 93% and that the PDL is negligible at λ_B. To our knowledge such a low PDL and high efficiency could not be obtained up to now with an equivalent surface-relief transmission grating.

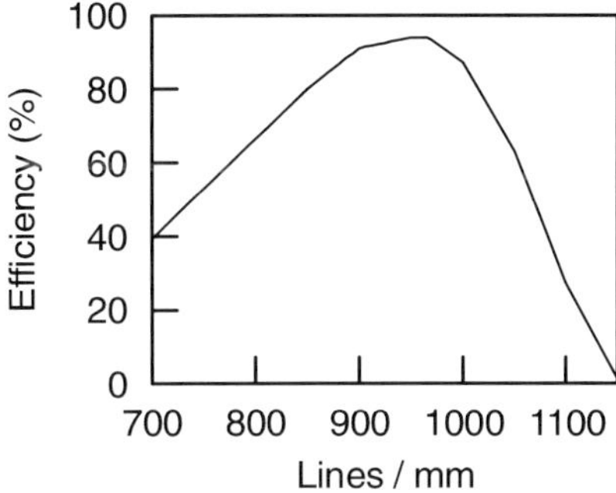

Fig. 3.19. Highest efficiency at Bragg wavelength of thin Dickson gratings optimized with $\eta_{TE} = \eta_{TM}$ ([42])

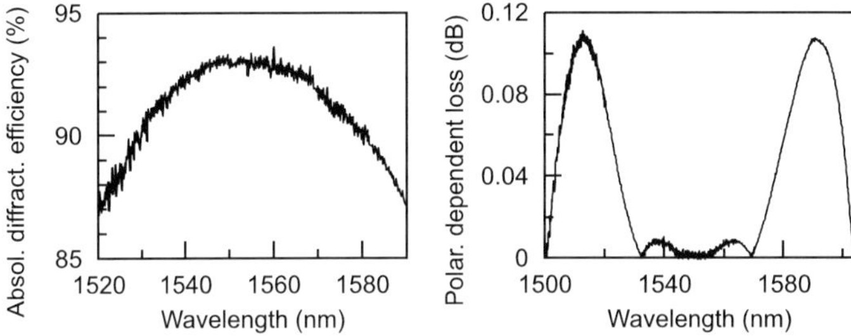

Fig. 3.20. Absolute diffraction efficiency (*left*) and polarisation-dependent loss (*right*) of a 940 lines /mm Dickson grating (after Ref. 32)

3.5 Bandwidth of Diffraction Grating Devices

3.5.1 Coupling without Aberration

Devices with Input and Output Single-mode Fibres

It was shown that in the ideal, zero aberration, all single-mode demultiplexer, the adjustments being assumed perfect, and in the Gaussian approximation of the single-mode fibre electric field description, the transmission function $F(\lambda)$ in each channel was also a Gaussian function [71, 98]. The ratio R_w between the width at half-maximum $\Delta\lambda_{FWHM}$ of the transmission function and the spectral distance between channels $\Delta\lambda$ is:

$$R_w = \Delta\lambda_{FWHM} / \Delta\lambda = 1.66 \; \omega'_0 / \Delta x', \qquad (3.25)$$

where $\Delta x'$ is the distance between fibres, and ω'_0 is defined as the mode radius corresponding to the half width of the amplitude distribution in the fibre core at $1/e$.

Devices with a Single-mode Entrance Fibre and an Exit Slit or a Diode Array with Rectangular Pixels

We chose a system with x',y' as the coordinates in the focal plane, and we further assume that the transmission inside the elementary slit, or the sensitivity inside the elementary rectangular pixel, $T(x',y')$ is uniform and equal to 1. The slit (or the pixel) of width $\Delta x'$ is considered to have a height $\Delta y'$ much larger than the fibre core diameter, corresponding to an intensity distribution I (x',y'). When λ varies, I is translated nearly linearly along x', and I is given by

$$I = I_0 \exp - \frac{2\,r^2}{\omega_0'^{\,2}} \tag{3.26}$$

The transmitted flux, $F(x'_0)$, is the modulus of the correlation of I with T:

$$F(x'_0) = K \cdot \iint I\left(x' - x'_0, y'\right) T(x', y')\, dx'dy' \tag{3.27}$$

The function $F(x'_0)$ shows a flat top depending on $\Delta x'/\omega_0'$. With a single-mode fibre of mode radius $\omega_0' = 4.57\ \mu$m and a pixel slit width of $20\ \mu$m, the full width of $F(x'_0)$ at -0.46 dB is $\Delta x'_{at\,-0.46\,dB} \approx 16\ \mu$m and the full width at -20 dB is $\Delta x'_{at\,-20\,dB} \approx 32\ \mu$m. $\Delta x'$ can be converted into $\Delta\lambda$ through the linear dispersion equation:

$$\mathrm{d}x' / \mathrm{d}\lambda = (f \times m) / (\Lambda \times \cos\theta') \tag{3.28}$$

where f is the focal length of the device, m is the diffraction order of the grating, Λ is the grating period, and θ' is the angle of the exit beam with respect to the normal to the grating surface.

Devices with a Single-mode Entrance Fibre and Multimode Step-index Exit Fibres

As in the previous case the bandpass is a flat top function. Typically the full width at -3.3 dB for a demultiplexer with a single-mode input fibre of $\omega_0' = 5.5\ \mu$m with jointed 50×125 step index output fibres is

$$\Delta\lambda_{-3.3\,dB} = 0.4\ \Delta\lambda, \tag{3.29}$$

$\Delta\lambda$ being the distance between channels.

Devices with a Single-mode Entrance Fibre and Multimode Graded-index Exit Fibres

Similar mathematical developments lead to the results shown in [Ref. 71, pp. 50–58].

3.5.2 Coupling with Small Aberration

It is very important to use diffraction limited optics on multi/demultiplexers using single-mode fibres. The first effect of aberration is an additional coupling loss. This loss is about 1 dB for a sum of aberrations of $\lambda/3$ measured in the exit pupil [99]. Of course, the loss depends on the aberration type. For fibres with numerical aperture NA = 0.099 (mode radius 5 μm at 1550 nm) a defocusing of 10 μm gives 0.81 dB loss, a spherical aberration of $\lambda/2.8$ or a coma of $\lambda/3.2$ or an astigmatism of $\lambda/4$ at the best focus also give a loss of about 1 dB. It can be useful to use a small astigmatism to enlarge the function $F(x'_0)$.

3.6 Grating Bulk Optic Devices

3.6.1 General Characteristics

Generally, grating multiplexers or demultiplexers consist of three main parts: entrance and exit elements (fibre array or transmission line fibre and emitters or receivers), focusing optics, and dispersive grating. The grating is often a plane grating [100–111]. Losses can be very small in components using diffraction limited optics and efficient gratings with monochromatic sources centred on the multiplexer transmission bands. For example, they were as low as 0.5 dB in the 1.54/1.56 μm single-mode device of [111].

The use of dedicated concave gratings simplifies the device [112–116]. For example, in the early Kita and Harada [114] configuration 2.6 dB losses are reached on a four-channel multiplexer at 821.5, 841.3, 860.9, and 871 nm with 60/125 μm multimode fibres. However, it is impossible to retain aberration-free focusing over a large spectral range with such configurations.

In order to obtain a large enough channel spectral width, as compared to channel spacing, the geometrical distance between fibre cores must be small. The pass bands are practically adjacent for a distance between fibres of 22 μm when an 11 μm diameter core is used. Fibres with a small cladding are necessary. This can be obtained by chemical etching of larger core fibres [117]. On practical stigmatic DWDMs using all single-mode fibres, the agreement between the theoretical value of R_w given in Sect. 3.5.1 (3.25) and the experimental results is quite good (within a few %) [71].

In order to obtain optimized spectral pass bands without requiring a small distance between fibre cores, as it is necessary in simple devices, different solutions have been proposed such as a spectral recombination within each channel via an intermediate spectrum image through micro prisms or through a micro lens array [118], where $R_w = 0.7$ is obtained with up to 32 channels. (In modern devices fibre lens sections operating as beam field expanders can advantageously replace the micro lens array). This method is interesting as it introduces only a limited amount of additional losses, but it corresponds to significant additional labour cost. An intermediate waveguiding fan-out section reducing the distance between the channels can also be used between the fibre array and the multiplexer, unfortunately at the expense of additional losses and cost. Another method is to introduce a spatial filtering in the pupil plane [119]. It can be, for instance, a variation in the optical path. Suppose a coupling device with a grating or a wavelength disperser R located at the pupil with coordinates (α',β'), the coordinate in the image (spectrum) plane being (x',y'). Let $A(x',y')$ correspond to the amplitude of the image of the input fibre in the image plane, and let us further suppose that a spatial filter is located in the pupil plane. If its filtering function is $S = \mathrm{sinc}\,(\pi\,a\,\alpha')$, where a is a numerical constant, its Fourier transform in the image plane will be:

$$P\,(x') = 1/a\;\mathrm{Rect}\,(x'/a). \tag{3.30}$$

The new amplitude distribution in the image is the convolution product A*P. So we get a transmission

$$F(x'_0) = \big|\,K\,[A{*}P{\otimes}A]\,](x'_0)\big|^2, \tag{3.31}$$

where K is a factor taking into account the losses of the filter. $F(x'_0)$, and consequently $F(\lambda)$, the wavelength transmission functions, become flat top functions. The width of the top of $F(\lambda)$ depends on the constant a that will be chosen according to its specification. In principle $F(\lambda)$ has no secondary lobes. This method is cheap but induces a theoretical additional loss of about 3 dB for an enlargement of the FWHM by a factor of two.

3.6.2 Devices for Use in WDM Systems

A schematic of a diffraction-grating based WDM multiplexer/demultiplexer is shown in Fig. 3.21. Corresponding devices have been commercialized for a wide range of channel numbers and for different channel separations as well: Typical channel numbers are 4, 8, ... up to 160, channel separations are 25, 50, 100, 200, or 400 GHz. The larger channel separations do normally go with the lower channel numbers and vice versa. Commercial devices cover the C-, the L-, or the C+L-bands, respectively, so far, but could be fabricated for all other bands as well.

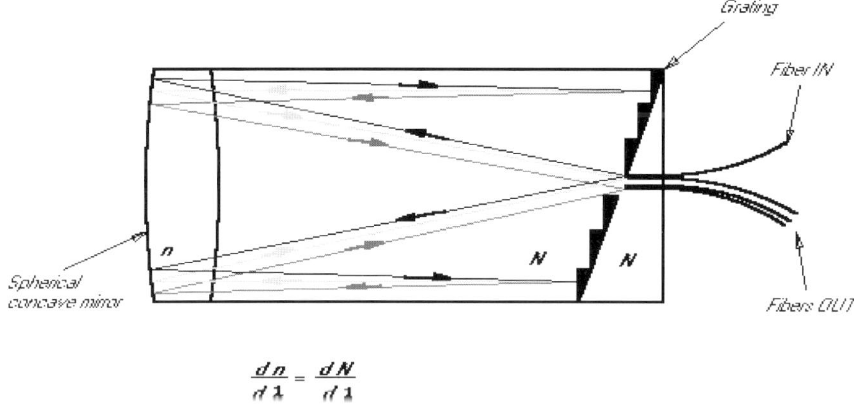

$$\frac{dn}{d\lambda} = \frac{dN}{d\lambda}$$

Fig. 3.21. Stimax® configuration with correction of the spherical aberration caused by the spherical concave mirror and of the residual chromatism by combination of two materials of index n and N with identical dispersion [71]

The normal transmission channel characteristics of diffraction gratings is Gaussian (See Chap. 3 in [71]), and the typical wavelength response of three adjacent channels with 100 GHz separation is shown in Fig. 3.22.

For telecom applications flat-top transmission channel characteristics is preferred to a Gaussian shape as the former is more tolerant to operating channel variations. This becomes particularly important, if signals travel through cascaded filters, and this will become more relevant as fibre optic networks develop from essentially point-to-point into more flexible optical networks.

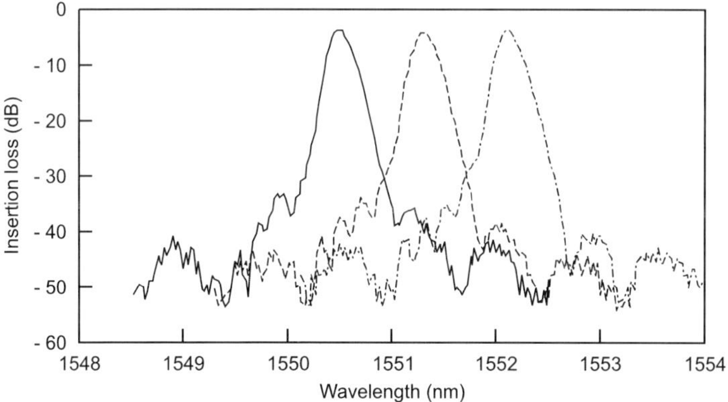

Fig. 3.22. Typical wavelength response of three adjacent channels with 100 GHz separation (Stimax® doc.)

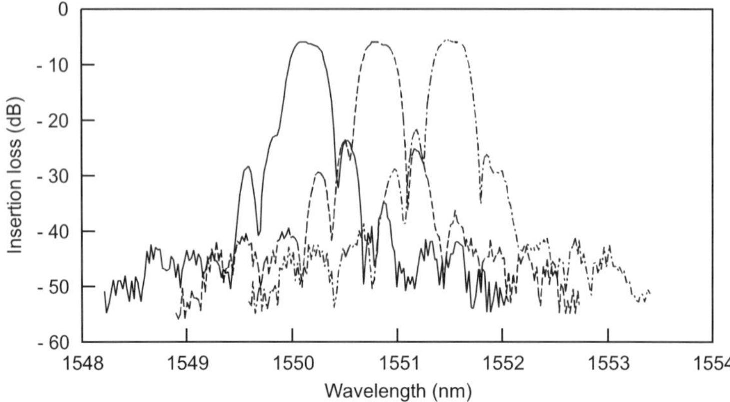

Fig. 3.23. Typical wavelength response of three adjacent channels with 100 GHz separation and flat-top characteristics (Stimax® doc.)

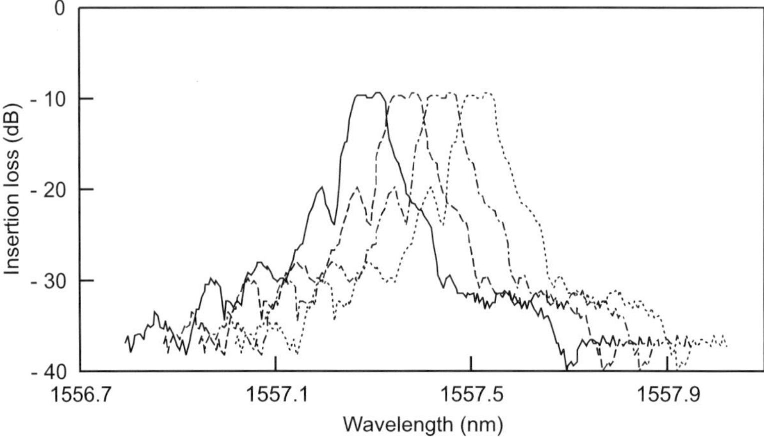

Fig. 3.24. Typical wavelength response of four adjacent channels with 19 GHz separation and flat-top characteristics (Presented by J. P. Laude et al., IEEE Workshop Vail USA Sept.15–17, 1999. No proceedings)

The channel half-width of transmission gratings can be enlarged (at the expense of a slightly higher insertion loss). Corresponding results are illustrated in Fig. 3.23 for 100 GHz spacing DWDM and in Fig. 3.24 for an ultra-dense UDWDM in Minilat® configuration (Fig. 3.39, p. 64, in [71].)

Other relevant characteristics of diffraction-grating based WDM multiplexers/demultiplexers are summarized in Table 3.1:

Table 3.1. Typical parameters of diffraction-grating based WDM mux/demux

Parameter	typical value	typical value	typical value
Channel shape	Gaussian	Gaussian	flat top
Number of channels	8	96	48
Channel separation (GHz)	100	100	100
Far end adjacent channel crosstalk (dB)	28	30	25
Near end adjacent channel crosstalk (dB)	40	50	40
Far end non-adjacent channel crosstalk (dB)	32	35	30
Near end non-adjacent channel crosstalk (dB)	40	50	40
Insertion loss (dB)	2.5	4	7
Loss uniformity (dB)	0.5	0.5	0.5
Channel width (3 dB FWHM) (GHz)	50	30	45
Frequency position accuracy (GHz)	± 5	± 5	± 5
Temperature range (°C)	$-20 \ldots +70$	$-20 \ldots +70$	$-20 \ldots +70$
Thermal drift (nm/°C)	0.0004	0.001	0.01
Polarisation-dependent shift (GHz)	0.1	0.1	0.1
Polarisation-dependent loss (dB)	0.5	0.5	0.5

3.7 Gratings on Planar Lightwave Circuits

Echelle grating designs for monolithic integration have been proposed. A packaged echelle grating device is usually about a quarter the size of a comparable AWG. Devices with an echelle grating etched on a concave surface with facets perpendicular to a silica-on-silicon or indium phosphide guide within $\pm 2°$ have been manufactured. For example, a 49 channel demux with a die size of $18 \times 20 \, \text{mm}^2$ with adjacent channel crosstalk 35 dB and background channel crosstalk 37 dB was described by Optenia in 2001 [120]. Unfortunatly, Optenia, a Mitel-founded company, closed doors in 2002 after failing to secure sufficient funding. Remarkable results were obtained later (e. g. by MetroPhotonics [121], Fig. 3.25). However, ultimately unsuccessful in its efforts to keep the company afloat, MetroPhotonics Inc., the Ontario-based company, closed down in December 2005.

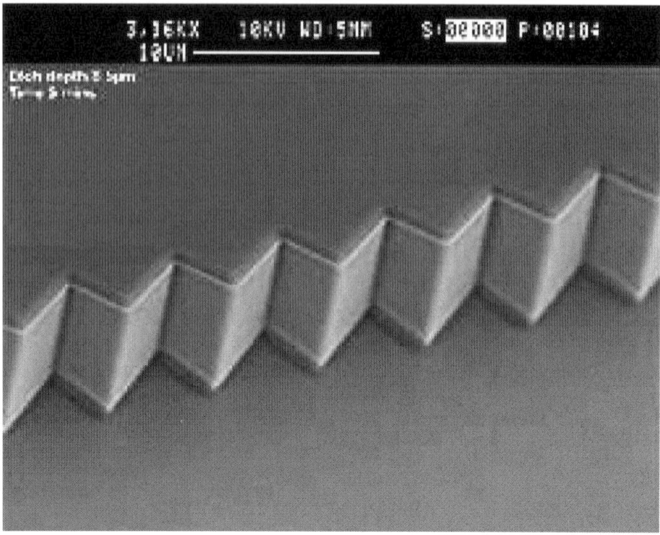

Fig. 3.25. Echelle grating on planar lightwave circuit (source: MetroPhotonics)

The shape of the grooves of the grating can be designed to take advantage of total internal reflection (TIR). TIR gratings with no facet metalisation can be used in place of echelle gratings [122]. In principle, in the absence of metallic absorption high-dispersion TIR gratings can reach more than 99% efficiencies for 15–20 nm spectral bandwidths [123]. The PDL of planar-integrated TIR concave grating structures can be very low [124].

3.8 Device Polarisation Sensitivity

Critical to all devices is polarisation sensitivity. It must be small since transmission lines generally do not maintain the polarisation state. In Integrated Optics devices birefringence may cause wavelength shifts: even for the best new germanium-doped waveguides from Nippon Telephone and Telegraph (NTT) the birefringence remains about 10^{-4}. However, polarisation compensation can be designed. Some 3-dimensional optics devices show very low birefringence, e. g. 5×10^{-7} in pure silica and 0 in air. Consequently, the Polarisation Mode Dispersion (PMD) can be very small and there is no wavelength shift of the channel centre with polarisation. But the grating can induce PDL. We have seen that the PDL of the surface-relief gratings can be lowered by the choice of a higher order of diffraction and specific groove profiles. In VPGs in Bragg condition the PDL is extremely low. If required, the PDL of any grating device can be eliminated by different means such as a second pass on the grating after 90° rotation of the electric field.

3.9 Thermal Drift

Thermal drift depends on the thermal expansion coefficients ε of the different materials and on the index variation dn/dt. For monoblock-grating WDM devices it has been shown [98] that the wavelength shift is given by:

$$\Delta\lambda/\lambda = (\varepsilon + 1/n\ dn/dT)\ \Delta T \qquad (3.32)$$

and a typical result is $\Delta\lambda = 12\,\text{pm/}°\text{C}$ with silica at $1550\,\text{nm}$.

For devices using a grating in air, the thermal drift can be reduced to about $1\,\text{pm/}°\text{C}$ with a grating on a low thermal expansion substrate. It can be less with a mechanical compensation of the expansion. For monoblock-grating DWDM devices the thermal expansion of the grating in silica can be compensated by a negative dn/dT of the optical block. Drifts lower than $0.4\,\text{pm/}°\text{C}$ of $25\,\text{GHz}$ spacing, 160 channel Very Dense WDM (VD-WDM) devices in the so-called Minilat configuration have been published [71, 125]. Similar results were also obtained on Minilat components with spacing lower than $20\,\text{GHz}$. (Today multiplexers in such a configuration are manufactured by Yenista (Lannion, France)).

3.10 Diffraction Grating Routers

3.10.1 Cyclic Free Space Diffraction Grating Router

A passive wavelength router (PWR) with N input ports to M outputs allows a wavelength-selective cross-connection in the optical domain from any input i to any output j at a wavelength λ_{ij} given for instance by a tuneable transmitter Tx (Fig. 3.26, [71]). Such a component will become particularly useful in full-mesh photonic networks, in reconfigurable networks, and on the longer term in all-optical packet routing systems. The cross-connection can be made by a single affordable component: an AWG router or a free space diffraction grating (FSDG) router. The optical path is determined by its input port and by λ_{ij}.

A standard grating multiplexer with a linear array of $2N$ fibres located in the focal plane can be used as an $N \times N$ router. (See Chap. 6: Routers, cross-connects, and add-drops, pp. 147–219 in [71]). An input fibre noted i

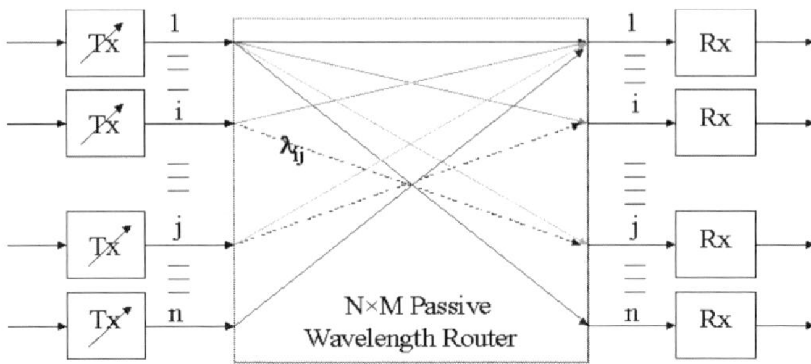

Fig. 3.26. Passive wavelength router with N input ports to M output ports used with tuneable sources and fixed receivers [71]

and an output fibre noted j, in angular locations θ_i and θ_j, respectively, are coupled through the grating equation:

$$\Lambda\,(\sin\theta_i + \sin\theta_j) = m\,\lambda_{ij} \tag{3.33}$$

With fibre positions corresponding to $\sin\theta_i$ and $\sin\theta_j$ in constant progression we get:

$$\lambda_{ij} = \lambda_0 + (i + j)\,\Delta\lambda \tag{3.34}$$

An $N\times N$ non-cyclic router (Fig. 3.27a) can be obtained with a multiplexer with $2N$ equidistant fibres. In CWDM International Telecommunication Union (ITU) standards, perfectly centred channels can be obtained for all input to output pairs. In DWDM systems, where the channels are defined by the ITU standards in constant progression of frequencies (cf. Appendix, Sect. A1), the centring of channels is obtained by approximation with small specific fibre distance shifts.

Out In	j=1	j=2	j=3	j=4
i=1	λ_1	λ_2	λ_3	λ_4
i=2	λ_2	λ_3	λ_4	λ_5
i=3	λ_3	λ_4	λ_5	λ_6
i=4	λ_4	λ_5	λ_6	λ_7

Out In	j=1	j=2	j=3	j=4
i=1	λ_1	λ_2	λ_3	λ_4
i=2	λ_2	λ_3	λ_4	λ_1
i=3	λ_3	λ_4	λ_1	λ_2
i=4	λ_4	λ_1	λ_2	λ_3

(a)   **(b)**

Fig. 3.27. (a) Position of channels in a non-cyclic router: $2N-1$ wavelengths are required for $N\times N$ non-blocking routers, **(b)** Position of channels in a cyclic router: only N wavelengths are required for $N\times N$ non-blocking routers

As an alternative to AWGs, FSDGs operating in 2 successive high orders m and $m + 1$ (for example, $m = 40$ for a typical DWDM router) can be used to design cyclic routers (Fig. 3.27b), but the deviations from the ITU channels, increasing with the spectral range, become too large in CWDM and remain a considerable drawback in DWDM.

3.10.2 From Non-cyclic to Cyclic FSDG Routers

To our knowledge, FSDG wavelength routers were exclusively non-cyclic (Fig. 3.27a) until 2001. Then it was shown that a traditional FSDG non-cyclic $2N \times N$ router could be transformed into an $N \times N$ cyclic router through a specific arrangement of the $2N$ inputs connected two by two by a duplexer array (Fig. 3.28a) [126–127]. In these cyclic routers the grating can operate in a single order instead of in the two adjacent high orders necessary in the classical Dragone configuration [128] generally used in AWG cyclic routers, and the diffraction order can be low enough to give

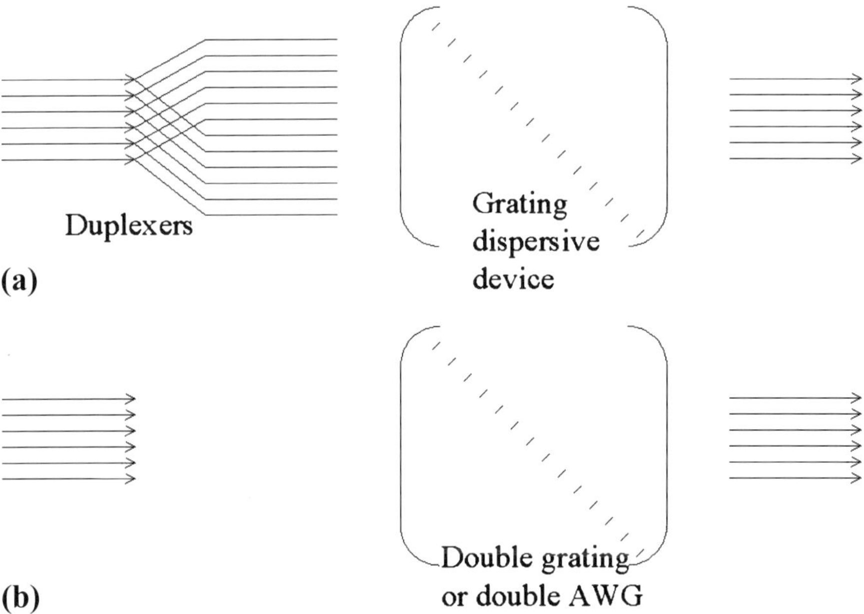

Fig. 3.28. (**a**) A non-cyclic $2N \times N$ router can be transformed into an $N \times N$ cyclic router through a specific arrangement of the $2N$ inputs connected two by two by a duplexer array. (**b**) It was known that an $N \times N$ cyclic router could be obtained coupling N inputs to N outputs through an AWG using 2 adjacent high orders. New solutions use a double dispersive device (i.e. a double diffraction grating or a double AWG; J.P. Laude [126–127])

a very large free spectral range. This is of special interest for Metro CWDM applications. But in DWDM this solution has also a great advantage as it gives much lower deviations from the ITU channels than the solution with 2 orders m and $m + 1$. Moreover, using a double grating (for example a VPG recorded one with two periodic structures operating in two adapted different Bragg conditions) it becomes possible to get directly an $N \times N$ cyclic router without the use of the duplexers (Fig. 3.28b). The FSDG devices with high efficiency, low crosstalk, and low PDL working in low and medium order, already proposed by different manufacturers of classical multi/demultiplexers, can be adapted to readily design efficient and reliable cyclic DWDM or CWDM routers accurately matched to ITU channels. It is only necessary to add a duplexer array in the wavelength dispersion plane (Fig. 3.29) or to use a double grating in them. Of course, the drawback is a 3 dB additional loss. However, in principle this additional loss can be avoided if the couplers are replaced by multiplexers or if the double grating is replaced by a multi VPG. It is worthwhile to note that in principle the solution can be applied to AWGs with an equivalent input/output coupling arrangement or with a specific double grating array.

For example, a 3×3 cyclic router, obtained from a Stimax FSDG noncyclic router [129] with 9 equidistant fibres (F1 to F9), is illustrated in Fig. 3.29. The three inputs are obtained from the six fibres F1 to F6 coupled

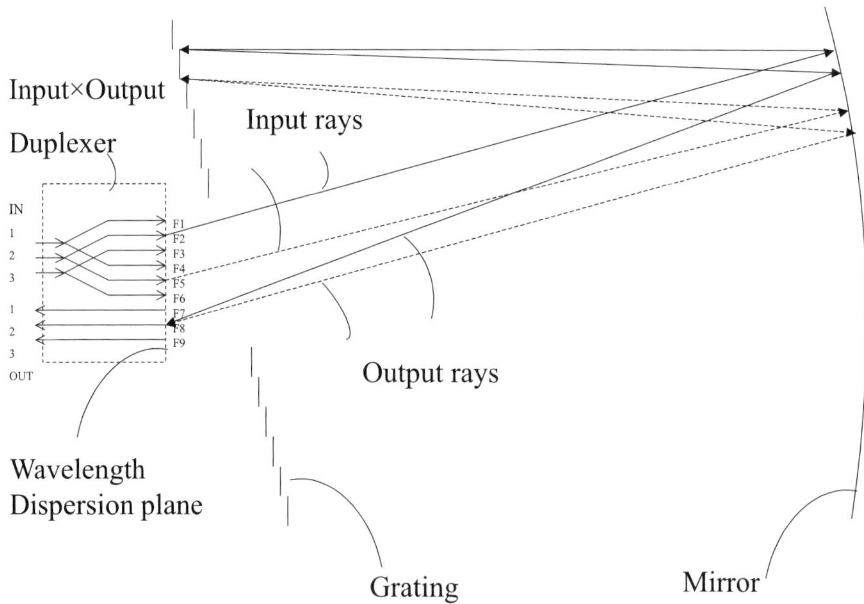

Fig. 3.29. A practical cyclic router design example

two by two (F1 to F4, F2 to F5, and F3 to F6), the three outputs are F7, F8, F9. More generally, a cyclic $N \times N$ router needs $3N$ fibres. Cyclic routers larger than 32×32 can be designed as more than 96 fibres FSDG multiplexers are commercially available. A non-cyclic router similar to the router described in [129] equipped with 3 dB couplers would give a total insertion loss of 5.4 dB and crosstalk between adjacent channels better than 40 dB.

3.10.3 An example: a 16×16 Cyclic Router Design for CWDM

A 16×16 cyclic router can be obtained using a classical grating multiplexer with an array of 32 equidistant input fibres and 16 equidistant output fibres. (For example, with a single linear row of 48 equidistant fibres).

In/out	1	2	3	4	5	6	7	8	9	10	11	12	13	14	15	16
1	990	1010	1030	1050	1070	1090	1110	1130	1150	1170	1190	1210	1230	1250	1270	1290
2	1010	1030	1050	1070	1090	1110	1130	1150	1170	1190	1210	1230	1250	1270	1290	1310
3	1030	1050	1070	1090	1110	1130	1150	1170	1190	1210	1230	1250	1270	1290	1310	1330
4	1050	1070	1090	1110	1130	1150	1170	1190	1210	1230	1250	1270	1290	1310	1330	1350
5	1070	1090	1110	1130	1150	1170	1190	1210	1230	1250	1270	1290	1310	1330	1350	1370
6	1090	1110	1130	1150	1170	1190	1210	1230	1250	1270	1290	1310	1330	1350	1370	1390
7	1110	1130	1150	1170	1190	1210	1230	1250	1270	1290	1310	1330	1350	1370	1390	1410
8	1130	1150	1170	1190	1210	1230	1250	1270	1290	1310	1330	1350	1370	1390	1410	1430
9	1150	1170	1190	1210	1230	1250	1270	1290	1310	1330	1350	1370	1390	1410	1430	1450
10	1170	1190	1210	1230	1250	1270	1290	1310	1330	1350	1370	1390	1410	1430	1450	1470
11	1190	1210	1230	1250	1270	1290	1310	1330	1350	1370	1390	1410	1430	1450	1470	1490
12	1210	1230	1250	1270	1290	1310	1330	1350	1370	1390	1410	1430	1450	1470	1490	1510
13	1230	1250	1270	1290	1310	1330	1350	1370	1390	1410	1430	1450	1470	1490	1510	1530
14	1250	1270	1290	1310	1330	1350	1370	1390	1410	1430	1450	1470	1490	1510	1530	1550
15	1270	1290	1310	1330	1350	1370	1390	1410	1430	1450	1470	1490	1510	1530	1550	1570
16	1290	1310	1330	1350	1370	1390	1410	1430	1450	1470	1490	1510	1530	1550	1570	1590
17	1310	1330	1350	1370	1390	1410	1430	1450	1470	1490	1510	1530	1550	1570	1590	1610
18	1330	1350	1370	1390	1410	1430	1450	1470	1490	1510	1530	1550	1570	1590	1610	1630
19	1350	1370	1390	1410	1430	1450	1470	1490	1510	1530	1550	1570	1590	1610	1630	1650
20	1370	1390	1410	1430	1450	1470	1490	1510	1530	1550	1570	1590	1610	1630	1650	1670
21	1390	1410	1430	1450	1470	1490	1510	1530	1550	1570	1590	1610	1630	1650	1670	1690
22	1410	1430	1450	1470	1490	1510	1530	1550	1570	1590	1610	1630	1650	1670	1690	1710
23	1430	1450	1470	1490	1510	1530	1550	1570	1590	1610	1630	1650	1670	1690	1710	1730
24	1450	1470	1490	1510	1530	1550	1570	1590	1610	1630	1650	1670	1690	1710	1730	1750
25	1470	1490	1510	1530	1550	1570	1590	1610	1630	1650	1670	1690	1710	1730	1750	1770
26	1490	1510	1530	1550	1570	1590	1610	1630	1650	1670	1690	1710	1730	1750	1770	1790
27	1510	1530	1550	1570	1590	1610	1630	1650	1670	1690	1710	1730	1750	1770	1790	1810
28	1530	1550	1570	1590	1610	1630	1650	1670	1690	1710	1730	1750	1770	1790	1810	1830
29	1550	1570	1590	1610	1630	1650	1670	1690	1710	1730	1750	1770	1790	1810	1830	1850
30	1570	1590	1610	1630	1650	1670	1690	1710	1730	1750	1770	1790	1810	1830	1850	1870
31	1590	1610	1630	1650	1670	1690	1710	1730	1750	1770	1790	1810	1830	1850	1870	1890
32	1610	1630	1650	1670	1690	1710	1730	1750	1770	1790	1810	1830	1850	1870	1890	1910

Fig. 3.30. Wavelengths used to connect any input 1 to 32 to any output 1 to 16 in a 32×16 non-cyclic router

In/out	1	2	3	4	5	6	7	8	9	10	11	12	13	14	15	16
1 + 17	1310	1330	1350	1370	1390	1410	1430	1450	1470	1490	1510	1530	1550	1570	1590	1610
2 + 18	1330	1350	1370	1390	1410	1430	1450	1470	1490	1510	1530	1550	1570	1590	1610	1310
3 + 19	1350	1370	1390	1410	1430	1450	1470	1490	1510	1530	1550	1570	1590	1610	1310	1330
4 + 20	1370	1390	1410	1430	1450	1470	1490	1510	1530	1550	1570	1590	1610	1310	1330	1350
5 + 21	1390	1410	1430	1450	1470	1490	1510	1530	1550	1570	1590	1610	1310	1330	1350	1370
6 + 22	1410	1430	1450	1470	1490	1510	1530	1550	1570	1590	1610	1310	1330	1350	1370	1390
7 + 23	1430	1450	1470	1490	1510	1530	1550	1570	1590	1610	1310	1330	1350	1370	1390	1410
8 + 24	1450	1470	1490	1510	1530	1550	1570	1590	1610	1310	1330	1350	1370	1390	1410	1430
9 + 25	1470	1490	1510	1530	1550	1570	1590	1610	1310	1330	1350	1370	1390	1410	1430	1450
10+26	1490	1510	1530	1550	1570	1590	1610	1310	1330	1350	1370	1390	1410	1430	1450	1470
11+27	1510	1530	1550	1570	1590	1610	1310	1330	1350	1370	1390	1410	1430	1450	1470	1490
12+28	1530	1550	1570	1590	1610	1310	1330	1350	1370	1390	1410	1430	1450	1470	1490	1510
13+29	1550	1570	1590	1610	1310	1330	1350	1370	1390	1410	1430	1450	1470	1490	1510	1530
14+30	1570	1590	1610	1310	1330	1350	1370	1390	1410	1430	1450	1470	1490	1510	1530	1550
15+31	1590	1610	1310	1330	1350	1370	1390	1410	1430	1450	1470	1490	1510	1530	1550	1570
16+32	1610	1310	1330	1350	1370	1390	1410	1430	1450	1470	1490	1510	1530	1550	1570	1590

Fig. 3.31. How to obtain a 16×16 cyclic router from the 32×16 non-cyclic router of Fig. 3.30

The wavelengths to be used to connect any input 1 to 32 to any output 1 to 16 of this multiplexer, initially used as a 32×16 non-cyclic router, are given in Fig. 3.30. If the input fibres 1 and 17, 2 and 18, ..., i and i + 16, ..., 16 and 32, are duplexed into 16 input fibres, a 16×16 cyclic router is obtained as shown in Fig. 3.31.

It is worth noting that this configuration allows an almost perfect matching of ITU wavelengths on all channels in CWDM and an amelioration of matching of ITU frequencies over classical AWG solutions in DWDM.

3.10.4 Practical Results

DWDM applications: All channels could be matched to the ITU frequency grid G.694.1 with a maximum deviation of 1 GHz on 32×32 FSDG cyclic routers with 25 GHz spacing [130].

CWDM applications: a maximum deviation of 0.57 nm from the wavelength grid ITU G.694.2 was calculated on the worst channel of a 16×16 FSDG cyclic router with 16 wavelengths at 20 nm spacing from 1310 nm to 1610 nm using the design characteristics of a basic Stimax® glass block configuration with an 80 lines/mm grating, a focal length of 40.102 mm, and a linear array of 48 fibres with a constant spacing of 42.5 µm ([127], Fig. 3.32). This small residual deviation comes from the dispersion of the material, here a high dispersion glass, that could be easily replaced by a lower dispersion glass material. With a similar component in aerial configuration the deviations become negligible: < 0.00005 nm (Fig. 3.33). For such devices, using available multiplexers and duplexers, the insertion loss

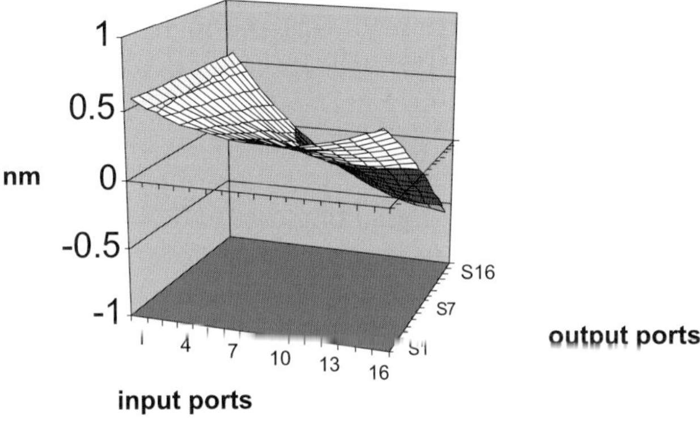

Fig. 3.32. Deviations from ITU wavelengths of glass block cyclic routers. The wavelength deviation does not depend on focal length or grating period, but only on glass dispersion: here high dispersion

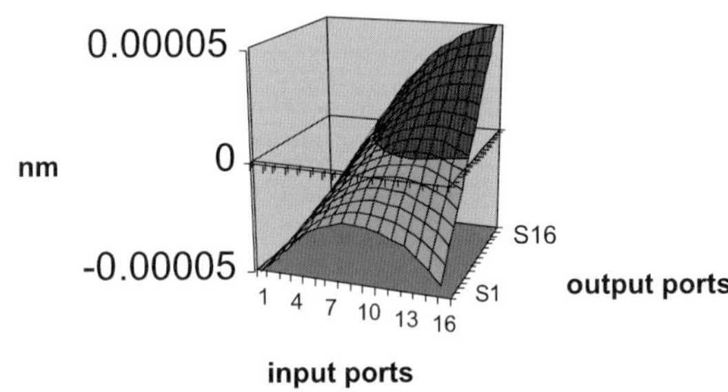

Fig. 3.33. Deviations from ITU wavelengths of aerial cyclic routers. The wavelength deviation does not depend on focal length or grating period

would be typically 5 to 7 dB. The use of a double or a multi-grating in place of the duplexers could reduce the loss by 3 dB.

3.11 Conclusion

Diffraction grating WDM components have tremendous capability in number of channels (160 channel components and more with bi-directionality are commercially available and components with two times more channels have been demonstrated experimentally). They have a high free spectral range (typically 775 nm with gratings working in the first order). Small spacing such as 25 GHz is commercially available and 5 GHz feasibility has been demonstrated. Athermal glass components down to 25 GHz spacing are manufactured with uncontrolled drifts as low as 0.35 pm/°C using commonly available glasses. These solutions are based on the traditional and mature optics technology used to manufacture millions of diffraction-limited reliable optics at low cost. The unique three-dimensional nature of these components can be exploited for advanced features requested by network designers such as add/drops, crossconnects, and routing devices. Cyclic passive routers accurately matched to the ITU frequencies in DWDM or to ITU wavelengths in CWDM have been proposed for advanced networks. Diffraction grating components can be used for high quality bi-directional links. For mono-directional or bi-directional links they have virtually unlimited numbers of channels and provide the lowest crosstalk for the higher number of channels. For low cost mass production the problems of fibre handling, that are the same with any solution, remain to be solved.

References

1.   M. C. Hutley: *Diffraction gratings* (Academic Press, London, 1982)
2.   J. P. Laude and J. Flamand: „Herstellung von Beugungsgittern für Spektrometrie und Optoelektronik," Feinwerktechnik und Meßtechnik (Carl Hanser Verlag, München) **94**, 5 (1986)
3.   J. M. Lerner, J. Flamand, J. P. Laude, G. Passereau, and A. Thevenon: "Diffraction gratings ruled and holographic – A review," *Proc. SPIE* **240–14**, 82–88 (1980)
4.   J. Cordelle, J. P. Laude, R. Petit, and G. Pieuchard: "Réseaux classiques, réseaux holographiques. Les questions d'efficacité et leurs conséquences en spectroscopie instrumentale," Nouv. Rev. d'Optique Appliquée **3**, 149–154 (1970)
5.   E. G. Loewen and E. Popov: *Diffraction gratings and applications* (Marcel Dekker, New York, Basel, 1997)
6.   A. Labeyrie and J. Flamand: "Spectrographic performance of holographically made diffraction gratings," Opt. Commun. **1**, 5–8 (1969)
7.   J. Cordelle, J. Flamand, G. Pieuchard, and A. Labeyrie: "Aberration-corrected concave gratings made holographically," *Proceedings ICO Conference,* Reading

1969; in *Optical Instruments and Techniques* (J. Home, ed.), (Dickson Oriel Press, 1970)

8.   Y. Fujii, K.-I. Aoyama, and J.-I. Minowa: "Optical demultiplexer using a silicon echelette grating," IEEE J. Quantum Electron. **QE–16**, 165–169 (1980)

9.   MicroSensonic: "IR and NIR silicon gratings," http://optics.org/press/5479 (2003)

10.   B. Wattelier, S. Norcia, J. P. Zou, and C. Sauteret: "Mesure de la tenue au flux de nouveaux matériaux pour les réseaux de compression," Rapport LULI, **D2**, 113–115 (1999)

11.   B. Wattelier, J. P. Zou, C. Sauteret, J. A. Reichart, N. Blanchot, P. Y. Baures, H. Bercegol, and J. Dijon: "Nouvelles technologies pour les réseaux de diffraction utilisés dans les compresseurs de chaines CPA," Rapport LULI, **E5**, 128–130 (2000)

12.   A. S. Svakhin, V. A. Sychugov, and A. E. Tikhomirov: "Diffraction gratings with high optical strength for laser resonators," Quantum Electron. **24**, 233–235 (1994)

13.   M. D. Perry, R. D. Boyd, J. A. Britten, D. Decker, B. W. Shore, C. Shannon, and E. Shults: "High-efficiency multilayer dielectric diffraction gratings," Opt. Lett. **20**, 940–942 (1995)

14.   M. D. Perry, J. A. Britten, H. T. Nguyen, R. D. Boyd, and B. W. Shore: "Multilayer dielectric diffraction gratings," US patent: 5,907,436 May 25 (1999)

15.   M. D. Perry: "Multilayer dielectric diffraction gratings increasing the power of light," Science and technology review, Sept. 1995, 24–33 (1995)

16.   B. Touzet and J. R. Gilchrist: "New Jobin Yvon multi-layer dielectric gratings will double the power of high energy laser," http://www.jyhoriba.jp/product_e/grating/grating/pulse.htm (2003)

17.   B. Touzet and J. R. Gilchrist: "Multi-layer dielectric gratings enable more-powerful high-energy laser," Photonics Spectra, Sept. 2003, 68–75 (2003)

18.   C. P. Barty: "Short pulse, high repetition rate laser technology issues for IFE FI," Snowmass 2002 Summer Fusion Study, July 10 (2002)

19.   Diffractive Optics Group at LLNL: "Multilayer dielectric diffraction grating," http://www.llnl.gov/nif/lst/diffractive-optics/multilayer.html, 1–3 (2003)

20.   D. M. Chambers and G. P. Nordin: "Stratified volume diffractive optical elements as high-efficiency gratings," J. Opt. Soc. Am. A **16**, 1184–1193 (1999)

21.   D. M. Chambers, G. P. Nordin, and S. Kim : "Fabrication and analysis of a three-layer stratified volume diffractive optical element high-efficiency grating," Opt. Express **11**, 27–38 (2003)

22.   G. Lippmann: "La photographie intégrale," C. R. Acad. Sci. **146**, 446–451 (1908)

23.   T. A. Shankoff: "Phase holograms in dichromated gelatin," Appl. Opt. **7**, 2101–2105 (1968)

24.   L. H. Lin: "Hologram formation in hardened dichromated gelatin films," Appl. Opt. **8**, 963–966 (1969)

25.   B. J. Chang: "Dichromated gelatin as a holographic storage medium," *Proc. SPIE* **177** "Optical Information Storage", 71–81 (1979)

26.   B. J. Chang and C. D. Leonardo: "Dichromated gelatin for the fabrication of holographic optical elements," Appl. Opt. **18**, 2407–2417 (1979)

27. B. J. Chang: "Dichromated gelatin holograms and their applications," Opt. Eng. **19**, 642–648 (1980)

28. J. R. Magarinos and D. J. Coleman: "Holographic mirrors," Opt. Eng. **24**, 769–780 (1985)

29. T. Kubota: "Control of the reconstruction wavelength of Lippmann holograms recorded in dichromated gelatin," Appl. Opt. **28**, 1845–1849 (1989)

30. R. D. Rallison: "DCG applied with a record player and broadband processed in 2 minutes," *Hughes Aircraft report*, Jan 1974 (never published)

31. M. Akagi, T. Kaneko, and T. Ishiba: "Electron micrographs of hologram cross sections," Appl. Phys. Lett. **21**, 93–95 (1972)

32. L. D. Dickson: "Method for making holographic optical elements with high diffraction efficiencies," US Patent 4,416,505 (1983)

33. Wasatch Photonics: "Dickson^{TM} Gratings and Grisms," http://www.wasatchphotonics.com/technical_info.htm (2004)

34. S. C. Barden, J. A. Arns, and W. S. Colburn: "Volume-phase holographic gratings and their potential for astronomical applications," *Proc. SPIE* **3355**, 866–876 (1998)

35. S. C. Barden, J. A. Arns, W. S. Colburn, and J. B. Williams: "Volume-phase holographic gratings and the efficiency of three simple VPH gratings," Pub. Astron. Soc. Pacific **112**, 809–820 (2000)

36. S. C. Barden, J. A. Arns, W. S. Colburn, and J. B. Williams: "Evaluation of volume-phase holographic grating technology," *Proc. SPIE* **4485**, 429–438 (2001)

37. G. M. Bernstein, A. Athey, R. Bernstein, S. Gunnels, D. Richstone, and S. Shectman: "Volume-phase holographic spectrograph for the Magellan telescope," *Proc. SPIE* **4485**, 453–459 (2002)

38. I. K. Baldry, J. Bland-Hawthorn, and J. G. Robertson: "Volume phase holographic gratings: polarisation properties and diffraction efficiency," Pub. Astron. Soc. Pacific **116**, Issue 819, 403–414, (2004)

39. B. L. Volodin, S. V. Dolgy, E. D. Melnik, and V. S. Ban: "Volume Bragg gratings^{TM}, a new platform technology for WDM applications," http://www.pd-ld.com/pdf/VBG_paper.pdf , 1–9 (2004)

40. Z. Yaqoob and N. A. Riza: "Low-loss wavelength-multiplexed optical scanners using volume Bragg gratings for transmit-receive lasercom systems," Opt. Eng. **43**, 1128–1135 (2004)

41. Y. Q. Lu, F. Du, and S.-T. Wu: "Polarisation switch using thick holographic polymer-dispersed liquid crystal grating," J. Appl. Phys. **95**, 810–815 (2004)

42. Wasatch Photonics: "Volume phase holographic gratings," http://www.wasatchphotonics.com/products.htm (2004)

43. Kaiser Inc.: "Telecom gratings," http:// www.kosi.com/gratings/telecom (2001)

44. R. D. Rallison: "Dense wavelength division multiplexing and the Dickson grating," White Paper, http://www.xmission.com/~ralcon/files/DWDM-Dickson_grating_white_paper.pdf~ralcon/ (2004)

45. http://soi.srv.pu.ru/developments/elem_basa/trans_grate.htm (2004)

46. D. Yu and W. Yang: "VPG-based DWDM mux/demux devices and specifications," http://www.bayspec.com/WP-DWDM.htm, (2002)

47. P. Boffi, M. C. Ubaldi, D. Piccinin, C. Frascolla, and M. Martinelli: "1550-nm volume holography for optical communication devices," IEEE Photon. Technol. Lett. **12**, 1355–1357 (2000)

48. P. Boffi: "Holography-based devices for WDM fiber communication," *Proc. Cost Meeting,* Berlin, Germany, 24–25 June (2002)

49. M. C. Ubaldi, P. Boffi, D. Piccinin, C. Frascolla, and M. Martinelli: "Volume holography for 1550 nm digital databases," *Proc. SPIE* **4089**, 625–629 (2000)

50. S. Han, B. Yu, S. Chung, H. Kim, J. Paek, and B. Lee: "Filter characteristics of a chirped volume holographic grating," Opt. Lett. **29**, 107–109 (2004)

51. J. A. Arns, W. S. Colburn, and S. C. Barden: "Volume phase gratings for spectroscopy, ultrafast laser compressors, and wavelength division multiplexing," *Proc. SPIE* **3779**, 313–323 (1999)

52. R. D. Rallison: "Phase material for HOE applications," http://www.xmission.com/~ralcon/phasemat.html (2004)

53. H. I. Bjelkhagen: *Silver-Halide recording materials for holography,* 2nd ed. (Springer, Heidelberg, 1996)

54. J. Liu, R. T. Chen, B. M. Davies, and L. Li: "Modeling and design of planar slanted volume holographic gratings for wavelength-division-multiplexing applications," Appl. Opt. **38**, 6981–6986 (1999)

55. R. K. Kostuk, A. Sato, and D. A. Waldman: "Volume holographic filters for 1550 nm dense wavelength division multiplexing applications," OSA Diffraction Optics and Micro-Optics Top. Meeting (DOMO), Tucson, AZ, June (2002)

56. D. M. Sykora and G. M. Morris: "Photopolymer characterization and grating playback at telecommunication wavelengths," OSA Diffraction Optics and Micro-Optics Top. Meeting (DOMO), Tucson, AZ, June (2002)

57. D. Sykora: "Volume gratings in photopolymer for playback at near infrared wavelengths: design and experiments," University of Rochester colloquia, http://www.optics.rochester.edu/events/colloquia/SykoraAbstract.html (2003).

58. D. A. Waldman, H.-Y. S. Li, and M. G. Horner: "Volume shrinkage in slant fringe gratings of a cationic ring-opening volume hologram recording material," J. Imaging Sci. Technol. **41**, 497–514 (1997)

59. R. T. Ingwall and D. A. Waldman: "CROP Photopolymers for holographic recording," SPIE, *Holography* **11**, 1 (2000)

60. R. T. Ingwall and D. A. Waldman: "Photopolymer systems," in *Holographic Data Storage* (H. J. Coufal, D. Psaltis, and G. Sincerbox, eds.), Springer Series in Optical Sciences, vol. **76**, 171–197 (Springer, New York, 2000)

61. C. Neipp, A. Beléndez, S. Gallego, M. Ortuño, and I. Pascual: "Angular responses of first and second diffracted orders in transmission diffraction grating recorded on photopolymer material," Opt. Express **11**, 1835–1843 (2003)

62. T. J. Trentler, J. E. Boyd, and V. L. Colvin: "Epoxy resin-photopolymer composites for volume holography ," Chem. Mater. **12**, 1431–1438 (2000)

63. R. Schriever, H. Franke, H. G. Festl, and E. Krätzig: "Optical waveguiding in doped poly(methyl methacrylate)," Polymer **26**, 1423–1427 (1985)

64. O. Beyer, I. Nee, F. Havermeyer, and K. Buse: "Holographic recording of Bragg gratings for wavelength division multiplexing in doped and partially polymerized poly(methyl metacrylate)," Appl. Opt. **42**, 30–37 (2003)

65. S. Breer and K. Buse: "Wavelength demultiplexing with volume phase holograms in photorefractive lithium niobate," Appl. Phys. B **66**, 339–345 (1998)
66. K. M. Davis, K. Miura, N. Sugimoto, and K. Hirao: "Writing waveguides in glass with femtosecond lasers," Opt. Lett. **21**, 1729–1731 (1996)
67. N. Takeshima, Y. Kuroiwa, Y. Narita, and S. Tanaka: "Fabrication of a periodic structure with a high refractive-index difference by femtosecond laser pulses," Opt. Express **12**, 4019–4024 (2004)
68. J. M. Lerner and J. P. Laude: "New vistas for diffraction gratings," Electro-Optics, May, 77–82 (1983)
69. P. Mouroulis, F. T. Hartley, D. W. Wilson, and V. E. White: "Blazed grating fabrication through gray-scale X-ray lithography," Opt. Express **11**, 270–281 (2003)
70. M. Detaille, M. Duban, and J. P. Laude: "Réalisation de réseaux de diffraction pour le satellite astronomique D2B," *Perspectives 91*, *Jobin Yvon documentation*, Feb. (1976)
71. J. P. Laude: *DWDM Fundamentals, Components, and Applications*, ISBN: 1-58053-177-6, (Artech House, Boston, London, 2002)
72. R. Petit: *Electromagnetic theory of gratings* (Springer, Heidelberg, Berlin, 1980)
73. L. I. Goray: "Rigorous integral method in application to computing diffraction on relief gratings working in wavelength range from microwaves to X-ray," *Proc. SPIE* **2532**, 427–433 (1995)
74. E. Popov, B. Bozhkov, D. Maystre, and J. Hoose: "Integral method for echelles covered with lossless or absorbing thin dielectric layers," Appl. Opt. **38**, 47–55 (1999)
75. L. I. Goray: "Modified integral method and real electromagnetic properties of echelles," in *Diffractive and Holographic Technologies for Integrated Photonic Systems* (R. I. Sutherland, D. W. Prather, and I. Cindrich, eds.), *Proc. SPIE* **4291**, 13–24 (2001)
76. M. G. Moharam and T. K. Gaylord: "Diffraction analysis of dielectric surface-relief gratings," J. Opt. Soc. Am. **72**, 1935 (1982)
77. M. G. Moharam and T. K. Gaylord: "Rigorous coupled-wave analysis of planar gratings," J. Opt. Soc. Am. **71**, 811 (1981)
78. M. G. Moharam and T. K. Gaylord: "Rigorous coupled-wave analysis of metallic surface- relief gratings," J. Opt. Soc. Am. A **3**, 1780–1787 (1986)
79. M. G. Moharam and T. K. Gaylord: "Three-dimensional vector coupled-wave analysis of planar grating diffraction," J. Opt. Soc. Am. **73**, 1105 (1983)
80. G. Cappiello: "DWDM and component technologies," *Proc. NEFC Meeting* Lexington MA 12/10/02, www.nefc.com (2002)
81. J. Qiao, F. Zhao, R. T. Chen, J. W. Horwitz, and W. W. Morey: "Athermalized low-loss echelle-grating-based multimode dense wavelength division demultiplexer," Appl. Opt. **41**, 6567–6575 (2002)
82. R. Ryf, Y. Su, L. Möller, S. Chandrasekhar, D. T. Neilson, and C. R. Giles: "Data rate and channel spacing flexible wavelength blocking filter," *Opt. Fiber Commun. Conf.* (OFC'2004), Techn. Digest (Los Angeles, CA, USA) PDP10; http://www.ofcconference.org/materials/pdp10-1489.PDF (2004)
83. E. Popov, J. Hoose, B. Frankel, C. Keast, M. Fritze, T. Y. Fan, D. Yost, and S. Rabe: "Low polarisation dependent diffraction grating for wavelength demultiplexing," Opt. Express **12**, 269–274 (2004)

84. M. Nevière and E. Popov: *Light propagation in periodic media, differential theory and design* (Marcel Dekker, New York, 2003)

85. D. Maystre, J. P. Laude, P. Gacoin, D. Lepere, and J. P. Priou: "Gratings for tuneable lasers: using multidielectric coatings to improve their efficiency," Appl. Opt. **19**, 3099–3102 (1980)

86. L. Weitzel, A. Krabbe, H. Kroker, N. Thatte, L. E. Tacconi-Garman, M. Cameron and R. Genzel: "3D: The next generation near-infrared imaging spectrometer," Astron. Astrophys. Suppl. Ser. **119**, 531–546 (1996)

87. J. P. Laude: brevet européen 332790, 18 mars (1988)

88. M. Nevière, D. Maystre, and J. P. Laude: "Perfect blazing for transmission gratings," J. Opt. Soc. Am. A **7**, 1736–1739 (1990)

89. Z. Shi, J.-J. He, and S. He: "Waveguide echelle grating with low polarisation-dependent loss using single-side metal-coated grooves," IEEE Photon. Technol. Lett. **16**, 1885–1887 (2004)

90. A. S. Svakhin, V. A. Sychugov, and A. E. Tikhomirov: "Diffraction gratings with high optical strength for laser resonators," Quantum Electron. **24**, 233–235 (1994)

91. J. A. Britten, M. D. Perry, B. W. Shore, R. D. Boyd, G. E. Loomis, and R. Chow: "High-efficiency dielectric multilayer gratings optimized for manufacturability and laser damage threshold," *Proc. SPIE* **2714**, 511–520 (1996)

92. J. P. Laude: "Guidage des ondes électromagnétique par les diélectriques," Perspective 91 Meeting, Jobin Yvon Application Note, F 2112, 1–12, Longjumeau, France, (1976)

93. H. Wei and L. Li: "All-dielectric reflection gratings: a study of the physical mechanism for achieving high efficiency," Appl. Opt. **42**, 6255–6260 (2003)

94. C. B. Burckhard: "Diffraction of a plane wave at a sinusoidally stratified dielectric grating," J. Opt. Soc. Am. **56**, 1502–1509 (1966)

95. H. Kogelnik: "Coupled wave theory for thick hologram gratings," Bell Syst. Technical J. **48**, 2909–2947 (1969)

96. R. Magnuson and T. K. Gaylord: "Analysis of multiwave diffraction by thick gratings," J. Opt. Soc. Am. **67**, 1165–1170 (1977)

97. T. K. Gaylord and M. G. Moharam: "Analysis and applications of optical diffraction by gratings," *Proc. IEEE* **73**, 894–937 (1985)

98. J. P. Laude: *Wavelength Division Multiplexing*, ISBN 0-13-489865-6 (hbk), (Prentice Hall, London, 1993) Out of print but still available at jplaude@libertysurf.fr

99. J. P. Laude, J. Flamand, J. C. Gautherin, D. Lepère, F. Bos, P. Gacoin, and A. Hamel: "Multiplexeur de longueurs d'onde à micro-optique pour fibres unimodales," *Sixième journées nationales d'optique guidée*, Issy, 20–21 mars (1985)

100. K. Aoyame and J. Minowa: "Low loss optical demultiplexer for WDM systems in the 0.8 μm wavelength region," Appl. Opt. **18**, 2854–2836 (1979)

101. R. Watanabe, K. Nosu, and Y. Fujii: "Optical grating multiplexer in the 1.1–1.5 μm wavelength region," Electron. Lett. **16**, 108–109 (1980)

102. W. J. Tomlinson: "Wavelength multiplexing in multimode optical fibers," Appl. Opt. **16**, 2180–2194 (1977)

103. B. D. Metcalf and J. F. Providakes: "High-capacity wavelength demultiplexer with a large-diameter Grin rod lens," Appl. Opt. **21**, 794–796 (1982)

104. M. Seki, K. Kobayashi, Y. Odagiri, M. Shikada, T. Tanigawa, and R. Ishikawa: "20-channel micro-optic grating demultiplexer for 1.1–1.6 µm band using a small focusing parameter graded-index rod lens," Electron. Lett. **18**, 257–258 (1982)

105. W. J. Tomlinson: "Wavelength division multiplexer," *US patent*, Doc. 4, III, 524, Sept. (1978)

106. G. Finke, A. Nicia, and D. Rittick: „Optische Kugellinsen Demultiplexer," *Optische Nachrichtentechnik*, Ntz Bd **37**, 346–351 (1984)

107. J. P. Laude: "High brightness monochromator using optical fibers," *Proc. Conference Opto* (ESI Ed. Paris, 1980) p. 60

108. J. P. Laude and J. Flamand: "Un multiplexeur-démultiplexeur de longueur d'onde (configuration Stimax)," Revue Opto **3**, 33–34 Février (1981)

109. L. Mannschke: "Microcomputer aided design and realization of low insertion loss wavelength multiplexer and demultiplexer," *Proc. SPIE* **399**, 92–97 (1983)

110. L. Mannschke: "A multiplexer/coupler with tapered graded-index glass fibers and a grin rod lens," *Proc. 10th Europ. Conf. Opt. Commun.* (ECOC'84), Berlin, Germany, 164–165 (1984)

111. J. P. Laude, J. Flamand, J. C. Gautherin, D. Lepere, P. Gacoin, F. Bos, and J. Lerner: "STIMAX, a grating multiplexer for monomode or multimode fibers," *Proc. 9th Europ. Conf. Opt. Commun.* (ECOC'83), Geneva, Switzerland, 417 (1983)

112. R. Watanabe, K. Nosu, T. Harada, and T. Kita: "Optical demultiplexer using concave grating in 0.7–0.9 µm wavelength region," Electron. Lett. **16**, 106–108 (1980)

113. Y. Fujii and J. Minowa: "Cylindrical concave grating utilising thin silicon chip," Electron. Lett. **17**, 934–936 (1981)

114. T. Kita and T. Harada: "Use of aberration corrected concave grating in optical demultiplexing," Appl. Opt. **22**, 819–825 (1983)

115. E. G. Churin and P. Bayvel: "Free-space, dense WDM router based on a new concave grating configuration," *Proc. 24th Europ. Conf. Opt. Commun.* (ECOC'98), Madrid, Spain, vol. **1**, 239–240 (1998)

116. A. Stavdas, P. Bayvel, and J. E. Midwinter: "Design and performance of concave holographic gratings for applications as multiplexers/demultiplexers for wavelength routed optical networks," Opt. Eng. **35**, 2816–2823 (1996)

117. J. Hegarty, S. D. Poulsen, K. A. Tackson, and P. Raminov: "Low-loss single-mode wavelength division multiplexer with etched fibre array," Electron. Lett. **20**, 685–686 (1984)

118. D. R. Wisely: "High performance 32 channels HDWDM multiplexer with 1 nm channel spacing and 0.7 nm bandwidth," *Proc. SPIE* **1578**, 170–176 (1991)

119. J. P. Laude and S. Louis: "A new method for broadening and flattening the spectral shape of the transmission channels of grating wavelength division multiplexers (WDM) and routers," *Proc. Optoelectron. Commun. Conf.* (OECC'98), Chiba, Japan, 522–532 (1998)

120. Optenia: "Silicon-based echelle grating technology for metropolitan and long-haul DWDM applications," www.optenia.com

121. Commercial document: "MetroPhotonics Technology Introduction," www.metrophotonics.com/html/tech.html

122. K. A. McGreer, Z. J. Sun, J. N. Broughton, and M. S. Smith: "Demultiplexing 120 optical channels with an integrated concave grating with total internal reflection facets," *Proc. SPIE* **3491**, 101–107 (1998)

123. J. R. Marciante and D. H. Raguin: "High-efficiency, high-dispersion diffraction gratings based on total internal reflection," Opt. Lett. **29**, 542–544 (2004)

124. Z. Shi, J-J. He, and S. He: "Analysis and design of a concave diffraction grating with total-reflection facets by hybrid diffraction method," J. Opt. Soc. Am. A **21**, 1198–1206 (2004)

125. J. P. Laude: "New athermal very dense wavelength division multiplexers," *Proc. 26th Europ. Conf. Opt. Commun.* (ECOC'2000), Munich, Germany, vol. **3**, 181–182 (2000)

126. J. P. Laude. Int. Patent, 27 nov. (2001)

127. J. P. Laude: "New cyclic wavelength routers using diffraction grating for CWDM and DWDM networks," *Opt. Fiber Commun. Conf.* (OFC'03), Techn. Digest (Atlanta, GA, USA) vol. **1**, 55–56 (2003)

128. C. Dragone: "An NxN optical demultiplexer using a planar arrangement of two star couplers," IEEE Photon. Technol. Lett. **3**, 812–815 (1991)

129. J. P. Laude, I. Long, and D. Fessard: "Very dense wavelength routers based on a new diffraction grating configuration," *Proc. 23rd Europ. Conf. Opt. Commun.* (ECOC'97), Edinburgh, UK, vol. **3**, 87–90 (1997)

130. J. P. Laude and C. Botton: "Cyclic wavelength routers using diffraction gratings accurately matched to ITU frequencies," *Proc. Optoelectron. Commun. Conf.* (OECC'02), Yokohama, Japan, 11B2–3 (2002)

4 Arrayed Waveguide Gratings

Xaveer J. M. Leijtens, Berndt Kuhlow and Meint K. Smit

4.1 Introduction

Arrayed Waveguide Grating (AWG) multiplexers/demultiplexers are planar devices which are based on an array of waveguides with both imaging and dispersive properties. They image the field in an input waveguide onto an array of output waveguides in such a way that the different wavelength signals present in the input waveguide are imaged onto different output waveguides. AWGs were first reported by Smit [1] (1988) and later by Takahashi [2] (1990) and Dragone [3] (1991). They are known under different names: Phased Arrays (PHASARs), Arrayed Waveguide Gratings (AWGs), and Waveguide Grating Routers (WGRs). The acronym AWG, introduced by Takahashi [2], is the most frequently used name today and will also be used in this text. Together with Thin-Film Filters and Fibre Bragg Gratings, AWGs are the most important filter type applied in WDM networks, and with the advance of Photonic Integrated Circuits technology they are expected to become the most important one. Their operation principles will be described in Sect. 4.2.

The most important technologies used for realisation of AWGs today are silica-on-silicon technology and Indiumphosphide (InP)-based semiconductor technology. In addition, research on silicon-based polymer technology [4, 5] and on lithium niobate [6] have been reported as well.

Silica-on-silicon (SoS) AWGs have been introduced to the market in 1994 and currently hold the largest share of the AWG market. Their modal field matches well with that of a fibre, and therefore it is relatively easy to couple them to fibres. They combine low propagation loss (< 0.05 dB/cm) with a high fibre-coupling efficiency (losses in the order of 0.1 dB). A disadvantage is that they are relatively large due to their fibre matched waveguide properties, which prohibit the use of short bends. This is presently being improved by using higher index contrast and spot-size converters to keep fibre coupling losses low. SoS-based AWGs will be described in detail in Sect. 4.3.

AWGs can be utilized to accomplish complex functionalities in fibre optic WDM networks. They are also increasingly used in other areas such as signal processing, measurement, and characterisation or sensing as well. Integration of AWGs enables compact, high functionality devices. Examples from the area of fibre optic communication will be presented in Sect. 4.4.

Semiconductor-based devices have the potential to integrate a wide variety of functions on a single chip; they are suitable for integration of passive devices such as AWGs with active ones such as electro-optical switches, modulators, and optical amplifiers, and also non-linear devices such as wavelength converters. The dominant technology for operation in the telecom window is based on InP. InP-based AWGs can be very compact due to the large index-contrast of InP-based waveguides. The market for integrated components is still small, but it is expected to become increasingly important in the coming years. InP-based technology will be described in Sect. 4.5, a brief description of other material systems is given in Sect. 4.6, and finally, characterisation of AWGs is discussed in Sect. 4.7.

4.2 Operation Principle and Device Characteristics

4.2.1 Principle

Figure 4.1 shows the schematic layout of an AWG-demultiplexer, and the operation can be understood as follows [7]. When a beam propagating through the transmitter waveguide enters the first Free Propagation Region (FPR) it is no longer laterally confined and becomes divergent. On arriving at the input aperture the beam is coupled into the waveguide array and propagates through the individual waveguides towards the output aperture. The length of the array waveguides is chosen such that the optical path length difference between adjacent waveguides equals an integer multiple of the central wavelength λ_c of the demultiplexer. For λ_c, the fields in the individual waveguides arrive at the output aperture with equal phase (mod. 2π), and thus the field distribution at the input aperture is reproduced at the output aperture. The divergent beam at the input aperture is thus transformed into a convergent one with equal amplitude and phase distribution, and the input field at the object plane gives rise to a corresponding image at the centre of the image plane. The spatial separation of different wavelengths is obtained by linearly increasing the lengths of the array waveguides, which introduces a wavelength-dependent tilt of the outgoing beam

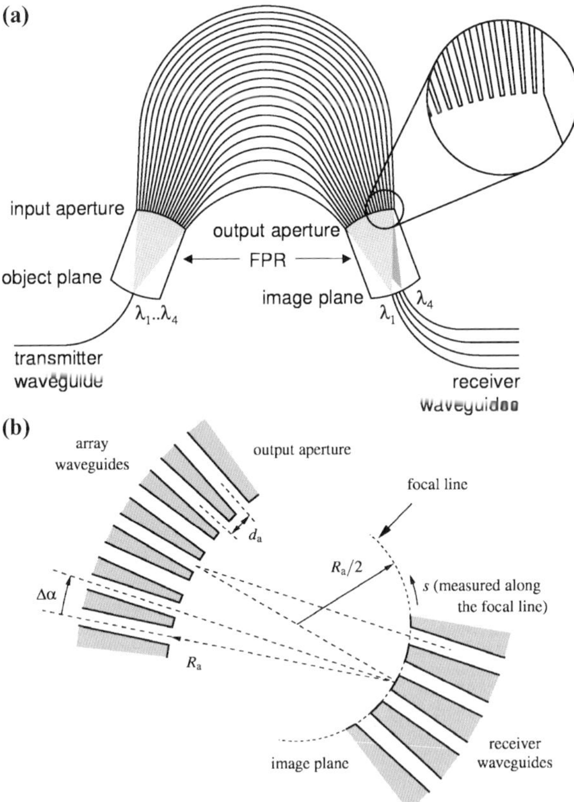

Fig. 4.1. (a) Geometry of an AWG demultiplexer, **(b)** beam focusing geometry in the free propagation region

associated with a shift of the focal point along the image plane. If receiver waveguides are placed at proper positions along the image plane, different wavelengths are led to different output ports.

4.2.2 Focusing, Spatial Dispersion, and Free Spectral Range

Focusing of the fields propagating in an AWG is obtained if the length difference ΔL between adjacent waveguides is equal to an integer number m of wavelengths inside the AWG:

$$\Delta L = m \cdot \frac{\lambda_c}{n_{\text{eff}}} \tag{4.1}$$

The integer m is called the order of the array, λ_c is the central wavelength (in vacuo) of the AWG, n_{eff} is the effective refractive (phase) index of the guided mode, and λ_c / n_{eff} corresponds to the wavelength inside the array waveguides. Under these circumstances the array behaves like a lens with image and object planes at a distance R_a of the array apertures. On the other hand, the focal line (which defines the image plain) follows a circle with radius $R_a / 2$, and transmitter and receiver waveguides should be located on this line. This geometry is equivalent to a Rowland-type mounting.

The length increment ΔL of the array gives rise to a phase difference according to

$$\Delta\phi = \beta\Delta L \tag{4.2}$$

where

$$\beta = 2\pi v n_{eff} / c \tag{4.3}$$

is the propagation constant in the waveguides, $v = c/\lambda$ is the frequency of the propagating wave, and c is the speed of light in vacuo. The wavelength-dependent phase difference $\Delta\phi$ introduces the wavelength-dependent tilt of the outgoing wave front associated with a wavelength-dependent shift of the corresponding image (as already mentioned above). The lateral displacement ds of the focal spot along the image plane per unit frequency change dv is called the spatial dispersion D_{sp} of the AWG, which is given by (for details see [8]):

$$D_{sp} = \frac{ds}{dv} = \frac{1}{v_c} \frac{n_g}{n_{FPR}} \cdot \frac{\Delta L}{\Delta\alpha} \tag{4.4}$$

where n_{FPR} is the (slab) mode index in the Free Propagation Region, $\Delta\alpha$ is the divergence angle between the array waveguides in the fan-in and the fan-out sections, and n_g is the group index of the waveguide mode:

$$n_g = n_{eff} + v \frac{dn_{eff}}{dv} \tag{4.5}$$

According to (4.1) and (4.4) the (spatial) dispersion is fully determined by the order m and the divergence angle $\Delta\alpha$, and as a consequence filling-in of the space between the array waveguides near the apertures due to finite lithographical resolution (cf. Sects. 4.3 and 4.4) does not affect the dispersive properties of the AWG.

If the input wavelength change is such that the phase difference $\Delta\phi$ between adjacent waveguides (4.2) has increased by 2π, the transfer will be the same as before, i.e. the response of the AWG is periodic. The period in the frequency domain is called the Free Spectral Range (FSR), and $\Delta\beta\Delta L = 2\pi$ in combination with (4.1) leads to:

$$\mathrm{FSR} = \frac{\nu_c}{m}\left(\frac{n_{eff}}{n_g}\right) \tag{4.6}$$

In order to avoid crosstalk problems with adjacent orders, the free spectral range should be larger than the whole frequency range spanned by all channels, and thus for a demultiplexer with 8 channels and 200 GHz channel spacing for example, the FSR should be at least 1600 GHz. If the channels are centred around 1550 nm, (4.6) yields that this requires an array with an order of about 120 (with $n_{eff}/n_g \approx 0.975$ for the SoS- and $n_{eff}/n_g \approx 0.9$ for the InP-material system).

If the device is to be used in combination with Erbium-Doped Fibre Amplifiers (EDFAs), the FSR should be chosen such that adjacent orders do not coincide with the peak of the EDFA gain spectrum in order to avoid accumulation of Amplified Spontaneous Emission (ASE).

4.2.3 Insertion Loss and Non-uniformity

Fields propagating through an AWG are attenuated due to various loss mechanisms. The most important contribution to this loss is coming from the junctions between the free propagation regions and the waveguide array. For low-loss the fan-in and fan-out sections should operate adiabatically, i.e. there should be a smooth transition from the guided propagation in the array to the free-space propagation in the FPRs and vice versa. This will occur if the divergence angle $\Delta\alpha$ between the array waveguides is sufficiently small and the vertex between the waveguides is sufficiently sharp. Due to the finite resolution of the lithographical process blunting of the vertex will occur. Junction losses for practical devices are between 1 and 2 dB per junction (i.e. between 2–4 dB for the total device). Propagation loss in the AWG and coupling losses due to a mismatch between the imaged field and the receiver waveguide mode (see below) are usually much smaller.

If the transmission for the central channel of the AWG is T_c, then the attenuation of the central channel is given by

$$A_0 = -10\log T_c \tag{4.7}$$

From Fig. 4.2a it is seen that the outer channels have more loss than the central ones. This reduction is caused by the fact that the far field of the

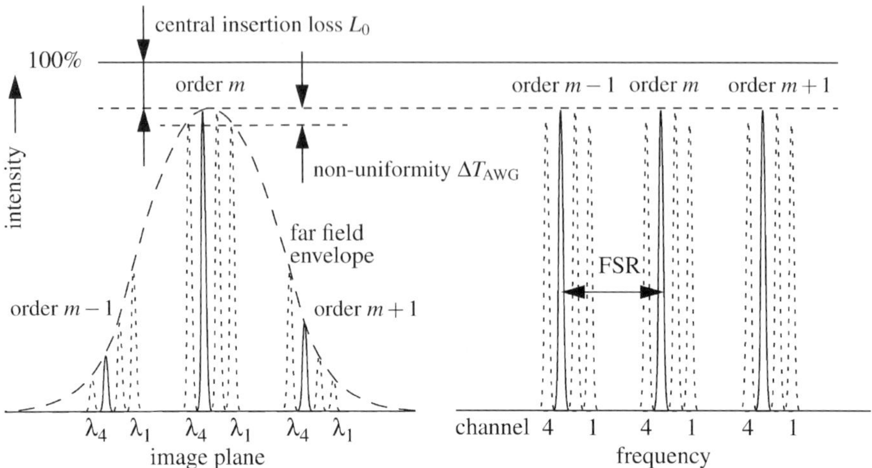

Fig. 4.2. (*Left*) The field in the image plane for different wavelengths, showing the influence of the far-field pattern of the individual array waveguide and the occurrence of different orders (*Right*) the corresponding frequency response curve for the different channels

individual array waveguides drops in directions different from the main optical axis. The envelope, as indicated in the figure, is mainly determined by the far field radiation pattern of the individual array waveguides. The non-uniformity ΔT_{AWG} is defined as the difference in transmission between the central channel(s) and the outermost channels (nrs. 1 and N):

$$\Delta T_{\mathrm{AWG}} = -10\log\frac{T_{1,N}}{T_{\mathrm{c}}} \tag{4.8}$$

in which T_{c} is the transmission of the central channel(s).

The power which is lost from the main lobe will appear in adjacent orders, as shown in the left part of Fig. 4.2. If the free spectral range is chosen N times the channel spacing, then the outer channels of the AWG will experience almost 3 dB more loss than the central channels, i. e. the non-uniformity is close to 3 dB. This is because at a deflection angle at half the angular distance between the orders of the array the power in the image is reduced by 50% because at that angle it will be equally divided over the two orders at both sides of the optical axis. In a periodical AWG the outer channels will be close to these 3–dB points.

From the above it is clear that the non-uniformity of an AWG can be reduced by increasing the FSR, however, at the expense of a larger device-size.

4.2.4 Bandwidth

As explained in the introduction, AWGs are lens-like imaging devices;
they form an image of the field in the object plane at the image plane. Because of the linear length increment of the array waveguides the lens exhibits dispersion: if the wavelength changes, the image moves along the
image plane without changing shape, in principle. Most of the AWG properties can be understood by considering the coupling behaviour of the focal
field in the image plane to the receiver waveguide(s). This coupling is described by the overlap integral of the normalised receiver waveguide mode
$U_r(s)$ and the normalised focal field $U_f(s)$ in the image plane, as illustrated
in Fig. 4.3:

$$\eta(\Delta s) = \left| \int U_f(s - \Delta s)\, U_r(s)\, ds \right|^2 \tag{4.9}$$

in which Δs is the displacement of the focal field relative to the receiver
waveguide centre. If $U_r(s)$ and the image field $U_f(s)$ have the same shape,
which will be the case if identical waveguides are used for the transmitter
and receiver waveguides, then the coupling efficiency η can be close to
100% on proper design of the AWG. The power transfer function $T_i(\nu)$
for the i-th receiver waveguide is found by substituting

$$\Delta s = D_{sp} \cdot (\nu - \nu_i) \tag{4.10}$$

in (4.9), which yields

$$T_i(\nu) = T_c \cdot \eta\{D_{sp} \cdot (\nu - \nu_i)\} \tag{4.11}$$

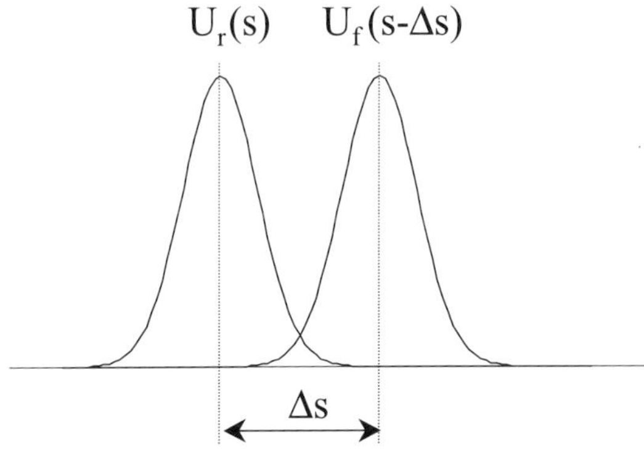

Fig. 4.3. The receiver waveguide mode profile $U_r(s)$ and the focal field $U_f(s)$

where v_i is the frequency corresponding to the i-th channel. The power transmission of the (virtual) central channel T_c is normally smaller than 1 due to the transmission losses in the AWG, and it is also worthwhile to note that for an even number of channels there will be no receiver waveguide at the centre of the image plane. Figure 4.2 shows an example of a demultiplexer response curve for different channels, in which the most important characteristics of the device are indicated.

To analyse the transmission loss at the input and output aperture of the array, which form the main contributions to the total insertion loss for the central channels of the device, the most usual approach is to overlap (4.9) the far field of the transmitter with the sum field of the individual array waveguides, with excitation coefficients proportional to the incident field strength. This analysis ignores the coupling between the waveguides, which will reduce the transition losses at the input and output aperture of the arrays. A more rigorous analysis of the array imaging properties is presented in [9].

4.2.5 Passband Shape

An important feature of the AWG-filter characteristics is the passband shape. Without special measures the channel response has a more or less Gaussian shape. This is because the mode profiles of the transmitter and receiver waveguides of the AWG can be described to a good approximation by a Gaussian function, so that their overlap (4.9) is also Gaussian. The 1–dB bandwidth is usually 25–30% of the channel spacing.

The Gaussian shape of the channel response imposes tight restrictions on the wavelength tolerance of the emitters (e. g. laser diodes) and requires accurate temperature control for both the AWGs and the laser diodes. Moreover, when signals are transmitted through several filters in a WDM network, the cumulative passband width for each channel narrows significantly. Therefore, flattened and broadened channel transmissions are an important requirement for AWG de/multiplexers. Different approaches to flatten the passbands of AWGs have been published. The simplest method is to use multimode waveguides at the receiver side of an AWG. If the focal spot moves at the AWG output along a broad waveguide, almost 100% of the light is coupled into the receiver to have a flat region of transmission [10]. However, this approach is unfavourable for single-mode systems.

Other methods convert the field at the transmitter or receiver into a double image as indicated in Fig. 4.4a. The wavelength response, which follows from the overlap of this field with the normal mode of a receiver/transmitter waveguide, will get a flat region as shown in Fig. 4.4b. The double image is

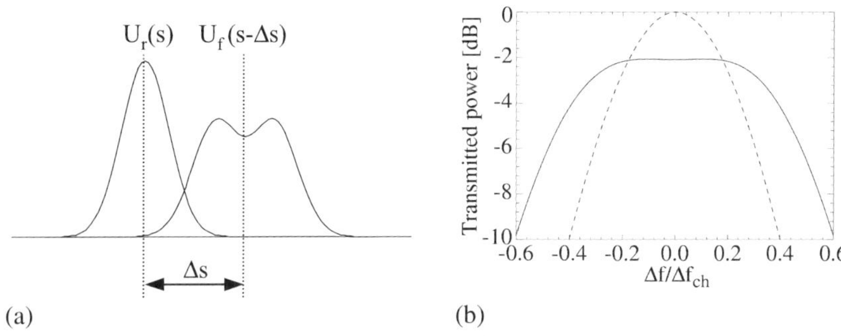

Fig. 4.4. Wavelength response flattening: **(a)** Shape of the focal field U_f required for obtaining a flat region in the overlap with the receiver mode U_r, **(b)** wavelength response obtained by applying a camel-shaped focal field (the dashed curve indicates the non-flattened response obtained by applying a non-modified focal field, i. e. $U_f = U_r$

created by use of a short MMI coupler [11], a Y-junction [12] or a non-adiabatic parabolic horn [13], where the best results are achieved with the parabolic horn. Other approaches based on spatial filtering use waveguide arm length changes inside the AWG that excite a sinc-like field distribution at the grating exit, which produces, via Fourier transform, a rectangular transmission shape [14, 15]. The use of interleaved gratings [16] or phase-dithering [17] for this purpose have also been published.

All these techniques fundamentally increase the insertion loss (on the order of 3 dB) because only a portion of the image is focused onto each waveguide output. This can be avoided by combining the AWG with a synchronized AWG or MZI-duplexer at the input, as is demonstrated, for example, in [18, 19]. A discussion of different techniques for passband flattening with a list of references is given in [15].

4.2.6 Crosstalk

One of the most important characteristics of the device is the inter-channel crosstalk. It is the contribution of unwanted signals, e. g. in the case of adjacent channels, the contribution of the (unwanted) signal at frequency v_{i+1} to the (detected) channel i. The theoretical adjacent-channel crosstalk A_x follows from the overlap of the focal field with the unwanted mode (4.9) as

$$A_x = \eta(d) \tag{4.12}$$

with d being the distance between adjacent receiver waveguides. From this formula it is seen that arbitrary large crosstalk attenuation is possible by positioning the receiver waveguides sufficiently far apart. Usually a gap of

1–2 times the waveguide width is sufficient for more than 40 dB adjacent inter-channel crosstalk attenuation. However, in practice other mechanisms appear to limit the crosstalk attenuation. The most important ones are errors in the phase transfer of the array waveguides. They are due to non-uniformities in layer thickness, waveguide width, and refractive index and cause a rather noisy "crosstalk floor", which is better than 35 dB for good devices. Semiconductor-based devices exhibit slightly inferior crosstalk figures compared to silica-based ones.

Crosstalk figures provided for experimental and commercial devices usually refer to single-channel crosstalk levels, i. e. the crosstalk resulting from a single channel. In an operational environment crosstalk contributions from all active channels will impair the crosstalk level compared to the single-channel crosstalk attenuation. This effect is discussed in [20] (Note: One should keep in mind that crosstalk is defined in such a way (cf. Glossary) that device behaviour is the better the higher the crosstalk which is somewhat counter-intuitive).

4.2.7 Wavelength Routing Properties

An interesting device is obtained if the AWG is designed with N input and N output waveguides and a free spectral range equalling N times the channel spacing. With such an arrangement the device behaves cyclical: a signal disappearing from output N will reappear at output 1, if the frequency is increased by an amount equal to the channel spacing. Such a device is called a cyclical wavelength router [3]. It provides an important additional functionality compared to multiplexers and demultiplexers and plays a key role in more complex devices as add-drop multiplexers and wavelength switches. Figure 4.5 illustrates its functionality. Each of the N input ports can carry N different frequencies. The N frequencies carried by input channel 1 (signals a_1–a_4 in Fig. 4.5) are distributed among output channels

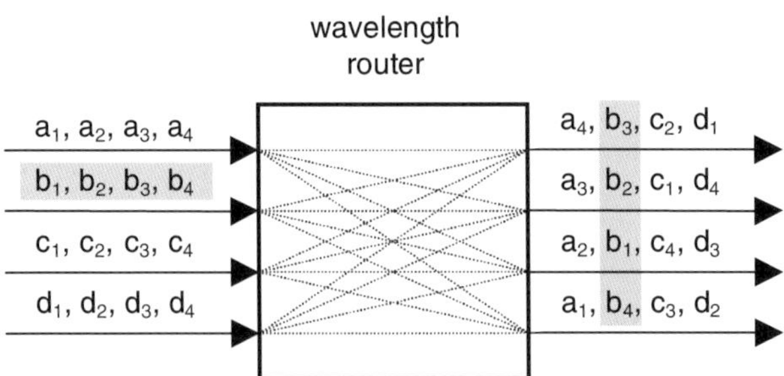

Fig. 4.5. Schematic diagram illustrating the operation of a wavelength router

1 to N in such a way that output channel 1 carries frequency N (signal a_4) and channel N frequency 1 (signal a_1). The N frequencies carried by input 2 (signals b_1–b_4) are distributed in the same way, but cyclically shifted by 1 channel in such a way that frequencies 1–3 are coupled to ports 3–1 and frequency 4 to port 4. In this way each output channel receives N different frequencies, one from each input channel. To realise such an interconnectivity scheme in a strictly non-blocking way using a single frequency, a large number of switches would be required. With a wavelength router, this functionality can be achieved using only a single component. Wavelength routers are key components also in multi-wavelength add-drop multiplexers and crossconnects (see below).

4.2.8 Configuration-dependent Crosstalk in Add-drop Multiplexers

One key application of AWGs is adding or dropping wavelengths from a multiplex. AWGs in different configurations can be used for this purpose, and the corresponding interchannel crosstalk is a parameter of particular importance. One standard configuration consists of two symmetrically connected $1 \times N$ AWGs with identical wavelength response (Fig. 4.6a). This architecture exhibits an almost perfect isolation between the add and the drop port, and spurious intensity of the dropped signal propagating towards the output is suppressed again by the multiplexer, so that the total crosstalk attenuation is essentially doubled. The latter is not the case for cheaper configuration illustrated in Fig. 4.6b which uses a power combiner instead of a multiplexing AWG.

An add-drop multiplexer as shown in Fig. 4.6a can be made configurable, i. e. the added and dropped wavelengths can be selected with an external control signal by combining the (de)multiplexers with switches as

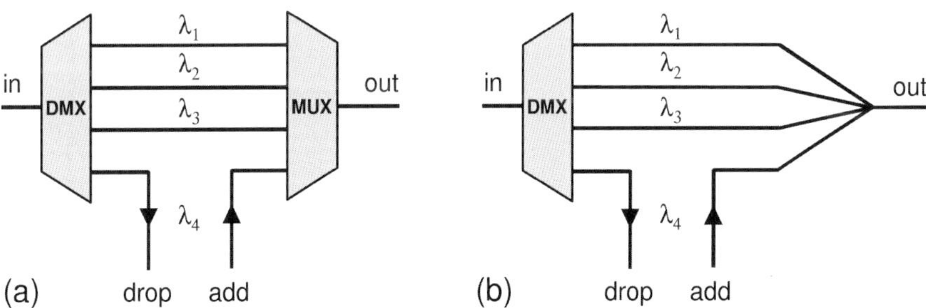

Fig. 4.6. Add-drop multiplexers with fixed add-drop channel. DMX: demultiplexer, MUX: multiplexer

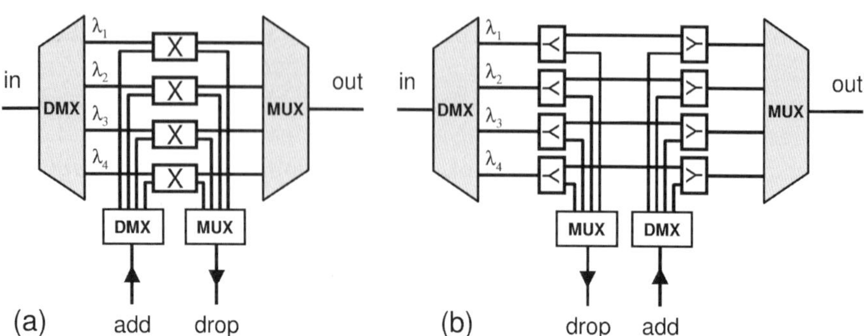

Fig. 4.7. Two reconfigurable add-drop multiplexers, based on one 2x2 switch (**a**) and two 1x2 switches (**b**) per channel. DMX: demultiplexer, MUX: multiplexer

shown in Fig. 4.7. In the configuration of Fig. 4.7a the add-signal may end up in the drop-port at an undesirably high level through-crosstalk in the switch. In the configuration of Fig. 4.7b, which requires two 1×2 switches per channel instead of one 2×2, the isolation is almost perfect.

A disadvantage of the configurations shown in Fig. 4.6 and Fig. 4.7 is that for low insertion loss operation they require two (de)multiplexers. Figure 4.8 shows an add-drop multiplexer realised with a single $(N+1) \times (N+1)$ AWG. In the first pass through the AWG the four wavelengths are demultiplexed. They are fed to four different switches with which they can

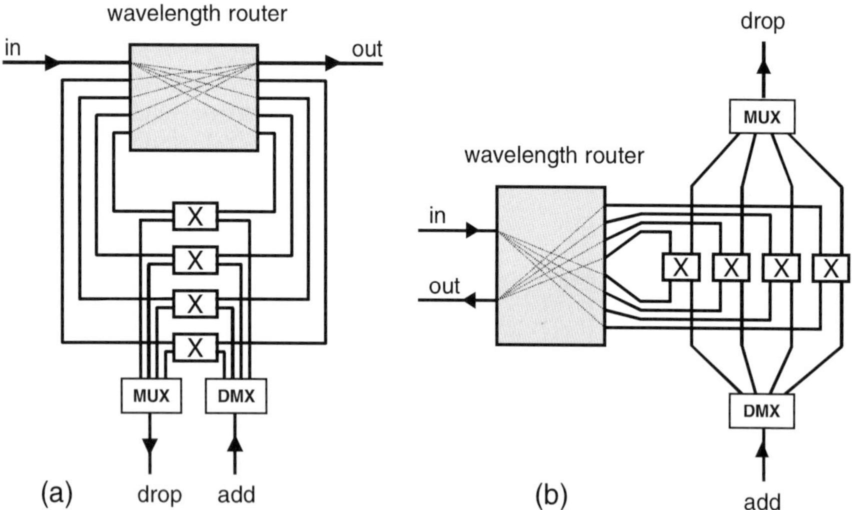

Fig. 4.8. Two add-drop multiplexer configurations based on a single wavelength router in a loop-back configuration (**a**) and in a fold-back configuration (**b**). MUX: multiplexer, DMX: demultiplexer

be switched to the drop port or be looped back to the wavelength router which multiplexes them to the output port. In the drop state a signal applied to the add port will be multiplexed to the output port. A disadvantage of this loop-back configuration is that crosstalk of the AWG is coupled directly into the main output port, where it competes with the transmitted signal, which is attenuated in the loop. This problem can be reduced by applying the AWG in a fold-back configuration as shown in Fig. 4.8b, which requires a larger AWG, however. One advantage of configurations using one AWG only is that the channel wavelengths automatically fit. A comparison of crosstalk properties of different configurations is given in [21].

4.2.9 Polarisation Dependence

If the propagation constants for the fundamental TE- and TM-modes (electrical field vector parallel and perpendicular to the wafer plane, respectively), of the array waveguides are equal, the AWG exhibits polarisation independence, i.e. identical spectral response for any polarisation. On the other hand, any waveguide birefringence, i.e. any difference in propagation constants for TE- and TM-polarised modes, results in a shift Δv_{pol} of the spectral response for TE- or TM-polarised modes, which is called polarisation dispersion. For semiconductor-based double-heterostructure waveguides, typical polarisation dispersion values are in the order of a few hundred GHz. In fibre-matched silica-on-silicon waveguide structures the birefringence should be zero, in principle. However, due to mechanical stress resulting from differences in thermal expansion coefficients of waveguide layers and substrate, polarisation dispersion of more than 20 GHz may still occur, which is too much for dense WDM systems applications.

Different concepts have been developed for reducing (residual) polarisation dependence. An elegant method applied in silica-based devices is the insertion of a $\lambda/2$-plate in the middle of the phased array [22]. Light entering the array in TE-polarised state will be converted by the $\lambda/2$-plate and travel through the second half of the array in TM-polarised state, and TM-polarised light will similarly traverse half the array in TE-state. As a consequence, TE- and TM-polarised input signals will experience the same phase change regardless of the birefringence properties of the waveguides. These and other approaches will be discussed in more detail in Sect. 4.3.

In semiconductor waveguides the birefringence is usually much larger due to the large refractive index contrast of these waveguides. The most straightforward approach for polarisation independence is by making the waveguide cross section square when the index contrast is the same in the vertical and lateral direction, as is the case, for example, in buried waveguide structures. In waveguides with a different lateral and vertical contrast, polarisation independence may be achieved by choosing a proper ratio between the height and the width of the waveguide. However, the restriction to waveguides supporting the propagation of a single mode only (monomode waveguides) in combination with typical material composition and index contrast leads to very small waveguide dimensions and tight tolerance requirements, which are extremely demanding to fabricate. Rather small deviations of layer thicknesses, composition, or waveguide width introduce a polarisation dependence again. Relaxed tolerances are obtained if low-contrast waveguides with a relatively large waveguide core are used, which is also advantageous for achieving low fibre coupling loss.

4.2.10 Temperature Dependence

Temperature independence is an important issue for passive devices that should operate in the field under varying temperature conditions. It is also important in integrated chips where chip temperatures may be dependent on bias currents of optical amplifiers. The main effect of temperature change in a temperature-dependent AWG is a shift of the wavelength response. In active devices temperature dependence is less important because power supply for active temperature control is available, although this raises the component cost.

InP-based AWGs have a temperature dependence in the order of 0.12 nm/°C, so they can be tuned over a wavelength range of a few nanometres with a temperature change of 30–40°C.

SoS-AWGs have a temperature dependence which is smaller by one order of magnitude, still too large for uncooled operation, but too small for active tuning.

A popular way to achieve temperature insensitivity is by insertion of triangular regions with material with a different temperature sensitivity in the free propagation region of the AWG in a similar way as applied for polarisation insensitivity [23]. Other methods have been reported too and are discussed in Sect. 4.3.3.

An important feature of temperature-insensitive devices is the accuracy with which their channel wavelength is controlled: if it deviates from the target value, one can no longer use temperature tuning to trim it to the right value. So temperature insensitive AWGs require a high degree of process control or the possibility to trim the devices, which is costly.

4.3 Silica-based AWG Devices

4.3.1 Silica-on-Silicon PLC Fabrication Techniques

As discussed in the introduction, AWGs are one of the most important filter types applied in WDM networks, and silica-on-silicon devices are the variety predominating in current optical networks.

The fabrication of SoS-AWGs relies essentially on the combination of techniques developed independently in two different fields: the well established deposition techniques for silica-based glasses and optical fibres plus etching and structuring developed for VLSI micro-fabrication [24].

AWGs are a special type of Planar Lightwave Circuit (PLC), and these are fabricated on planar substrates, typically a crystalline flat silicon wafer. Alternatively, optically flat fused silica substrates are also used, which results in reduced polarisation-dependence of the waveguide characteristics and enables the fusing of optical fibres to the waveguides [25].

Favourable properties of silicon wafers are their high degree of planarity, excellent heat dissipation property, and the potential for hybridisation of optical and electronic components onto a common substrate. The use of large size wafers, typically 6" or 8" in diameter, is cost effective as it enables multiple devices to be fabricated simultaneously on one wafer. Silicon wafers are normally pre-oxidized resulting in an oxide layer of typically several µm thickness. This layer facilitates the adhesion of deposited silica and forms at the same time part of a buffer layer.

Usually, the fabrication of silica PLCs starts with the deposition of a stack of glass layers of high silica content onto the wafer, where the glass composition is similar to that of an optical fibre. Their simple waveguide core structure, low propagation loss, and an almost perfect field match to optical fibres are the key factors which have given silica PLCs their predominance.

For the deposition of silica waveguide layers various processes have been explored, and two of them have gained major importance: flame hydrolysis, a method originally developed for fibre pre-form fabrication, and chemical vapour deposition (CVD), a common process in micro-fabrication.

The PLC base layer is usually made of undoped SiO_2, while the other glass layers are made of doped silica and are flowed during annealing, a process that helps to form homogeneous low loss material. Typical dopants for the silica core layer are Ge, P, or Ti, which slightly raise the refractive index of the core with respect to buffer and cladding. In order to enable filling-in between closely spaced core sections the upper cladding must exhibit easy flow characteristics, while the core and buffer layer have to remain rigid. Furthermore, the refractive index of the cladding layer should closely match that of the buffer layer. For this purpose dopants such as B and P are added simultaneously to the cladding layer. While the addition of B lowers both, the flow temperature and the refractive index, P raises the refractive index and thus enables index compensation. As an alternative F has also been used to reduce the refractive index [26].

The standard fabrication of silica PLCs uses optical lithography, and this relies on a chromium mask on a quartz blank which carries the waveguide structure. The mask is designed by CAD and subsequently written by electron-beam lithography with sub-µm accuracy. The pixel size during mask writing has a strong influence on the AWG's performance [27]. Waveguide cross sections are quadratic in most cases, however, their width is sometimes enlarged in waveguide bends.

Waveguide forming starts by deposition of the etching mask layer on top of the core layer, and the mask structure is then transferred into the silica core layer by reactive ion etching (RIE) or similar techniques. It is of key importance that the etching process provides waveguides with sufficiently smooth side-walls since the side-wall roughness induces scattering losses which should be kept as low as possible. In order to get embedded waveguides it is necessary to deposit a cladding layer with a refractive index similar to that of the buffer layer in a final processing step after mask removal.

Flame Hydrolysis

Flame hydrolysis (FHD) for silica PLCs has mainly been developed by Nippon Telephone and Telegraph Corporation (NTT) in Japan [24]. Fine glass particles produced in an O_2/H_2 torch fed by a mixture of vapours such as $SiCl_4$ and $GeCl_4$ or $TiCl_4$ (in the case of the core) are deposited onto the substrate, while the torch is traversed repeatedly over the moving substrate held at lower temperature. The refractive index of the core layer is determined by the relative content of Ge or Ti, while the base layer is usually made of undoped silica. The porous glass layers produced by FHD are consolidated by annealing due to heating up to around 1200°C to 1300°C. Small amounts of BCl_3 and PCl_3 are added to the gas mixture to build the upper cladding layer which is consolidated at a somewhat lower temperature. In the case of PLCs on silicon substrates, a 10 to 15 µm thick SiO_2 buffer layer is normally used in order to avoid leakage from the silica waveguide core to the silicon substrate which has a much higher refractive index.

The thickness of the doped core layer is normally determined by the requirement of single mode propagation for a quadratic core cross section. Differences Δn of the effective refractive index of the core and that of the buffer can be adjusted from about 0.25% to 1.5–2.0%. In the case of $\Delta n = 0.75\%$ silica, which is commonly used in 1.55 µm wavelength PLCs, the core layer is 6 µm thick.

Chemical Vapour Deposition

Chemical vapour deposition (CVD) and related processes such as low pressure chemical vapour deposition (LPCVD) [28] and plasma-enhanced chemical vapour deposition (PECVD) are also often used for silica PLC fabrication [29]. These methods enable the deposition of thin layers of undoped or doped silica by using admixtures of either vapours or gases [30]. Phosphorous is a frequently used dopant for the core layer. In a typical CVD process the substrate is exposed to one or more volatile precursors which react and/or decompose on the substrate surface to produce the desired deposition. PECVD and its variants, on the other hand, rely on ionic catalysts.

LPCVD is a method of film deposition primarily used in silicon technology which has also been adapted to silica device fabrication [31]. LPCVD works at sub-atmospheric pressure which tends to reduce unwanted gas phase reactions and improves film uniformity across the wafer. Several manufacturers use this deposition method with great success, and sometimes a combination of both, FHD and CVD, is also used. Although deposition rates of ~1 µm/h are rather low, the total wafer throughput is normally high enough because 50 or even more wafers can be processed simultaneously in a CVD reactor. Deposition temperatures range from 400 to 800°C, depending on the reactants. Annealing is carried out at temperatures of ~900–1100°C. In a CVD process layer growth takes place on both sides of the wafer which minimizes wafer bowing.

One advantage of PECVD, compared to both FHD and CVD, is that it is a low-temperature process, compatible with temperatures occurring in microelectronics processing. However, this potential advantage is lost if high-temperature annealing is needed for smoothing out irregularities, which might have been introduced during the waveguide etching process. In addition, the PECVD process offers the possibility to monitor the refractive index of the deposited layers using in-situ ellipsometry.

Other advantages of CVD processes in silica waveguide fabrication are both lower stress (reduced birefringence) in waveguide layers and higher layer uniformity over large wafer areas, and these features can be of high relevance for large channel count, high resolution AWGs designed for DWDM applications.

4.3.2 Birefringence

Glass layers of silica-on-silicon wafers tend to exhibit compressive stress after they have been cooled down to room temperature as a consequence of different thermal expansion coefficients of glass and silicon, and this effect is particularly pronounced if the layer deposition has been made by FHD. The stress induces birefringence (cf. Sect. 4.2.9), and as a consequence standard silica-based AWGs exhibit polarisation-dependent optical properties. The corresponding wavelength separation $\Delta\lambda$ between the peak transmission for TE- and TM-polarisation is given by

$$\Delta\lambda = 1/m \cdot \Delta L \cdot (n_{TE} - n_{TM}), \qquad (4.13)$$

where m is the waveguide grating order, ΔL is the path difference between neighbouring arrayed waveguides, and n_{TE} and n_{TM} are the effective refractive indices for TE- and TM-polarised guided waves, respectively. A typical value for silica-on-silicon at 1.55 µm wavelength is

$$(n_{TE} - n_{TM}) = 2.4 \cdot 10^{-4}.$$

For the case of an AWG with 100 GHz channel spacing such a refractive index difference induces a TE-TM shift of $\Delta\lambda = 0.26$ nm (33 GHz), which is not acceptable in many cases, in particular for DWDM applications [32].

Several methods have been reported which compensate this polarisation dependence, and these include TE/TM mode conversion [33], dopant-rich cladding [32, 34, 35] and undercladding [36], groove-assisted waveguides [37], width-controlled waveguides [38], and core-over-etching methods [39].

In the case of TE/TM mode conversion (cf. Sect. 4.2.9), a thin half-wave plate (being a 14.5 µm thick stretched polyimide foil) is inserted into a slit in the centre of the arrayed waveguide grating and fixed with UV-curable adhesive [33]. In the case of a conventional AWG a typical excess loss of 0.4 dB is introduced by this method. Using dopant-rich cladding a TE-TM shift $\Delta\lambda < 0.02$ nm (2.5 GHz) was obtained without significant excess loss. This corresponds to a reduced birefringence of $(n_{TE} - n_{TM}) < 1.8 \cdot 10^{-5}$ [32].

4.3.3 Silica Waveguide Features and their Influence on AWGs

Three waveguide parameters are particularly essential for PLC-based AWGs and are relevant for the AWG design. These parameters are:

1. the waveguide loss (over the wavelength range of interest),
2. the waveguide coupling characteristics to other components of a network, e. g. fibres,
3. the lowest possible bend radius which still assures acceptable radiation loss.

ad 1: As known from optical fibres, silica itself has extremely low optical loss in the wavelength range throughout the visible and near infrared region which covers all wavelengths utilized in optical communication.

However, the preferred spectral transmission region is between about 1.3 to 1.6 µm. This is essentially due to the fact that Rayleigh scattering losses strongly increase towards shorter wavelengths, and the same is true for planar silica waveguides as a consequence of inevitable residual sidewall roughness induced by technological processes such as lithography and etching. The associated losses are approximately 10^3 times higher than those of a fibre. Furthermore, the effect is more pronounced with waveguides of higher index contrast, provided the roughness is comparable.

ad 2: For low-contrast silica devices (see below), the waveguide mode matches with that of a standard single mode fibre (SMF), resulting in very low coupling losses. For higher contrast waveguides, a high coupling efficiency requires mode matching, and it is necessary to integrate a taper to the input/output waveguides of the AWG. Coupling losses well below 0.5 dB are feasible, even for high contrast waveguides, if a proper mode-converter is applied.

ad 3: The choice of a waveguide design with a given Δn value affects the minimum possible bending radius and therefore the compactness and size of a device. Since the typical bending radius R of a silica waveguide is around 2–25 mm, the chip size of large-scale circuits can amount to several cm^2. Therefore, reduction of propagation loss and uniformity of refractive indices throughout the wafer are strongly required. Increasing Δn of a waveguide allows for smaller waveguide bending radii and enables the realization of high-density integrated circuits. Low-Δn waveguides are superior to the high-Δn waveguides in terms of fibre coupling losses with standard single mode fibres SMF-28. On the other hand, the much smaller minimum bending radii of high-Δn waveguides enable the fabrication of highly integrated and large scale optical circuits such as AWGs with high channel numbers and with high resolution.

Four different (standard) Δn values are typically used for silica waveguides, and each Δn value corresponds to a specific core width which assures single-mode propagation. Lowest loss down to 0.017 dB/cm has been achieved in low-Δn waveguides with $\Delta n = 0.30\%$ (core size $8 \times 8\,\mu\text{m}^2$, $R = 25$ mm) and also with $\Delta n = 0.45\%$ (core size $7 \times 7\,\mu\text{m}^2$, $R = 15$ mm) [40]. These waveguides match best to standard SMFs, but their large bending radii lead to rather large PLCs. Waveguides with "standard" $\Delta n = 0.75\%$-silica (core size $6 \times 6\,\mu\text{m}^2$, $R = 5$ mm) are more practical for compact PLCs and have still tolerable mode match to SMFs. Corresponding minimum waveguide propagation losses of 0.035 dB/cm have been reported [41]. A propagation loss of about 0.1 dB/cm was obtained for super-high $\Delta n = 1.5 - 2\%$-silica (core size 4.5×4.5 to $3 \times 3\,\mu\text{m}^2$, $R = 2$ mm) [42], which enables high density PLCs, but requires special tapers for coupling to SMFs.

Slightly higher losses are observed for the TM modes in comparison to TE modes, and this is attributed to the roughness of the waveguide sidewalls caused by the RIE process (cf. above). However, this effect is usually negligible in loss considerations of AWGs.

A conventional 16-channel 100-GHz-spacing AWG fabricated with 0.75%-Δn waveguides may have a size of about 26×21 mm^2. This circuit size increases rapidly when the channel count rises beyond 100. For example, a 200 channel AWG can be more than $100\,\text{cm}^2$ in size, which makes its fabrication difficult. Moving to 1.5%-Δn waveguides with corresponding lower bending radius allows to fabricate such a 200 channel AWG on a 4-inch wafer [43].

Another high-contrast low-loss "silica-like" material is Hydex™ with pre-adjustable index contrast between 1.5% and 25%, developed by Little Optics. Bending radii in the sub-mm range and correspondingly dense PLCs are possible. In addition to different passive optical PLCs such as µ-ring filters (cf. Chap. 8), star couplers, delay lines and many others, a 40-channel AWG has also been demonstrated with more than 200 pieces integrated on a single 6-inch wafer [44]. However, efficient coupling to the optical fibre is not at all an easy task in that case because of strongly different optical modes of the AWG input waveguides and the optical fibre.

4.3.4 AWG Development Trends

AWGs are available as wavelength multiplexers and demultiplexers to be used in WDM networks for more than a decade. Since then they have steadily been improved, and this process is still going on. In the following we will discuss current development trends, but we will start with a description of typical characteristics of state-of-the-art AWGs.

Various kinds of de/multiplexers relying on "standard" $\Delta n = 0.75\%$ silica and exhibiting good performance have been demonstrated in the laboratory rather early [45–47]. They range from 50 nm spacing, 8-channel AWGs for CWDM to 25 GHz spacing, 128-channel AWGs for DWDM applications or even AWGs with 10 GHz spacing.

Table 4.1 shows characteristic parameters of corresponding typical AWG multiplexers with Gaussian passband fabricated in the standard silica-on-silicon technique [48]. Table 4.1 also illustrates a number of trends and dependences as the number of channels increases and the channel spacing gets lower simultaneously. For example, clearly to be seen is the proper choice of grating order m yielding a FSR value, according to (4.6), capable to capture the full channel number N. Moreover, the spectral resolution of the AWGs is essentially determined by the product $m{\cdot}n$, and the number of grating arms n is typically chosen to be about three to four times that of the number of output channels (over-sampling), which is also illustrated in the table.

Table 4.1. Experimental performance examples of fabricated AWG multiplexers

Parameter	Layout specifications and experimental results				
Number of channels N	8	16	32	64	128
Centre wavelength	1.30 µm	1.55 µm	1.55 µm	1.55 µm	1.55 µm
Channel spacing $\Delta\lambda$	50 nm	2 nm	0.8 nm	0.4 nm	0.2 nm
(Δf)	(6.25 THz)	(250 GHz)	(100 GHz)	(50 GHz)	(25 GHz)
Grating order m	3	47	59	59	59
Number of arrayed waveguides n	28	60	100	160	388
On-chip loss for λ_c	2.2 dB	2.3 dB	2.1 dB	3.1 dB	3.5 dB
Channel crosstalk	> 30 dB	> 35 dB	> 35 dB	> 30 dB	> 20 dB

In general the insertion loss increases from central to peripheral output ports by a certain amount, referred to as non-uniformity, (see (4.8)). Typical values of this variation are 2 to 3 dB, and there is an additional dependence on the FSR and on the corresponding grating order m of the AWG under consideration. In addition, the insertion loss tends to increase slightly with increasing channel number and device size. One option to reduce the channel-dependent loss is choosing the number N of AWG channels (or accordingly the FSR value) larger than the number of channels required.

The channel centre wavelengths of AWGs to be used in telecommunication systems have to be aligned exactly to the ITU grid (cf. Appendix, Sect. A.1), and this can be assured by a Vernier design of the input waveguide geometry and temperature tuning of the AWG module [49].

In the early days the total insertion loss of AWGs was typically in the range of 3 to 4 dB and channel crosstalk in the range of 20 to 35 dB. Both values have been considerably improved since then by the use of special layout, trimming methods, and advanced filtering methods.

Improving the characteristics of existing AWGs has focused on lower insertion loss, improved crosstalk, higher channel resolution, and higher total channel count, and corresponding work is still going on. The priority among these parameters is essentially application- and cost dependent.

A silica-based AWG with particularly low insertion loss of only 0.75 dB has been achieved using a structure with vertically tapered waveguides in the intersection region between the arrayed waveguides and the slab which reduces the transition loss at that interface. In addition, a spot-size converter was incorporated into the AWG in order to reduce the fibre-to-waveguide coupling loss [50].

Another highly effective method to reduce the insertion loss of an AWG, which is based on the same idea of tapering, has been patented by Lucent: A segmented transition region is inserted between the slab and the arrayed waveguides, which comprises a number of paths intersecting the waveguide array and exhibiting progressively decreasing width as they depart further away from the slab [51].

Besides lowering the overall insertion loss of AWGs the improvement of crosstalk is an issue of high relevance. One source of crosstalk is phase error. It can be compensated by trimming all waveguides in the AWG individually, but it is laborious and time consuming [52, 53]. An alternative method is to use additional bandpass filters at the output ports of a conventional AWG. This approach has been demonstrated for a 64 channel, 50 GHz spaced silica AWG integrated with a secondary element consisting of 64 1×1-channel-type AWGs (bandpass filters), each of them with a different centre wavelength which corresponds to the centre wavelength of the respective output of the main AWG. Silica with $\Delta n = 0.75\%$ was used for the main AWG, and $\Delta n = 1.5\%$ silica for the AWG bandpass filters, respectively. This device exhibited an excellent crosstalk of 82 dB and a moderate insertion loss of 3.3 dB [54].

Different AWGs with particularly high channel number and very close channel spacing (relying on 1.5%-Δn waveguides in silica) have been published by NTT. Examples are 25 GHz spacing, 256-channel AWGs with a circuit size of about $75 \times 55 \, \text{mm}^2$ and fabricated on a 4-inch wafer [55, 56] or a 400-channel AWG with the same channel spacing that covers the full C- and L-bands from 1530 to 1610 nm [57]. The corresponding chip size was about $124 \times 64 \, \text{mm}^2$, fabricated on a 6-inch wafer. On-chip losses ranged from 3.8 to 6.4 dB from the central to the outer output ports, and adjacent and far-end crosstalk were about 20 dB and better than 30 dB, respectively.

Furthermore, demultiplexers with 10 GHz channel separation and 320 channels or even more than 1000 channels have been realized in a two-stage architecture using auxiliary 10 GHz-spaced AWGs connected to each output of a coarse filter AWG [58, 59]. In the latter device one 10-channel AWG and ten 160-channel AWGs have been connected in a tandem configuration.

Recently, even a 5 GHz spaced, two-stage tandem demultiplexer with 4200 channels that covers the complete S-, C- and L-bands from 1460 nm to 1620 nm has been reported [60]. In this case the configuration was a 20-channel AWG with 1 THz channel spacing and Gaussian passband as a primary filter and twenty 5 GHz-spaced AWGs as secondary filters. Primary and secondary AWGs have been fabricated in $\Delta n = 0.75\%$ and $\Delta n = 1.5\%$ silica, respectively. The secondary AWGs have been fabricated on 4-inch wafers, and UV laser trimming has been used for phase error correction. The loss values of the whole device ranged from 7 to 12 dB and the adjacent crosstalk ranged from 40 to 14 dB. The crosstalk worsened periodically with channel number since side lobes in the transmission spectrum of each secondary AWG increased as the channel deviated from the centre channel.

Even a 16 channel AWG with 1 GHz channel separation has been realized already which uses a grating order as high as $m = 11,818$. The corresponding device layout had to be folded in order to fit onto a 4-inch wafer, and in addition, phase errors had to be compensated in each grating arm in order to make the device work [61].

On the other hand, AWG based de/multiplexers for CWDM applications have been demonstrated which are typically dimensioned for 10 or 20 nm channel spacing. Also very broad-band low channel count AWGs of low grating order have been reported using a special AWG design [62]. Here e. g. devices with one input and two output channels have been fabricated operating at 1.0–1.55 µm, 1.31–1.53 µm, and 1.47–1.55 µm, respectively.

Athermal AWGs

In standard silica-based AWGs the centre wavelength of any output channel shifts by about

$$\mathrm{d}\lambda/\mathrm{d}T = 1.2 \cdot 10^{-2}\,\mathrm{nm/K} \text{ or } \mathrm{d}\nu/\mathrm{d}T = -1.5\,\mathrm{GHz/K}$$

at 1.55 µm. This is mainly determined by the temperature-induced variation of the refractive index of the silica glass, which is

$$\mathrm{d}n/\mathrm{d}T = 1.1 \cdot 10^{-5}\,\mathrm{K}^{-1}.$$

For many applications, e. g. in WDM networks, the AWG-channel wavelengths are stabilised by temperature control, either by use of a Peltier element or more simply by a heater if the operating temperature of the component is set to a higher temperature. The control requires additional equipment and the availability of electrical power at the operation place. The desire to get rid of this extra effort has spurred the development of athermal (i. e. temperature insensitive) AWGs which exhibit stable filter response over a certain temperature range, e. g. 0–85°C.

One realisation concept of athermal AWGs relies on replacing a section of the standard waveguides by waveguides made of a material with a negative dn/dT value in order to compensate the temperature dependent optical path length variation [63]. A material commonly used for this purpose is silicone which has a rather large value

$$dn/dT = -37 \cdot 10^{-5}\,\mathrm{K}^{-1}.$$

In this case only about 3% of the waveguide path in an essentially silica-based AWG has to be replaced by silicone. However, due to additional light scattering at the interfaces between silica and silicone and a slightly higher light loss in silicone the total AWG insertion loss is slightly enhanced by this procedure, typically by 2 dB. However, this excess loss can be reduced to about 0.4 dB by segmenting a single trapezoidal silicone region into multiple groove regions [64]. Polymer is another material utilized for the compensation of temperature-induced optical path variations of silicon [65].

A different approach to achieve an athermal AWG is based on tracking the AWG focal points, which exhibit a temperature-dependent move at the slab output (or vice versa at the input) related to the optical path length variation in the arrayed grating. Tracking is accomplished by fixing the input or output waveguides to a metal arm (copper or aluminium) which exhibits the appropriate thermal contraction or expansion [66, 67].

4.3.5 Interleave Filters

The combination of an AWG with an interleave filter (cf. Chap. 9) is an efficient method to double the channel count of AWG-based de/multiplexers without changing the corresponding AWG-design. The interleave filter separates WDM channels, which have equal spacing, into two groups with twice the spacing, and therefore, one interleaver filter plus two AWGs with N channels is equivalent to one AWG with $2N$ output channels and half the channel separation of each individual AWG. The interleaver is basically a Mach–Zehnder-interferometer, but more sophisticated designs offer spectral characteristics which are more favourable than a simple $\sin^2(\lambda)$ power-dependence. For example, flat-top characteristics have been demonstrated by a Fourier transform filter composed of directional couplers and delay lines [68]. The advantage of this filter is that it exhibits no excess loss other than fibre coupling and waveguide propagation loss.

Thus in addition to doubling the total channel number, the integration of interleave filters and AWGs enables the construction of compact WDM filters with wide pass-band and low insertion loss as well. Ref. [69] illustrates the combination of a 50 GHz interleave filter with two identical 100 GHz spacing 51-channel AWGs with their centre wavelengths shifted by 50 GHz with respect to each other which provides a 102-channel 50 GHz-spacing demultiplexer.

4.3.6 Commercially available AWGs

AWGs have become one of the standard filter types used in WDM networks, and they are available from a number of different suppliers. These include (at the time of writing this text) ANDevices, Gemfire, JDS Uniphase, and Wavesplitter from the US, Hitachi-Cable, NEC, and NEL from Japan, PPI from South Korea, and NKT from Denmark. There have been additional suppliers in the past, in particular around the year 2000, which have either closed down their business or have undergone mergers with other companies, and similar developments are likely to happen in the future as well.

Typical characteristics of commercially available AWGs are compiled in Table 4.2, where the data in brackets refer to less common devices. Offered low channel-spacing filters, e. g. those with 25 GHz, are primarily accomplished using an additional interleaver filter. The channel characteristics of AWGs have been primarily Gaussian a number of years ago, however, more recently flat-top characteristics have become predominant.

Two examples of filter characteristics of commercially available AWGs in SoS-technology are shown in Figs. 4.9 and 4.10, one with Gaussian and the other with flat-top shape.

Table 4.2. Typical characteristics of commercially available AWGs in SoS-technology

Quantity	Typical characteristics	Options
Maximum channel number	up to 40	80
channel spacing	100, 50 GHz	25, 200 GHz
Mode	Gaussian and flat-top	
Wavelength range	C- and L-band	S-band
Temperature dependence	active temperature-control	athermal
Typical insertion loss		
Gaussian:	< 3.5 ... < 5 dB	< 2.5 dB
flat-top:	< 4.5 ... < 8 dB	
Typical crosstalk		
adjacent channel crosstalk	> 25 ... > 30 dB	> 33 dB
non-adjacent channel X-talk	> 30 ... > 35 dB	
Polarisation-dependent loss	< 0.35 ... < 0.50 dB	< 0.2 dB
Added interleaving filter	2 and 4 channels, 25–100 GHz	

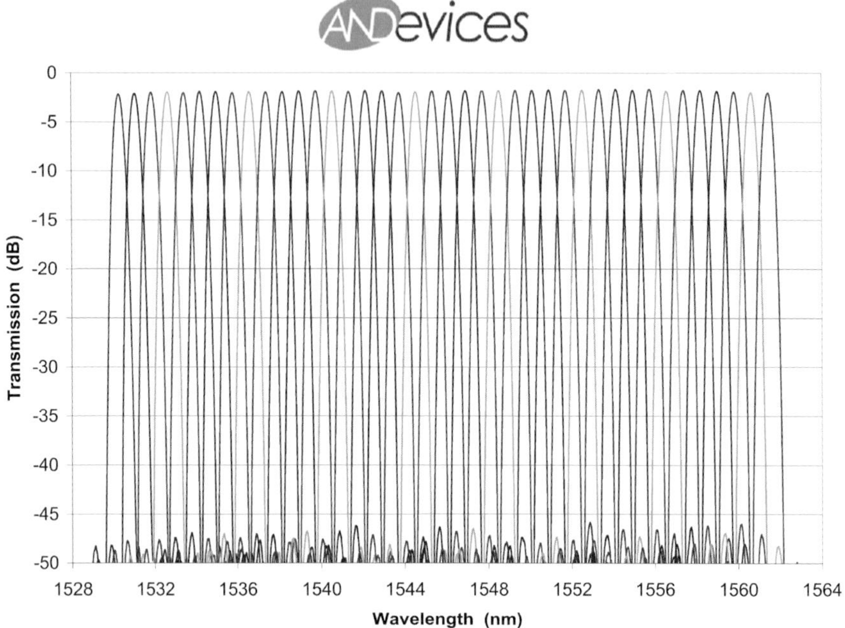

Fig. 4.9. Wavelength characteristics of a 40 channel AWG with Gaussian passband

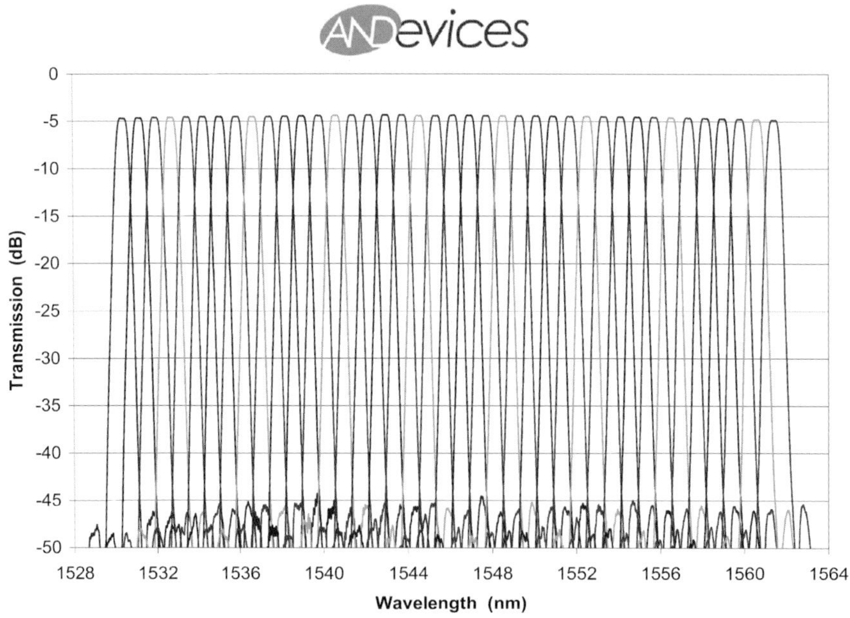

Fig. 4.10. 40 channel flat-top passband AWG

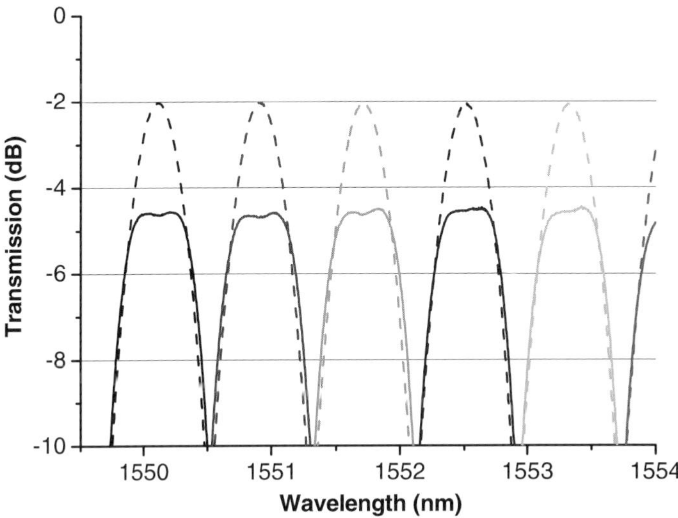

Fig. 4.11. Comparison of Gaussian and flat-top AWGs (courtesy of ANDevices)

The differences between Gaussian and flat-top characteristics are illustrated in Fig. 4.11. As can be seen, the more favourable channel shape (i. e. higher tolerance to wavelength variations) is accomplished at the expense of a somewhat higher overall attenuation.

4.4 Applications

AWGs have found a large number of applications ranging from simple add-drop multiplexers to complex-functionality crossconnects in telecommunications, and corresponding typical examples will be illustrated in the following, while other applications such as in signal processing, measurement and characterisation, and sensing for example, are beyond the scope of this chapter.

4.4.1 Add-drop Multiplexer

Generic add-drop multiplexer (ADM) configurations have been illustrated in Sect. 4.2.8. A device consisting of three AWGs connected by 16 thermo-optic switches has been the first fully integrated silica PLC-based ADM that enabled full access to 16 individual wavelength channels [70]. In a second version [71] double gate switches were used and four AWGs were installed. This double-gate configuration enabled a remarkable improvement

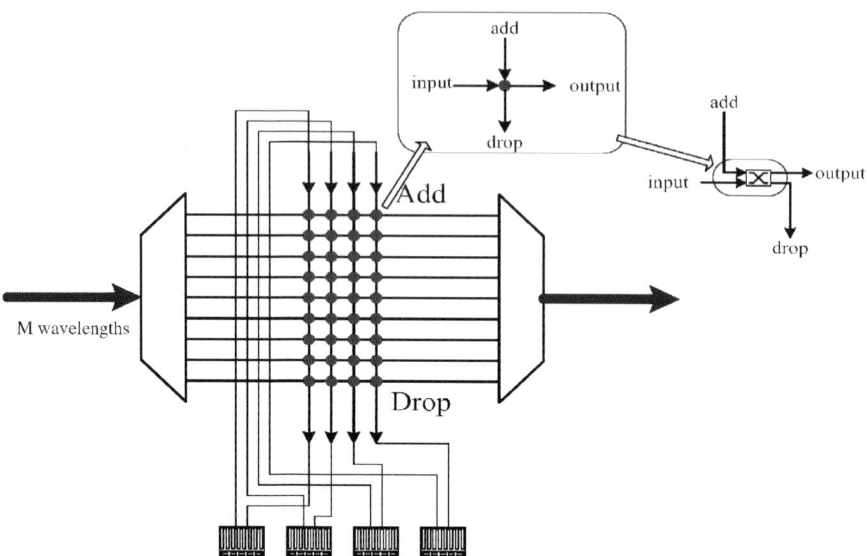

Fig. 4.12. Structure of a client-reconfigurable ADM, after [73]

in crosstalk: The on-off crosstalk from the main input to the main output or any drop port was larger than 28 dB, and the on-chip insertion loss ranged between 7.8 and 10.3 dB. Similar configurations using athermal AWGs and TO switches in silica can be operated without temperature control of the AWGs [72].

Other integrated optical add-drop multiplexers on silica PLCs reported in the literature include a 32-channel ADM which can be re-configured to add-drop any of the 32 input channels from/to any of the 4 add-drop ports as illustrated in Fig. 4.12 [73, 74] and a 40-channel programmable add-drop filter with flattened passbands [75]. For other examples see also Sect. 4.5.4.

4.4.2 Equalizer

Equalizers for power and dispersion are often needed in WDM networks, especially in ultralong-haul (> 100 km) systems.

The equalization of channel power in order to compensate for residual gain ripple in optical amplifiers, sudden channel power changes, and incorrectly added channel powers in optical add-drop multiplexers normally requires dynamic gain equalizers. On the other hand, fixed dispersion equalizers are usually adequate for the compensation of large delay time differences. A large variety of compensation schemes have been developed for these purposes including mechanical [76], acoustooptic [77], and planar lightwave circuits containing AWGs as key elements [78, 79, 80]. These conventional equalizers consist of a demultiplexer-multiplexer pair and, in between, attenuators or amplifiers and delay lines, respectively (see also Sect. 4.5.4, Fig. 4.25).

The introduction of erbium-doped fibre amplifiers has greatly increased the transmission distance in optical fibre communications, and chromatic dispersion has now become the main factor limiting the maximum repeater span. Several techniques have been reported for the compensation of the corresponding signal distortion in optical links [81, 82] including lattice-form dispersion equalizers made in silica PLCs [83]. However, it is rather difficult to fabricate a dispersion equalizer by the lattice-form configuration having both wide operational bandwidth and large delay compensation capability.

A PLC dispersion equalizer, which satisfies both requirements, has been realized utilizing two AWGs (25 GHz channel spacing, 3200 GHz (25.6 nm) free spectral range, $\lambda_c = 1.55\,\mu$m centre wavelength) as a demultiplexer-multiplexer pair and multiple delay arms in between [84]. In this special case two diffraction orders $m = 59$ and 60 had to be used by constraints of the delay arm layout. The delay arm lengths were designed to compensate for the delay time $\tau(\lambda)$ introduced by a specific length of a given dispersion-shifted fibre (DSF). The waveguide length Δl of the i-th delay arm is given by

$$\Delta l(\lambda_i) = l_{\max} - [\tau(\lambda_i) - \tau(\lambda_c)] \cdot \frac{cL}{n_c} \qquad (4.14)$$

where $l_{\max}$ is the maximum arm length for the dispersion-zero centre wavelength λ_c, $\tau(\lambda_i)$ is the delay for each AWG channel wavelength λ_i along the fibre span with length $L\,(= 400\,$km). It has been shown that over 10 nm wavelength range this equalizer can almost completely compensate the delay accumulated over 400 km of DSF.

4.4.3 WDM-PON Overlay Device

Passive optical networks (PON), which use passive, wavelength-independent optical splitters for power splitting/branching and time division multiplexing for upstream and downstream signalling, constitute a low cost solution for many applications.

The upgrade of existing broadcast PONs with broadband interactive services by employing high density wavelength division multiplexing (WDM) represents an attractive way of exploiting the huge optical fibre capacity more efficiently and of adding additional features to the access network such as privacy for example, since individual subscribers can be addressed by separate WDM channels. For a WDM/PON with well defined wavelength channels, the broadcast signal and the WDM channels can be distributed over the same optical network, each using separate wavelength bands [85, 86], see also Appendix, Sect. A.2.

A number of integrated WDM-PON devices fabricated in silica waveguide technology have been reported which utilize AWGs as de/multiplexers for specific wavelength bands. One solution, which enables a 1.31 μm wavelength broadcast overlay over eight WDM channels in the 1.55 μm wavelength band, uses individual coarse wavelength selective 1×2 couplers for overlay in each channel [87]. Waveguide crossings are inevitable in such a structure. Another solution that is suitable for much smaller separation of both wavelength bands uses a specially designed AWG with chirped grating which acts like an optical multiplexer in one wavelength band and as an optical power splitter in another one [88].

Other solutions for WDM-PON overlay devices have been demonstrated that make use of the inherent overlay features of AWGs with their different grating orders and which utilize wavelength dispersion and imaging properties between input and output [89–95] as described in Sect. 4.2.1. This concept can be adapted to a wide range of band separations and can even enable an overlay of more than two bands.

In one corresponding solution the device is composed of an AWG demultiplexer for the WDM channels, a $1/N$ power splitter for the broadcast wavelength, and an overlay-AWG of low order. The latter provides the two-band multiplexing, essentially due to its imaging properties. A 1.31 μm-broadcast/ 1.55 μm-WDM overlay and a 1.50 μm-broadcast/ 1.55 μm-WDM overlay have been realized as well (cf. column 1 in Table 4.3). In another approach the required demux/overlay-mux functionality was implemented by one specially designed AWG only which again was combined with a distributing star-coupler attached in front of it [91, 92] (columns 2 and 3 in Table 4.3). All these devices carried 8 WDM channels in the 1.55 μm band with 200 GHz and 100 GHz spacing, respectively,

The general AWG behaviour is described by (4.15), where θ_{IN} and θ_{OUT} represent the angles between the respective slab centre axis and the attached input/output waveguides.

$$\theta_{i,IN} = \frac{m}{n_s d}\left\{\lambda_i - \lambda_C\left[1 + \frac{1}{n_{wg}}\frac{dn_{wg}}{d\lambda}(\lambda_i - \lambda_C)\right]\right\} - \theta_{i,OUT} \qquad (4.15)$$

In this equation, λ_i is the signal wavelength and m is the grating order at λ_i, λ_c is the design centre wavelength at order m, d is the grating pitch, and n_s and n_{wg} are the effective refractive indices at λ_c of the slab and the grating waveguides, respectively. This equation takes into account also the

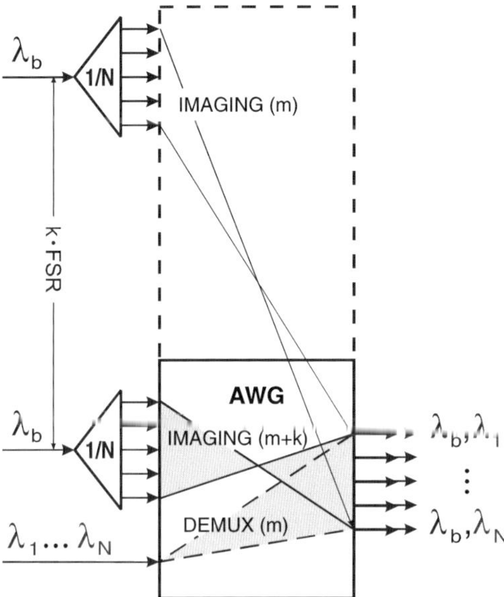

Fig. 4.13. Design principle of a broadcast WDM/PON overlay device using demultiplexing and imaging properties of an AWG, using different grating orders m and $m+k$

chromatic material dispersion by a linear fit of n_{wg} around λ_c (given in square brackets). The input/output waveguide positions at both slabs are $x = \theta f$, where f is the slab focal length.

The method of operation is illustrated schematically in Fig. 4.13. The typical AWG demux functionality for WDM wavelengths $\lambda_1 ... \lambda_N$ between the common WDM input port and N output ports is obtained using grating order m. The overlay functionality of a broadcast signal at wavelength λ_b onto these N output ports utilizes imaging from properly set N input ports. If the imaging position of the λ_b-comb from the star-coupler at the AWG input were $k \cdot FSR$ away from the centre axis with respect to order m, then one uses an order $m \pm k$, depending on the sign of the wavelength shift $\Delta\lambda_{band}$. Obviously a multiple choice of the broadcast wavelength in such a device is possible, spaced by the FSR of the AWG. The FSR determined by the chosen order m should be large enough to cover all inputs without overlap.

Furthermore, a 3-band PON-AWG configuration has been realized (column 3 in Table 4.3) which combines the mux- and demux capabilities of 2×8 WDM up- and downstream channels and the overlay of 8 distributed broadcast channels onto the WDM downstream channels (see Fig. 4.14). Two bands of them are identical in their functionality to the previous ones (orders $m = 100$ and $m = 102$). The additional third band is

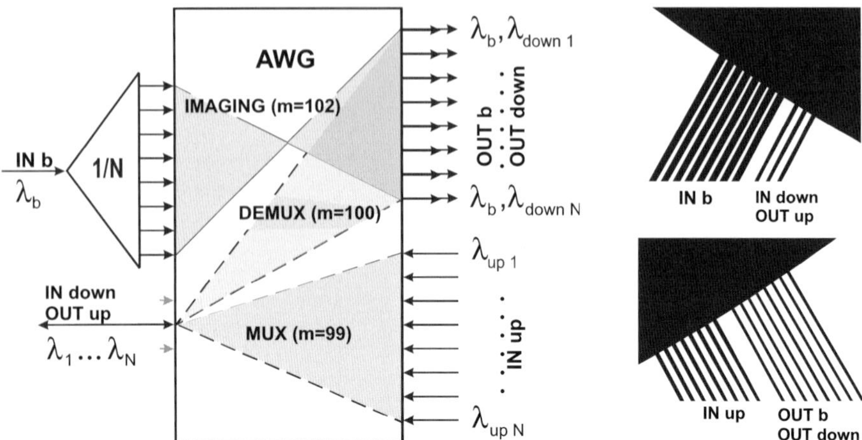

Fig. 4.14a. Principle of a broadcast PON / WDM overlay device using multi-/demultiplexing and imaging properties of an AWG, consulting different grating orders m

Fig. 4.14b. Input and output sections of the corresponding AWG device

used for 8 upstream WDM channels from the users to the central office (order $m = 99$). All these 8 upstream channels are multiplexed by the AWG and directed to one output port. Corresponding experimental results are shown in Fig. 4.15.

Table 4.3. Measured features of three of the above-mentioned WDM-PON overlay devices fabricated as silica PLCs

	Overlay device	Type 1	Type 2	Type 3
	λ_b / λ_{WDM}	1.31 / 1.55 µm	1.50 / 1.55 µm	1.50 / 1.55 µm
	diffraction order		70 (λ_b)	102 (λ_b)
	m	3(λ_b)	68 (λ_{WDM})	99 ($\lambda_{WDM\,/up}$)
				100 ($\lambda_{WDM\,/down}$)
	WDM mode	pass through	DEMUX, m = 68	DEMUX, m = 99,100
λ_b	no. of channels	8	8	8
	excess loss	11 dB ± 2 dB	4.5 dB ± 1.2 dB	5.5 dB ± 1.2 dB
	$\Delta\lambda_{3dB}$	50 nm	0.8 nm	0.4 nm
	method	horn	star	star
λ_{WDM}	channel spacing	1.6 nm (200 GHz)	1.6 nm (200 GHz)	0.8 nm (100 GHz)
	insertion loss	3.8 dB – 5.2 dB	3 dB	3 dB
	$\Delta\lambda_{3dB}$	8 nm	0.65 nm	0.35 nm
	crosstalk	> 25 dB	> 25 dB	> 27 dB

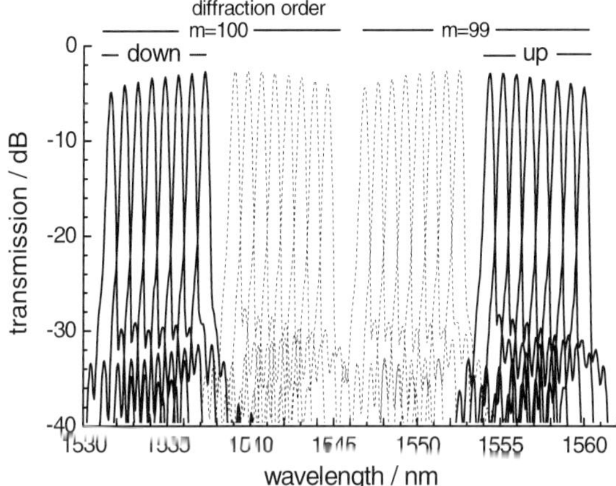

Fig. 4.15. (a) Experimental results for the 2×8 ports of type 3 WDM-PON overlay device. Downstream channels around 1533 nm at order $m = 100$ and upstream channels around 1557 nm at order $m = 99$ (*bold curves*). The centre loss amounts to about -2.5 dB. *Dotted curves:* additional WDM bands due to periodicity of AWG characteristics

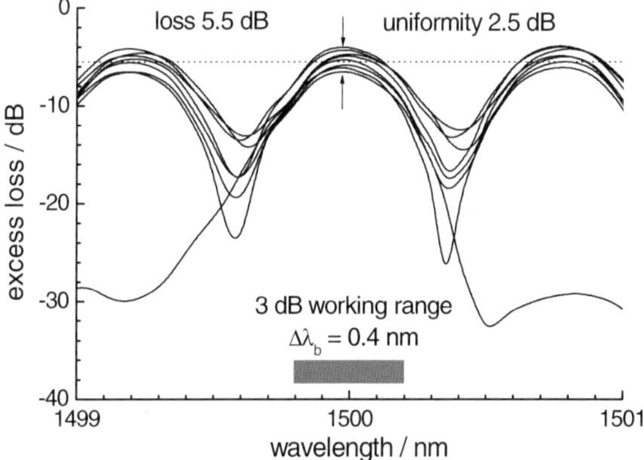

Fig. 4.15. (b) Measured excess loss of the distributing system for the broadcast wavelength λ_b around 1500 nm at the ports $\mathrm{OUT_b}$ ($= \mathrm{OUT_{down}}$). The passband is broadened to 0.4 nm by horn structures at the input

4.5 InP-based Devices

4.5.1 Introduction

The characteristics of InP-based AWGs are rather different compared to silica-based ones, and in some respects InP is superior, in others silica-on-silicon is more favourable. The most important advantage of InP-based AWGs is that they can be monolithically integrated with devices like lasers, semiconductor optical amplifiers (SOA), RF-modulators and switches, wavelength converters, signal regenerators or detectors, for example. Moreover, due to the high index contrast that is possible in semiconductor waveguides, InP-based AWGs can be smaller than their silica-based counterparts by more than two orders of magnitude. Insertion-losses of InP and silica AWGs are comparable. Crosstalk figures for InP-based AWGs are lagging behind by 5–10 dB compared to silica devices (> 35 dB), however, considerable improvement can be expected in the coming years. The most important disadvantage of InP-based AWGs is their coupling losses to single-mode fibres which are in the order of 10 dB because of the large difference in waveguide mode size. Coupling losses can be reduced by using lensed fibres, but at the price of a reduced alignment tolerance and, as a consequence, higher packaging costs. For efficient and tolerant fibre coupling integration of fibre-mode adaptors (also called spot-size converters or tapers) on the chip is mandatory. So far, this has prevented InP-based AWGs to become a competitor of silica-based AWGs for use as stand-alone devices. The real advantage of InP-based AWGs lies in their potential for integration in circuits with an increased functionality, such as WDM transmitters and receivers, both for Metro and Access Networks, and optical add-drop multiplexers. So far the market for this kind of functionalities has been too small to enable a real breakthrough of InP-based Photonic Integrated Circuits (PICs), but at the time of writing this chapter the recovery of the WDM market is creating the conditions for broader application of InP-based Photonic ICs.

The power of micro-electronic integration technology is that a broad class of electronic functionalities can be synthesised from a small set of elementary components such as transistors, resistors, and capacitors. A technology that supports integration of these elementary components can, therefore, be used for a broad class of applications, and investments made in its development are paid back by a large market. Although photonic integration has much in common with micro-electronic integration, a major difference is the variety of devices and device-principles in photonics. For couplers, filters, multiplexers, lasers, optical amplifiers, detectors, switches, modulators, etc. a broad variety of different operation principles and materials have been reported. It is impossible to develop a monolithic technology which is capable of realising even a modest subset of all these devices. The key for the success of integration in photonics is, therefore, the reduction of the broad variety of optical functionalities to a few elementary components such as waveguides, couplers, and generic active components.

4.5.2 Fabrication

The first step in the fabrication of InP-based PICs is epitaxial growth of the layer stacks required in the PIC. The most widely applied technology is Metal-Organic Vapour Phase Epitaxy (MOVPE, mostly called Metal-Organic Chemical Vapour Deposition, MOCVD, which is a less correct name, however). In this technology the materials (Indium, Phosphorous, Gallium, and Arsenic) are carried to the reactor in gaseous form coupled to methyl- or ethyl-groups (In, Ga) or as hydrides (As, P). In the reactor they are thermally cracked and deposited onto the heated substrate, where they crystallise in a composition that is dependent on the concentrations of the constituent elements. During the epitaxy, both the composition of the crystal (the amounts of In, Ga, P, and As) and its lattice constant have to be controlled, the latter to avoid formation of large numbers of crystal defects. This can be achieved by tight control of the temperature and the gas flows. By controlling the composition the properties of the material can be chosen to be transparent or active (amplifying or absorbing), and also proper materials for use in modulators can be grown.

A major issue for photonic integration is the way in which active sections (laser and amplifier) are integrated with transparent sections (waveguide or modulator). In principle three different schemes can be distinguished, each with a number of variants. They are illustrated in Fig. 4.16. In the first one active and transparent waveguides are in the same plane and are coupled via a so-called butt-joint. The structure is realized by first growing the active layer stack, selectively removing it while covering the active regions with a mask, selectively re-growing the transparent regions, while covering the active regions with a mask, and covering the whole structure with a cladding layer in a third epitaxial growth step. The whole process requires good equipment and skills because all layers should get an almost perfect crystalline structure with well-controlled composition, morphology, and dopant level. In the second and third scheme the whole structure is grown in a single epitaxy growth, and the active layers are removed afterwards while covering the active regions with a mask. In the second scheme a thin active layer is applied on top of the transparent waveguide in

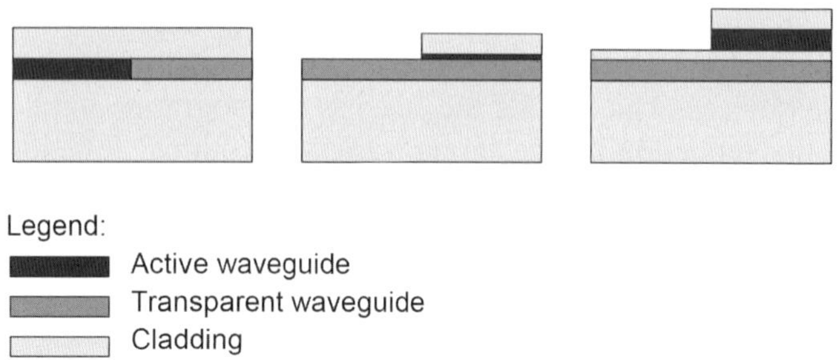

Fig. 4.16. Three different schemes for integration of active (laser and amplifier) and passive waveguides

such a way that the optical fields at both sides of the junction are not too different so that the transition loss at the junction is small. Special provisions are necessary to avoid reflections. In the third scheme the light is coupled from the lower transparent layer to the upper active layer. This scheme needs special provisions for efficient coupling from the lower to the upper layer. The single-step epitaxial growth schemes have the advantage of being simpler from a fabrication point of view. Both schemes have been successfully applied for development of commercial products. A disadvantage is the restricted flexibility: because the whole structure is grown in one step, it is not possible to use different compositions and doping levels in different regions. This reduces the design freedom and the number of different components that can be integrated.

Etching of InP-based waveguides is commonly done with Reactive Ion Etching (RIE) with methane/hydrogen (CH_4/H_2) plasma [96]. Excellent anisotropy and smooth surfaces can be obtained. During etching a polymer film deposits that can cause waveguide roughness and therefore optical losses. The polymer can be removed during etching by using an etch process where the CH_4/H_2 etching step is alternated with an O_2 descumming step [97], resulting in optical waveguide propagation losses below 1 dB/cm.

Another common etching method for InP-based materials is using inductively-coupled plasma RIE with $Cl_2/CH_4/H_2$ chemistry. This results in high etching rates while maintaining smooth morphology. Moreover, the chlorine prevents the formation of polymers [98].

4.5.3 InP-based AWGs

The first InP-based AWG (de-)multiplexer was reported by Zirngibl et al. [99]. This 15×15 channel device measured 1 cm^2 and had an insertion loss of 2–7 dB and a crosstalk > 18 dB. The InP-based device with the largest

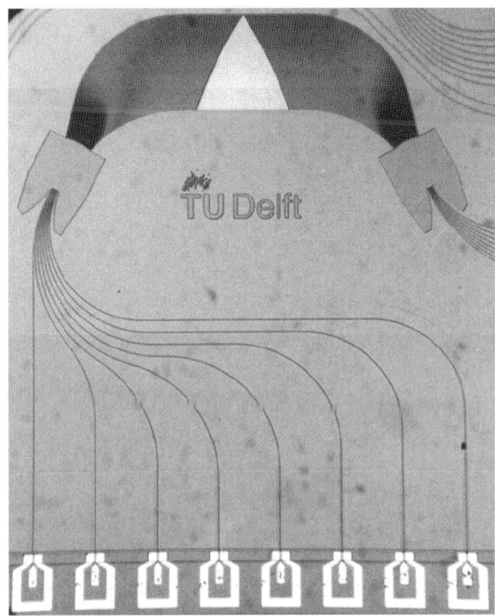

Fig. 4.17. WDM-receiver chip $(3.1 \times 3.9\,\mathrm{mm}^2)$ containing an AWG-demultiplexer integrated with 8 detector diodes at the bottom [104]

number of channels so far is a 64-channel AWG with dimensions $3.6 \times 7\,\mathrm{mm}^2$ reported by NTT [100], and it has a crosstalk level better than 20 dB. Small device dimensions are achieved by applying deeply etched waveguides with a high lateral index contrast which allow for small bending radii. Disadvantages of this solution are the high coupling losses at the discontinuity between the array and the free propagation regions at both sides and the increased propagation losses of deeply etched waveguides. Loss figures for compact AWGs reported [100–102] range from 7 to 14 dB. These losses have been reduced to 3 dB for an 8-channel AWG with dimensions $700 \times 750\,\mu\mathrm{m}^2$ using a double etch process [103].

Polarisation sensitivity is an important issue. GaInAsP/InP DH-waveguide structures inherently have a strong birefringence which translates into a TE-TM shift of several nanometres. A solution for InP-based AWGs is found in compensating the polarisation dispersion of the phased array by inserting a waveguide section with a different birefringence [104–106]. Figure 4.17 shows a polarisation-independent demultiplexer based on this principle. In the triangular section the waveguide structure has been changed by partial removal of the top layer and reduction of the waveguide width which creates a local increase in birefringence. Through proper design of the triangle this increased birefringence can compensate the birefringence in the rest of the array.

Another way of reducing the polarisation dependence is by adapting the waveguide width of deeply etched waveguides, where the birefringence of the DH-structure is compensated by the birefringence induced by the vertical waveguide walls. For single-mode waveguides with a core of quaternary material with a bandgap-equivalent wavelength around $1.3\,\mu m$ ("Q1.3") the waveguides become rather small, around $1.5 \times 0.6\,\mu m^2$. To relax the fabrication tolerance, the waveguide size can be increased, when lowering the index contrast by using, for example, Q1.0 material for the waveguide core.

Another point of relevance is the temperature sensitivity. In silica-based AWGs several techniques have been proposed to reduce the temperature sensitivity (cf. Sect. 4.3.4). As InP-based AWGs usually operate in an actively controlled temperature-stabilised environment the need for temperature insensitivity is less urgent. NTT has reported a temperature insensitive AWG using a temperature compensation triangle [107].

A major issue in the development of photonic integration technology is reduction of device dimensions. The key to reduction of device dimensions is the application of high lateral index contrast (i. e. the index contrast in the plane of the waveguide layer) realised by deep etching of the waveguides. Because of the high index contrast between the semiconductor waveguide (typical refractive index values around 3.3) and air, deeply etched waveguides provide strong light confinement. This allows for application of

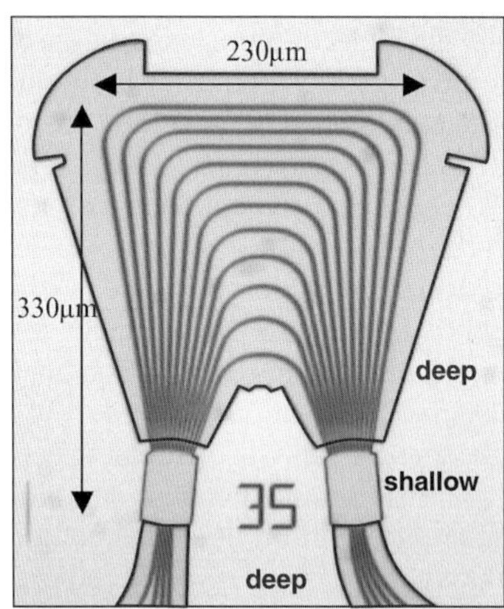

Fig. 4.18. Deeply etched InP-based AWG wavelength demultiplexer with record small dimensions

small waveguide widths, small bending radii, compact waveguide couplers, and compact AWGs.

The price that is paid for the high contrast is increased propagation losses due to scattering at rough waveguide walls and increased insertion losses at junctions between slabs and waveguide arrays as used in an AWG where the finite resolution of the lithography causes abrupt closure of the tapered inter-waveguide gaps. With a low index contrast the optical discontinuities caused by this closure are small; with deeply etched waveguides they can introduce a few dB loss per junction. A solution is the use of a technology that supports a combination of low-loss shallowly etched waveguides for interconnect purposes and at waveguide junctions with locally deeply etched regions where short bends or other high-contrast functions are required. Losses smaller than 0.1 dB per junction have been shown at transitions between deeply and shallowly etched regions, so that several transitions can be included in a circuit. Using such a technology AWG-dimensions have been reduced below 0.1 mm^2 [108], see Fig. 4.18. This is more than two orders of magnitude smaller than the smallest silica-based devices reported so far.

An important feature of AWGs is their crosstalk performance. A major cause of crosstalk are random errors in the phase transfer of the array waveguides, due to random variations in waveguide width, layer thickness, or composition. As far as these random variations occur on a spatial scale comparable to the operation wavelengths they will also contribute to propagation losses (scattering). Improvement of uniformity and smoothness will, therefore, affect both crosstalk and propagation loss in a positive way. This is mainly a matter of lithographic and etching technology and it will become better and better with progress in these technologies.

A second cause of crosstalk is the generation of higher order modes in the array at junctions between straight and curved waveguides or other irregularities. Two approaches have been proposed to filter out higher order modes: application of mode filters halfway the waveguide array [109, 110] and application of a special waveguide type that is leaky for higher order modes [111]. Crosstalk levels around 30 dB have been demonstrated for InP-based AWGs using these approaches.

4.5.4 AWG-based Circuits

Multi-wavelength Transmitters

Today's WDM systems use wavelength-selected or tuneable lasers as sources. Multiplexing a number of wavelengths into one fibre is done using a power combiner or a wavelength multiplexer. A disadvantage of this

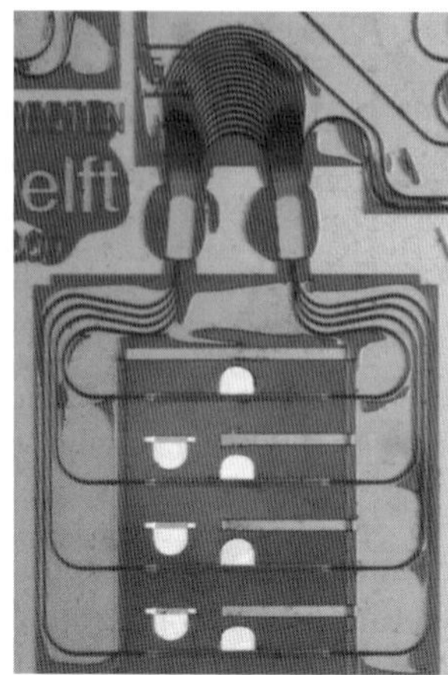

Fig. 4.21. Monolithically integrated multi-wavelength ring laser measuring $1.0 \times 1.5\,\mathrm{mm}^2$ [123, 124]

This device, which measures only $1.5\,\mathrm{mm}^2$, showed lasing behaviour at 7 different wavelengths with four amplifiers [123, 124]. Since ring lasers do not need mirrors, they are particulary suitable for integration into larger circuits.

Multi-wavelength Receivers

The most straightforward way to realise a multi-wavelength receiver is by connecting N detectors to the output ports of a wavelength demultiplexer. Integration of an AWG-based demultiplexer with detectors on a single chip

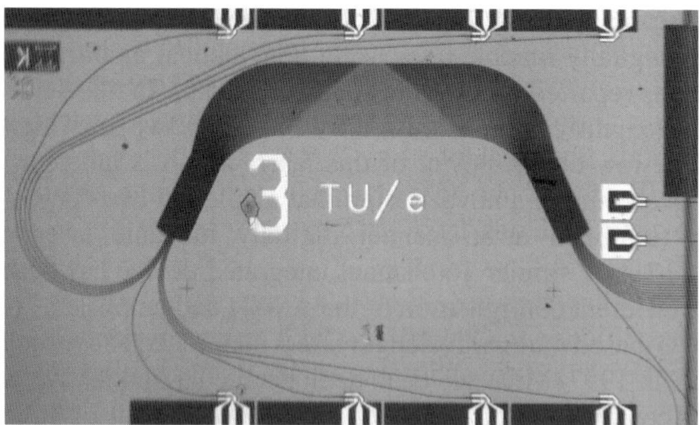

Fig. 4.22. WDM-receiver chip containing an AWG-demultiplexer integrated with 8 photodiodes (*top* and *bottom*), the chip measures $3 \times 5\,\mathrm{mm}^2$ [127]

has been reported by several authors [104, 127–130], and one of these devices is shown in Fig. 4.22. The photodetectors are twin guide structures grown in a single epitaxial growth on top of the waveguiding layer and have been placed around the chip in order to reduce the electrical crosstalk between the channels. The device has eight channels with a 3-dB electrical bandwidth of 25 GHz and a responsivity around 0.2 A/W at a wavelength of 1550 nm. Optical crosstalk between the channels is > 18 dB. The device reported in [130] has similar characteristics, but integrates a spot-size converter, resulting in an increased responsivity of 0.46 A/W. A 4-channel device, similar to these, hybridly integrated with 4 frontend amplifiers and packaged in a 26-pin butterfly package, operating at 10 Gbit/s is commercially available from ThreeFive Photonics. A 10-channel multi-wavelength (MW)-receiver operating at 10 Gbit/s per channel was reported by Infinera, matching the 10×10 Gbit/s transmitter module described above [114]. This device has the highest aggregated throughput of 100 Gbit/s demonstrated to date. An example of a highly sophisticated commercially available device is the 40-channel optical monitor shown in Fig. 4.23. The device integrates 9 AWGs with 40 detectors on a chip area of less than 5×5 mm^2, thus combining high integration density with good performance: 4 dB total on-chip loss, 0.4 A/W responsivity, and a crosstalk level > 35 dB. These devices illustrate the volume reduction that can be achieved using monolithic integration.

In InP the receiver amplifiers can also be integrated; this was first reported by Chandrasekhar et al. who realized an 8×2.5 Gbit/s multi-wavelength-receiver with integrated heterojunction bipolar transistor (HBT) preamplifiers [131]. Monolithic integration has the potential to bring down

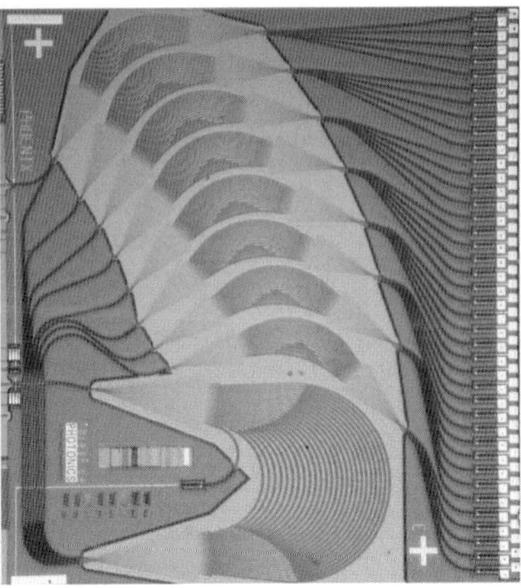

Fig. 4.23. 40-channel optical monitor chip containing 9 AWGs and 40 detectors (courtesy ThreeFive Photonics)

the cost and increase the reliability of MW-receiver modules. It further leads to a drastic volume reduction with an increased functionality. Main challenges to exploit the advantage of integration are the problems related to electrical crosstalk inside the package. For devices with larger channel count special care has to be taken to reach acceptable crosstalk figures at data rates higher than 1 Gbit/s.

Channel Selectors and Equalizers

Arrayed waveguide gratings can be used as tuneable filters or channel selectors. In their simplest form, a pair of AWGs is used back-to-back, the first AWG demultiplexes the N channels and the second multiplexes them again. When the AWGs are connected by an array of N amplifiers (see Fig. 4.24), one or more wavelength channels can be selected and coupled to the outputs. Wavelength selectors have been reported both in hybrid and monolithic form. Monolithically integrated devices combine a small device

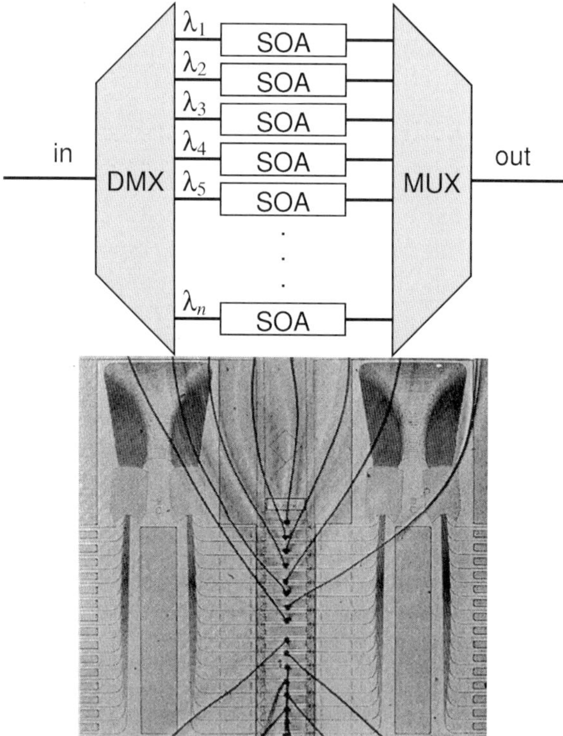

Fig. 4.24. Channel selector capable of selecting any combination of channels by activating one or more Semiconductor Optical Amplifiers. Schematic (*top*) and photograph of realized device [132] (*bottom*)

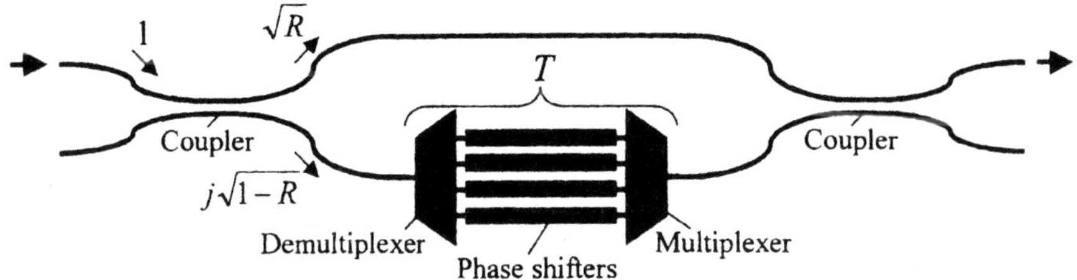

Fig. 4.25. Schematic layout of a power equalizer [80]

size with on-chip loss-compensation (zero-loss-between-fibres) at a potentially lower cost. In the basic configuration [132], the number of SOAs is equal to the number of channels. The same device can be used as channel power equalizer. For the power equalization function, attenuators may also be used, instead of amplifiers. However, since the loss of such a passive power equalizer using attenuators is too high for many networks, sophisticated solutions have been introduced.

One of these [80] uses a Mach–Zehnder interferometer (MZI), where one arm is a simple waveguide, while the other arm contains a demultiplexer-multiplexer pair interconnected by an array of programmable phase shifters whose effective path lengths can be controlled externally, as illustrated in Fig. 4.25. A proper splitting ratio in the MZI couplers assures the required dynamic range for power equalization. Fractions R and $1-R$ of the total power are sent to the "non-filtered" straight arm and the "filtered" arm, respectively, and are then combined in the second coupler with the same splitting ratio. In Fig. 4.25 the optical field strengths are indicated instead of the optical power levels. Interference occurring for each channel wavelength leads to the final power level. This kind of equalizer can have significantly lower loss than the conventional channel-by-channel equalizer.

For channel selectors that make use of amplifiers, various solutions to reduce the number of amplifiers have been demonstrated. Kikuchi et al. [133] reported a device in which one out of 64 channels can be selected using 16 SOAs that are operated as optical gates. In this device the reduction in the number of gates from N to $2\sqrt{N}$ was achieved with a two-step selection using two AWGs and a power combiner (see Fig. 4.26). Note that with this device, although any single wavelength can be selected, not every combination of wavelengths can be selected, as it is the case with the general device. A variant with a two-step selection has been reported by Lucent and uses a combination of a star coupler and a single AWG to select one out of 32 channels with 12 SOA gates [134].

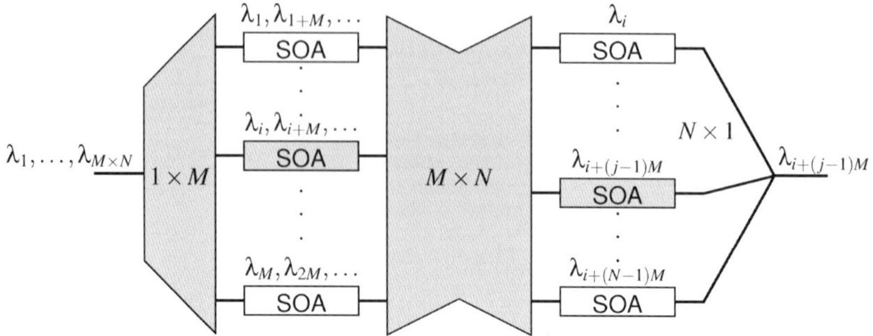

Fig. 4.26. Channel selector selecting one out of $M \times N$ channels using a combination of two AWGs and a power combiner

Multi-wavelength Add-drop Multiplexers and Crossconnects

The generic structure of reconfigurable add-drop multiplexers has been illustrated in Sect. 4.2.8, and an early example of an InP based reconfigurable add-drop multiplexer can be found in [135]. This device integrates 4 Mach–Zehnder interferometer switches with a 5×5 AWG in a loop-back configuration (cf. Fig. 4.8a).

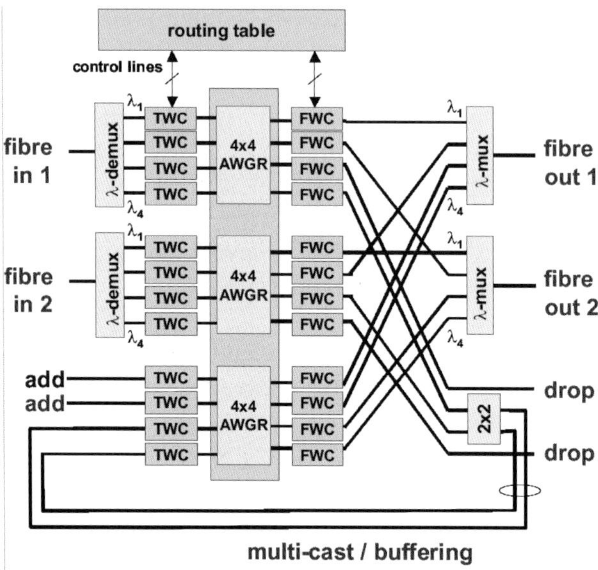

Fig. 4.27. IST-STOLAS label-controlled optical crossconnect with non-blocking node architecture

Another key device for advanced multi-wavelength networks is an optical crossconnect. Figure 4.27 shows an example of an experimental crossconnect which has been investigated in a European research project (IST-STOLAS). In this project, data packets carry a label that is modulated orthogonally to the data payload: the data is intensity-modulated on a specific wavelength channel (at e. g. 10 Gbit/s), and the label is frequency- or phase-modulated on the same channel (at e. g. 155 Mbit/s). In the crossconnect, the packets are routed by their wavelength, which is changed by a tuneable wavelength converter according to the destination encoded in the label. Then the packets are routed by means of a passive wavelength router, after which the wavelength is converted to a wavelength selected for the desired output. At the same time, collisions between two packets for the same destination are thus avoided, making the node non-blocking.

The cost presently involved in crossconnects like these is prohibitive for large-scale application of optical cross-connected networks. Application will only be economical at the highest levels of the network where cost can be shared by many users. For broader application, volume of devices like these has to be ramped up while cost has to be reduced at the same time. Present progress in photonic integration holds a great promise that this goal can be achieved. Figure 4.28 shows the function, the circuit scheme, and a photograph of a PIC which may be used to cross-connect two fibre links. The device, which was monolithically integrated in InP-based

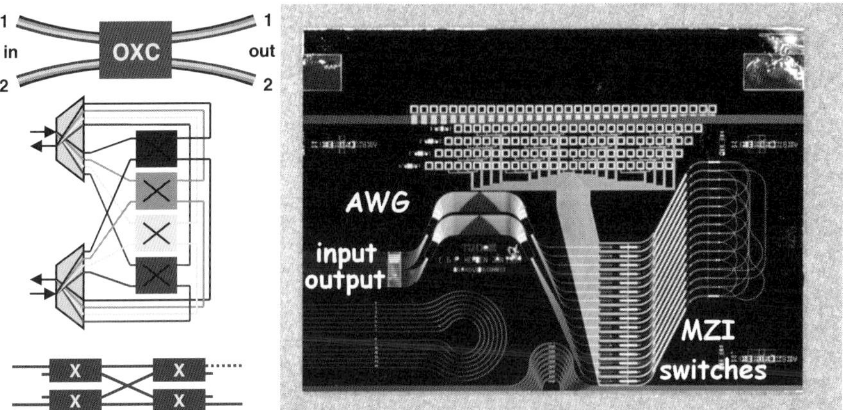

Fig. 4.28. Functional diagram (*upper left*), circuit scheme (*middle left*), dilated switch layout (*lower left*) and microscope photograph (*right*) of a compact photonic integrated crossconnect in a double AWG configuration. The AWGs are seen in the centre; the triangular structures make them polarisation independent. The sixteen switches at the right are positioned under a specific angle with the crystal axis to make them polarisation independent. The chip measures $8 \times 12 \, \mathrm{mm}^2$

semiconductor material, measures only $8 \times 12\,\text{mm}^2$ and contains two polarisation-independent AWGs, and 16 Mach–Zehnder interferometric switches (4 for each wavelength, in a dilated scheme) that are positioned under an angle for polarisation-independent operation. The insertion loss was $< 16\,\text{dB}$ and the inter-channel crosstalk $> 20\,\text{dB}$. An even more compact crossconnect with the same functionality and similar performance, but measuring only $3.3 \times 1.5\,\text{mm}^2$, was also realized [136], demonstrating the potential of this very compact technology. At present the performance of PICs like this is not yet sufficient for operational applications. Silica-based devices show better performance, but have a much larger device size. With the ongoing progress in integration technology such crossconnects can already be realised with much improved performance. When there is sufficient demand for such devices, this may give rise to a new generation of highly sophisticated multi-wavelength networks.

4.6 Other Material Systems

AWGs have been realized primarily in silica-on-silicon (SiO_2/Si) and on indium phosphide (InP), but to a smaller extent in other material systems such as polymers, silicon-oxinitride (SiON), and silicon-on-insulator (SOI) as well.

Polymer optical waveguides and related PLCs have been fabricated using different materials, which all rely on a relatively low index contrast comparable to that of silica-on-silicon, and as a consequence the device properties are rather similar as well, which also applies to AWGs [137–140]. Some interesting features have been obtained with polymer material, e. g. athermal AWG behaviour [141] or low power needed for switching [142, 143], but the widespread expectation of particularly low cost AWG devices has not been fulfilled so far. The main reasons for that are a comparable device fabrication effort and in addition even higher pigtailing and packaging cost (in order to meet the Telcordia (Belcore) requirements), which essentially compensate or even go beyond the savings in material cost. In addition, polymer AWGs do not yet meet the outstanding performance of present silica AWGs with respect to insertion loss, crosstalk, and PDL.

On the other hand, silicon-oxinitride is a high-index contrast material suitable for highly compact devices. Several reports on silicon-oxinitride PLCs [144–148] including small size AWG realizations have been published. But none of them has reached commercialisation up to now.

Silicon waveguide technology is another promising field of photonics [149–152]. It is based on the use of silicon-on-insulator wafers, offering a platform for diverse waveguide structures such as slab-, strip- , and rib-waveguides. All-Si waveguides rely on the very high index contrast, too. But, especially rib waveguides offer single-mode behaviour up to rather large dimensions, e. g. 10 μm, enabling low coupling loss to the fibre and low propagation loss as well [153]. Therefore, SOI rib waveguides have been preferred for AWG implementation. In addition, electrically controlled optical elements monolithically incorporated into the Si waveguide have been demonstrated, namely μs-optical switches [154, 155], variable optical attenuators (VOA) [156], and (slow) photodetectors [157]. Also, SOI-based optical devices can take advantage of the highly-developed silicon technology including the availability of high quality SOI wafers, the applicability of CMOS compatible processing steps, and the possibility to monolithically integrate Si-electronics. A number of promising AWG devices have been shown in this material system [158–160].

The SOI-based optical components technology had been pioneered by Bookham Technology (UK). Bookham's ASOC ('application-specific optical circuit') platform enabled the monolithic integration of passive optical devices with other functionalities such as switching or channel monitoring [155–157] onto a single SOI chip, and one example is a 40-channel 100 GHz multiplexer arrayed-waveguide grating for WDM applications monolithically integrated with variable optical attenuators [160]. Commercialisation of ASOC-platform devices started in 2002, but was discontinued only one year later as a consequence of the downturn in the telecom industry. In 2005 these activities were renewed by Kotura Inc. who licenses silicon photonic patents from Bookham and appointed one of Bookham's founders to its board of directors. These patents cover the design and manufacture of silicon waveguide devices such as VOAs and modulators and the construction of multi-functional optical silicon chips. As a first product, this company brought a high-speed multi-channel VOA component in SOI technology to market.

4.7 Methods for AWG Characterisation

Testing of fibre optic components such as de/multiplexers for DWDM applications has become a great challenge due to high accuracy demands versus constraints with respect to the affordable cost to generate the data [161]. In many cases a limited set of data has to be determined experimentally, while other parameters are derived from these parameters (see below).

4.7.1 Insertion Loss

One of the most important characteristics of passive optical components is the insertion loss value, and for WDM applications insertion loss has to be determined as a function of wavelength. Other parameters such as cross-talk between different channels or uniformity, for example, can be derived comparing the insertion loss values at different wavelengths.

Since insertion loss is the ratio of two power levels (cf. Glossary), two measurement steps are required for each specific wavelength:

1. The reference power, i. e. the power P_{ref} without the device under test (DUT), must be recorded.
2. Then the DUT is inserted and the output power P_{out} after the DUT is measured.

In the case of a pigtailed device the insertion loss, essentially given by $P_{\mathrm{out}} / P_{\mathrm{ref}}$, does not only include any errors of the measurement system, but the uncertainty related to the loss of one connector pair. The latter problem is frequently circumvented by splicing the DUT into the signal path and cutting the fibre before taking the reference reading.

In contrast to absolute power measurements, the uncertainty of an insertion loss measurement is not affected by the uncertainty of the detector sensitivity, and one has to ensure only that the detector's power linearity is sufficient.

4.7.2 Wavelength-dependent Measurements

One key issue in wavelength-dependent testing is (in addition to absolute wavelength accuracy) the selection of the most appropriate method to do swept-wavelength testing. Most frequently one chooses between one of the following options:

a) A **broadband light source** is used and the wavelength discrimination is done at the detector site by sweeping the narrow detector pass band over the whole wavelength range of interest. This measurement principle is realized in optical spectrum analysers (OSA). Typical sources used are either a white light source (halogen lamp), or one or more edge emitting LEDs (EELED). The proper choice of source depends on the wavelength measurement range and the output power needed.
b) A **narrow-band tuneable light source** (frequently an external cavity tuneable laser) is used to excite the DUT and the receiver is a non-selective detector, usually a semiconductor photo-detector.

Both methods introduce characteristic accuracy and sensitivity limitations. In the first case (using an OSA) resolution bandwidth (typically > 50 pm) is normally the most severe limitation, and as a consequence narrow features cannot be spectrally resolved.

On the other hand, if the light source is a tuneable laser, the so-called source spontaneous emission (SSE) is a limiting factor. The SSE appears as a broad continuum below the laser emission peak, and if a broadband detector is used, the SSE gives rise to a receiver signal even in the stop band region of the filter, which limits the achievable dynamics.

Consequently, best results are achieved by combining a tuneable laser source and an OSA (i. e. combining a narrow source and a narrowband receiver). However, as this solution is rather expensive, it is only used for the most demanding applications.

4.7.3 Polarisation-dependent Loss

Optical signals propagating in fibre networks comprising a large number of active (e. g. optical amplifiers) and passive components do usually not have a well-defined state of polarisation. As a consequence, modern communication systems require components with low polarisation dependence (or low polarisation-dependent loss (PDL)). For this reason the evaluation of PDL has become important, and two methods are commonly used in evaluating PDL.

a) All polarisation states are applied to the DUT, and the states of maximum and minimum transmittance are determined. In the case of an AWG demultiplexer with low or moderate PDL under test, a convenient and fast procedure is the following: One first chooses a wavelength with arbitrary polarisation that lies at the steep edge of a filter peak. This can be for example a wavelength at the −10 dB decay point from the filter peak. Subsequently one changes the incoming polarisation, e. g. by twisting the fibre paddles in a convenient apparatus or by means of a polarisation controller. These changes do normally induce variations of the output signal intensity, and the two states with maximum and minimum transmittance usually correspond to the TE- and TM-polarisation states of the inserted light. The wavelength-dependent device characterisation is then performed using these two settings. Systematic errors may arise from the PDL and the reflection of the test equipment itself, which does consequently need appropriate attention.

b) The so-called **Mueller/Stokes method** is based upon the application of four defined polarisation states to the DUT and subsequent calculation of polarisation dependence using Mueller mathematical formulas.

Application of this method requires a programmable polarisation controller (PC) that is able to generate polarisation states which uniformly cover the entire Poincaré sphere [162]. In addition, the optical source power at the DUT should be time-invariant and exhibit no polarisation dependence. The procedure consists of the following steps: 1. maximize power throughput by adjusting the polariser, 2. measure the output power of the PC at four different polarisation states: horizontal, vertical, +45°, and right-hand circular, 3. insert the test device and measure again under the same polarisation states, 4. calculate max / min transmissions using the 8 measurement results and Mueller mathematics, 5. calculate the PDL.

For all PDL measurements a detector with low polarisation dependence is mandatory.

References

1.   M. K. Smit: "New focusing and dispersive planar component based on an optical phased array," Electron. Lett. **24**, 385–386 (1988)
2.   H. Takahashi, S. Suzuki, K. Kato, and I. Nishi: "Arrayed-waveguide grating for wavelength division multi/demultiplexer with nanometer resolution," Electron. Lett. **26**, 87–88 (1990)
3.   C. Dragone: "An N×N optical multiplexer using a planar arrangement of two star couplers," IEEE Photon. Technol. Lett. **3**, 812–815 (1991)
4.   M. B. J. Diemeer, L. H. Spiekman, R. Ramsamoedj, and M. K. Smit: "Polymeric phased array wavelength multiplexer operating around 1550 nm," Electron. Lett **32**, 1132–1133 (1996)
5.   J. T. Ahn, S. Park, J. Y. Do, J.-M. Lee, M.-H. Lee, and K. H. Kim: "Polymer wavelength channel selector composed of electrooptic polymer switch array and two polymer arrayed waveguide gratings," IEEE Photon. Technol. Lett. **16**, 1567–1569 (2004)
6.   H. Okayama and M. Kawahara: "Waveguide array grating demultiplexer on LiNbO$_3$," *Integrated Photonics Research* (IPR'95), Techn. Digest (Dana Point, CA, USA, 1995), 296–298 (1995)
7.   C. van Dam: *"InP-based polarization independent wavelength demultiplexers,"* PhD thesis, Delft University of Technology, Delft, The Netherlands (1997) ISBN 90-9010798-3
8.   M. K. Smit and C. van Dam: "PHASAR-based WDM-devices: principles, design and applications," J. Select. Topics Quantum Electron. **2**, 236–250 (1996)
9.   A. Klekamp and R. Münzner: "Calculation of imaging errors of AWG," J. Lightwave Technol. **21**, 1978–1986 (2003)
10.   M. R. Amersfoort, C. R. de Boer, F. P. G. M. van Ham, M. K. Smit, P. Demeester, J. J. G. M. van der Tol, and A. Kuntze: "Phased-array wavelength demultiplexer with flattened wavelength response," Electron. Lett. **30**, 300–302 (1994)

11. M. R. Amersfoort, J. B. D. Soole, H. P. LeBlanc, N. C. Andreadakis, A. Rajhel, and C. Caneau: "Passband broadening of integrated arrayed waveguide filters using multimode interference couplers," Electron. Lett. **32**, 449–451 (1996)

12. C. Dragone. US Patent No. 5,412,744 (1995)

13. K. Okamoto and H. Yamada: "Flat spectral response arrayed waveguide grating multiplexer with parabolic waveguide horns," Electron. Lett. **32**, 1661–1662 (1996)

14. K. Okamoto and H. Yamada: "Arrayed-waveguide grating multiplexer with flat spectral response," Opt. Lett. **20**, 43–45 (1995)

15. C. Dragone: "Efficient techniques for widening the passband of a wavelength router," J. Lightwave Technol. **16**, 1895–1906 (1998)

16. A. Rigny, A. Bruno, and H. Sik: "Multigrating method for flattened spectral response wavelength multi/demultiplexer," Electron. Lett. **33**, 1701–1702 (1997)

17. J.-J. He, E. S. Koteles, and B. Humphreys: "Passband flattening via waveguide grating devices using phase-dithering," *Integrated Photonics Research* (IPR '02), Techn. Digest, (Vancouver, Canada, 2002), paper IFE2 (2002)

18. G. H. B. Thompson, R. Epworth, C. Rogers, S. Day, and S. Ojha: "An original low-loss and pass-band flattened SiO_2 on Si planar wavelength demultiplexer," *Opt. Fiber Commun. Conf.* (OFC'98), Techn. Digest, (San Jose, CA, USA, 1998), paper TuN1 (1998)

19. C. R. Doerr, L. W. Stulz, and R. Pafchek: "Compact and low-loss integrated box-like passband multiplexer," IEEE Photon. Technol. Lett. **15**, 918–920 (2003)

20. H. Takahashi, K. Oda, H. Toba, and Y. Inoue: "Transmission characteristics of arrayed waveguide N×N wavelength multiplexer," J. Lightwave Technol. **13**, 447-455 (1995)

21. C. G. P. Herben, X. J. M. Leijtens, D. H. P. Maat, H. Blok, and M. K. Smit: "Crosstalk performance of integrated optical cross-connects," J. Lightwave Technol. **17**, 1126–1134 (1999)

22. H. Takahashi, Y. Hibino, Y. Ohmori, and M. Kawachi: "Polarization-insensitive arrayed-waveguide wavelength multiplexer with birefringence compensating film," IEEE Photon. Technol. Lett. **5**, 707–709 (1993)

23. K. Maru, K. Matsui, H. Ishikawa, Y. Abe, S. Kashimura, S. Himi, and H. Uetsuka: "Super-high-Δ athermal arrayed waveguide grating with resin-filled trenches in slab region," Electron. Lett. **40**, 374–375 (2004)

24. M. Kawachi: "Silica waveguides on silicon and their application to integrated-optic components," Opt. and Quantum Electron. **22**, 391–416 (1990)

25. K. Imoto: "Progress in high silica waveguide devices," *Integrated Photonics Research* (IPR'94), Techn. Digest (1994), 62–64 (1994)

26. Y. Saito: "Optical fibre technology - key to undersea cable development," J. Asia Electron. Union **5**, 63–67 (1989)

27. C. D. Lee, W. Chen, Y.-J. Chen, W. T. Beard, D. Stone, R. F. Smith, R. Mincher, and I. R. Stewart: "The role of photomask resolution on the performance of arrayed-waveguide grating devices," J. Lightwave Technol. **19**, 1726–1733 (2001)

28. C. H. Henry, G. E. Blonder, and R. F. Kazarinov: "Glass waveguides on silicon for hybrid optical packaging," J. Lightwave Technol. **7**, 1530–1539 (1989)

29. Q. Lai, J. S. Gu, M. K. Smit, J. Schmid, and H. Melchior: "Simple technologies for fabrication of low-loss silica waveguides," Electron. Lett. **28**, 1000–1001 (1992)

30. S. Valette, J. P. Jadot, P. Gidon, S. Renard, A. Fournier, A. M. Grouillet, H. Dennis, P. Philippe, and E. Desgranges: "Si-based integrated optics technologies," Solid State Technol. **32**, 69–74 (1989)

31. Y. P. Li and C. H. Henry: "Silica-based optical integrated circuits," IEE Proc. Optoelectron. **143**, 263–280 (1996)

32. A. Kilian, J. Kirchhof, B. Kuhlow, G. Przyrembel, and W. Wischmann: "Birefringence free planar optical waveguide made by flame hydrolysis deposition (FHD) through tailoring of the overcladding," J. Lightwave Technol. **18**, 193–198 (2000)

33. Y. Inoue, H. Takahashi, S. Ando, T. Sawada, A. Himeno, and M. Kawachi: "Elimination of polarization sensitivity in silica-based wavelength division multiplexer using a polyimide half waveplate," J. Lightwave Technol. **15**, 1947-1957 (1997)

34. S. Suzuki, S. Sumida, Y. Inoue, M. Ishii, and Y. Ohmori: "Polarization-insensitive arrayed-waveguide gratings using dopant-rich silica-based glass with thermal expansion asjusted to Si substrate," Electron. Lett. **33**, 1173–1174 (1997)

35. S. M. Ojha, C. G. Cureton, T. Bricheno, S. Day, D. Moule, A. J. Bell, and J. Taylor: "Simple method of fabricating polarisation-insensitive and very low crosstalk AWG grating devices," Electron. Lett. **34**, 78–79 (1998)

36. H. H. Yaffe, C. H. Henry, R. F. Kazarinov, and M. A. Milbrodt: "Polarization-independent silica-on-silicon Mach–Zehnder interferometers," J. Lightwave Technol. **12**, 64–67 (1994)

37. C. K. Nadler, E. K. Wildermuth, M. Lanker, W. Hunziker, and H. Melchior: "Polarization insensitive, low-loss, low crosstalk wavelength multiplexer modules," IEEE J. Select. Topics Quantum Electron. **5**, 1407–1412 (1999)

38. Y. Inoue, M. Itoh, Y. Hibino, A. Sugita, and A. Himeno: "Novel birefringence compensating AWG design," *Conf. Opt. Fiber Commun.* (OFC'01), Techn. Digest (Anaheim, CA, USA, 2001), paper WB4 (2001)

39. R. Kasahara, M. Itoh, Y. Hida, T. Saida, Y. Inoue, and Y. Hibino: "Novel polarization-insensitive AWG with undercladding ridge structure," *Integr. Photon. Res.* (IPR'02) Technical Digest (Vancouver, Canada, 2002), paper Ife5 (2002)

40. Y. Hida, Y. Hibino, H. Okazaki, and Y. Ohmori: "10 m-long silica-based waveguide with a loss of 1.7 dB/m," *Integrated Photonics Research* (IPR'95), Techn. Digest, (Dana Point, CA, USA, 1995), 49–51 (1995)

41. Y. Hibino, H. Okazaki, Y. Hida, and Y. Ohmori: "Propagation loss characteristics of long silica-based optical waveguides on 5-inch Si wafers," Electron. Lett. **29**, 1847-1848 (1993)

42. S. Suzuki, K. Shuto, H. Takahashi, and Y. Hibino: "Large scale and high-density planar lightwave circuits with high-Δ GeO$_2$-doped silica waveguides," Electron. Lett. **28**, 1863–1864 (1992)

43. Y. Hibino: "Recent advances in high-density and large-scale AWG multi/-demultiplexers with higher index-contrast silica-based PLCs," IEEE J. Select. Topics Quantum Electron. **8**, 1090–1101 (2002)

44. B. Little: "A VLSI photonics platform," *Opt. Fiber Commun. Conf.* (OFC'03). Techn. Digest (Atlanta, GA, USA, 2003), 444–445 (2003)

45. K. Okamoto, K. Moriwaki, and S. Suzuki: "Fabrication of 64 × 64 arrayed-waveguide grating multiplexer on silicon," Electron. Lett. **31**, 184–186 (1995)

46. K. Okamoto, K. Shuto, H. Takahashi, and Y. Ohmori: "Fabrication of 128-channel arrayed-waveguide grating multiplexer with 25 GHz spacing," Electron. Lett. **32**, 1474–1475 (1996)

47. H. Takahashi, I. Nishi, and Y. Hibino: "10 GHz spacing optical frequency division multiplexer based on arrayed-waveguide grating," Electron. Lett. **28**, 380–382 (1992)

48. K. Okamoto: *Fundamentals of Optical Waveguides* (Academic Press, San Diego, CA, 2000)

49. H. Uetsuka, L. L. Marra, K. Akiba, K. Morosawa, H. Okano, S. Takasugi, and K. Inaba: "Recent improvements in arrayed waveguide grating dense wavelength division multi/demultiplexers," *Proc. 8th Europ. Conf. Integr. Optics* (ECIO'97), Stockholm, Sweden, 76–79 (1997)

50. A. Sugita, A. Kaneko, K. Okamoto, M. Itoh, A. Himeno, and Y. Ohmori: "Very low insertion loss arrayed-waveguide grating with vertically tapered waveguides," J. Lightwave Technol. **12**, 1180–1182 (2000)

51. Y. P. Li. US Patent No. 5,745,618 (1998)

52. K. Takada, H. Yamada, and Y. Inoue: "Optical low coherence method for characterizing silica-based arrayed-waveguide grating multiplexers," J. Lightwave Technol. **14**, 1677–1689 (1996)

53. E. Pawlowski, B. Kuhlow, G. Przyrembel, and C. Warmuth: "Arrayed-waveguide grating demultiplexer with variable center frequency and transmission characteristic," *Proc. 9th Eur. Conf. on Integr. Opt.* (ECIO'99) Torino, Italy, 207–210 (1999)

54. S. Kamei, M. Ishii, T. Kitagawa, M. Itoh, and Y. Hibino: "64-channel ultra-low crosstalk arrayed-waveguide grating multi/demultiplexer module using cascade connection technique," Electron. Lett. **39**, 81–82 (2003)

55. Y. Hibino, Y. Hida, A. Kaneko, M. Ishii, M. Itoh, T. Goh, A. Sugita, T. Saida, A. Himeno, and Y. Ohmori: "Fabrication of silica-on-Si waveguide with higher index difference and its application to 256 channel arayed-waveguide multi/-demultiplexer," *Opt. Fiber Commun. Conf.* (OFC'2000), Techn. Digest, (Baltimore, MD, USA, 2000), 127–129 (2000)

56. Y. Hida, Y. Hibino, M. Itoh, A. Sugita, A. Himeno, and Y. Ohmori: "Fabrication of low-loss and polarisation-insensitive 256 channel arrayed-waveguide grating with 25 GHz spacing using 1.5% Δ waveguides," Electron. Lett. **36**, 820–821 (2000)

57. Y. Hida, Y. Hibino, T. Kitoh, Y. Inoue, T. Shibata, and A. Himeno: "400-channel 25-GHz spacing arrayed-waveguide grating covering a full range of C- and L-bands," *Opt. Fiber Commun. Conf.* (OFC'01), Techn. Digest (Anaheim, CA, USA, 2001), paper WB2 (2001)

58. K. Takada, H. Yamada, and K. Okamoto: "320-channel multiplexer consisting of 100 GHz-spaced parent AWG and 10 GHz-spaced subsidiary AWGs," Electron. Lett. **35**, 824–826 (1999)

59. K. Takada, M. Abe, T. Shibata, M. Ishii, Y. Inoue, H. Yamada, Y. Hibino, and K. Okamoto: "10 GHz spaced 1010-channel AWG filters achieved by tandem connection of primary and secondary AWGs," *Proc. 26th Europ. Conf. Opt. Commun.* (ECOC'2000), Munich, Germany, paper PD3.8 (2000)

60. K. Takada, M. Abe, T. Shibata, and K. Okamoto: "5 GHz-spaced 4200-channel two-stage tandem demultiplexer for ultra-multi-wavelength light source using supercontinuum generation," Electron. Lett. **38**, 572–573 (2002)

61. K. Takada, M. Abe, T. Shibata, and K. Okamoto: "1-GHz-spaced 16-channel arrayed-waveguide grating for a wavelength reference standard in DWDM network systems," J. Lightwave Technol. **20**, 850–853 (2002)

62. R. Adar, C. H. Henry, C. Dragone, R. C. Kistler, and M. A. Milbrodt: "Broadband array multiplexers made with silica waveguides on silicon," J. Lightwave Technol. **11**, 212-219 (1993)

63. Y. Inoue, A. Kaneko, F. Hanawa, H. Takahashi, K. Hattori, and S. Sumida: "Athermal silica-based arrayed-waveguide grating multiplexer," Electron. Lett. **33**, 1945–1947 (1997)

64. A. Kaneko, S. Kamei, Y. Inoue, H. Takahashi, and A. Sugita: "Athermal silica-based arrayed-waveguide grating (AWG) multiplexers with new low loss groove design," Electron. Lett. **36**, 318–319 (2000)

65. D. Kim, Y. Han, J. Shin, S. Park, Y. Park, H. Sung, S. Lee, Y. Lee, and D. Kim: "Suppression of temperature and polarization dependence by polymer overcladding in silica-based AWG multiplexer," *Opt. Fiber Commun. Conf.* (OFC'03), Techn. Digest (Atlanta, GA, USA, 2003), paper MF50 (2003)

66. G. Heise, H. W. Schneider, and P. C. Clemens: "Optical phased array filter module with passively compensated temperature dependence," *Proc. 24th Europ. Conf. Opt. Commun.* (ECOC'98), Madrid, Spain, 319–320 (1998)

67. T. Saito, K. Nara, Y. Nekado, J. Hasegawa, and K. Kashihara: "100 GHz-32ch athermal AWG with extremely low temperature dependency of center wavelength," *Opt. Fiber Comm. (OFC '03)* vol. **1**, Techn. Digest (Atlanta, Georgia, USA, 2003), pp. MF47 pp. 57–59 (2003)

68. T. Chiba, H. Arai, K. Ohira, H. Nonen, H. Okano, and U. Uetsuka: "Novel achitecture of wavelength interleaving filter with Fourier transorm-based MZIs," *Opt. Fiber Commun. Conf.* (OFC'01), Techn. Digest (Anaheim, CA, USA, 2001), paper WB5 (2001)

69. M. Oguma, T. Kitoh, K. Jinguji, T. Shibata, A. Himeno, and Y. Hibino: "Flat-top and low loss WDM filter composed of lattice-form interleave filter and arrayed-waveguide grating on one chip," *Opt. Fiber Commun. Conf.* (OFC'01), Techn. Digest (Anaheim, CA, USA, 2001), paper WB3 (2001)

70. K. Okamoto, K. Takiguchi, and Y. Ohmori: "16-channel optical add/drop multiplexer using silica-based arrayed-waveguide gratings," Electron. Lett. **31**, 723–724 (1995)

71. K. Okamoto, M. Okuno, A. Himeno, and Y. Ohmori: "16-channel optical add/-drop multiplexer consisting of arrayed waveguide gratings and double-gate switches," Electron. Lett. **32**, 1471–1472 (1996)

72. T. Saida, T. Goh, M. Okuno, A. Himeno, K. Takiguchi, and K. Okamoto: "Athermal silica-based optical add/drop multiplexer consisting of arrayed waveguide gratings and double gate thermo-optical switches," Electron. Lett. **36**, 528–529 (2000)

73. W. Chen, Z. Zhu, and Y. Chen: "Monolithically integrated 32-channel client reconfigurable optical add/drop multiplexer on planar lightwave circuit," *Opt. Fiber Commun. Conf.* (OFC'03), Techn. Digest (Atlanta, GA, USA, 2003), paper TuE5 (2003)

74. W. Chen, Z. Zhu, Y. J. Chen, J. Sun, B. Grek, and K. Schmidt: "Monolithically integrated $32\times$ four-channel client reconfigurable optical add/drop multiplexer on planar lightwave circuit," IEEE Photon. Technol. Lett. **15**, 1413–1415 (2003)

75. C. R. Doerr, L. W. Stulz, M. Cappuzzo, E. Laskowski, A. Paunescu, L. Gomez, J. V. Cates, S. Shunk, and A. E. White: "40-wavelength add-drop filter," IEEE Photon. Technol. Lett. **11**, 1437–1439 (1999)

76. J. E. Ford and J. A. Walker: "Dynamic spectral power equalization using micro-opto-mechanics," IEEE Photon. Technol. Lett. **10**, 1440–1442 (1998)

77. S. H. Huang, X. Y. Zou, S.-M. Hwang, A. E. Willner, Z. Bao, and D. A. Smith: "Experimental demonstration of dynamic network equalization of three 2.5-Gb/s WDM channels over 1000 km using acoustooptic tunable filters," IEEE Photon. Technol. Lett. **8**, 1243-1245 (1996)

78. K. Inoue, T. Kominato, and H. Toba: "Tunable gain equalization using a Mach–Zehnder optical filter in multistage fiber amplifiers," IEEE Photon. Technol. Lett. **3**, 718–720 (1991)

79. M. Zirngibl, C. H. Joyner, and B. Glance: "Digitally tunable channel dropping filter/equalizer based on waveguide grating router and optical amplifier integration," IEEE Photon. Technol. Lett. **6**, 513–515 (1994)

80. C. R. Doerr, C. H. Joyner, and L. W. Stulz: "Integrated WDM dynamic power equalizer with potentially low insertion loss," IEEE Photon. Technol. Lett. **10**, 1443–1445 (1998)

81. A. M. Vengsarkar, A. E. Miller, and W. A. Reed: "Highly efficient single-mode fiber for broadband dispersion compensation," *Opt. Fiber Commun. Conf.* (OFC'93) *Vol. 4*, Techn. Digest (San Jose, CA, USA, 1993), post-deadline paper PD13 (1993)

82. K. O. Hill, S. Theriault, B. Malo, F. Bilodeau, T. Kitagawa, D. D. Johnson, J. Albert, K. Takiguchi, T. Kataoka, and K. Hagimoto: "Chirped in-fiber Bragg grating dispersion compensators; linearization of dispersion characteristics and demonstration of dispersion compensation in 100 km, 10 Gbit/s optical fiber link," Electron. Lett. **30**, 1755–1756 (1994)

83. K. Takiguchi, S. Kawanishi, H. Takara, K. Okamoto, K. Jinguji, and Y. Ohmori: "Higher order dispersion equaliser of dispersion shifted fibre using a lattice-form programmable optical filter," Electron. Lett. **32**, 755–757 (1996)

84. A. Kaneko, T. Goh, H. Yamada, T. Tanaka, and I. Ogawa: "Design and application of silica-based planar lightwave circuits," IEEE J. Select. Topics Quantum Electron. **5**, 1227–1235 (1999)

85. U. Hilbk, T. Hermes, J. Saniter, and F.-J. Westphal: "High capacity WDM overlay on a passive optical network," Electron. Lett. **32**, 2162–2163 (1996)

86. C. R. Giles, R. D. Feldman, T. H. Wood, M. Zirngibl, G. Raybon, T. Strasser, L. Stulz, A. McCormick, C. H. Joyner, and C. R. Doerr: "Access PON using downstream 1550 nm WDM routing and upstream through a fiber-grating router," IEEE Photon. Technol. Lett. **8**, 1549–1551 (1996)

87. Y. P. Li, L. G. Cohen, C. H. Henry, E. J. Laskowski, and M. A. Capuzzo: "Demonstration and application of a monolithic two-PONs-in-one device," *Proc. 22nd Europ. Conf. Opt. Commun.* (ECOC'96) Oslo, Norway, paper TuC.3.4 (1996)

88. M. Zirngibl, C. R. Doerr, and C. H. Joyner: "Demonstration of a splitter/router based on a chirped waveguide grating router," IEEE Photon. Technol. Lett. **10**, 87–89 (1998)

89. G. Przyrembel, B. Kuhlow, E. Pawlowski, M. Ferstl, W. Fürst, H. Ehlers, and R. Steingrüber: "Multichannel 1.3 µm/1.55 µm AWG multiplexer/demultiplexer for WDM-PONs," Electron. Lett. **34**, 263–264 (1998)

90. B. Kuhlow, G. Przyrembel, E. Pawlowski, M. Ferstl, and W. Fürst: "AWG-based device for a WDM overlay PON in the 1.5-µm band," IEEE Photon. Technol. Lett. **11**, 218–220 (1999)

91. G. Przyrembel and B. Kuhlow: "AWG based device for a WDM/PON overlay in the 1.5 µm fiber transmission window," in *Opt. Fiber Commun. Conf.* (OFC'99), Techn. Digest, (San Diego, CA, USA, 1999), 207–209 (1999)

92. B. Kuhlow and G. Przyrembel: "WDM/PON overlay devices in silica technology for the 1.5 µm band," *Proc. 9th Europ. Conf. Integr. Optics* (ECIO'99), Torino, Italy, 215–218 (1999)

93. G. Przyrembel and B. Kuhlow: "AWG-based devices for a WDM overlay PON," in *WDM Components,* Chap. 29, pp. 87–94 (Opt. Soc. America TOP, 2005), ISBN 1-55752-610-9

94. B. Kuhlow and G. Przyrembel: "Silica based PON-AWG with 2×8 up- and downstream channels," *Opt. Fiber Commun. Conf.* (OFC'02), Techn. Digest (Anaheim, CA, USA, 2002), 695-697 (2002)

95. B. Kuhlow and G. Przyrembel: "Device for superposing optical signals with different wavelengths," US-Patent No. 6 347 166.

96. T. R. Hayes, M. A. Dreisbach, P. M. Thomas, W. C. Dautremont-Smith, and L. A. Heimbrook: "Reactive ion etching of InP using CH_4/H_2 mixtures: Mechanisms of etching and anisotropy," J. Vac. Sci. Technol. B. **7**, 1130–1140 (1989)

97. Y. S. Oei, L. H. Spiekman, F. H. Groen, I. Moerman, E. G. Metaal, and J. W. Pedersen: "Novel RIE-process for high quality InP-based waveguide structures," *Proc. 7th Europ. Conf. Integr. Optics* (ECIO'95), Delft, The Netherlands, 205–208 (1995)

98. T. Yoshikawa, S. Kohmoto, M. Anan, N. Hamao, M. Baba, N. Takado, Y. Sugimoto, M. Sugimoto, and K. Asakawa: "Chlorine-based smooth reactive ion beam etching of indium-containing III-V compound semiconductor," Jpn. J. Appl. Phys. **31**, 4381–4386 (1992)

99. M. Zirngibl, C. Dragone, and C. H. Joyner: "Demonstration of a 15×15 arrayed waveguide multiplexer on InP," IEEE Photon. Technol. Lett. **4**, 1250–1253 (1992)

100. M. Kohtoku, H. Sanjoy, S. Oku, Y. Kadota, Y. Yoshikuni, and Y. Shibata: "InP-based 64-channel arrayed waveguide grating with 50 GHz channel spacing and up to −20 dB crosstalk," Electron. Lett. **33**, 1786–1787 (1997)

101. C. G. P. Herben, X. J. M. Leijtens, F. H. Groen, I. Moerman, and M. K. Smit: "Ultra-compact polarisation independent PHASAR demultiplexer," *Proc. 24th Europ. Conf. Opt. Commun.* (ECOC'98), Madrid, Spain, 125–126 (1998)

102. H. Bissessur, P. Pagnod-Rossiaux, R. Mestric, and B. Martin: "Extremely small polarization independent phased-array demultiplexers on InP," IEEE Photon. Technol. Lett. **8**, 554–556 (1996)

103. C. G. P. Herben, X. J. M. Leijtens, F. H. Groen, and M. K. Smit: "Low-loss and compact phased-array demultiplexers using a double etch process," *Proc. 9th Europ. Conf. Integr. Optics* (ECIO'99), Torino, Italy, 211–214 (1999)

104. C. A. M. Steenbergen, C. van Dam, A. Looijen, C. G. P. Herben, M. de Kok, M. K. Smit, J. W. Pedersen, I. Moerman, R. G. Baets, and B. H. Verbeek:

"Compact low-loss 8×10 GHz polarization independent WDM receiver," *Proc. 22nd Europ. Conf. Opt. Commun.* (ECOC'96), Oslo, Norway, 1.129–1.131 (1996)

105. M. Zirngibl, C. H. Joyner, and P. C. Chou: "Polarisation compensated waveguide grating router on InP," Electron. Lett. **31**, 1662–1664 (1995)

106. C. G. M. Vreeburg, C. G. P. Herben, X. J. M. Leijtens, M. K. Smit, F. H. Groen, J. J. G. M. van der Tol, and P. Demeester: "A low-loss 16-channel polarization dispersion-compensated PHASAR demultiplexer," IEEE Photon. Technol. Lett. **10**, 382–384 (1998)

107. H. Tanobe, Y. Kondo, Y. Kadota, K. Okamoto, and Y. Yoshikuni: "Temperature insensitive arrayed waveguide gratings on InP substrates," IEEE Photon. Technol. Lett. **10**, 235–237 (1998)

108. Y. Barbarin, X. J. M. Leijtens, E. A. J. M. Bente, C. M. Louzao, J. R. Kooiman, and M. K. Smit. "Extremely small AWG demultiplexer fabricated on InP by using a double-etch process," IEEE Photon. Technol. Lett. **16**, 2478–2480 (2004)

109. M. Kohtoku, Y. Shibata, and Y. Yoshikuni: "Evaluation of the rejection ratio of an MMI-based higher order mode filter using optical low-coherence reflectometry," IEEE Photon. Technol. Lett. **14**, 968–970 (2002)

110. N. Kikuchi, Y. Shibata, H. Okamoto, Y. Kawaguchi, S. Oku, Y. Kondo, and Y. Tohmori: "Monolithically integrated 100-channel WDM channel selector employing low-crosstalk AWG," IEEE Photon. Technol. Lett. **16**, 2481–2483 (2004)

111. M. Kohtoku, T. Hirono, Member, S. Oku, Y. Kadota, Y. Shibata, and Y. Yoshikuni: "Control of higher order leaky modes in deep-ridge waveguides and application to low-crosstalk arrayed waveguide gratings," J. Lightwave Technol. **22**, 499–508 (2004)

112. M. G. Young, U. Koren, B. I. Miller, M. Chien, T. L. Koch, D. M. Tennant, K. Feder, K. Dreyer, and G. Raybon: "Six wavelength laser array with integrated amplifier and modulator," Electron. Lett. **31**, 1835–1836 (1995)

113. C. E. Zah, F. J. Favire, B. Pathak, R. Bhat, C. Caneau, P. S. D. Lin, A. S. Gozdz, N. C. Andreadakis, M. A. Koza, and T. P. Lee: "Monolithic integration of multi-wavelength compressive-strained multiquantum-well distributed-feedback laser array with star coupler and optical amplifiers," Electron. Lett. **28**, 2361–2362 (1992)

114. R. Nagarajan, C. H. Joyner, R. P. Schneider, Jr., J. S. Bostak, T. Butrie, A. G. Dentai, V. G. Dominic, P. W. Evans, M. Kato, M. Kauffman, D. J. H. Lambert, S. K. Mathis, A. Mathur, R. H. Miles, M. L. Mitchell, M. J. Missey, S. Murthy, A. C. Nilsson, F. H. Peters, S. C. Pennypacker, J. L. Pleumeekers, R. A. Salvatore, R. K. Schlenker, R. B. Taylor, H.-S. Tsai, M. F. van Leeuwen, J. Webjorn, M. Ziari, D. Perkins, J. Singh, S. G. Grubb, M. S. Reffle, D. G. Mehuys, F. A. Kish, and D. F. Welch: "Large-scale photonic integrated circuits," IEEE J. Select. Topics Quantum Electron. **11**, 50–65 (2005)

115. M. Zirngibl and C. H. Joyner: "12 frequency WDM laser based on a transmissive waveguide grating router," Electron. Lett. **30**, 701–702 (1994)

116. M. Zirngibl, B. Glance, L. W. Stulz, C. H. Joyner, G. Raybon, and I. P. Kaminow: "Characterization of a multiwavelength waveguide grating router laser," IEEE Photon. Technol. Lett. **6**, 1082–1084 (1994)

117. C. H. Joyner, M. Zirngibl, and J. C. Centanni: "An 8-channel digitally tunable transmitter with electroabsorption modulated output by selective area epitaxy," IEEE Photon. Technol. Lett. **7**, 1013–1015 (1995)

118. A. A. M. Staring, L. H. Spiekman, J. J. M. Binsma, E. J. Jansen, T. van Dongen, P. A. J. Thijs, M. K. Smit, and B. H. Verbeek: "A compact nine-channel multi-wavelength laser," IEEE Photon. Technol. Lett. **8**, 1139–1141 (1996)

119. R. Monnard, A. K. Srivastava, C. R. Doerr, C. H. Joyner, L. W. Stulz, M. Zirngibl, Y. Sun, J. W. Sulhoff, J. L. Zyskind, and C. Wolf: "16-channel 50 GHz channel spacing long-haul transmitter for DWDM systems," Electron. Lett. **34**, 765–767 (1998)

120. C. R. Doerr, C. H. Joyner, and L. W. Stulz: "40-wavelength rapidly digitally tunable laser," IEEE Photon. Technol. Lett. **11**, 1348–1350 (1999)

121. D. Van Thourhout, L. Zhang, W. Yang, B. I. Miller, N. J. Sauer, and C. R. Doerr: "Compact digitally tunable laser," IEEE Photon. Technol. Lett. **15**, 182–184 (2003)

122. J. H. den Besten, R. G. Broeke, M. van Geemert, J. J. M. Binsma, F. Heinrichs-dorff, T. van Dongen, E. A. J. M. Bente, X. J. M. Leijtens, and M. K. Smit: "An integrated coupled-cavity 16-wavelength digitally tunable laser," IEEE Photon. Technol. Lett. **14**, 1653–1655 (2002) erratum: **15**, 353 (2003)

123. J. H. den Besten, R. G. Broeke, M. van Geemert, J. J. M. Binsma, F. Heinrichs-dorff, T. van Dongen, E. A. J. M. Bente, X. J. M. Leijtens, and M. K. Smit: "A compact digitally tunable seven-channel ring laser," IEEE Photon. Technol. Lett. **14**, 753–755 (2002)

124. E. A. J. M. Bente, Y. Barbarin, J. H. den Besten, M. K. Smit, and J. J. M. Binsma: "Wavelength selection in an integrated multiwavelength ring laser," IEEE J. Quantum Electron. **40**, 1208-1216 (2004)

125. T. Miyazaki, N. Edagawa, S. Yamamoto, and S. Akiba: "A multiwavelength fiber ring-laser employing pair of silica-based arrayed-waveguide-gratings," IEEE Photon. Technol. Lett. **9**, 910–912 (1997)

126. N. Pleros, C. Bintjas, M. Kalyvas, G. Theophilopoulos, K. Vlachos, and H. Avramopoulos: "50 channel and 50 GHz multiwavelength laser source," *Proc. 27*th *Europ. Conf. Opt. Commun.* (ECOC'01), Amsterdam, The Netherlands, 410–411 (2001)

127. M. Nikoufard, X. J. M. Leijtens, Y. C. Zhu, T. J. J. Kwaspen, E. A. J. M. Bente, and M. K. Smit: "An 8×25 GHz polarization-independent integrated multi-wavelength receiver," *Integrated Photonics Research* (IPR'04), Techn. Digest (San Francisco, CA, USA, 2004), paper IThB2 (2004)

128. M. R. Amersfoort, J. B. D. Soole, H. P. Leblanc, N. C. Andreakakis, A. Rajhel, and C. Caneau: "8×2 nm polarization-independent WDM detector based on compact arrayed waveguide demultiplexer," *Integrated Photonics Research* (IPR'95), Techn. Digest (Dana Point, CA, USA, 1995), post-deadline paper PD3 (1995)

129. M. Zirngibl, C. H. Joyner, and L. W. Stulz: "WDM receiver by monolithic integration of an optical preamplifier, waveguide grating router and photodiode array," Electron. Lett. **31**, 581–582 (1995)

130. W. Tong, V. M. Menon, X. Fengnian, and S. R. Forrest: "An asymmetric twin waveguide eight-channel polarization-independent arrayed waveguide grating

with an integrated photodiode array," IEEE Photon. Technol. Lett. **16**, 1170–1172 (2004)

131. S. Chandrasekhar, M. Zirngibl, A. G. Dentai, C. H. Joyner, F. Storz, C. A. Burrus, and L. M. Lunardi: "Monolithic eight-wavelength demultiplexed receiver for dense WDM applications," IEEE Photon. Technol. Lett. **7**, 1342–1344 (1995)

132. R. Mestric, C. Porcheron, B. Martin, F. Pommereau, I. Guillemont, F. Gaborit, C. Fortin, J. Rotte, and M. Renaud: "Sixteen-channel wavelength selector monolithically integrated on InP," *Opt. Fiber Commun. Conf.* (OFC'2000), Techn. Digest (Baltimore, MD, USA, 2000), 81–83 (2000)

133. N. Kikuchi, Y. Shibata, H. Okamoto, Y. Kawaguchi, S. Oku, H. Ishii, Y. Yoshikuni, and Y. Tohmori: "Monolithically integrated 64-channel WDM channel selector with novel configuration," Electron. Lett. **38**, 331–332 (2002)

134. D. Van Thourhout, P. Bernasconi, B. Miller, W. Yang, L. Zhang, N. Sauer, L. Stulz, and S. Cabot: "Novel geometry for an integrated channel selector," IEEE J. Select. Topics Quantum Electron. **8**, 1211–1214 (2002)

135. C. G. M. Vreeburg, T. Uitterdijk, Y. S. Oei, M. K. Smit, F. H. Groen, E. G. Metaal, P. Demeester, and H. J. Frankena: "First InP-based reconfigurable integrated add-drop multiplexer," IEEE Photon. Technol. Lett. **9**, 188–190 (1997)

136. C. G. P. Herben, X. J. M. Leijtens, M. R. Leys, F. H. Groen, and M. K. Smit: "Extremely compact WDM cross connect on InP," *Proc. IEEE/LEOS Symposium (Benelux Chapter)*, Delft, The Netherlands, 17–20 (2000)

137. T. Watanabe, Y. Inoue, A. Kaneko, N. Ooba, and T. Kurihara: "Polymeric arrayed-waveguide grating multiplexer with wide tuning range," Electron. Lett. **33**, 1547–1548 (1997)

138. J. Kobayashi, Y. Inoue, T. Matsuura, and T. Maruno: "Tunable and polarization-insensitive arrayed-waveguide grating multiplexer fabricated from fluorinated polyimides," IEICE Trans. Electron. **E81-C**, 1020–1026 (1998)

139. J.-H. Ahn, H.-J. Lee, W.-Y. Hwang, M.-C. Oh, M. H. Lee, S. G. Han, H.-G. Kim, and C. H. Yim: "Polymeric 1×16 arrayed waveguide grating multiplexer using fluorinated poly(acrylene ethers) at 1550 nm," IEICE Trans. Electron. **E82-B**, 406–408 (1999)

140. C. L. Callender, J.-F. Viens, J. P. Noad, and L. Eldada: "Compact low-cost tunable acrylate polymer arrayed-waveguide grating multiplexer," Electron. Lett. **35**, 1839–1840 (1999)

141. N. Keil, H. H. Yao, C. Zawadski, J. Bauer, M. Bauer, C. Dreyer, and J. Schneider: "Athermal all-polymer arrayed waveguide grating multiplexer," Electron. Lett. **37**, 579–580 (2001)

142. N. Keil, H. H. Yao, C. Zawadzki, and B. Strebel: "Re-arrangeable nonblocking polymer waveguide thermo-optic 4×4 switching matrix with low power consumption at 1.55 μm," Electron. Lett. **31**, 403–404 (1995)

143. N. Keil, C. Weinert, W. Wirges, H. H. Yao, C. Zawadzki, J. Schneider, J. Bauer, M. Bauer, K. Lösch, K. Satzke, W. Wischmann, and J. v. Wirth: "Thermo-optic switches using vertically coupled polymer/silica waveguides," *Proc. 26th Europ. Conf. Opt. Commun.* (ECOC'2000), Munich, Germany, 117–120 (2000)

144. G. L. Bona, E. Flück, F. Horst, B. J. Offrein, H. W. M. Salemink, and R. Germann: "Flexible thermo-optical switch in SiON planar waveguides," *Proc. 9th Europ. Conf. Integr. Optics* (ECIO'99), Torino, Italy, 65–68 (1999)

145. T. Shimoda, K. Suzuki, S. Takaesu, and A. Furukawa: "Low-loss, polarization-independent silicon-oxynitride waveguides for high-density integrated planar lightwave circuits," *Proc. 28th Europ. Conf. Opt. Commun.* (ECOC'02), Copenhagen, Denmark, paper 4.2.2 (2002)

146. F. Horst, R. Beyeler, G. L. Bona, E. Flück, R. Germann, B. Offrein, H. Salemink, and D. Wiesmann: "Compact, tunable optical devices in silicon-oxynitride waveguide technology," *Integrated Photonics Research* (IPR'2000), Techn. Digest (Quebec, Canada, 2000), paper IThF1 (2000)

147. B. Schauwecker, M. Arnold, C. Radehaus, G. Przyrembel, and B. Kuhlow: "Optical waveguide components with high refractive index difference in silicon-oxynitride for application in integrated optoelectronics," Opt. Engineering **41**, 237–243 (2002)

148. B. Schauwecker, G. Przyrembel, B. Kuhlow, and C. Radehaus: "Small-size silicon-oxynitride AWG demultiplexer operating around 725 nm," IEEE Photon. Technol. Lett. **12**, 1645–1646 (2000)

149. A. P. Harpin: "Integrated optics in silicon: coming of age," Silicon-based monolithic and hybrid optoelectronic devices," *Proc. SPIE* **3007**, 128–135 (1997)

150. C. Z. Zhao, G. Z. Li, E. K. Liu, Y. Gao, and X. D. Liu: "Silicon on insulator Mach–Zehnder waveguide interferometer operating at 1.3 μm," Appl. Phys. Lett. **67**, 2448–2449 (1995)

151. I. Day, I. Evans, A. Knights, F. Hopper, S. Roberts, J. Johnston, S. Day, J. Luff, H. Tsang, and M. Asghari: "Tapered silicon waveguides for low insertion loss highly-efficient high-speed electronic variable optical attenuators," *Opt. Fiber Commun. Conf.* (OFC'03), Techn. Digest (Atlanta, GA, USA, 2003), 249–251 (2003)

152. U. Fischer, T. Zinke, B. Schüppert, and K. Petermann: "Singlemode optical switches based on SOI waveguides with large cross-section," Electron. Lett. **30**, 406–408 (1994)

153. R. A. Soref, J. Schmidtchen, and K. Petermann: "Large single-mode rib waveguides in GeSi-Si and Si-on-SiO$_2$," IEEE J. Quantum Electron. **27**, 1971–1974 (1991)

154. T. Aalto, M. Kapulainen, S. Yliniemi, P. Heimala, and M. Leppihalme: "Fast thermo-optical switch based on SOI waveguides," *Photonics West, Proc. SPIE: Integrated Optics: Devices, Materials, and Technologies VII*, vol. **4987**, 149–159 (2003)

155. A. House, R. Whiteman, L. Kling, S. Day, A. Knights, F. H. D. Hogan, and M. Asghari: "Silicon waveguide integrated optical switching with microsecond switching speed," *Opt. Fiber Commun. Conf.* (OFC'03), Techn. Digest (Atlanta, GA, USA, 2003), vol. **2**, 449–450 (2003)

156. I. E. Day, S. W. Roberts, R. O. Caroll, A. Knights, P. Sharp, G. F. Hopper, B. J. Luff, and M. Asghari: "Single-chip variable optical attenuator and multiplexer subsystem integration," *Opt. Fiber Commun. Conf.* (OFC/IOOC'02), Techn. Digest (Anaheim, CA, USA, 2002), 72–73 (2002)

157. A. Knights, A. House, R. MacNaughton, and F. Hopper: "Optical power monitoring function compatible with single chip integration on silicon-on-insulator," *Opt. Fiber Commun. Conf.* (OFC'03), Techn. Digest (Atlanta, GA, USA, 2003), vol. **2**, 705–706 (2003)

158. B. Jalali, S. Yegnanarayanan, T. Yoon, T. Yoshimoto, I. Rendina, and F. Coppinger: "Advances in silica-on-insulator optoelctronics," IEEE J. Select. Topics Quantum Electron. **4**, 938–947 (1998)
159. P. D. Trinh, S. Yegnanarayanan, F. Coppinger, and B. Jalali: "Silicon-on-insulator (SOI) phased-array wavelength multi-demultiplexer with extremely low-polarization sensitivity," IEEE Photon. Technol. Lett. **9**, 940–942 (1997)
160. G. T. Reed and A. P. Knights: *Silicon Photonics: An Introduction* (Wiley, New York, 2004)
161. A. Gerster, M. Maile, and C. Hentschel: "State-of-the-art characterization of optical components for DWDM applications," Agilent Technologies, Application Brief (2001)
162. *Handbook of Optics*, vol. **I** (McGraw-Hill, New York, 1994) Chap. 5 "Polarization," ISBN 007047740X.

5 Fibre Bragg Gratings

Andreas Othonos, Kyriacos Kalli, David Pureur and Alain Mugnier

5.1 Introduction

The discovery of fibre optics has revolutionized the field of telecommunications making possible high-quality, high-capacity, long distance telephone links. Over the past three decades the advancements in optical fibre have undoubtedly improved and reshaped fibre optic technology so that optical fibres plus related components have become synonymous with "telecommunication". In addition to applications in telecommunications, optical fibres are also utilized in the rapidly growing field of fibre sensors. Despite the improvements in optical fibre manufacturing and advancements in the field in general, it has remained challenging to integrate basic optical components such as mirrors, wavelength filters, and partial reflectors with fibre optics. Recently, however, all this has changed with the ability to alter the core index of refraction in a single-mode optical fibre by optical absorption of UV light. The photosensitivity of optical fibres allows the fabrication of phase structures directly into the fibre core, called *fibre Bragg gratings* (FBG), Fig. 5.1. Photosensitivity refers to a permanent change in the index of refraction of the fibre core when exposed to light with characteristic wavelength and intensity that depend on the core material. The fibre Bragg grating can perform many primary functions, such as reflection and filtering for example, in a highly efficient, low loss manner. This versatility has stimulated a number of significant innovations [1–3].

For a conventional fibre Bragg grating the periodicity of the index modulation has a physical spacing that is one half of the wavelength of light propagating in the waveguide (phase matching between the grating planes and incident light results in coherent back reflection). Reflectivities approaching 100% are possible, with the grating bandwidth tailored from typically 0.1 nm to more than tens of nanometres. These characteristics make Bragg gratings suitable for telecommunications, where they are used to reflect, filter or disperse light [1]. Fibre lasers capable of producing light at telecommunications windows utilize Bragg gratings for forming both, the high-reflectivity end mirror and output coupler to the laser cavity, resulting

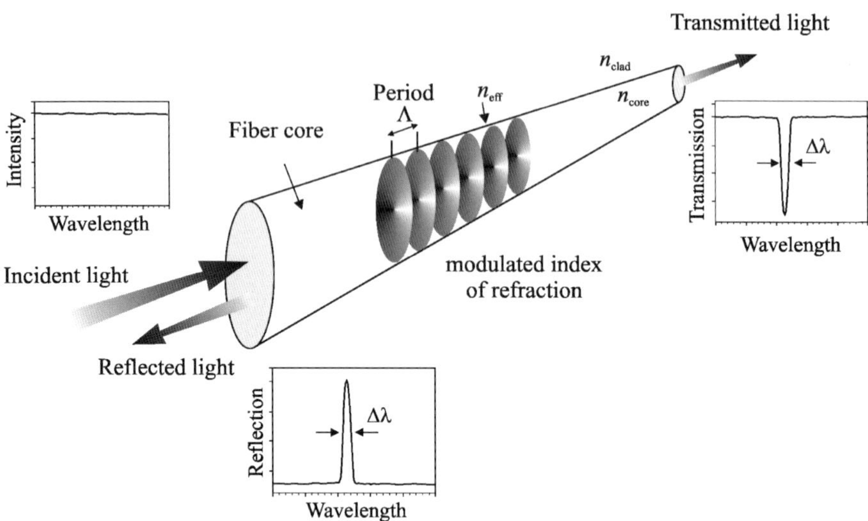

Fig. 5.1. Schematic representation of a Bragg grating inscribed into the core of an optical fibre. The period of the index of refraction variation is represented by Λ. A broadband light is coupled into the core of the fibre. Part of the input light is reflected (at the Bragg condition) and the rest is transmitted. The bandwidth of the reflected and transmitted light depends on the characteristics of the Bragg grating, its length and modulation depth

in an efficient and inherently stable source. Moreover, the ability of gratings with non-uniform periodicity to compress or expand pulses is particularly important to high-bit-rate, long-haul communication systems. Grating-based tuneable dispersion compensator devices can be used to alleviate non-linear signal distortion resulting from optical power variations. A multitude of grating-based transmission experiments has been reported [4], including 10 Gbit/s over 400 km of non-dispersion-shifted fibre with fixed dispersion compensation using chirped fibre Bragg gratings [5]. Power penalties are routinely less than 1 dB [6, 7]. In optical systems with bit rates of 40 Gbit/s and higher, tuneable dispersion compensation becomes necessary to maintain system performance. Tuneable chirp devices through uniform tuning of nonlinearly chirped Bragg gratings [8], or a non-uniform gradient via temperature [9] or strain gradients [10–12] along the grating length have been demonstrated. Systems employing Bragg gratings have demonstrated in excess of 100 km–40 Gbit/s transmission [4]. Given that future systems will operate at bit rates of 160 Gbit/s accurate dispersion maps are required for the fibre network, furthermore at this bit rate there are no electronic alternatives and dispersion compensation must be all-optical and tuneable in nature. There are demonstrations to 100 km at this repetition rate using tuneable, chirped gratings [13]. Furthermore, the Bragg grating meets the demands of dense wavelength division multiplexing, which requires narrowband, wavelength-selective components, offering very high extinction between different information channels. Numerous applications exist for such low loss, fibre optic filters, examples of which are ASE noise suppression in amplified systems, pump recycling in fibre amplifiers, and soliton pulse control.

The grating planes are subject to temperature and strain perturbations, as is the host glass material, modifying the phase matching condition and leading to wavelength dependent reflectivity. Typically, at $1.5\,\mu$m, the wavelength-strain responsivity is ~ 1 pm/$\mu\varepsilon$, with a wavelength shift of about 10 to 15 pm/$^\circ$C for temperature excursions (strain ε defined as Δ-length/length). Therefore by tracking the wavelength at which the Bragg reflection occurs the magnitude of an external perturbation may be obtained. This functionality approaches the ideal goal of optical fibre sensors: to have an intrinsic in-line, fibre-core structure that offers an absolute readout mechanism. The reliable detection of sensor signals is critical and spectrally encoded information is potentially the simplest approach, offering simple decoding that may even be facilitated by another grating. An alternative approach is to use the grating as a reflective marker, mapping out lengths of optical fibre. Optical time domain measurements allow for accurate length or strain monitoring.

The grating may be photo-imprinted into the fibre core during the fibre manufacturing process, with no measurable loss to the mechanical strength of the host material. This makes it possible to place a large number of Bragg gratings at predetermined locations into the optical fibre to realize a quasi-distributed sensor network for structural monitoring, with relative ease and low cost. Importantly, the basic instrumentation applicable to conventional optical fibre sensor arrays may also incorporate grating sensors, permitting the combination of both sensor types. Bragg gratings are ideal candidates for sensors, measuring dynamic strain to nε-resolution in aerospace applications and as temperature sensors for medical applications. They also operate well in hostile environments such as high pressure, borehole-drilling applications, principally as a result of the properties of host glass material.

Fibre optic photosensitivity has indeed opened a new era in the field of fibre optic based devices [1], with innovative new Bragg grating structures finding their way into telecommunication and sensor applications. Devices like fibre *Fabry–Perot Bragg gratings* for band-pass filters, *chirped gratings* for dispersion compensation and pulse shaping in ultra-short work, and *blazed gratings* for mode converters are becoming routine applications. Fibre optics sensing is an area that has embraced Bragg gratings since the early days of their discovery, and most fibre optics sensor systems today make use of Bragg grating technology.

Within a few years from the initial development, fibre Bragg gratings have moved from laboratory interest and curiosity to implementation in optical communication and sensor systems. In a few years, it may be as difficult to think of fibre optic systems without fibre Bragg gratings as it is to think of bulk optics without the familiar laboratory mirror.

5.2 Fundamentals of Fibre Bragg Gratings

In this section we will describe in detail the various properties that are characteristic of fibre Bragg gratings and this will involve the discussion of a diverse range of topics. We will begin by examining the measurable wavelength-dependent properties, such as the reflection and transmission spectral profiles, for a number of simple and complex grating structures. The dependence of the grating wavelength response to externally applied perturbations, such as temperature and strain, is also investigated.

5.2.1 Simple Bragg Grating

A fibre Bragg grating consists of a periodic modulation of the refractive index in the core of a single-mode optical fibre. These types of uniform fibre gratings, where the phase fronts are perpendicular to the fibre's longitudinal axis with grating planes having constant period (Fig. 5.2), are

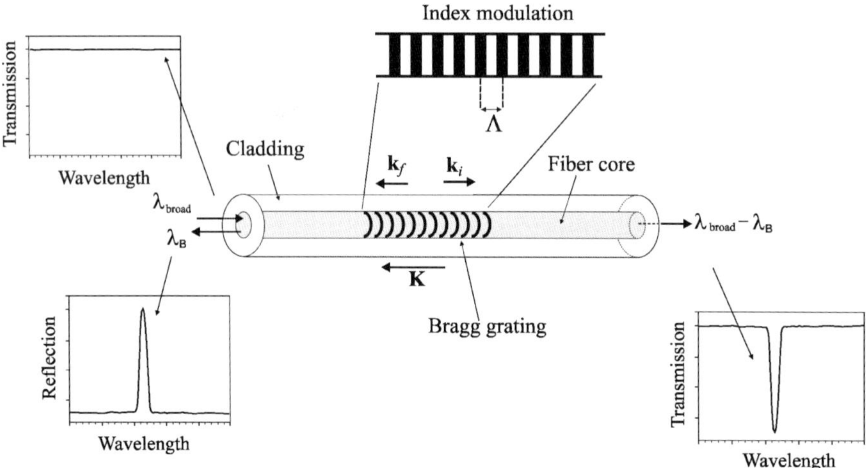

Fig. 5.2. Illustration of a uniform Bragg grating with constant index of modulation amplitude and period. Also shown are the incident, diffracted, and grating wave vectors that have to be matched for momentum conservation

considered the fundamental building blocks for most Bragg grating structures. Light guided along the core of an optical fibre will be scattered by each grating plane. If the Bragg condition is not satisfied, the reflected light from each of the subsequent planes becomes progressively out of phase and will eventually cancel out. Additionally, light that is not coincident with the Bragg wavelength resonance will experience very weak reflection at each of the grating planes because of the index mismatch; this reflection accumulates over the length of the grating. As an example, a 1 mm grating at 1.5 µm with a strong Δn of 10^{-3} will reflect ~0.05% of the off-resonance incident light at wavelengths sufficiently far from the Bragg wavelength. Where the Bragg condition is satisfied the contributions of reflected light from each grating plane add constructively in the backward direction to form a back-reflected peak with a centre wavelength defined by the grating parameters.

The Bragg grating condition is simply the requirement that satisfies both energy and momentum conservation. Energy conservation ($\hbar\omega_f = \hbar\omega_i$) requires that the frequency of the incident and the reflected radiation is the same. Momentum conservation requires that the wavevector of the incident wave, k_i, plus the grating wavevector, K, equal the wavevector of the scattered radiation k_f, which is simply stated as

$$k_i + K = k_f \qquad (5.1)$$

where the grating wavevector, K, has a direction normal to the grating planes with a magnitude $2\pi/\Lambda$ (Λ is the grating spacing shown in Fig. 5.2). The diffracted wavevector is equal in magnitude, but opposite in direction, to the incident wavevector. Hence the momentum conservation condition becomes

$$2\left(\frac{2\pi n_{eff}}{\lambda_B}\right) = \frac{2\pi}{\Lambda} \qquad (5.2)$$

which simplifies to the first order Bragg condition

$$\lambda_B = 2n_{eff}\Lambda \qquad (5.3)$$

where the Bragg grating wavelength, λ_B, is the free space centre wavelength of the input light that will be back-reflected from the Bragg grating, and n_{eff} is the effective refractive index of the fibre core at the free space centre wavelength.

5.2.2 Uniform Bragg Grating

Consider a uniform Bragg grating formed within the core of an optical fibre with an average refractive index n_0. The index of the refractive profile can be expressed as

$$n(z) = n_0 + \Delta n \cos\left(\frac{2\pi z}{\Lambda}\right) \tag{5.4}$$

where Δn is the amplitude of the induced refractive index perturbation (typically 10^{-5} to 10^{-3}) and z is the distance along the fibre longitudinal axis. Using coupled-mode theory [14] the reflectivity of a grating with constant modulation amplitude and period is given by the following expression

$$R(l,\lambda) = \frac{\kappa^2 \sinh^2(sl)}{\Delta\beta^2 \sinh^2(sl) + s^2 \cosh^2(sl)} \tag{5.5}$$

where $R(l, \lambda)$ is the reflectivity, which is a function of the grating length l and wavelength λ. κ is the coupling coefficient, $\Delta\beta = \beta - \pi/\Lambda$ is the detuning wavevector, $\beta = 2\pi n_0/\lambda$ is the propagation constant and finally $s^2 = \kappa^2 - \Delta\beta^2$. For sinusoidal variations of the index perturbation the coupling coefficient, κ, is given by

$$\kappa = \frac{\pi \Delta n}{\lambda} M_{power} \tag{5.6}$$

where M_{power} is the fraction of the fibre mode power contained by the fibre core. In the case where the grating is uniformly written through the core, M_{power} can be approximated by $1 - V^{-2}$, where V is the normalized frequency of the fibre, given by

$$V = (2\pi/\lambda)a\sqrt{n_{co}^2 - n_{cl}^2} \tag{5.7}$$

where a is the core radius, and n_{co} and n_{cl} are the core and cladding indices, respectively. At the centre wavelength of the Bragg grating the wavevector detuning is $\Delta\beta = 0$, therefore the expression for the reflectivity becomes

$$R(l,\lambda) = \tanh^2(\kappa l) \tag{5.8}$$

The reflectivity increases as the induced index of refraction change gets larger. Similarly, as the length of the grating increases, so does the resultant reflectivity. Figure 5.3 shows a calculated reflection spectrum as a function of wavelength of a uniform Bragg grating. The side lobes of the resonance are due to multiple reflections to and from opposite ends of the grating region. The sinc spectrum arises mathematically through the

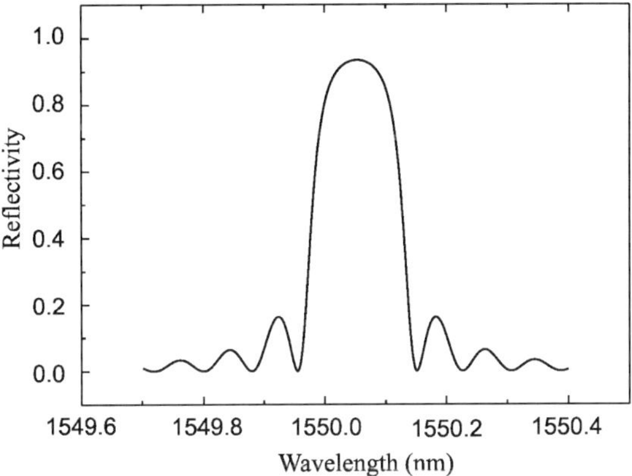

Fig. 5.3. Typical wavelength-dependent reflection spectrum of a Bragg grating with centre wavelength around 1550 nm

Fourier transform of a harmonic signal having finite extent, while an infinitely long grating would transform to an ideal delta function response in the wavelength domain.

A general expression for the approximate full-width-half maximum bandwidth of a grating is given by [15]

$$\Delta\lambda = \lambda_B S \sqrt{\left(\frac{\Delta n}{2n_0}\right)^2 + \left(\frac{1}{N}\right)^2} \tag{5.9}$$

where N is the number of grating planes. For strong gratings (with near 100% reflection) $S \approx 1$ holds, while $S \approx 0.5$ for weak gratings.

5.2.3 Phase and Group Delay of Uniform Gratings

Figure 5.4 shows the phase response of two uniform-period Bragg gratings (λ_B around 1550 nm) as a function of wavelength. The two gratings have the same length (1 cm), however, they have different index perturbation change, namely a "strong" grating with $\delta n_{eff} = 3 \times 10^{-4}$ and a "weak" grating with $\delta n_{eff} = 5 \times 10^{-5}$. It appears that the phase change around the Bragg wavelength decreases with higher index of refraction perturbation.

The group delay of the same two gratings is shown in Fig. 5.5. Strong dispersion (change of group delay with wavelength) is clearly seen at the edge of the stop band and it increases with increasing index perturbation change, although it is limited to a small bandwidth. The group delay is minimum at the centre of the band (see also Chap. 2).

The second term in (5.10) represents the effect of temperature on an optical fibre. A shift in the Bragg wavelength due to thermal expansion changes the grating spacing and the index of refraction. This fractional wavelength shift for a temperature change ΔT may be written as [16]

$$\Delta\lambda_B = \lambda_B \left(\alpha_\Lambda + \alpha_n \right) \Delta T \tag{5.12}$$

where $\alpha_\Lambda = (1/\Lambda)(\partial\Lambda/\partial T)$ is the thermal expansion coefficient for the fibre (approximately 0.55×10^{-6} for silica). The quantity $\alpha_n = (1/n)(\partial n/\partial T)$ represents the thermo-optic coefficient and its approximately equal to 8.6×10^{-6} for a germanium-doped, silica-core fibre. Clearly the index change is by far the dominant effect. From (5.12) the expected sensitivity for a ~1550 nm Bragg grating is approximately 14 pm/°C, in close agreement with the data shown in Fig. 5.6, which illustrate results of a Bragg grating centre wavelength shift as a function of temperature. It is apparent that any change in wavelength, associated with the action of an external perturbation to the grating, is the sum of strain and temperature terms. Therefore, in sensing applications where only one perturbation is of interest, the deconvolution of temperature and strain becomes necessary.

5.2.5 Other Properties of Fibre Gratings

When a grating is formed under conditions for which the modulated index change is saturated under UV exposure, then the effective length will be reduced as the transmitted signal is depleted by reflection. As a result, the spectrum will broaden appreciably and depart from a symmetric sinc or Gaussian shape spectrum, whose width is inversely proportional to the grating length. This is illustrated in Figs. 5.7 (a) and (b). In addition, the cosine-like shape of the grating will change into a waveform with steeper sides, and second-order Bragg lines (Fig. 5.7(c)) will appear due to the new harmonics in the Fourier spatial spectrum of the grating [17].

The presence of higher order grating modes has been utilised as a means of separating temperature and strain measurements using a single grating device, as the grating response to external perturbations is wavelength dependent [18]. Another interesting feature, which is observed in strongly reflecting gratings with large index perturbations, is the small-shape spectral resonance on the short wavelength side of the grating centre line. This is due to self-chirping from $\Delta n_{eff}(z)$. Such features do not occur if the average index change is held constant or adjusted to be constant by a second exposure of the grating. A Bragg grating will also couple dissimilar modes in reflection and transmission, provided the following two conditions are

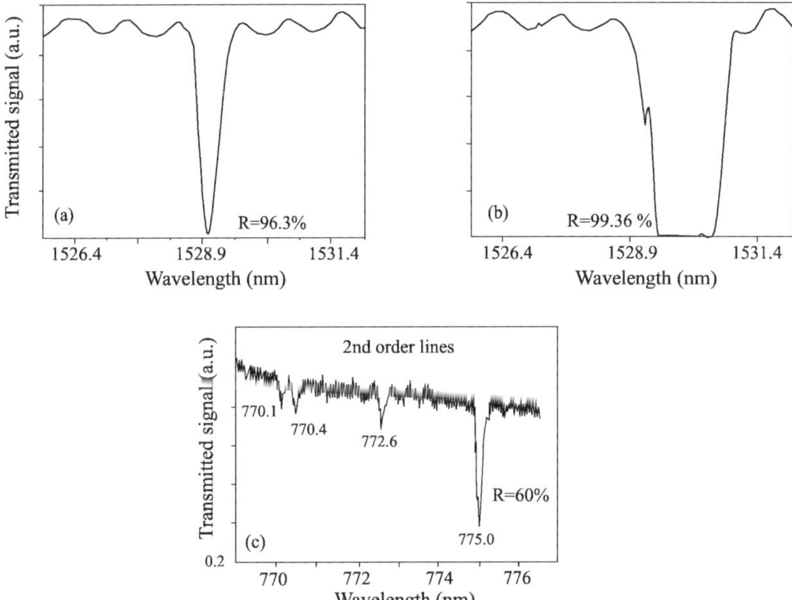

Fig. 5.7. Transmission of standard Bragg grating (**a**) and of Bragg grating with large index change due to saturation under UV exposure (**b**). The spectrum (**b**) broadens under continuous exposure because the incident wave is completely reflected before reaching the end of the grating. The strongly saturated grating is no longer sinusoidal, and the peak index regions are flattened, whereas the valleys in the perturbation index distribution are sharpened. As a result second order Bragg reflection lines (**c**) are observed at about one-half the fundamental Bragg wavelength and at other shorter wavelengths for higher order modes (after [17])

satisfied, namely phase matching and sufficient mode overlap in the region of the fibre that contains the grating. The phase matching condition, which ensures a coherent exchange of energy between the modes, is given by [17]

$$n_{eff} - \frac{\lambda}{\Lambda_z} = n_e \tag{5.13}$$

where n_{eff} is the modal index of the incident wave and n_e is the modal index of the grating-coupled reflected or transmitted wave. It should be pointed out that the above equation allows for a tilted or blazed grating by adjusting the grating pitch along the fibre axis Λ_z.

Normally Bragg gratings do not only reflect radiation into back-travelling guided modes, but also into cladding and radiation modes at wavelengths shorter than the Bragg wavelength. Since these modes are not guided, they are not observed in reflection, but in transmission only.

A schematic illustration of the combined effect of cladding and radiation mode coupling is given in Fig. 5.8.

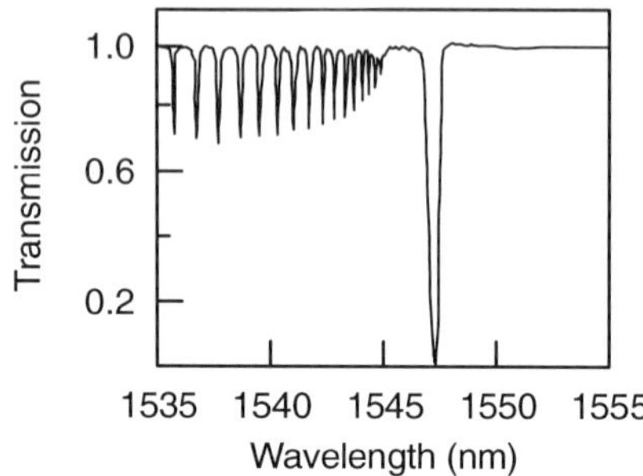

Fig. 5.8. Transmission of strong fibre Bragg grating (schematic) showing loss to radiation modes plus sharp lines due to coupling to distinct cladding modes

Such extra short-wavelength transmission structures are particularly pronounced in highly photosensitive fibres or in hydrogenated ones. In general these losses are unwanted and different methods have been developed in order to suppress them. One approach relies on having a uniform photosensitive region all over the cross-section plane of the optical fibre [19], an alternative variant is the use of fibres with high numerical aperture. For more details including specific references see [1].

5.2.6 Bragg Grating Types

Bragg gratings grow differently in response to particular inscription conditions and the laser used, in addition to the optical fibre type and photosensitivity conditioning prior to inscription. The gratings are characterized by four distinct dynamical regimes known as Type I, Type IA, Type IIA, and Type II. The key differences are highlighted below, bearing in mind that the mechanisms responsible for these types are different. The physical properties of these grating types can be inferred through their growth dynamics and by measurement of thermally induced decay. Broadly speaking Type IA are the least and Type II the most stable gratings with increasing temperature. This is not surprising given that Type IA appears to be a true colour centre grating and purely related to local electronic defects, Type I has both a colour centre and densification element, Type IIA is related to compaction, and Type II is related to fusion of the glass matrix.

Type I Fibre Bragg Gratings

Type I Bragg gratings refer to gratings that are formed in normal photosensitive fibres under moderate intensities. The growth dynamics of the Type I grating is characterized by a power law with time of the form $\Delta n \propto t^{\alpha}$ [20]. It is interesting to point out that the reflection spectra of the guided mode are complementary to the transmission signal, implying that there is negligible loss due to absorption or reflection into the cladding. This is a fundamental characteristic of a Type I Bragg grating. Furthermore, due to the photosensitivity type of the Bragg grating, the grating itself has a characteristic behaviour with respect to temperature erasure. Type I gratings can be erased at relatively low temperatures, approximately 200°C. Nevertheless, Type I gratings are the most utilized Bragg gratings and operate effectively from −40 to +80°C, a temperature range that satisfactorily covers most telecommunications and some sensor applications.

Type IA Fibre Bragg Gratings

Type IA fibre Bragg gratings are the most recently revealed grating type and may be considered a subtype of Type I gratings. (The transmission and reflection spectra are complementary, thus this grating type is indistinguishable from Type I in a static situation.) They are typically formed after prolonged UV exposure of a standard grating in hydrogenated germanosilicate fibre [21, 22], although recent improvements in their inscription have shown that they can be readily inscribed in a suitably prepared optical fibre [23]. The spectral characteristics of Type IA gratings are unique; they are distinct from other grating types as they exhibit a large increase in the mean core index that is identifiable as a large red shift seen in the Bragg wavelength λ_B of the grating during inscription. The mean wavelength change is characterised by three distinct regimes, with the Type I grating growth being superseded by a quasi-linear region followed by saturation. This saturated red shift is dependent on fibre type and hydrogenation conditions, but for a highly doped fibre (either high Ge dopant or B/Ge co-doped fibre) is typically in the order of 15–20 nm, and 5–8 nm for SMF-28 fibre. The maximum wavelength shift translates to an increase in the mean index of up to 2×10^{-2}. More importantly, IA gratings have been shown to exhibit the lowest temperature coefficient of all grating types reported to date, which makes them ideal for use in a temperature compensating, dual grating sensor, as has recently been demonstrated by Kalli and co-workers [24, 25]. Recent studies by Kalli et al. have also shown that their primary limitation of having to work at low temperatures (80°C) can be greatly mitigated if inscribed under strain (stability to 200°C) [26].

Type IIA Fibre Bragg Gratings

Type IIA fibre Bragg gratings appear to have the same spectral characteristics as Type I gratings. The transmission and reflection spectra are again complementary, also rendering this type of grating indistinguishable from Type I in a static situation. However, due to the different mechanism involved in fabricating these gratings, there are some distinguishable features that are noticeable under dynamic conditions either in the initial fabrication or in the temperature erasure of the gratings. Type IIA gratings are inscribed through a long process, following Type I grating inscription [27]. After approximately 30 min of exposure (depending on the fibre type and exposure fluence), the Type IIA grating is fully developed. Clearly, Type IIA gratings are not very practical to fabricate. Although the mechanism of the index change is different from Type I, occurring through compaction of the glass matrix, the behaviour subject to external perturbations is the same for both grating types. Irrespective of the subtleties of the index change on a microscopic scale, the perturbations act macroscopically and, therefore, the wavelength response remains the same. However, when the grating is exposed to high ambient temperature, a noticeable erasure is observed only at temperatures as high as 500°C. A clear advantage of the Type IIA gratings over the Type I is the dramatically improved temperature stability of the grating, which may prove very useful, if the system has to be exposed to high ambient temperatures (as may be the case for sensor applications).

Type II Fibre Bragg Gratings

A single excimer light pulse of fluence $>0.5\,\mathrm{J/cm^2}$ can photoinduce large refractive-index changes in small, localized regions at the core-cladding boundary, resulting in the formation of the Type II grating [28]. This change results from physical damage through localized fusion that is limited to the fibre core, and it produces very large refractive-index modulations estimated to be close to 10^{-2}. The reflection spectrum is broad and several features appear over the entire spectral profile due to non-uniformities in the excimer beam profile that are strongly magnified by the highly non-linear response mechanism of the glass core. Type II gratings pass wavelengths longer than the Bragg wavelength, whereas shorter wavelengths are strongly coupled into the cladding, as is observed for etched or relief fibre gratings, permitting their use as effective wavelength-selective taps. Results of stability tests have shown Type II gratings to be extremely stable at elevated temperatures [28], surviving temperatures in excess of 800°C for several hours; this superior temperature stability can be utilized for sensing applications in hostile environments.

5.3 Spectral Response from Bragg Gratings

Many models have been developed to describe the behaviour of Bragg gratings in optical fibres [1]. The most widely used technique has been coupled-mode theory, where the counter-propagating fields inside the grating structure, obtained by convenient perturbation of the fields in the unperturbed waveguide, are related by coupled differential equations. Here a simple T-Matrix formalism will be presented for solving the coupled-mode equations for a Bragg grating structure [1] thus obtaining its spectral response.

5.3.1 Coupled-mode Theory and the T-Matrix Formalism

The spectral characteristics of a Bragg grating structure may be simulated using the T-Matrix formalism. For this analysis two counter-propagating plane waves are considered confined to the core of an optical fibre, in which a uniform intra-core Bragg grating of length l and uniform period Λ exists. This is illustrated in Fig. 5.9. The electric fields of the backward- and forward-propagating waves can be expressed as

$$E_a(x,t) = A(x)\exp[i(\omega t - \beta x)] \qquad \text{and} \qquad (5.14a)$$

$$E_b(x,t) = B(x)\exp[i(\omega t + \beta x)] \qquad (5.14b)$$

respectively, where β is the wave propagation constant. The complex amplitudes $A(x)$ and $B(x)$ of these electric fields obey the coupled-mode equations [29]

$$\frac{dA(x)}{dx} = i\kappa B(x)\exp[-i2(\Delta\beta)x] \qquad (0 \leq x \leq l) \qquad (5.15a)$$

$$\frac{dB(x)}{dx} = -i\kappa^* A(x)\exp[i2(\Delta\beta)x] \qquad (0 \leq x \leq l) \qquad (5.15b)$$

where $\Delta\beta = \beta - \beta_0$ is the differential propagation constant ($\beta_0 = \pi/\Lambda$, and Λ is the grating period) and κ is the coupling coefficient. For uniform gratings, κ is constant and it is related to the index modulation depth. For a sinusoidally-modulated refractive index the coupling coefficient is real and it is given by (5.6).

Assuming that there are both, forward and backward inputs to the Bragg grating, and boundary conditions $B(0) = B_0$ and $A(l) = A_l$, closed-form solutions for $A(x)$ and $B(x)$ are obtained from (5.15). Following these assumptions, the closed-form solutions for x-dependencies of the two waves are

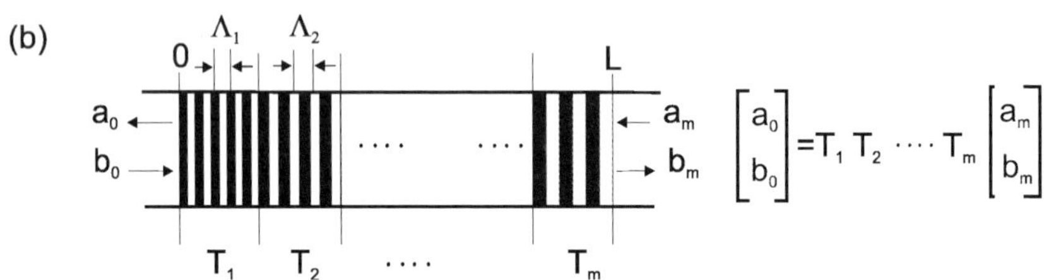

Fig. 5.9. Illustration of T-matrix model: (a) single uniform Bragg grating and (b) series of gratings with different periods back to back

$a(x) = A(x)\exp(-i\beta x)$ and $b(x) = B(x)\exp(i\beta x)$. Therefore, the backward output (reflected wave), a_0, and the forward output (transmitted wave), b_1, from the grating can be expressed by means of the scattering matrix

$$\begin{bmatrix} a_0 \\ b_1 \end{bmatrix} = \begin{bmatrix} S_{11} & S_{12} \\ S_{21} & S_{22} \end{bmatrix} \cdot \begin{bmatrix} a_1 \\ b_0 \end{bmatrix} \tag{5.16}$$

with $a_1 = A_1\exp(i\beta l)$ and $b_0 = B_0$, and

$$S_{11} = S_{22} = \frac{is\exp(-i\beta_0 l)}{-\Delta\beta\sinh(sl) + is\cosh(sl)} \tag{5.17a}$$

$$S_{12} = S_{21}\exp(2i\beta_0 l) = \frac{\kappa\sinh(sl)}{-\Delta\beta\sinh(sl) + is\cosh(sl)} \tag{5.17b}$$

where $s = \sqrt{|\kappa|^2 - \Delta\beta^2}$. Based on the scattering-matrix expression in (5.17), the T-matrix for the Bragg grating is [30]:

$$\begin{bmatrix} a_0 \\ b_0 \end{bmatrix} = \begin{bmatrix} T_{11} & T_{12} \\ T_{21} & T_{22} \end{bmatrix} \cdot \begin{bmatrix} a_1 \\ b_1 \end{bmatrix} \tag{5.18}$$

where

$$T_{11} = T_{22}^* = \exp(-i\beta_0 l)\frac{\Delta\beta\sinh(sl) + is\cosh(sl)}{is} \tag{5.19a}$$

$$T_{12} = T_{21}^* = \exp(-i\beta_0 l)\frac{\kappa\sinh(sl)}{is} \tag{5.19b}$$

The T-matrix relates the input and output of the Bragg grating and is ideal for analyzing a cascade of gratings (Fig. 5.9). Figure 5.9(b) shows a series of gratings back to back with a total length L. This grating structure is made up of "m" Bragg grating segments. Each segment has a different period Λ_m and has its own T-matrix T_m. The total grating structure may be expressed as

$$\begin{bmatrix} a_0 \\ b_0 \end{bmatrix} = T_1 \cdot T_2 \cdots T_{m-1} \cdot T_m \cdot \begin{bmatrix} a_m \\ b_m \end{bmatrix}, \tag{5.20}$$

and the spectral reflectivity of the grating structure is given by $|a_0(\lambda)/b_0(\lambda)|^2$. From the phase information one may also obtain the delay for the light reflected back from the grating [1]. It should be noted that this model does not take into account cladding mode-coupling losses.

Grating-length Dependence

The reflection spectral response for uniform Bragg gratings is calculated using the T-matrix formalism described above. The objective of this set of simulations is to demonstrate how the spectral response of a grating is affected as the length of the grating is altered. The index of refraction change is assumed uniform over the grating length, however, the value of the change is reduced with increasing grating length in such a way that the maximum grating reflectivity remains constant. Figure 5.10 shows the spectral profile of three uniform Bragg gratings.

The various plots clearly demonstrate that the bandwidth of the gratings decreases with increasing length. The 1-cm long uniform grating has a bandwidth of approximately 0.15 nm, that of the 2-cm long grating is 0.074 nm and finally the 4-cm long grating exhibits 0.057 nm bandwidth. Theoretically Bragg gratings may be constructed with extremely small bandwidths by simply increasing the grating length. However, in practice such devices are not easy to manufacture. The error associated with the spacing between the periods of a grating (during manufacturing) is cumulative, therefore, with increasing grating length the total error increases, resulting in out-of-phase periods and leading to broadening of the Bragg

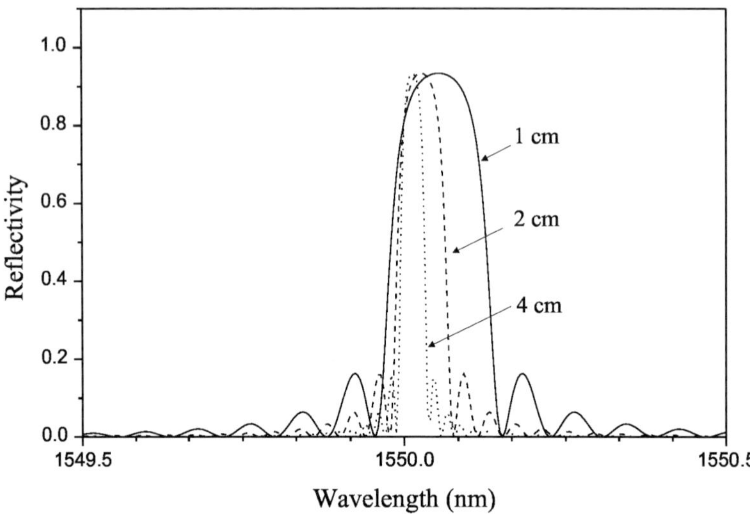

Fig. 5.10. Spectral reflectivity response from uniform Bragg gratings. The various spectral profiles correspond to different grating lengths: 1 cm (*solid-*), 2 cm (*dashed-*), and 4 cm (*dotted curve*)

grating reflection. Furthermore, if a long perfect Bragg grating is constructed, the effects of the environment have to be considered very carefully. For example, any strain or temperature fluctuations on any part of the grating will cause the periods to move out of phase resulting in broadening of the Bragg grating reflection.

Index of Refraction Dependence

Figure 5.11 shows a set of simulations assuming a uniform Bragg grating of 2 cm length and different index of refraction changes. For the first grating with $\Delta n = 0.5 \times 10^{-4}$ the reflectivity is 90% and the bandwidth is approximately 0.074 nm. If the change of the index of refraction is reduced to half the value of the first grating ($\Delta n = 0.25 \times 10^{-4}$), the reflectivity decreases to 59% and the bandwidth to 0.049 nm. A further decrease in the index of refraction change ($\Delta n = 0.1 \times 10^{-4}$) results in a reflectivity of 15% and a bandwidth of 0.039 nm. It appears that the bandwidth approaches a minimum value and remains constant for further reductions in the index of refraction change.

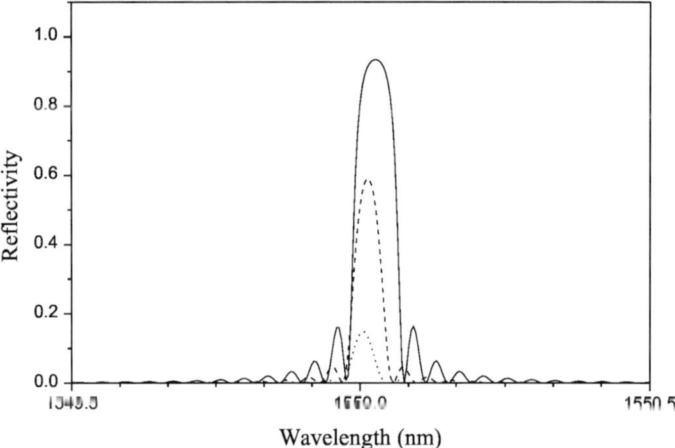

Fig. 5.11. Spectral reflectivity response from uniform Bragg grating 2 cm in length for different refraction indices. The solid, dashed, and dotted lines correspond to $\Delta n = 0.5 \times 10^{-4}$, $\Delta n = 0.25 \times 10^{-4}$, and $\Delta n = 0.1 \times 10^{-4}$ index of refraction change, respectively

Time Delay Dependence

Figure 5.12 shows the delay τ calculated from the derivative of the phase with respect to the wavelength for a uniform grating length L of about 10 mm. The design wavelength for this grating was 1550 nm and the index of refraction of the fibre was set at $n_{eff} = 1.45$. Figure 5.12 also shows the reflectivity spectral response of the same Bragg grating. Clearly both,

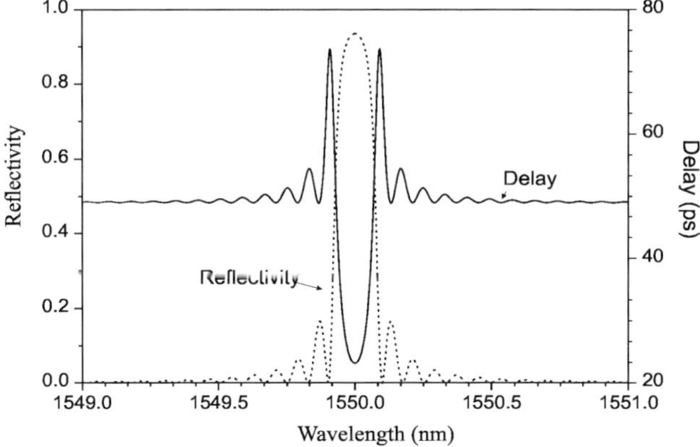

Fig. 5.12. Calculated group delay (*solid line*) and reflectivity (*dotted line*) for uniform weak Bragg grating ($v*\delta n_{eff} = 1 \times 10^{-4}$ and $L = 10$ mm). Design wavelength of the grating: 1550 nm, fringe visibility $v = 100\%$

reflectivity and delay, are symmetric about the peak wavelength λ_{max}. The dispersion is zero near λ_{max} for uniform gratings and becomes appreciable near the band edges and side lobes of the reflection spectrum, where it tends to vary rapidly with wavelength.

5.3.2 Chirped Bragg Gratings

One of the most interesting Bragg grating structures with immediate applications in telecommunications is the chirped Bragg grating. This grating has a monotonically varying period, as illustrated schematically in Fig. 5.13. There are certain characteristic properties offered by monotonically varying the period of gratings that are considered advantages for specific applications in telecommunication and sensor technology, such as dispersion compensation and the stable synthesis of multiple-wavelength sources. These types of gratings can be realized by axially varying either the period of the grating Λ or the index of refraction of the core, or both. From (5.3) we have

$$\lambda_B(z) = 2n_{eff}(z)\Lambda(z) \tag{5.21}$$

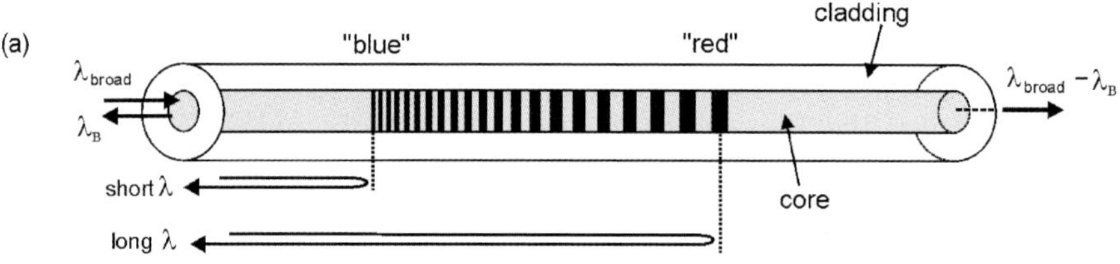

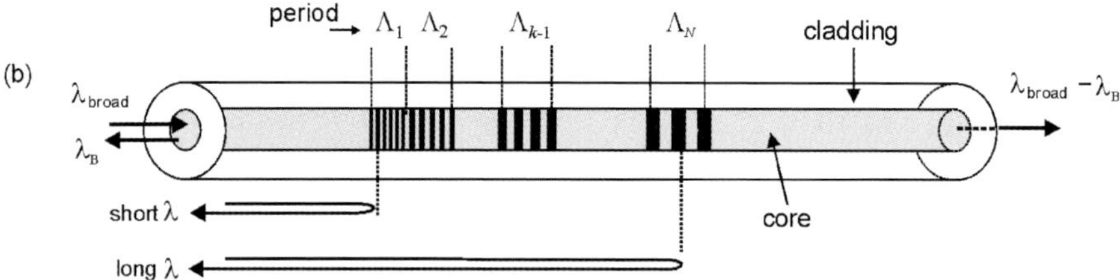

Fig. 5.13. (a) Schematic diagram of a chirped grating with an aperiodic pitch. For forward-propagating light as shown, long wavelengths travel further into the grating than shorter wavelengths before being reflected. **(b)** Schematic diagram of a cascade of several gratings with increasing period that are used to simulate long, chirped gratings

The simplest type of chirped grating structure is one with a linear variation of the grating period

$$\Lambda(z) = \Lambda_0 + \Lambda_1 z \qquad (5.22)$$

where Λ_0 is the starting period and Λ_1 is the linear change (slope) along the length of the grating. One may consider such a grating structure made up of a series of smaller length uniform Bragg gratings increasing in period. If such a structure is designed properly one may realize a broadband reflector. Typically the linear chirped grating has associated with it a chirp value/unit length ($chirp_\lambda = 2n_0\Lambda_1$) and the starting period. For example, a chirped grating 2 cm in length may have a starting wavelength at 1550 nm and a chirp value of 1 nm/cm. This implies that the end of the chirped grating will have a wavelength period corresponding to 1552 nm.

The simulation results shown in Fig. 5.14 illustrate the characteristics of chirped Bragg grating structures. The three different reflection spectra in the left part of Fig. 5.14 correspond to chirp values 0, 0.2, and 0.4 nm over the entire length of the grating. In these calculations all gratings are assumed to be 10 mm long with a constant index of refraction change $\delta n_{eff} = 1 \times 10^{-4}$. With increasing chirp value the reflectivity response becomes broader and the reflection maximum decreases. In these simulations the chirped gratings are approximated by a number of progressively increasing period gratings, whose total length amounts to the length of the chirped grating. The number of "steps" (the number of smaller gratings) assumed in the calculations is 100 (simulations indicated that calculations

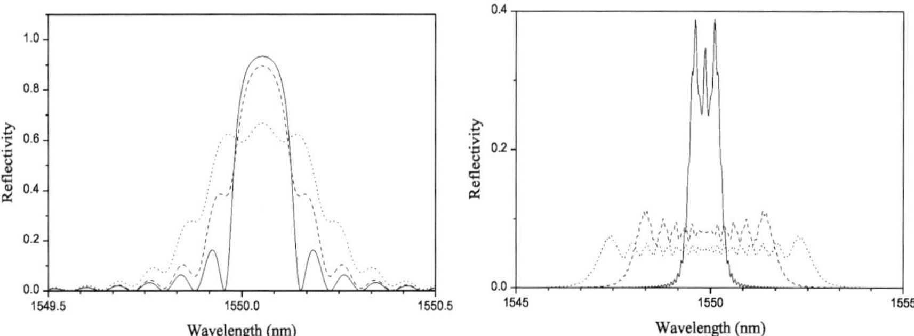

Fig. 5.14. Spectral reflectivity response from different Bragg gratings showing the effect of chirping. All gratings are 10 mm long and the index of refraction change is assumed to be $\delta n_{eff} = 1 \times 10^{-4}$ for all of them. Left part: The solid curve corresponds to 0 chirp, the dashed and dotted curves correspond to 0.2 and 0.4 nm chirp, respectively (where the chirp value is over the length of the grating). Right part: Spectral reflectivity response from highly chirped Bragg gratings for chirp values of 1 nm, 4 nm, and 8 nm over the 10 mm length of the gratings

with more than 20 steps will give approximately the same result). The spectral response from Bragg gratings with very large chirp values (1, 4, and 8 nm over the 10 mm length of the grating) is shown in the right part of Fig. 5.14. As can be seen it is possible to span a very large spectral area with increasing chirp value, however, with a reduction in the maximum reflectivity of the grating. This problem may be overcome by increasing the index of refraction modulation.

5.3.3 Apodisation of Spectral Response of Bragg Gratings

The reflection spectrum of a finite length Bragg grating with uniform modulation of the index of refraction gives rise to a series of side lobes at adjacent wavelengths (cf. Figs. 2.1 (log scale!), 5.3 or 5.12). It is very important to minimize and if possible eliminate the reflectivity of these side lobes, (or apodize the reflection spectrum of the grating) in devices where high rejection of the non-resonant light is required. An additional benefit of apodization is the improvement of the dispersion compensation characteristics of chirped Bragg gratings [31]. In practice apodization is accomplished by varying the amplitude of the coupling coefficient along the length of the grating. One method used to apodize an FBG consists in exposing the optical fibre with the interference pattern formed by two non-uniform ultraviolet light beams [32]. In the phase mask technique, apodization can be achieved by varying the exposure time along the length of the grating, either from a double exposure, by scanning a small writing beam, or by using a variable diffraction efficiency phase mask. In all these apodization techniques, the variation in coupling coefficient along the length of the grating comes from local changes in the intensity of the UV light reaching the fibre.

Figure 5.15 demonstrates the characteristics of apodized Bragg gratings, to be compared for example with Figs. 2.1, 5.3, or 5.12, which illustrate the typical side lobes of uniform Bragg gratings. For both gratings of Fig. 5.15 the magnitude of the index variation and the extension of the apodized regions are the same, but in the first case (left) the average refractive index changes also along the apodized region, while it remains constant for the second example (right part of Fig. 5.15). It is obvious that the latter approach results in a significantly stronger side lobe suppression.

Apodization of the fibre Bragg grating spectral response has been reported by Albert et al. using a phase mask with variable diffraction efficiency [33]. Bragg gratings with side lobe levels 26 dB lower than the peak reflectivity have been fabricated in standard telecommunication fibres [34]. This represents a reduction of 14 dB in the side lobe levels compared to uniform gratings with the same bandwidth and reflectivity.

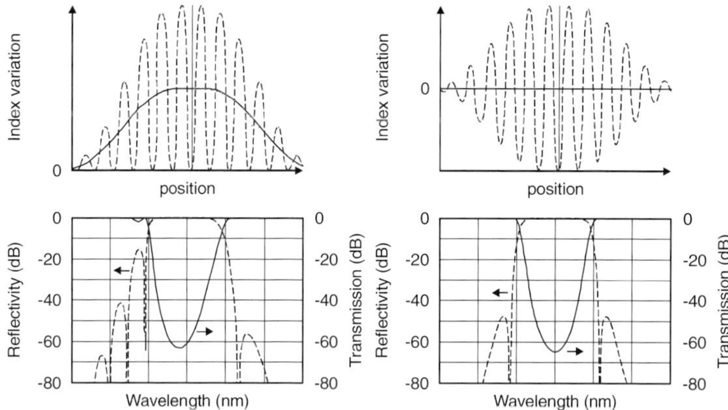

Fig. 5.15. Index variation and corresponding reflectivity and transmission of apodised gratings with varying (*left*) and constant (*right*) average effective refractive index

A technique for cosine apodization that was obtained by repetitive, symmetric longitudinal stretching of the fibre around the centre of the grating while the grating was written has also been demonstrated [35]. This apodization scheme is applicable to all types of fibre gratings, written by direct replication by a scanning or a static beam, or by use of any other interferometer and is independent of length. The simplicity of this technique allows the rapid production of fibre gratings required for wavelength-division-multiplex (WDM) systems and for dispersion compensation.

5.3.4 Fibre Bragg Gratings with Other Types of Mode Coupling

Tilting or blazing the Bragg grating at angles with respect to the fibre axis can couple light out of the fibre core into cladding modes or to radiation modes outside the fibre. This wavelength-selective tapping occurs over a rather broad range of wavelengths that can be controlled by the grating and waveguide design. One of the advantages is that the signals are not reflected into the fibre core, and thus the tap forms an absorption type of filter. An important application is gain flattening filters for erbium-doped fibre amplifiers (cf. Sect. 5.5.5).

With a small tilt of the grating planes to the fibre axis (~1°), one can make a reflecting spatial mode coupler such that the grating reflects one guided mode into another. It is interesting to point out that by making long period gratings, one can perturb the fibre to couple to other forward going modes. A wavelength filter based on this effect has been demonstrated by Hill et al. [36]. The spatial mode-converting grating was written using the point-by-point technique (cf. Sect. 5.4.4) with a period of $590\,\mu$m over a length of

60 cm. Using mode strippers before and after the grating makes a wavelength filter. In a similar manner, a polarisation mode converter or rocking filter in polarisation-maintaining fibre can be made. A rocking filter of this type, generated with the point-by-point technique, was also demonstrated by Hill and co-workers [37]. In their work they demonstrated an 87 cm long, 85 step rocking filter that had a bandwidth of 7.6 nm and a peak transmission of 89%.

5.4 Fabrication of Fibre Bragg Gratings

In the following section we will describe various techniques used in fabricating standard and complex Bragg grating structures in optical fibres. Depending on the fabrication technique Bragg gratings may be labelled as internally or externally written. Although internally written Bragg gratings may not be considered very practical or useful, nevertheless it is important to consider them, thus obtaining a complete historical perspective. Externally written Bragg gratings, that is gratings inscribed using techniques such as interferometric, point-by-point, and phase-mask overcome the limitations of internally written gratings and are considered far more useful. Although most of these inscription techniques were initially considered difficult due to the requirements of sub-micron resolution and thus stability, they are well controlled today and the inscription of Bragg gratings using these techniques is considered routine.

5.4.1 Internally Inscribed Bragg Gratings

Internally inscribed Bragg gratings were first demonstrated in 1978 by Hill and co-workers [38, 39] in a simple experimental set-up as shown in Fig. 5.16. An argon ion laser was used as the source, oscillating on a single longitudinal mode at 514.5 nm (or 488 nm) exposing the photosensitive fibre by coupling light into its core. Isolation of the argon ion laser from the back-reflected beam was necessary to avoid instability. Furthermore, the pump laser and the fibre were placed in a tube for thermal isolation. The incident laser light interfered with the 4% reflection (from the cleaved end of the fibre) to initially form a weak standing wave intensity pattern within the core of the fibre. At the high intensity points the index of refraction in the photosensitive fibre changed permanently. Thus a refractive index perturbation having the same spatial periodicity as the interference pattern was formed. These types of gratings normally have a long length (tens of centimetres) in order to achieve useful reflectivity values due to the small index of refraction changes.

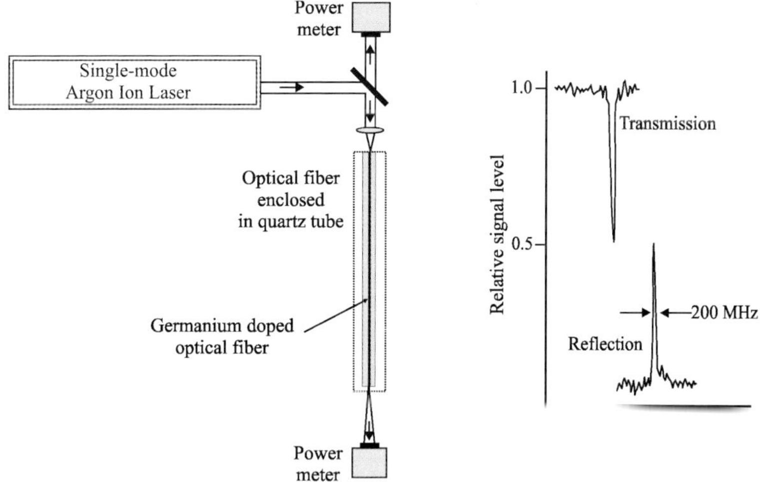

Fig. 5.16. A typical apparatus used in generating self-induced Bragg gratings using an argon ion laser. Typical reflection and transmission characteristics of these types of gratings are shown in the graph

5.4.2 Interferometric Inscription of Bragg Gratings

Amplitude-splitting Interferometer

The interferometric fabrication technique, which is an external writing approach for inscribing Bragg gratings in photosensitive fibres, was first demonstrated by Meltz and coworkers [40], who used an amplitude-splitting interferometer to fabricate fibre Bragg gratings in an experimental arrangement similar to the one shown in Fig. 5.17. An excimer-pumped dye laser operating at a wavelength in the range of 486–500 nm was frequency doubled using a non-linear crystal. This provided a UV source in the 244-nm band with adequate coherence length (a critical parameter in this inscription technique). The UV radiation was split into two beams of equal intensity that were recombined to produce an interference pattern, normal to the fibre axis. A pair of cylindrical lenses focused the light onto the fibre and the resulting focal line was approximately 4-mm long by 124-μm wide. A broadband source was also used in conjunction with a high-resolution monochromator to monitor the reflection and transmission spectra of the grating. The graph in Fig. 5.17 shows the reflection and complementary transmission spectra of the grating formed in a 2.6-μm diameter core, 6.6-mol % GeO_2-doped fibre after 5 minutes exposure to a 244-nm interference pattern with an average power of 18.5 mW. The length of the exposed region was estimated to be between 4.2 and 4.6 mm.

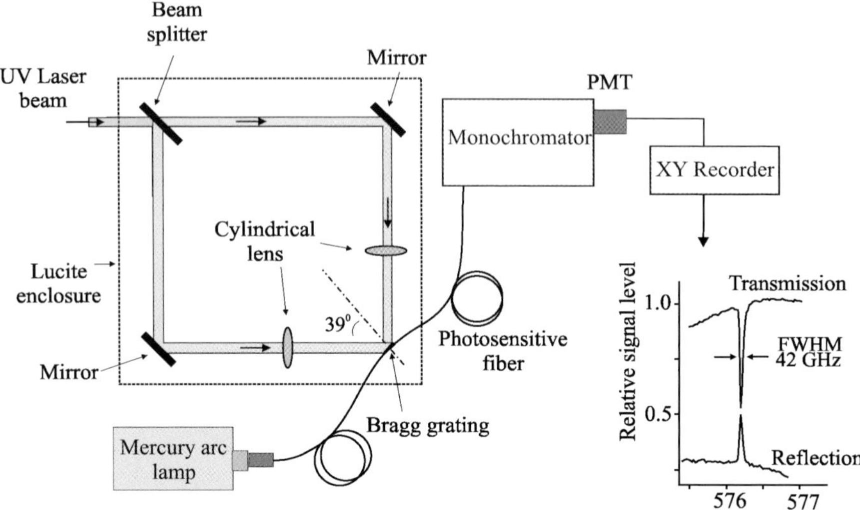

Fig. 5.17. An amplitude-splitting interferometer used by Meltz et al. [40], which demonstrated the first externally fabricated Bragg grating in optical fibre. The reflection and transmission spectra of a 4.4-mm long Bragg grating fabricated with this apparatus are also shown

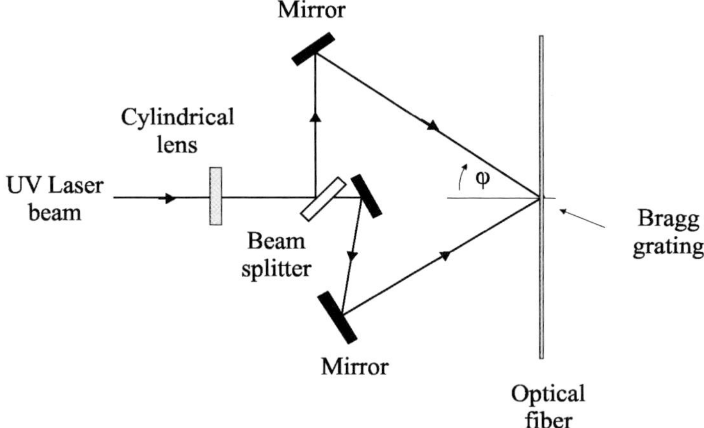

Fig. 5.18. An improved version of the amplitude-splitting interferometer, where an additional mirror is used to achieve an equal number of reflections, thus eliminating the different lateral orientations of the interfering beams. This type of interferometer is applicable to fabrication systems where the source spatial coherence is low, such as with excimer lasers

In a conventional interferometer such as the one shown in Fig. 5.17 the UV writing laser light is split into equal intensity beams that subsequently recombine after having undergone a different number of reflections in each optical path. Therefore, the interfering beams (wave fronts) acquire different (lateral) orientations. This results in a low quality fringe pattern for laser beams having low spatial coherence. This problem is eliminated by including a second mirror in one of the optical paths, as shown in Fig. 5.18, which in effect compensates for the beam splitter reflection. Since the total number of reflections is now the same in both optical arms, the two beams interfering at the fibre are identical.

The interfering beams are normally focused to a fine line matching the fibre core using a cylindrical lens placed outside the interferometer. This results in higher intensities at the core of the fibre, thereby improving the grating inscription. In interferometer systems as the ones shown in Figs. 5.17 and 5.18 the interference fringe pattern period Λ depends on both the irradiation wavelength λ_w and the half angle between the intersecting UV beams φ (cf. Fig. 5.18). Since the Bragg grating period is identical to the period of the interference fringe pattern, the fibre grating period is given by

$$\Lambda = \frac{\lambda_w}{2 \sin \varphi}.$$
(5.23)

Given the Bragg condition, $\lambda_B = 2n_{eff}\Lambda$, the Bragg resonance wavelength, λ_B, can be represented in terms of the UV writing wavelength and the half angle between intersecting UV beams as

$$\lambda_B = \frac{n_{eff}\lambda_w}{\sin \varphi}$$
(5.24)

where n_{eff} is the effective core index. From (5.24) one can easily see that the Bragg grating wavelength can be varied either by changing λ_w [41] and/or φ. The choice of λ_w is limited to the UV photosensitivity region of the fibre; however, there is no restriction for the choice of the angle φ.

One of the advantages of the interferometric method is the ability to introduce optical components within the arms of the interferometer, allowing for the wavefronts of the interfering beams to be modified. In practice, incorporating one or more cylindrical lenses into one or both arms of the interferometer produces chirped gratings with a wide parameter range [2]. The most important advantage offered by the amplitude-splitting interferometric technique is the ability to inscribe Bragg gratings at any wavelength desired. This is accomplished by changing the intersecting angle between the UV beams. This method also offers complete flexibility for producing gratings of various lengths, which allows the fabrication of wavelength-narrowed or broadened gratings. The main disadvantage of this approach is a susceptibility to mechanical vibrations. Sub-micron displacements in the position of mirrors, the beam splitter, or other optical mounts in the interferometer during UV irradiation will cause the fringe pattern to drift, washing out the grating from the fibre. Furthermore, because the laser light travels long optical distances, air currents, which affect the refractive index locally, can become problematic, degrading the formation of a stable fringe pattern. In addition to the above-mentioned shortcomings, quality gratings can only be produced with a laser source that has good spatial and temporal coherence and excellent wavelength and output power stability.

Wavefront-splitting Interferometers

Wavefront-splitting interferometers are not as popular as the amplitude-splitting interferometers for grating fabrication, however, they offer some useful advantages. Two examples of wavefront-splitting interferometers used to fabricate Bragg gratings in optical fibres are the prism interferometer [42, 43] and the Lloyd interferometer [44]. The experimental set-up for fabricating gratings with the Lloyd interferometer is shown in Fig. 5.19. This interferometer consists of a dielectric mirror, which directs half of the UV beam to a fibre that is perpendicular to the mirror. The writing beam is centred at the intersection of the mirror surface and fibre. The overlap of the direct and the deviated portions of the UV beam creates interference fringes normal to the fibre axis. A cylindrical lens is usually placed in front

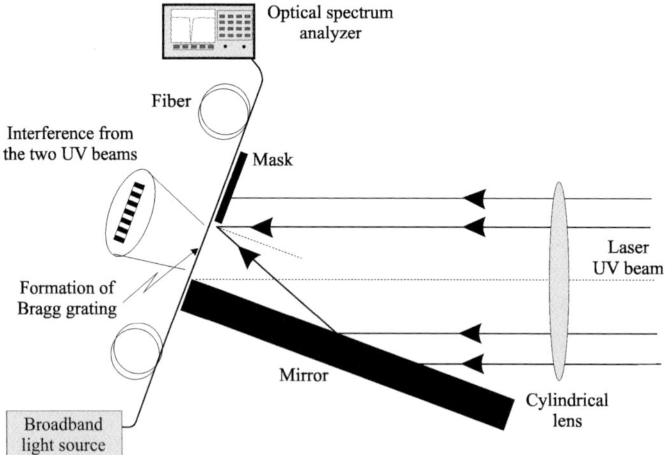

Fig. 5.19. Schematic of the Lloyd wavefront-splitting interferometer

of the system to focus the fringe pattern along the core of the fibre. Since half of the incident beam is reflected, interference fringes appear in a region of length equal to half the width of the beam. Secondly, since half the beam is folded onto the other half, interference occurs, but the fringes may not be of high quality. In the Lloyd arrangement, the folding action of the mirror limits what is possible. It requires a source with a coherence length equal to at least the path difference introduced by the fold in the beam. Ideally the coherence and intensity profile should be constant across the writing beam, otherwise the fringe pattern and thus the inscribed grating will not be uniform. Furthermore, diffraction effects at the edge of the dielectric mirror may also cause problems with the fringe pattern.

A schematic of the prism interferometer is shown in Fig. 5.20. The prism is made from high homogeneity, ultraviolet-grade, fused silica allowing for good transmission characteristics. In this set-up the UV beam is spatially bisected by the prism edge and half the beam is spatially reversed by total internal reflection from the prism face. The two beam halves are then recombined at the output face of the prism, giving a fringe pattern parallel to the photosensitive fibre core. A cylindrical lens placed just before the set-up helps in forming the interference pattern on a line along the fibre core. The interferometer is intrinsically stable as the path difference is generated within the prism and remains unaffected by vibrations. Writing times of over 8 hours have been reported with this type of interferometer. One disadvantage of this system is the geometry of the interference. Folding the beam onto itself forms the interferogram; hence different parts of the beam must interfere, which requires a UV source with good spatial coherence.

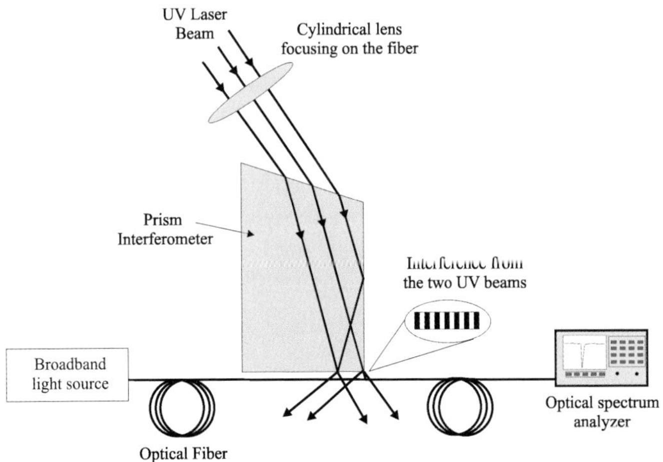

Fig. 5.20. Schematics of the prism wavefront-splitting interferometer

A key advantage of the wavefront-splitting interferometer is the requirement for only one optical component, greatly reducing sensitivity to mechanical vibrations. In addition, the short distance where the UV beams are separated reduces the wavefront distortion induced by air currents and temperature differences between the two interfering beams. Furthermore, this assembly can be easily rotated to vary the angle of intersection of the two beams for wavelength tuning. One disadvantage of this system is the limitation on the grating length, which is restricted to half the beam width. Another disadvantage is the range of Bragg wavelength tuneability, which is restricted by the physical arrangement of the interferometers. As the intersection angle increases, the difference between the beam path lengths increases as well, therefore, the beam coherence length limits the Bragg wavelength tuneability.

Laser Source Requirements

Laser sources used for inscribing Bragg gratings via the above interferometric techniques must have good temporal and spatial coherence. The spatial coherence requirements can be relaxed in the case of the amplitude-splitting interferometer by simply making sure that the total number of reflections are the same in both arms. This is especially critical in the case where a laser with low spatial coherence, like an excimer laser, is used as the source of UV light. The temporal coherence should correspond to a coherence length at least equal to the length of the grating in order for the interfering beams to have a good contrast ratio thus resulting in good quality Bragg gratings. The above coherence requirement together with the UV wavelength range needed (240–250 nm) forced researchers to initially use very complicated laser systems.

One such system consists of an excimer pumped tuneable dye laser, operating in the range of 480 to 500 nm. The output from the dye laser is focused onto a non-linear crystal to double the frequency of the fundamental light (Fig. 5.21). Typically this arrangement provides approximately 3–5 mJ, 10–20 nsec pulses (depending on the excimer pump laser) with excellent temporal and spatial coherence. An alternative to this elaborate and often troublesome set-up is a specially designed excimer laser that has a long temporal coherence length. These spectrally narrow linewidth excimer lasers may operate for extended periods of time on the same gas mixture with little changes in their characteristics. Commercially available narrow linewidth excimer systems are complicated oscillator amplifier configurations, which make them extremely costly. Othonos and Lee [45] developed a low cost simple technique, where existing KrF excimer lasers may be retrofitted with a spectral narrowing system for inscribing Bragg gratings in

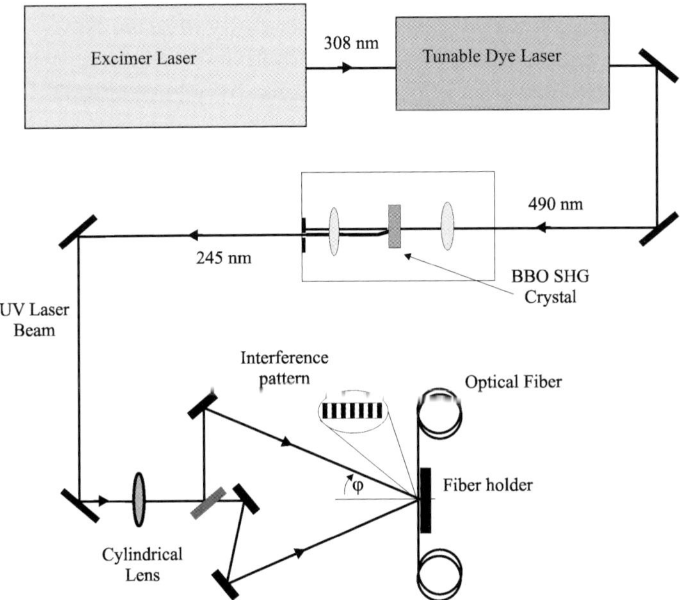

Fig. 5.21. Experimental set-up of an excimer pump dye laser with a frequency-doubled BBO crystal for generating UV light at 245 nm for inscribing Bragg gratings in an interferometer

a side-written interferometric configuration. In that work a commercially available KrF excimer laser (Lumonics Ex-600) was modified to produce a spectrally narrow laser beam (Fig. 5.22) with a linewidth of approximately

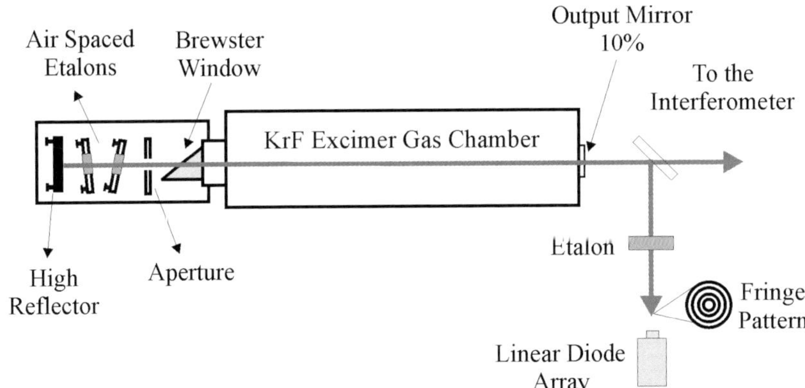

Fig. 5.22. Schematic of a narrow linewidth excimer laser system (KrF) consisting of two air-spaced etalons and an intracavity aperture placed between the KrF excimer gas chamber and the high reflector (after [45])

4×10^{-12} m. This system was used to successfully inscribe Bragg gratings in photosensitive optical fibres [45]. An alternative to the above system, which is becoming very popular, is the intracavity frequency-doubled argon ion laser that uses Beta-Barium Borate (BBO). This system efficiently converts high-power visible laser wavelengths into deep ultraviolet (244 and 248 nm). The characteristics of these lasers include unmatched spatial coherence, narrow linewidth and excellent beam pointing stability, which make such systems very successful in inscribing Bragg gratings in optical fibres [46].

5.4.3 Phase-mask Technique

One of the most effective methods for inscribing Bragg gratings in photosensitive fibre is the phase-mask technique [47]. This method employs a diffractive optical element (phase mask) to spatially modulate the UV writing beam (Fig. 5.23). Phase masks may be formed holographically or by electron-beam lithography. Holographically induced phase masks have no stitch error, which is normally present in the electron-beam phase masks. However, complicated patterns can be written into the electron beam-fabricated masks (quadratic chirps, Moiré patterns etc.). The phase mask grating has a one-dimensional surface-relief structure fabricated in a high quality fused silica flat transparent to the UV writing beam. The

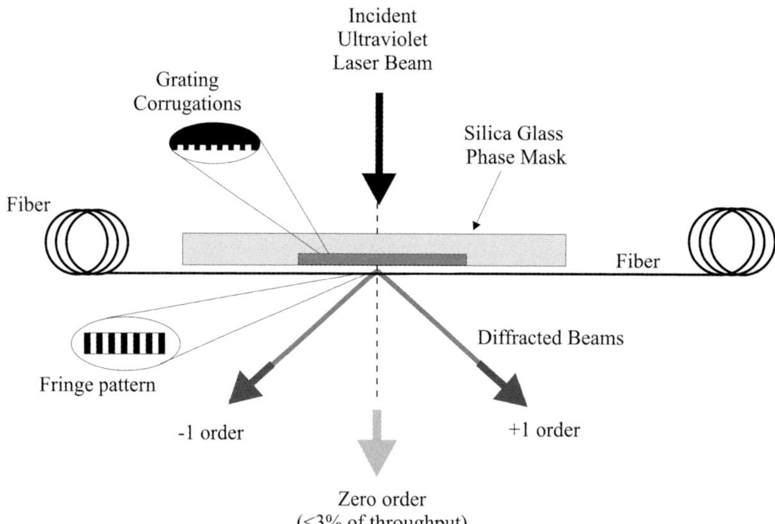

Fig. 5.23. Phase-mask geometry for inscribing Bragg gratings in optical fibres (see also Fig. 5.24)

profile of the periodic surface-relief gratings is chosen such that when a UV beam is incident on the phase mask, the zero-order diffracted beam is suppressed to less than a few percent (typically less than 5%) of the transmitted power. In addition, the diffracted plus and minus first orders are maximized each containing typically more than 35% of the transmitted power. A near field fringe pattern is produced by the interference of the plus and minus first-order diffracted beams. The period of the fringes is one half that of the mask. The interference pattern photo-imprints a refractive index modulation into the core of a photosensitive optical fibre placed in contact with or in close proximity immediately behind the phase mask (Fig. 5.23). A cylindrical lens may be used to focus the fringe pattern along the fibre core.

The phase mask greatly reduces the complexity of the fibre grating fabrication system. The simplicity of using only one optical element provides a robust and inherently stable method for reproducing fibre Bragg gratings. Since the fibre is usually placed directly behind the phase mask in the near field of the diffracting UV beams, sensitivity to mechanical vibrations and therefore stability problems are minimized. Low temporal coherence does not affect the writing capability (as opposed to the interferometric technique) due to the geometry of the problem.

KrF excimer lasers are the most common UV sources used to fabricate Bragg gratings with a phase mask. The UV laser sources typically have low spatial and temporal coherence. The low spatial coherence requires the fibre to be placed in near contact to the grating corrugations on the phase mask in order to induce maximum modulation in the index of refraction. The further the fibre is placed from the phase mask, the lower the induced index modulation, resulting in lower reflectivity Bragg gratings. Clearly, the separation of the fibre from the phase mask is a critical parameter in producing high quality gratings. However, placing the fibre in contact with the fine grating corrugations is not desirable due to possible damage to the phase mask. Othonos and Lee [48] demonstrated the importance of spatial coherence of UV sources used in writing Bragg gratings using the phase-mask technique. Improving the spatial coherence of the UV writing beam does not only improve the strength and quality of the gratings inscribed by the phase-mask technique, it also relaxes the requirement that the fibre has to be in contact with the phase mask.

To understand the significance of spatial coherence in the fabrication of Bragg gratings using the phase-mask technique it is helpful to consider a simple schematic diagram (Fig. 5.24). Consider the fibre core to be at a distance h from the phase mask. The transmitted *plus* and *minus* first orders that interfere to form the fringe pattern on the fibre emanate from different parts of the mask (referred to as distance d in Fig. 5.24). Since the

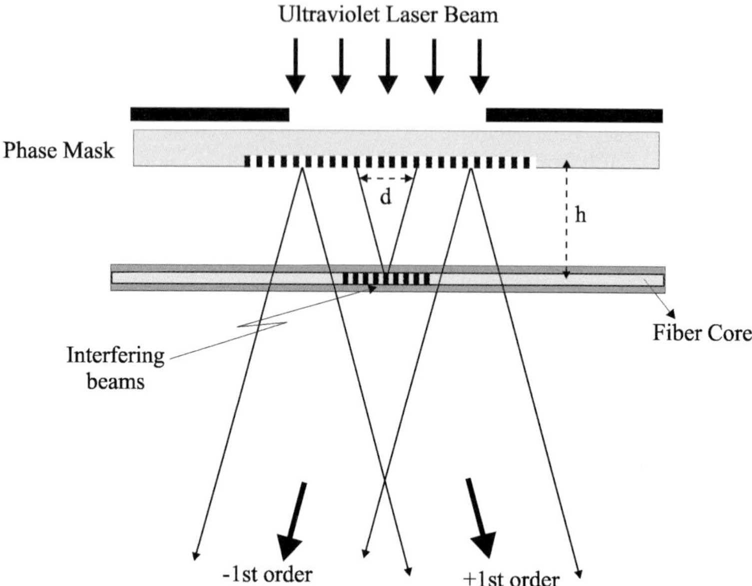

Fig. 5.24. Simple schematic of phase-mask geometry for inscribing Bragg gratings in optical fibres. The plus and minus first-order diffracted beams interfere at the fibre core, placed at distance *h* from the mask

distance of the fibre from the phase mask is identical for the two interfering beams, the requirement for temporal coherence is not critical for the formation of a high contrast fringe pattern. On the other hand, as the distance *h* increases, the separation *d* between the two interfering beams emerging from the mask, increases as well. In this case, the requirement for good spatial coherence is critical for the formation of a high contrast fringe pattern. As the distance *h* extends beyond the spatial coherence of the incident UV beam, the interference fringe contrast will deteriorate, eventually resulting in no interference at all. The importance of spatial coherence was also demonstrated by Dyer et al. [49], who used a KrF laser irradiated phase mask to form gratings in polyimide films. It should also be noted that if the zeroth order beam is not significantly suppressed, interference will occur between 0th- and 1st-order diffracted beams; in this case the interference pattern will change as a function of the fibre-phase mask separation resulting in fringes that vary from half the phase-mask period to one period of the mask.

5.4.4 Point-by-point Fabrication of Bragg Gratings

The point-by-point technique [50] for fabricating Bragg gratings is accomplished by inducing a change in the index of refraction a step at a time along the core of the fibre. A focused single pulse from an excimer laser produces each grating plane separately. A single pulse of UV light from an excimer laser passes through a mask containing a slit. A focusing lens images the slit onto the core of the optical fibre from the side as shown in Fig. 5.25, and the refractive index of the core increases locally in the irradiated fibre section. The fibre is then translated through a distance Λ corresponding to the grating pitch in a direction parallel to the fibre axis, and the process is repeated to form the grating structure in the fibre core. Essential to the point-by-point fabrication technique is a very stable and precise submicron translational system.

The main advantage of the point-by-point writing technique lies in its flexibility to alter the Bragg grating parameters. Because the grating structure is built up a point at a time, variations in grating length, grating pitch, and spectral response can easily be incorporated. Chirped gratings can be produced accurately simply by increasing the amount of fibre translation each time the fibre is irradiated. The point-by-point method allows the fabrication of spatial mode converters [51] and polarisation mode converters or rocking filters [37], that have grating periods, Λ, ranging from tens of micrometres to tens of millimetres. Because the UV pulse energy can be varied between points of induced index change, the refractive index profile of the grating can be tailored to provide any desired spectral response.

One disadvantage of the point-by-point technique is that it is a tedious process. Because it is a step-by-step procedure, this method requires a relatively long process time. Errors in the grating spacing due to thermal effects and/or small variations in the fibre's strain can occur. This limits

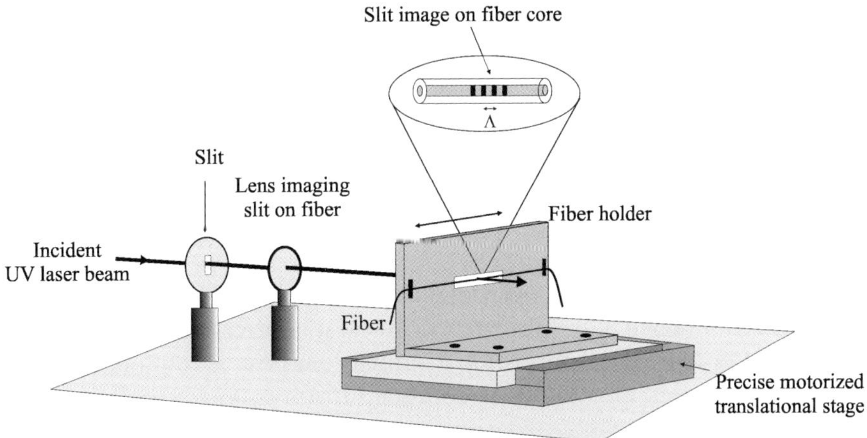

Fig. 5.25. Schematic of set-up for fabricating Bragg gratings using the point-by-point technique

the gratings to very short lengths. Typically, the grating period required for first order reflection at 1550 nm is approximately 530 nm. Because of the submicron translation and tight focusing required, first order 1550 nm Bragg gratings have yet to be demonstrated using the point-by-point technique. Malo et al. [50] have only been able to fabricate Bragg gratings, which reflect light in the 2nd and 3rd order, that have a grating pitch of approximately 1 μm and 1.5 μm, respectively.

5.4.5 Direct-writing Technique

The direct-writing technique [52] is an alternative process for the flexible inscription of high quality gratings in fibre or optical waveguide circuits with integral Bragg gratings. The direct writing-process uses two focused UV laser beams that are overlapped to give a micron-sized, circular spot with an intrinsic linear interference pattern in one dimension, much as a conventional amplitude-splitting interferometer. The interference beam is focused onto the photosensitive fibre or waveguide and translated by one grating period, re-exposed and moved to a new position, systematically building up a grating structure. The piecewise process is where the direct-writing technique differs from the traditional amplitude-splitting arrangement. The sample is moved continuously relative to a modulated laser beam that is synchronised to the sample motion, thereby ensuring a smooth inscription process. Changing the time for which the laser is switched on during the exposure cycle controls the grating contrast. This technique allows for the fabrication of any Bragg grating profile at any point in the structure. The process is computer controlled, providing a straightforward method of producing complex waveguide circuits and Bragg gratings with arbitrary chirp and apodization. Direct writing proves advantageous, as it eliminates the usual need for Bragg grating phase masks. As direct writing is essentially a modified interferometric set-up it has similar limitations regarding system stability requirements.

5.4.6 Femtosecond Laser Inscription of Bragg Gratings

The use of high power femtosecond laser sources for inscribing Bragg gratings has attained significant recent interest. Corresponding laser systems come in myriads of forms: their primary purpose is to produce high-energy laser pulses of 5 to 500 fs duration (150 fs is typical), offering the tempting prospect of extraordinarily high laser intensity when focused onto the sample, 200 GW/cm^2, for example, is not uncommon. The high-energy pulses and wavelength of operation both vary depending on the specific laser design, and there are, in general, two categories i) oscillator systems without extra optical amplification, where high energy pulses build up in an extended cavity and ii) amplified systems that use external optical amplifiers to enhance oscillator pulse energies. Values of 100 nJ/pulse at 10 MHz repetition rate are typical for extended oscillator cavities; 500 nJ/pulse at 1 MHz have been achieved for cavity-dumped oscillators, and finally 1 mJ/pulse to 10 µJ/pulse for repetition rates of 1 kHz to 500 kHz, respectively, for regeneratively amplified systems. Prior to recent developments, the amplified laser system was economically prohibitive and complex, with separate modules coupled by careful alignment. However, recent developments have seen this scheme replaced by less expensive, diode pumped amplified systems operating in the near infrared that can accept harmonic generators for frequency doubling, tripling, and quadrupling, in a very compact form factor, with minimal infrastructure needs.

The principal advantage of high-energy pulses is their ability of grating inscription in any material type without pre-processing, such as hydrogenation or special core doping with photosensitive materials – the inscription process is controlled multi-photon absorption, void generation and subsequent local refractive index changes. Furthermore, the use of an infrared source or its second harmonic removes the requirement to strip the optical fibre as gratings can readily be written through the buffer layer. However, one must consider that the very nature of the short duration pulse poses several technical difficulties that must be accounted for in choosing a suitable inscription method. Interferometric set-ups must have path lengths matched to within the physical location of the fs pulse; for a 150 fs pulse this corresponds to 45 µm. If a phase mask is utilised the issue of temporal coherence is resolved, but one must now consider the large spectral content of the pulse and its subsequent dispersion through and energy spread beyond the mask. There is also the potential risk of optically damaging the mask, if pulse energies or focusing conditions are not carefully constrained. With respect to the grating inscription schemes discussed so far, the femtosecond laser has been utilised as follows: Mihailov and co workers [53–55] have developed phase masks designed for use at 800 nm for the inscription of quality, higher-order Bragg gratings in optical fibres. The use of phase masks with larger pitch mitigates pulse spreading, and gratings were produced having the dual advantage of good spectral quality and high thermal stability to 950°C. Martinez and co-workers [56, 57] have opted for point-by-point inscription of multiple-order gratings, with the third-order giving the best spectral quality. Computer-controlled systems are also under development for the formation of arbitrary and multiple Bragg gratings [58].

5.5 Fibre Bragg Gratings in Optical Communication Systems

The unique filtering properties of fibre Bragg gratings and their versatility as in-fibre devices have made FBGs one of the key components in fibre optic networks [1], and FBGs are used in ultra long haul (ULH), long haul (LH), and Metropolitan dense WDM (DWDM) telecommunication networks as well. A schematic representation of a network is depicted in Fig. 5.26. Bragg gratings are located in optical erbium-doped fibre amplifiers (EDFA) for pump wavelength stabilization and erbium-gain flattening as well as in add-drop nodes for wavelength filtering and multiplexing. They are also used in line or at the emission-reception side for channel and band dispersion compensation.

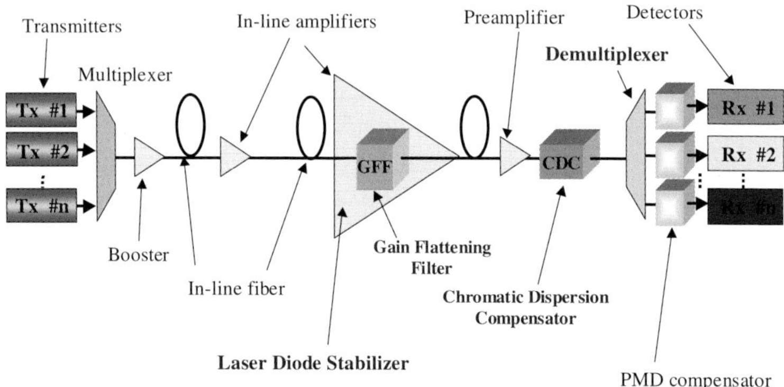

Fig. 5.26. Schematic representation of telecommunication network with preferred Bragg grating locations

Wavelength Channel Filters

Band-pass filters are key components of WDM systems, and many solutions utilizing fibre Bragg gratings for implementing such filters have been developed including Sagnac [59], Michelson [60, 61] or Mach–Zehnder [62, 63] interferometric configurations. Another approach, based on the principle of the Moiré grating resonator [64], has been applied with uniform period [65] and chirped [66] gratings. Resonant filter structures have been fabricated by introducing a phase shift into the grating by an additional UV exposure, or by using a phase-shifted phase mask. In general the resonant type transmission filters are capable of large wavelength selectivity and are, in principle, simple to manufacture and do not require carefully balanced arms or identical gratings, as in the case of interferometric filters.

Fibre Bragg 100-, 50- and 25-GHz bandwidth reflective filters are commercial products with low out-of-band crosstalk (better than 25 dB for the adjacent channel) and low chromatic dispersion (<30 ps/nm for a 50 GHz grating). The realization of specific fibres (low cladding mode coupling fibre or high numerical aperture (HNA) fibre, cf. Sect. 5.2.5) is important in order to get these specifications with lowest possible losses (<0.2 dB). One key application is channel selection in optical add-drop multiplexers.

Chromatic Dispersion Compensation

Chromatic dispersion is one of the key factors limiting transmission distances of 10 and 40 Gbit/s optical systems, and FBGs are widely used to mitigate these adverse effects. The phase response of a fibre Bragg grating can be designed to compensate many wavelengths and chirped fibre gratings are a well known (commercially available) solution for single channel and multi-channel dispersion compensation. The phase of the FBG is tailored during the photo-inscription process in order to get a linear variation of the delay as a function of wavelength. Dispersion compensation ranging from 100 to 2000 ps/nm has been demonstrated using chirped gratings. Many studies have focused on controlling the linear delay and suppression of the phase ripple inherent to multi-interference occuring inside the grating ($<\pm5$ ps).

Tuneable dispersion gratings for 40 Gbit/s systems enable the compensation of slow dispersion variations in telecommunication networks (temperature and vibration effects), for example by the application of an external chirp to the grating by means of a mechanical strain or a temperature gradient. This external chirp can then be added to the one created during photo-inscription and leads to a dispersion tuning over a few hundred of ps/nm. Bragg gratings constitute the most efficient current technology for dispersion tuning. The main alternative technologies are planar devices and thin-film filters exploiting the Gires–Tournois effects (cf. Chaps. 6 and 7).

Gain-flattening Filters

In Metro, LH or ULH transmission systems optical amplifiers (EDFA) are necessary to compensate the link losses, and good overall amplifier flatness is required for low optical signal-to-noise ratio at all wavelengths. Different variants of Bragg gratings can be used to achieve this purpose. The FBGs can have non-uniform period (mostly linearly varying) to cover the full amplifier bandwidth, and these chirped gratings (cf. Sect. 5.3.2) are used in reflection with isolators on both sides. The major competitive technology is dielectric thin film filters (cf. Chap. 7), which have inferior optical performance, but are available at lower cost, since they do not require optical isolators as FBGs do. To overcome this drawback slanted gratings have been developed [67]. They exhibit a tilt between the fibre axis and that of the grating in order to couple the forward-propagating fundamental mode (LP_{01}) to radiative modes which makes the filter dissipative in transmission. Different fibres have been designed to improve this coupling and suppress the residual backward wave (LP_{01-}). Total C-band transmission error functions lower than ± 0.2 dB over the temperature range $-10°C... +70°C$ have been demonstrated with induced losses smaller than 1 dB.

Pump Laser Stabilisation

Finally, fibre Bragg gratings are also used for the stabilisation of pump laser diodes in optical amplifiers. In EDFAs used in DWDM configurations, the precise stabilization of the pump wavelength is crucial. A small reflective grating is directly photo-written into the core of the laser pigtailed fibre. The feedback provided by the grating causes the laser to oscillate at the reflected wavelength, and most of today's pump laser diodes use a stabilizing grating written in standard or polarisation-maintaining fibre.

Gain flattening filters as well as Bragg grating-based add-drop multiplexers and dispersion compensators will be described in more detail in the following sections.

5.5.1 Add-drop Multiplexers

Fixed or reconfigurable optical add-drop multiplexers ((R)OADM) enable static or dynamic routing in wavelength division multiplexing networks, and AWGs (cf. Chap. 4), dielectric multilayer filters (cf. Chap. 7) or FBGs are most frequently used for the realisation of OADMs. The main advantages of FBG-based WDM filters are very high selectivity, flat top spectral response, steep spectral roll-off, and low insertion loss. As a consequence, FBG filters are widely used in low channel spacing (< 100 GHz) OADMs. The reflectivity and transmission spectra of a 50 GHz FBG are shown in Fig. 5.27. The grating is photo-written in a specific low-cladding-mode-loss fibre to avoid extra transmission losses out of the FBG spectral band. The complex apodization applied during the writing process assures an in-band dispersion lower than 10 ps/nm.

OADMs can be implemented in different configurations. One example is a passive temperature-compensated FBG sandwiched between two optical circulators (Fig. 5.28). If different wavelengths are injected at the input

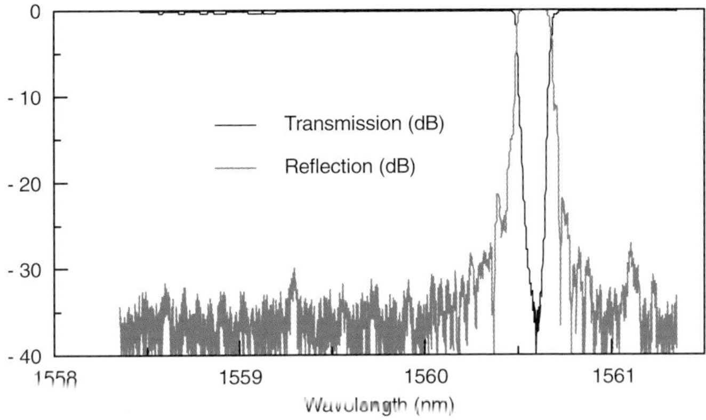

Fig. 5.27. Characteristics of 50 GHz fibre Bragg grating filter

port (1), the Bragg wavelength (λ_B) is reflected by the FBG and routed by the optical circulator to the drop port, while all other signals ($\lambda_i \neq \lambda_B$) pass through the FBG and exit via the output port (2). Since the component is symmetric, a new optical channel can be added (at λ_B) to the network through the add port, and the inserted wavelength exits the OADM from the output port (2) as well. Furthermore, if the FBG is made tuneable (thermally or by applying a mechanical force, cf. Sect. 5.2.4), its Bragg wavelength can change and the device becomes reconfigurable. Temperature-induced tuning over 8 nm with < 50 ms switching time has been demonstrated [68].

Placing Bragg gratings between optical circulators in a hybrid fashion is the simplest way of making FBG-based OADMs, but such devices exhibit high insertion loss and they are expensive. One option with no need of (expensive) circulators is a Mach–Zehnder interferometer (all-fibre solution) configuration with photo-imprinted Bragg gratings in each arm as illustrated schematically in Fig. 5.29. The device consists of two identical

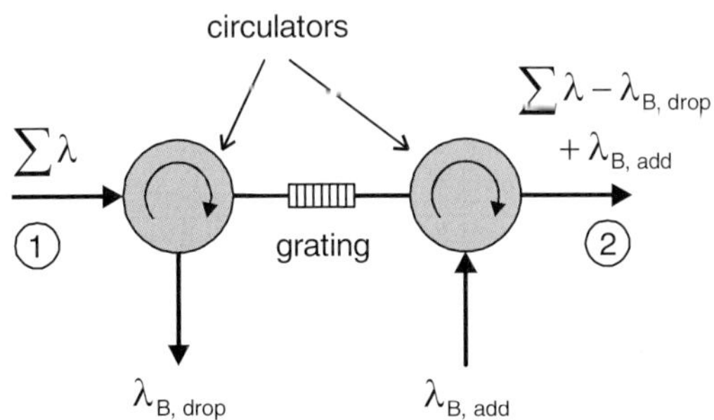

Fig. 5.28. Fibre Bragg grating-based add-drop multiplexer

FBGs and 3-dB couplers fused on both sides of the gratings, which makes the device a MZI for all wavelengths but the Bragg wavelength for which it is a Michelson interferometer. A signal launched into the input port is equally split through the first 3-dB coupler, and the Bragg wavelength is back-reflected on each arm of the MZI, while all other wavelengths pass through the FBG. Phase matching by UV trimming assures efficient extraction of the Bragg wavelength via the drop port. All other phase-matched wavelengths ($\lambda_i \neq \lambda_B$) leave the device via the output port, and a signal at the channel wavelength (λ_B) launched into the add port emerges at the output port as well (Fig. 5.29).

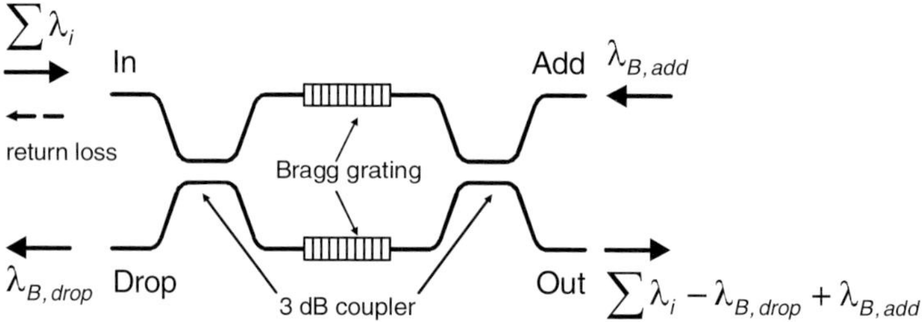

Fig. 5.29. Twin core fibre-based Mach–Zehnder interferometer add-drop multiplexer

The signal intensity propagating to the output and the add-port, respectively ($I_{in>out}$, $I_{in>add}$) as a function of the wavelength-dependent phase difference $\varphi(\lambda)$ between the arms, is given for the lossless case by (5.25) and (5.26), where $t_i(\lambda)$ is the complex transmission coefficient of the FBGs and $x_i(\lambda)$ the power coupling ratio of the couplers:

$$I_{in>out}(\lambda) = \left| \sqrt{(1-x_1)x_2} \cdot t_1(\lambda) \cdot \exp(i\varphi(\lambda)) + \sqrt{(x_1(1-x_2))} \cdot t_2(\lambda) \right|^2 \cdot I_{in}(\lambda)$$

$$(5.25)$$

$$I_{in>add}(\lambda) = \left| \sqrt{(1-x_1)(1-x_2)}\, t_1(\lambda) \cdot \exp(i\varphi(\lambda)) - \sqrt{x_1 x_2}\, t_2(\lambda) \right|^2 \cdot I_{in}(\lambda)$$

$$(5.26)$$

Similarly, $I_{in>drop}$ and the return loss $I_{in>in}$ can easily be obtained by (5.25) and (5.26), if all subscripts 2 are replaced by 1 and the reflection coefficients are used instead of the transmission coefficients.

Such MZIs can be realized using standard single-mode fibre or twin core fibre (TCF), where the TCF has a number of advantages compared to the SMF. Since the cores are embedded within the same cladding, the effective index difference between the arms is very small, and consequently the chromatic dispersion of the MZI is essentially that of the couplers. TCFs allow a very reproducible fusion-tapering process as well, which means that the centre wavelength of the couplers can be well controlled. For example, for a 30 dB isolation between $I_{add>out}$ and $I_{add>add}$, the centre wavelength of couplers must be controlled within 5 nm.

The performance of a 200 GHz MZI OADM is illustrated in Fig. 5.30: The Bragg wavelength is extracted from the drop port with more than 30 dB adjacent crosstalk, the in-band rejection of the output port is better than 30 dB, and the crosstalk isolation between signals passing through the add and the output port is better than 30 dB over a 25 nm range. These results confirm that the couplers exhibit very low chromatic dispersion and are well centred at the Bragg wavelength. The return loss reaches 18 dB in-band, while it is better than 25 dB over 30 nm out-of-band. It is worthwhile to note that the measurements have been performed on SMFs coupled to the TCFs, and total insertion loss of any path is better than 1.5 dB. Bit error rate measurements at 10 Gbit/s using this device did not show any power penalty at all [69].

Passive athermal add-drop devices have been demonstrated with less than 1 pm/°C deviation of the Bragg wavelength over the −10°C to +70°C temperature range [70–72]. Theoretical analyses of MZI-OADMs have demonstrated that the phase mismatch of the Bragg gratings (leading to Bragg wavelength detuning) is the most critical parameter determining in-band return loss.

Recent studies focused on the realization of MZI-based add-drop multiplexers in planar waveguide architecture with large negative index

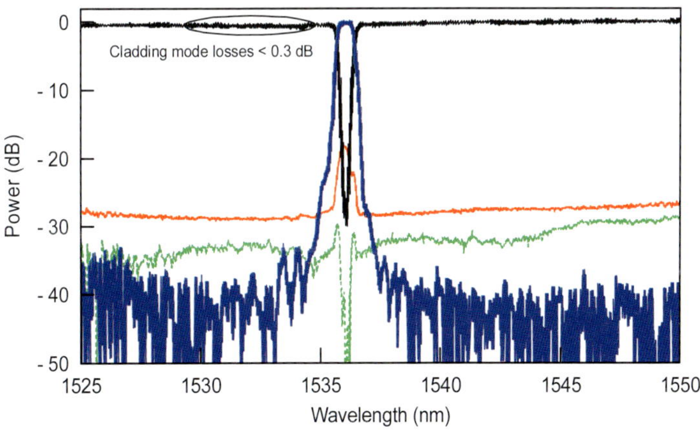

Fig. 5.30. Optical performance of packaged 200 GHz twin core fibre-based OADM. Transmission: black; drop (reflection): blue; return loss: red; input to add channel: green

modulation. Moreover, efficient thermal tuning of a Mach–Zehnder-based OADM has been demonstrated over 1.7 nm without significant modification of the drop directivity and the throughput signals.

Table 5.1 illustrates the performance of FBG-based hybrid and all-fibre fixed 100 GHz optical add-drop multiplexers.

Table 5.1. Characteristics of FBG-based hybrid and in-fibre, fixed 100 GHz OADMs

Optical parameter	Hybrid OADM	In-fibre OADM
Insertion loss (dB)	1.5	1.5
Isolation (dB)	30	30
Rejection (dB)	30	30
In-band return loss (dB)	30	20
Out-of-band return loss (dB)	30	30

Sub-band and single wavelength Bragg grating-based OADMs in fibre or/and planar waveguide technology are still topics of current research, and in particular SiO_2/Si offers the potential for the realization of complex multi-wavelength functionalities.

A device rather similar to the add-drop multiplexer is a single Bragg grating in a single-mode fibre acting as a wavelength selective distributed reflector or a band-rejection filter by reflecting wavelengths around the Bragg resonance. By placing identical gratings in two lengths of a fibre coupler, as in a Michelson arrangement, one can make a band-pass filter [73]. This filter, shown in Fig. 5.31, passes only wavelengths in a band around the Bragg resonance and discards other wavelengths without reflections. If the input port is excited by broadband light and the wavelengths reflected by the gratings arrive at the coupler with identical optical delays, then this wavelength simply returns to the input port. If, however, a path-length difference of $\pi/2$ is introduced between the two arms, then it is possible to steer the reflected wavelength to arrive at the second input port, creating a bandpass filter. In principle, this is a low-loss filter, although there is a 3 dB loss penalty for the wavelengths that are not reflected, unless a Mach–Zehnder interferometer is used to recombine the signal at the output [60]. An efficient band-pass filter was demonstrated by Bilodeau et al. [61] using a scheme identical to that presented in reference [60]. The device had back reflection of -30 dB (cf. Glossary for definition). However, all wavelengths out of the pass-band suffered from the 3-dB loss associated with the Michelson interferometer.

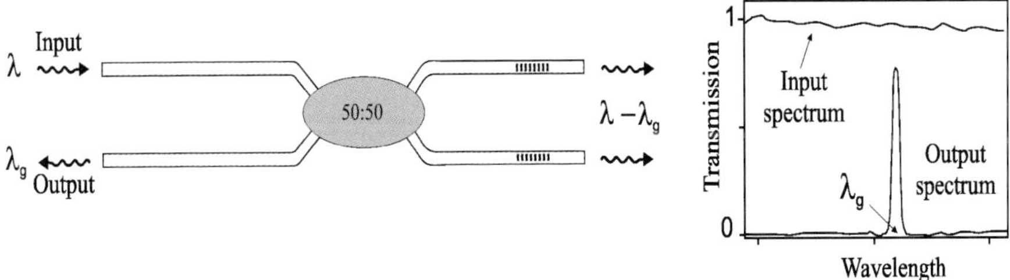

Fig. 5.31. Fibre-optic bandpass filter using Bragg reflectors arranged in a Michelson-type configuration

5.5.2 FBG based Chromatic Dispersion Compensators

The group velocity of optical signals travelling along a fibre is wavelength-dependent due to material and waveguide dispersion, and this phenomenon is usually called chromatic dispersion (CD). Waveguide dispersion is essentially due to a wavelength-dependent distribution of light between the core and the cladding of a single-mode fibre. Due to CD pulses propagating along an optical fibre experience a temporal broadening which limits the maximum transmission distance since subsequent pulses become overlapping, i. e. one gets inter-symbol interference. Tolerable chromatic dispersion depends on the modulation format, the channel wavelength, the spectral width of the pulses, the fibre type, and on the bit rate. To a first approximation tolerable chromatic dispersion decreases proportional to the square of the bit rate, so its effect becomes significant for 10 Gbit/s and beyond, and different techniques have been developed in the past to compensate unwanted chromatic dispersion, so that the maximum optical span length is extended. An estimate of dispersion-limited maximum transmission lengths (1 dB power penalty) for standard transmission formats and without dispersion compensation is given by [74]

$$B^2 DL \leq 10^5 \qquad (5.27)$$

where B is the bit rate, L the fibre length, and $D(\lambda)$ the fibre chromatic dispersion. For a Corning SMF-28e fibre, for example, $D(\lambda)$ (in [ps/(nm·km)]) is given by

$$D(\lambda) \approx \frac{S_0}{4}\left[\lambda - \frac{\lambda_0^4}{\lambda^3}\right] \qquad (5.28)$$

for the wavelength range $1200\,\text{nm} \leq \lambda \leq 1625\,\text{nm}$, where λ_0 is the zero dispersion wavelength ($1302\,\text{nm} \leq \lambda_0 \leq 1322\,\text{nm}$) and S_0 is the zero dispersion slope: $S_0 \leq 0.092\,\text{ps}/(\text{nm}^2 \cdot \text{km})$ [75]. Table 5.2 illustrates that dispersion poses severe limits on the maximum transmission distances at higher bit rates and consequently dispersion compensation is of high relevance at higher bit rates.

Table 5.2. Dispersion-limited maximum transmission distance for standard single-mode (SMF-28) and non-zero dispersion shifted fibre (1 dB penalty, no dispersion compensation)

Bit rate	Transmission over SMF-28	Transmission over NZ-DSF
2.5 Gbit/s	1000 km	6000 km
10 Gbit/s	60 km	400 km
40 Gbit/s	4 km	25 km

One well established, broadband solution for compensating CD is the use of dispersion-compensating fibre (DCF). A DCF has a high negative dispersion (at 1.5 µm) which is several times larger than the positive dispersion of a standard SMF. However, DCF exhibits rather high losses (typically 10 dB for 80 km fibre), high non-linearity due to a small mode effective area, and the use of DCF adds extra cost because extra amplifiers are needed to compensate the DCF loss. Moreover, for certain system configurations only single wavelengths or sub-band dispersion compensation is needed (4, 8, or 16 wavelengths), and DCF is not particularly suited to accomplish this task. Finally, in particular at 40 Gbit/s, tuneable dispersion compensation is requested in order to precisely compensate the dispersion for each channel and to correct for environmentally-induced variations, and DCF cannot meet these requirements.

On the other hand, chirped fibre Bragg gratings enable versatile and efficient dispersion compensation. The first practical demonstrations of dispersion compensation by means of chirped FBGs used short pulses of either 1.8 ps [76] or 21 ps [77] duration.

The design of a dispersion-compensating module (DCM) consisting of a 3-port circulator and a chirped grating and operating in reflection is illustrated schematically in Fig. 5.32.

The index modulation of the grating is given by

$$n(z) = n_{eff} + \Delta n_{mod} \times f_{apod}(z) \times \cos\left(\frac{2\pi}{\Lambda} z - \frac{\pi}{2\,n_{eff}\,\Lambda^2}\,\Lambda_1\,z^2 \right) \qquad (5.29)$$

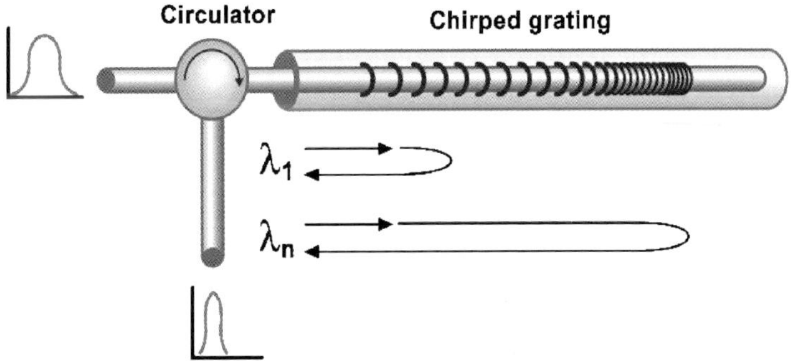

Fig. 5.32. Dispersion compensating module (schematic, $\lambda_1 > \lambda_n$)

where n_{eff} is the mean effective index, $n(z)$ is the index variation along the fibre axis, Λ is the grating mean pitch, Δn_{mod} is the index modulation amplitude, $f_{apod}(z)$ is the apodization function, and Λ_l is the so-called chirp parameter of the Bragg wavelength (see (5.22)) which ensures a linear variation of the local Bragg wavelength along the grating via the variation of the pitch or/and of the effective index. The spectral bandwidth of the grating increases with the grating length, and different wavelengths are reflected at different spatial locations inside the grating. Shorter wavelengths are reflected towards the end of the grating and are thus delayed with respect to longer wavelengths which are reflected at the beginning of the grating.

The delay τ, the bandwidth $\Delta\lambda$, and the dispersion D associated with the chirped grating are linked by the following set of equations :

$$\tau = \frac{2n_g L_g}{c} \tag{5.30a}$$

$$\Delta\lambda = \Lambda_l \times L_g \tag{5.30b}$$

so that

$$D = \frac{\tau}{\Delta\lambda} = \frac{2n_g}{c \times \Lambda_l} \tag{5.31}$$

where n_g is the fiber group index, L_g the grating length, and Λ_l the chirp parameter.

Chirped gratings can be realized in different ways. The most popular approach uses a chirped phase mask and a standard photo-inscription set-up. An alternative is to use a standard uniform phase mask and then apply a specific back and forth movement of the fibre in front of the phase mask [78]. A further option is to taper the fibre continuously (by fusion) in order to generate a continuous change of the refractive index along the fibre in the region where the grating is to be inscribed [79]. A uniform phase mask is then used to photo-write the grating.

Figure 5.33 shows the reflectivity and the resulting group delay of a linearly chirped FGB-based DCM. This component has been written according to the second method using an excimer laser (10 mJ, 200 Hz) and a standard uniform phase mask ($\Lambda = 1066.0$ nm). Moving the fibre in front of the mask introduced a chirp of 0.185 nm/cm, and according to (5.31) the resulting dispersion is equal to -530 ps/nm. The grating length was 70 mm and the apodization function an hyperbolic tangent.

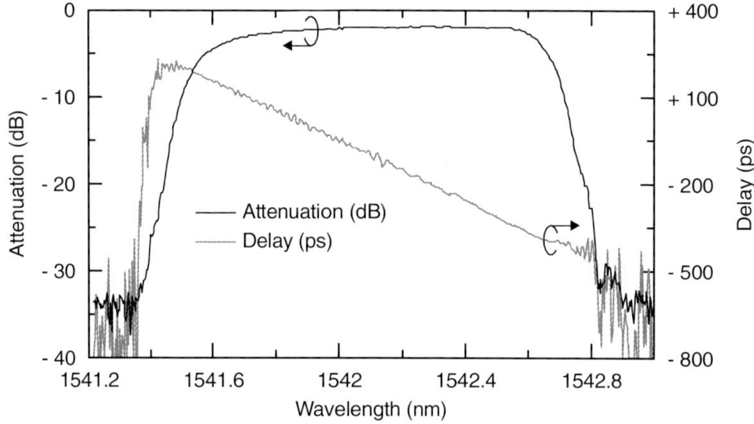

Fig. 5.33. Reflectivity and group delay spectrum of a chirped FBG-based DCM

Figure 5.33 shows a linear variation of the group delay over the filter bandwidth (constant dispersion), and the existence of a residual ripple on the delay curve can also be seen which will be explained below. In principle, the dispersion can be positive or negative, depending on the propagation direction of the beam in the grating region. Dispersion ranging from -2000 ps/nm to $+2000$ ps/nm has been demonstrated with module insertion loss < 2 dB.

FBG-based dispersion compensators are not restricted to operate over narrow wavelength bands only. As the delay increases with the grating length, one possibility to achieve broadband operation is making long FBGs (length > 1 meter). However, such devices are very difficult to realize and the group delay ripple is generally too high for network applications.

Another option is phase- or amplitude-sampled gratings [80]. Such components require a large number of phase shifts along the grating length. The expression of the index modulation $\Delta n_{mod}(z)$ in sampled gratings is given by

$$\Delta n_{\text{mod}}(z) = \Delta n_0 \, f_{\text{apod}}(z) \sum_{m=1}^{M} \cos\left(\frac{4\pi \, n_{\text{eff}} z}{\lambda_0 + \Delta\lambda_{\text{m}}} + \phi_{\text{m}} \right)$$

$$= \Delta n_0 \, f_{\text{apod}}(z) \, A_{\text{sampl}}(z) \cos\left(\frac{4\pi \, n_{\text{eff}} z}{\lambda_0} + \phi_{\text{sampl}}(z) \right)$$

(5.32)

where Δn_0 is the index modulation amplitude for a single channel, M is the number of channels, $\Delta\lambda_m$ is the spectral spacing between the mth channel and the central wavelength λ_0, and ϕ_m is the phase of the mth channel.

In the case of amplitude sampling (i. e. the case when $\phi_{sampl}(z) = \text{const}$) there are dead zones, where no gratings are photo-written, and there lower the reflectivity of the FBG as the number of WDM channels increases.

The phase-sampling method (i. e. the case when $A_{sampl}(z) = \text{const}$) is an alternative that overcomes these difficulties [81]. Since the refractive index modulation Δn required to maintain the reflectivity is much lower than that for the amplitude-sampling method, phase-sampled gratings can be fabricated more easily. Without using specially designed phase masks, phase-sampled FBGs have been realized by dithering and displacing standard phase masks. Phase-sampled gratings represent the most promising approach to realize multichannel DCM, but a very precise photo-inscription set-up and processing are mandatory [82].

A further variant of multi-channel DCM relies on the superposition of multiple Bragg gratings implemented over an arbitrary large range of wavelengths [83]. This third alternative, called superimposition, involves inefficient fabrication processes, because several FBGs are overwritten in the same location of a fibre, and each FBG corresponds to one channel. On the other hand, an advantage of the superimposed approach is that the separate FBGs are independent from each other. This allows the design of complex structures including channel to channel varying dispersion, which is useful for dispersion-slope compensation.

Multichannel FBG-based DCM offer many advantages compared to DCF, e. g. small footprint, low insertion loss, dispersion-slope compensation, and negligible non-linear effects. Table 5.3 compiles characteristic parameters of single- and 32-channel dispersion-compensating modules.

Table 5.3. Characteristics of single- and 32-channel chirped FBG-based dispersion-compensating modules

Parameter	Single-λ DCM	32-λ DCM
Channel bandwidth (GHz)	20 to 80	30
Insertion loss (dB)	1.5 (with circulator)	<3.5
Dispersion (ps/nm)	-2000 to $+2000$	-2000 to $+2000$
Group delay ripple (ps)	$<\pm 10$	$<\pm 40$
PMD (ps)	<0.5	<0.5
PDL (dB)	<0.1	<0.3
Athermal package size	168 mm length, 12 mm diameter	209 mm length, 14 mm diameter
Operating temperature (°C)	-5 to $+70$	-5 to $+70$

5.5.3 Tuneable Dispersion-compensating Module

Dynamic dispersion compensation is essential in DWDM optical communication systems operating at 40 Gbit/s and beyond. At these high bit rates, dispersion tolerances become so small that variations in dispersion can severely influence network performance. In such systems the amount of dispersion compensation required at the receiver to maintain optimum system performance may vary in time due to impairments which exhibit temporal variations. Factors which contribute to total dispersion are temperature fluctuations along the fibre, component dispersion, and dispersion variations in the transmission fibre. For example, for a 2000 km non-zero dispersion shifted fibre (NZDSF) span, a 10 to 20°C temperature change is sufficient to introduce measurable system impairments at 40 Gbit/s. Furthermore, in reconfigurable networks the total accumulated dispersion can experience significant sudden changes.

Chirped FBGs combined with an appropriate tuning platform are well suited for single or multiple channel dispersion compensation. Tuning can be achieved by creating a temperature gradient along the fibre [84] or by applying a linear strain gradient which can be provided by the S-bending beam technique for example [85]. Most commercially available tuneable dispersion-compensating modules (TDCM) use two (or more) thermoelectric coolers (TEC) located at each end of the grating. A thermal gradient is then created along the fibre, while the temperature at the centre is kept unchanged. This controllable temperature gradient generates chirp and thus dispersion variation. The chirp can be adjusted reversibly by varying the TEC current. If the grating in a DCM has an intrinsic chirp, the total chirp is the algebraic sum of both chirp contributions (intrinsic and reversible, TEC-induced one). As the total chirp is increased, the component

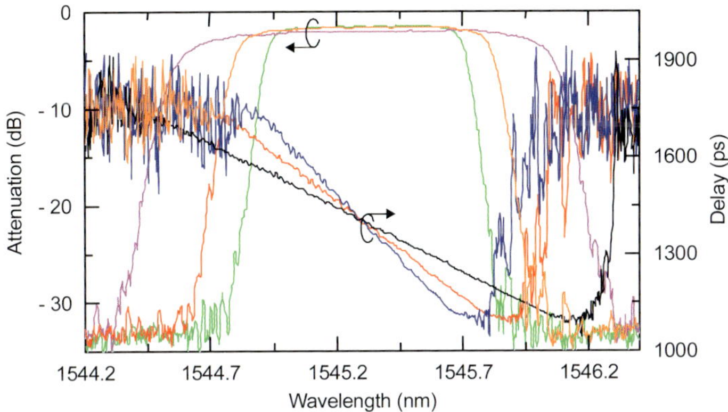

Fig. 5.34. Evolution of the spectral response of a dispersion-compensating module with tuning

bandwidth is increased, so that the delay slope, i. e. the dispersion, is reduced (in absolute terms). Corresponding integrated devices satisfy important requirements such as power efficiency, small size, and ease of fabrication. Figure 5.34 shows the evolution of the spectral bandwidth and the delay of a tuneable DCM.

It is important to note that with this architecture the Bragg wavelength is fixed, and that tuneability around a zero dispersion value is not possible. These spectra also show that the maximum reflectivity is not the same as the temperature gradient is changed. For the case of thermal tuning Fig. 5.35a illustrates the resulting non-linear dispersion variation for a grating with an initial dispersion value of -533 ps/nm.

As the index modulation is constant, the dispersion tuning is associated with a variation of the grating reflectivity. In the example of Fig. 5.35a it corresponds to 0.5 dB insertion loss variation.

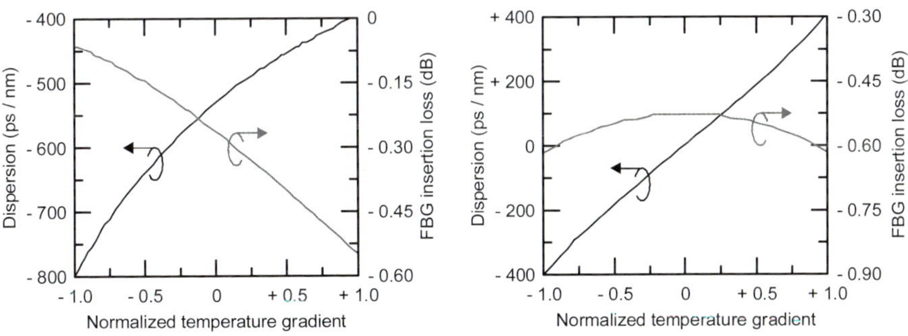

Fig. 5.35. Tuning characteristics examples of (**a**) single grating and (**b**) twin gratings configuration (after [86])

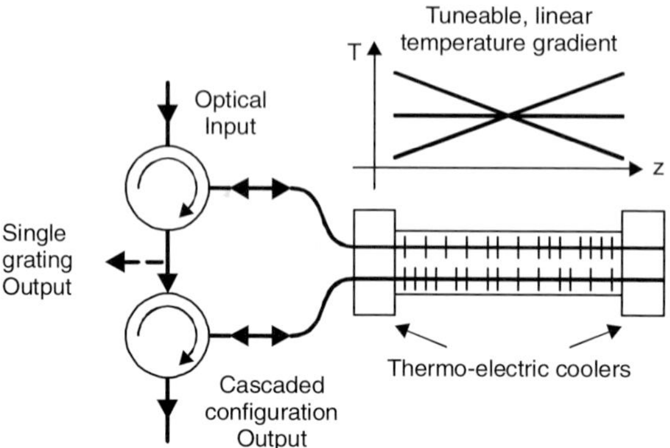

Fig. 5.36. Temperature tuning principle scheme (after [86])

Another configuration also using thermal tuning is illustrated in Fig. 5.36. Key elements are two identical, chirped cascaded FBGs and two 3-port circulators. Both gratings are simultaneously tuned (similar to the case of a single FBG), but the key point is that the chirp of the two FBGs varies in opposite directions with respect to the temperature gradient. As the sign of the dispersion of a chirped FBG changes with propagation direction, the total dispersion is zero without applied temperature gradient [86]. A temperature gradient induces opposite bandwidth variations for each grating. The tuning characteristics can simply be deduced from the single grating case (one has to keep in mind that transmission losses are expressed in decibels and dispersion values simply add). The tuning range (Fig. 5.35b) is now centred around zero and is doubled without any additional power consumption. The insertion loss variation with tuning is considerably reduced (less than 0.1 dB in this example) because both reflectivity variations cancel each other. Another interesting point is the quasi-linear relation between the applied gradient and the resulting dispersion, i. e. one gets a quasi-constant tuning sensitivity.

With a maximum temperature variation of 60°C the dispersion tuning range extends from −400 to +400 ps/nm. The total electrical power consumption of this component is 5 watts during the transient regime and 3 watts in the stationary state.

Mechanical designs are also very attractive [87], but they are only suited for applications with no need to permanently adjust the dispersion. The advantage of such a 'set and forget'-device is that there is no electrical power consumption except when it is tuned to the proper dispersion compensating value.

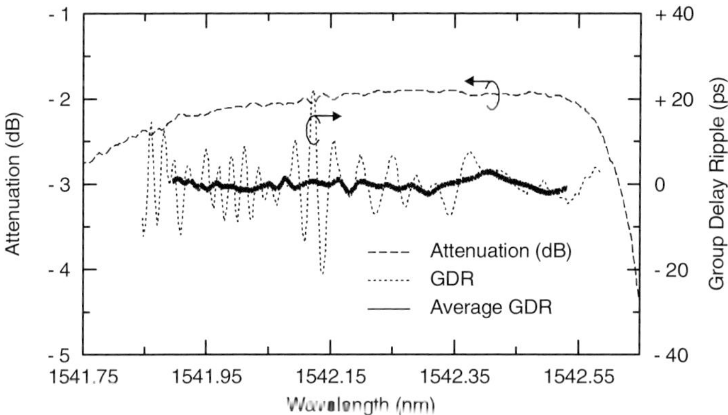

Fig. 5.37. Reflectivity (*dashed*), group delay ripple after subtracting a polynomial fit of the delay over the 0.5 dB bandwidth (*dotted*), and mean group delay ripple with a 100 pm resolution (*heavy line*)

5.5.4 Ripple Analysis in Dispersion-compensating Modules

The main limitation of chirped fibre Bragg grating (CFBG) technology is currently given by the well-known phase ripple (cf. also Chap. 2, Sect. 2.4.3) which is due to a non-optimum apodization function and/or imperfections related to the photo-inscription process. Although ripples can be equivalently characterized by the dispersion, the group delay, or the phase spectrum, respectively, there was until recently an implicit preference for directly measured group delay spectra (Fig. 5.37). However, very recent work showed that characterising the ripple over the phase spectrum instead of the group delay spectrum provides significant advantages. In addition, there is clear evidence that minimizing the group delay ripple (GDR) amplitude is not enough to reduce the impact on system performance, but the GDR period is also of high importance [88].

The worst case eye-opening penalty (EOP) associated with the use of a DCM can be calculated from the period and the amplitude of the group delay ripples [89] which can be separated into high and low frequency ones (Fig. 5.37). The latter are associated with a slowly varying curve, which gives a residual dispersion and leads to a broadening of the input pulse. The corresponding system impact can be estimated with conventional techniques. High frequency ripples are much more complicated to understand and to correct as well. They create new pulses which interfere with adjacent pulses. For a given GDR amplitude highest system impairments occur when the GDR period, expressed in optical frequency units,

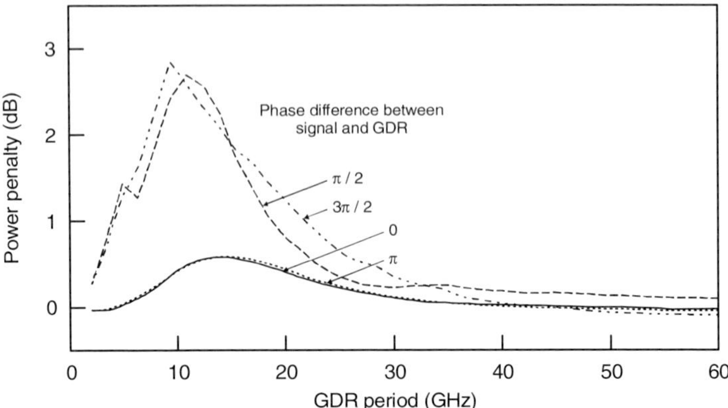

Fig. 5.38. Penalty versus group delay ripple (GDR) period for different phase differences between signal and ripple (NRZ modulation format, 10 Gbit/s)

equals the bit rate (10, 40 Gbit/s). For a high frequency sinusoidal ripple the EOP is approximately proportional to the product of the GDR amplitude and period [90], (cf. Fig. 5.38), i. e. to the phase-ripple amplitude.

One approach to determine the impact of GDR ripple on a telecommunication system is known as wavelength-frequency signal analysis based on spectrogram distributions (a method more commonly used in mechanical vibration analysis). The method reveals the variation of the frequency spectrum with respect to wavelength, and it provides useful information on the presence or absence of specific periods (ripple with 80 pm or 320 pm period is particularly detrimental for 10 or 40 Gbit/s systems, respectively) and on their localization inside the component bandwidth. It also helps to differentiate between apodization-induced and index variation noise-induced GDR [91], and it enables linking the measured penalty spectrum with the spectrogram and to correlate the peak penalty and critical frequencies in the GDR at the same wavelength as well.

An alternative approach to determine the impact of GDR on system performance is based on phase measurements since the phase ripple magnitude is a good indicator of system performance as long as it includes high frequency ripple components only. This is outlined in more detail in recent publications [92–94].

5.5.5 Gain-flattening Filters

Gain-equalized optical amplifiers have allowed the deployment of new DWDM systems. Power uniformity over the communication bandwidth is an important requirement in such systems since flatter optical amplifier gain enables longer fibre spans between regenerators which is more cost effective. Moreover, gain flatness of EDFAs is necessary to improve the optical signal-noise ratio. The straightforward option for gain equalization is the insertion of a gain flattening filter (GFF) with a spectral response precisely tailored to the inverse gain profile. In addition to exhibiting a spectral shape as close as possible to the inverse EDFA gain curve, GFFs should have low polarisation dependence, small footprint, and low loss.

Several technologies have been developed to perform gain equalization. The most common choice is dielectric thin film filters (cf. Chap. 7) which are particularly attractive with respect to cost due to high volume production. Long period gratings (LPG) [95] would in principle be a second possibility, however, they did not find any significant use with EDFAs due to their poor ageing characteristics (in particular, UV LPG). Finally, chirped and slanted/chirped short period FBGs form the last technology option, and best specifications have been obtained so far with this approach. Their superior performance combined with their good ageing behaviour and package reliability have led the main sub-marine carriers to use this technology.

Characteristic data of GFFs realized using different technologies are listed in Table 5.4.

Table 5.4. Comparison of gain-flattening filters based on different technologies (after [96])

Characteristics over full T range	Thin-film Filter	Long-period Grating	Grade A Chirped GFF	Slanted/ Chirped GFF
Error function	$\pm 0.5\,$dB	$\pm 0.5\,$dB	$\pm 0.1\,$dB	$\pm 0.2\,$dB
PDL	$<0.1\,$dB	$<0.2\,$dB	$<0.1\,$dB	$<0.1\,$dB
PMD	n.a.	n.a.	$<0.1\,$ps	$<0.1\,$ps
Insertion loss (out of band)	1 dB	0.3 dB	0.3 dB	0.5 dB
Back reflection	low	low	high	low
Temperature dependence	$<5\,$pm$/\,^{\circ}$C	$<3\,$pm$/\,^{\circ}$C	$<1\,$pm$/\,^{\circ}$C	$<1\,$pm$/\,^{\circ}$C
Dimensions	small	large	small	small

The fabrication of short period chirped FBGs for GFFs starts by using a chirped phase mask to change the period of the fringe pattern along the length of the grating and by adjusting the induced index modulation to match the required transmission. This adjustment can be made by blurring the fringes with a displacement of the phase mask in front of the fibre during the photo-inscription process. These filters are compatible with

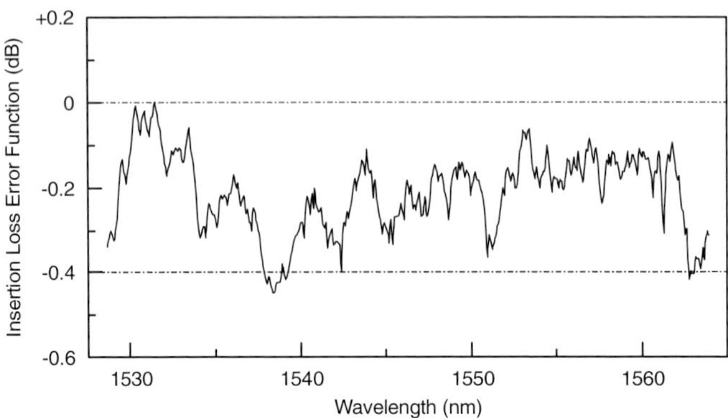

Fig. 5.39. Residual insertion loss variation of a gain-compensated EDFA

standard SMF-28 fibre and in most cases no particular fibre is needed, although fibres with a specific refractive index and photosensitivity profile may more efficiently suppress perturbations due to cladding mode coupling. The short piece of (specific) fibre with the grating is then spliced and sandwiched between two standard fibres. The GFF component itself is incorporated into the EDFA mostly in the middle of the erbium-doped fibre in order to avoid noise degradation. Losses are generally low and adequate annealing and packaging (hermetic sealing or fibre metallization) assure long term stability (> 25 years).

The main drawback of these filters, which are reflective filters used in transmission, is the need of optical isolators in order to avoid feedback inside the amplifier stage, which results in additional losses and higher overall cost. Furthermore, even if the induced dispersion and group delay ripple in transmission are very small ($< \pm 2$ ps around zero dispersion), the CFBG can sometimes lead to multi-path interference phenomena and pulse echoes shifted in time by a quantity equal to the reciprocal of the ripple modulation period. However, experiments show that group delay ripple of gain-flattening filters based on chirped FBGs do not cause significant system impairment at 10 and 40 Gbit/s [97]. Residual gain variations can be reduced to ± 0.2 dB as illustrated in Fig. 5.39, and one of the key parameters to get ultra low flatness over the whole gain curve is the quality of the phase mask.

An alternative to standard GFFs are slanted chirped short period gratings (SCG) which offer a number of advantages compared to standard CFBG gain-flattening filters. Instead of coupling the power into the counter-propagating mode (reflective filter), slanted gratings direct the energy into the radiative modes (cf. Sect. 5.3.4). In order to minimize the residual coupling into the counter-propagating mode, a specific index profile-fibre is

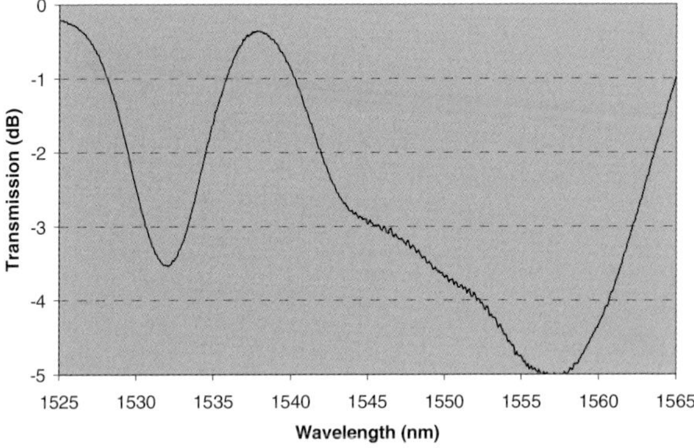

Fig. 5.40. Spectral shape of gain-flattening filter based on a slanted chirped grating

used with an adequate photosensitive profile. These two profiles are different, and different photosensitive species (like germanium or boron) can be used to obtain the required photosensitivity. Similar to standard chirped gratings, these slanted gratings are also chirped, i. e. the period of the fringe pattern changes along the grating (typical chirp is 10 nm/cm). The induced index modification is longitudinally adjusted to match the required transmission, and the spectral shape of the filter characteristics is the convolution of this blurred profile with the slanted grating elementary response. The final result is a smooth large band filter with a global shape very close to the complementary EDFA shape (Fig. 5.40). Using this component flatness smaller than 0.1 dB peak to peak has been demonstrated. The residual flatness is determined by the grating angle θ (larger cell for higher angle), the fringe pattern visibility, and the UV spot size.

The high frequency ripple which is relatively important in standard chirped GFFs (±0.1 dB) is reduced to less than ±0.02 dB for SCGs. Such low ripple is obtained by recoating the component with a polymer with an index greater than that of silica which suppresses cladding modes associated with high frequency ripple. The size and packaging of SCGs are similar to those of standard chirped gratings. Athermal Telcordia-qualified products for terrestrial and sub-marine applications are commercially available.

The residual reflectivity of slanted chirped FGBs can be made smaller than −25 dB, thus the requirements on optical isolation are relaxed compared to standard gratings, and in some EDFA configurations one single-stage isolator is sufficient instead of two two-stage isolators.

In telecommunication systems it is not only the gain flatness of a single EDFA which matters, but also the behaviour of cascaded EDFAs. If the GFF technology used exhibits systematic errors (such as dielectric multilayer filters, for example), the total error accumulates linearly with the number of amplifiers, and the resulting error at the end of a chain of EDFAs has to be corrected by a clean-up filter or by signal regeneration. Chirped (slanted) GFFs exhibit a small systematic error and a high degree of random error (manifesting itself primarily by the high frequency ripple), and cascading these components causes a statistical error accumulation which is less serious than linear accumulation. For example, a peak-to-peak flatness per GFF better than 0.1 dB for a chain of 10 cascaded EDFAs has been demonstrated [98].

The polarisation-dependent loss of chirped or SCGs is small (< 0.1 dB), but it can sometimes be detrimental for a system. The PDL spectral shape can be explained by considering that it comes from a birefringence-induced spectral splitting of the two polarisation axes [99]. This effective index birefringence can be explained by an asymmetric UV-induced refractive-index profile across the fibre core due to the absorption of the writing UV-laser beam. This residual PDL can be reduced by double-side exposure of the fibre.

5.5.6 Wavelength Tuning and Chirping of Bragg Gratings Using Infrared Lasers

Kalli and co-workers reported the first experimental measurements on the spectral modification of fibre Bragg gratings (Type IA), resulting from high-power, near infrared laser irradiation [100]. The grating properties were modified in a controlled manner by exploiting the characteristics of the inherent 1400 nm absorption band of the optical fibre, which grows during the grating inscription and is due to the optical fibre pre-conditioning. Therefore, the increase in absorption occurs in the grating region. Illuminating the area with a high power laser having an emission wavelength coincident with the absorption band, produced reversibly modified centre wavelength and chirp. Furthermore, partial and permanent grating erasure was demonstrated.

A low power (10 mW) laser source coinciding with the absorption peak at 1400 nm induced small but significant wavelength shifts of approximately 100 pm. This wavelength shift would change the grating's spectral response adversely affecting the filter performance and reducing the isolation between wavelength channels, transmission properties, and effective bandwidth. A second source operating at 1425 nm and far from the absorption peak induced wavelength shifts in excess of 750 pm and a 30% increase in FWHM for a pump power of 350 mW. This has serious implications on all grating types when the fibre undergoes photosensitivity pre-conditioning, as their spectrum can be modified using purely optical methods (no external heat source acts on the fibre), and it is also of relevance for long-term grating stability. It should be noted that high power lasers are increasingly being used in optical networks and this study may have greater implications to all grating types, as laser powers and the useable wavelength spectrum increase. This may result from the presence of absorption features in the visible and near infrared that are produced due to the fibre being pre-conditioned, prior to grating inscription (as in this case). Absorption features are observable at shorter wavelengths for conventional Type I gratings inscribed in hydrogenated fibre. Although there is no evidence that they have consequences for grating lifetime, their impact (or negligible impact) has still to be established conclusively.

Conversely, there are applications where suitably stabilized Type IA gratings can be spectrally tailored for tuning fibre lasers or modifying edge filters in sensing applications. The latter results from the non-uniform absorption of the pump laser source as it traverses the Type IA grating. The fact that this type of spectral tuning can be realized by the use of an additional laser source can be advantageous, as no special coatings to the fibre are necessary, and all degrees of tuning can be set during the grating manufacturing process which offers great flexibility at the design stage. Since all grating types can be written in a section of pre-exposed fibre, this method of optical tuning could be used for all existing Bragg grating applications making the technique invaluable to a multitude of applications. Finally we note that it is possible to tailor the absorption of the pre-exposed section to mirror the decay in intensity resulting in a uniform heating of the grating. However, this would alter the mean fibre index along the pre-exposed section, inducing a potentially large (up to 20 nm) chirp across the grating.

5.6 Other Applications of Fibre-based Bragg Gratings

Beyond their use in telecommunication systems fibre Bragg gratings have emerged as important components in a variety of other lightwave applications such as wavelength stabilized lasers, fibre lasers, remotely pumped amplifiers, Raman amplifiers, phase conjugators, or wavelength converters for example, and they are also considered excellent sensor elements, suitable for measuring static and dynamic fields such as temperature, strain, and pressure [101]. The principal advantage is that the information to be determined is wavelength-encoded (an absolute quantity) thereby making the sensor self-referencing, rendering it independent of fluctuating light levels and the system immune to source power and connector losses that plague many other types of optical fibre sensors. It follows that any system incorporating Bragg gratings as sensor elements is potentially interrupt-immune. Their very low insertion loss and narrowband wavelength reflection offers convenient serial multiplexing along a single monomode optical fibre. There are further advantages of the Bragg grating over conventional electrical strain gauges, such as linearity in response over many orders of magnitude, immunity to electromagnetic interference (EMI), light weight, flexibility, stability, high temperature tolerance, and even durability against high radiation environments (darkening of fibres). Moreover, Bragg gratings can easily be embedded into materials to provide local damage detection as well as internal strain field mapping with high localization, strain resolution, and measurement range. The Bragg grating is an important component for the development of smart structure technology, with applications also emerging in process control and aerospace industries. This section will describe in brief some applications of fibre gratings.

5.6.1 Fibre Bragg Grating Diode Lasers

A fibre Bragg grating may be coupled to a semiconductor laser chip to obtain a fibre Bragg diode laser [102]. A semiconductor laser chip is anti-reflection coated on the output facet and coupled to a fibre with a Bragg grating as illustrated in Fig. 5.41. If this Bragg grating reflects at the gain bandwidth of the semiconductor material it is possible to obtain lasing at

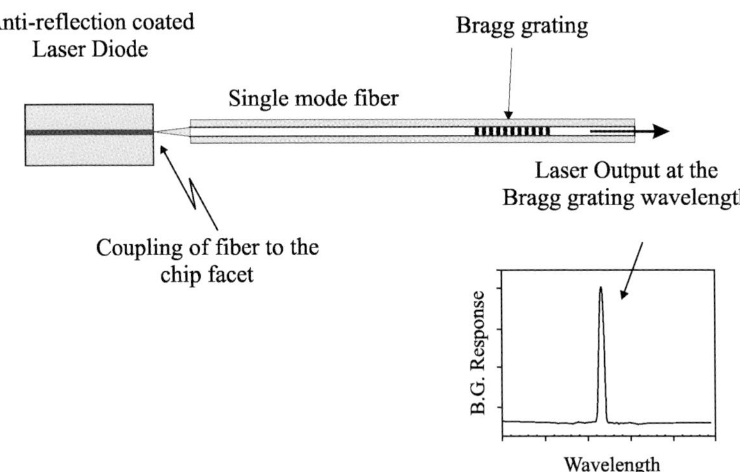

Fig. 5.41. External cavity fibre grating semiconductor laser. Semiconductor laser chip is antireflection coated on the output facet and coupled to a fibre with a Bragg grating, forcing oscillation at the Bragg grating wavelength

the Bragg grating wavelength. The grating bandwidth can be narrow enough to force single frequency operation with a linewidth of much less than a GHz. High output powers up to tens of mW have been obtained with these types of lasers. An added advantage of these systems is their temperature sensitivity, which is approximately 10% of that of a semiconductor laser, thus reducing temperature induced wavelength drift. These fibre grating semiconductor laser sources have been used to generate ultrashort mode-locked soliton pulses up to 2.9 GHz [103].

A manufacture problem in DBR lasers is the precise control of the laser wavelength. Routine production of DBR lasers with wavelength specified to better than 1 nm is difficult. On the other hand, Bragg gratings can be manufactured precisely (better than 0.1 nm) to the wavelength required. With anti-reflection coating on the semiconductor chip, the lasing wavelength may be selected from anywhere in the gain bandwidth by choosing the appropriate fibre Bragg grating. Clearly, such an approach will increase the yield from semiconductor wafers. In addition, since each laser has to be coupled to a fibre, the Bragg grating may be written after the packaging process has proved to be successful, thus reducing the time spent on unsuccessful products.

5.6.2 Fibre Bragg Grating Lasers

The majority of Bragg grating fibre laser research has been on erbium-doped lasers due to their potential in communication and sensor applications. The characteristic broadband gain profile of the erbium-doped fibre around the 1550 nm region makes it an extremely useful tuneable light source. Employing this doped fibre in an optical cavity as the lasing medium, along with some tuning element, results in a continuously tuneable laser source over its broad gain profile. In fact, a tuneable erbium-doped fibre with an external grating was reported by Reekie et al. [104] in 1986. Since then several laser configurations have been demonstrated with two or more intracavity gratings [105–109].

A simple Bragg grating tuneable Er-doped fibre laser was demonstrated, where a broadband Bragg mirror and a narrow Bragg grating served as the high reflector and the output coupler, respectively [110]. The broadband mirror was constructed from a series of Bragg gratings resulting in the broadband reflector with a bandwidth of approximately 4 nm. It should be noted that with today's advancements in photosensitivity and writing techniques such a broadband mirror may have any shape and bandwidth desired. The fibre laser consisted of a two-meter long erbium-doped fibre with Bragg gratings at each end (broadband and narrowband) providing

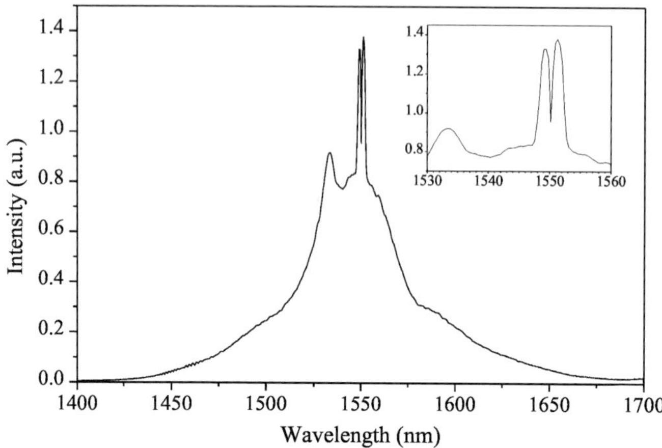

Fig. 5.42. Broadband fluorescence of an erbium-doped fibre laser. Broadband peak at 1550 nm due to broadband Bragg grating, notch within peak corresponding to Bragg grating (after [110])

feedback to the laser cavity. The output coupler to the fibre laser cavity was a single grating with approximately 80% reflectivity and 0.12 nm linewidth. Figure 5.42 shows the broadband fluorescence obtained from the Er-doped fibre laser system before lasing threshold is reached. The spectrum is the characteristic broadband gain profile from an erbium-doped fibre spanning a range of several tens of nanometres, namely between 1.45 and 1.65 μm. Superimposed on the gain profile is a broadband peak at 1550 nm corresponding to the reflection of the fluorescence from the broadband Bragg mirror and within this peak there is a notch at 1550 nm corresponding to the narrow Bragg grating. With increasing incident pump power, the losses in the fibre laser cavity are overcome and lasing begins. At pump powers just above threshold the notch due to the Bragg grating begins to grow in the positive direction and as the pump power increases further, the laser grows even stronger by depleting the broadband fluorescence.

Single frequency Er^{3+}-doped Fabry–Perot fibre lasers using fibre Bragg gratings as the end mirrors [111, 112] are emerging as an interesting alternative to distributed feedback (DFB) diode lasers for use in future optical cable television (CATV) networks and high capacity WDM communication systems [113]. They are fibre compatible, simple, scaleable to high output powers, and have low noise and kilohertz linewidth. In addition, the lasing wavelength can be determined to an accuracy of better than 0.1 nm, which is very difficult to achieve for DFB diode lasers.

Fibre lasers can operate in a single frequency mode provided that the grating bandwidth is kept below the separation between the axial mode spacings. Furthermore, it is necessary to keep the erbium concentration low enough (a few 100 ppm) to reduce ion-pair quenching, which causes a reduction in the quantum efficiency and in addition may lead to strong self-pulsation of the laser [112, 113]. The combination of these practical limits implies that the pump absorption of an erbium-doped fibre system can be as low as a few percent resulting in low output lasing power. One solution to this problem is to use the residual pump power to pump an erbium-doped fibre amplifier following the fibre laser. However, in such cases the amplified spontaneous emission from the amplifier increases the output noise. Another way to overcome the problem of low pump absorption is by codoping the erbium-doped fibre with Yb^{+3}. This increases the absorption at the pump wavelength by more than two orders of magnitude and enables highly efficient operation of centimetre long lasers with relatively low Er^{3+} concentration. Kringlebotn et al. [114] reported a highly-efficient, short, robust single-frequency and linearly polarised Er^{3+}:Yb^{3+}-codoped fibre laser with fibre grating Bragg reflectors, an output power of 19 mW, and a linewidth of 300 kHz for 100 mW of 980 nm diode pump power.

One other interesting application of Bragg gratings in fibre lasers makes use of stimulated Raman scattering. The development of fibre Bragg gratings has enabled the fabrication of numerous highly reflecting elements directly in the core of germanosilicate fibres. This technology coupled with that of cladding-pumped fibre lasers has made fibre Raman lasers possible. The pump light is introduced through one set of highly reflecting fibre Bragg gratings. The cavity consists of several hundred metres to a kilometre of germanosilicate fibre. The output consists of a set of highly reflecting gratings through Raman-order n-1 and the output wavelength of Raman-order n is coupled out by means of a partially reflecting fibre grating ($R \sim 20\%$). The intermediate Raman Stokes orders are contained by sets of highly reflecting fibre Bragg gratings and this power is circulated until it is nearly entirely converted to the next successive Raman Stokes order [115].

5.6.3 Fibre Bragg Grating Sensors

Fibre Bragg gratings are excellent fibre optic sensing elements. They are integrated into the light guiding core of the fibre and are wavelength encoded, eliminating the problems of amplitude or intensity variations that plague many other types of fibre sensors. Due to their narrow band wavelength reflection they are also conveniently multiplexed in a fibre optic network. Fibre gratings have been embedded into composite materials for smart structure monitoring and tested with civil structures to monitor load levels. They have also been successfully tested as acoustic sensing arrays. Applications for fibre grating sensors should also be emerging in process control and aerospace industries in the near future.

The temperature sensitivity of a Bragg grating occurs principally through the effect on the index of refraction and to a smaller extent through the expansion coefficient (Sect. 5.2.4). It is noteworthy that temperature sensitivity can be enhanced or eliminated by proper bonding to other materials. The maximum operating temperatures may be around 500 °C, however, this may depend on the fabrication condition of the Bragg grating. For example, Type II gratings may operate at higher temperatures than Type I gratings.

Strain affects the Bragg response directly through the expansion or contraction of the grating elements and through the strain optic effect. Many other physical parameters other than tension can also be measured such as pressure, flow, vibration acoustics, acceleration, electric, magnetic fields, and certain chemical effects. Therefore, fibre Bragg gratings can be thought off as generic transducer elements. There are various schemes for detecting the Bragg resonance shift, which can be very sensitive. One such scheme involves the injection of a broadband light (generated e. g. by a super-luminescent diode, an edge-emitting LED, or an erbium-doped fibre super-fluorescent source) into the fibre and determining the peak wavelength of the reflected light. Another way involves the interrogation of the Bragg grating with a laser tuned to the sensor wavelength, or by using the sensor as a tuning element in a laser cavity. Detecting small shifts in the Bragg wavelength of fibre Bragg grating sensor elements, which corresponds to changes of the sensing parameter is important. In a laboratory environment this can be accomplished using a high precision optical spectrum analyzer. In practical applications, this function must be performed using compact, low cost instrumentation. Schemes based on simple broadband optical filtering, interferometric approaches, and fibre-laser approaches allow varying degrees of resolution and dynamic range and should be suitable for most applications.

The most straightforward means for interrogating an FBG sensor is using a passive broadband illumination of the device, and several options exist for measuring the wavelength of the optical signal reflected from the Bragg grating element, for example a miniaturized spectrometer, passive optical filtering, tracking using a tuneable filter, and interferometric detection. The optical characteristics of these filtering options are as shown in Fig. 5.43. Filtering techniques based on the use of broadband filters allow the shift in the Bragg grating wavelength of the sensor element to be assessed by comparing the transmittance through the filter compared to

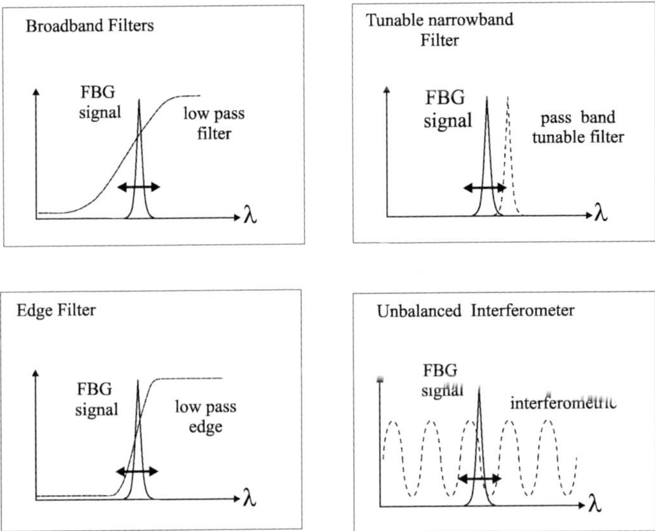

Fig. 5.43. Diagram of basic filtering function for processing fibre Bragg grating return signals

a direct 'reference' path [116]. A relatively limited sensitivity is obtained using this approach due to problems associated with the use of bulk-optic components and alignment stability. One way to improve this sensitivity is to use a fibre device with a wavelength-dependent transfer function, such as a fibre WDM coupler. Fused WDM couplers for 1550/1570 nm operation are commercially available. This coupler will provide a monotonic change in coupling ratio between two output filters for an input optical signal over the entire optical spectrum of an erbium broadband source, and has thus a suitable transfer function for wavelength discrimination over this bandwidth. An alternate means to increasing the sensitivity is to use a filter with a steeper cut-off such as an edge filter. However, this can limit the dynamic range of the system. One of the most attractive filter-based techniques for interrogating Bragg grating sensors is based on the use of a tuneable passband filter for tracking the Bragg grating signal. Examples of these types of filter include Fabry–Perot filters [117], acousto-optic filters [118], and fibre Bragg grating-based filters [119].

Tuneable Filter Interrogation

Figure 5.44 shows the configuration used to implement a tuneable filter (such as a fibre Fabry–Perot filter) to interrogate a Bragg grating sensor. The fibre Fabry–Perot can be operated in either a tracking or scanning mode for addressing a single or multiple grating element(s), respectively.

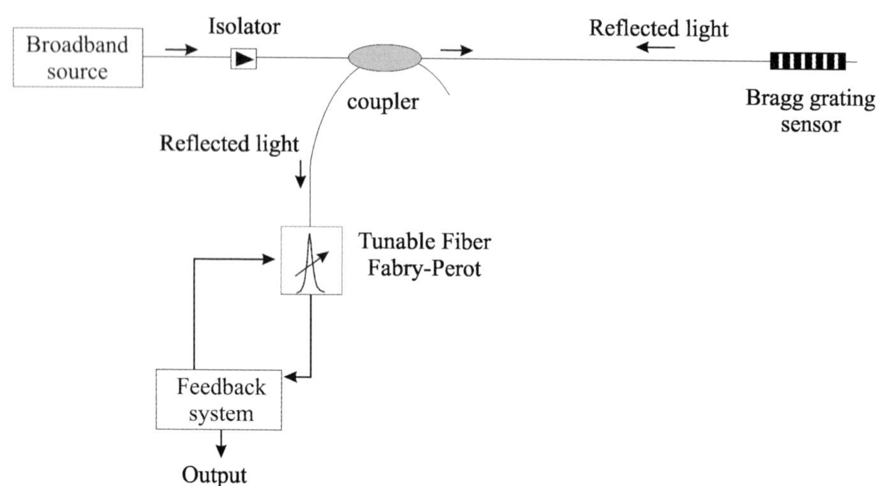

Fig. 5.44. Schematic of tuned filter-based interrogation technique for fibre Bragg grating sensors

In a single sensor configuration the Fabry–Perot filter with a bandwidth of 0.1 nm is locked to the Bragg grating reflected light using a feedback loop. This is accomplished by dithering the fibre Fabry–Perot resonance wavelength by a small amount (typically 0.01 nm) and using a feedback loop to lock to the Bragg wavelength of the sensor return signal. The fibre Fabry–Perot control voltage is a measure of the mechanical or thermal perturbation of the Bragg grating sensor.

Operating the fibre Fabry–Perot filter in a wavelength-scanning mode provides a means for addressing a number of fibre Bragg grating elements placed along a fibre path (Fig. 5.45). In this mode, the direct Bragg grating sensor spectral returns are obtained from the photodetector output. If the minimum resolvable Bragg wavelength shift that can be detected by simple scanning is insufficient, the resolution can be enhanced by dithering the Fabry–Perot filter transmission, which provides the derivative response of the spectral components in the array, i.e. a zero crossing at each of the Bragg grating centre wavelengths. This technique improves the accuracy in determining the wavelength shifts and hence the strain (or any other sensing parameter the transducer is made for).

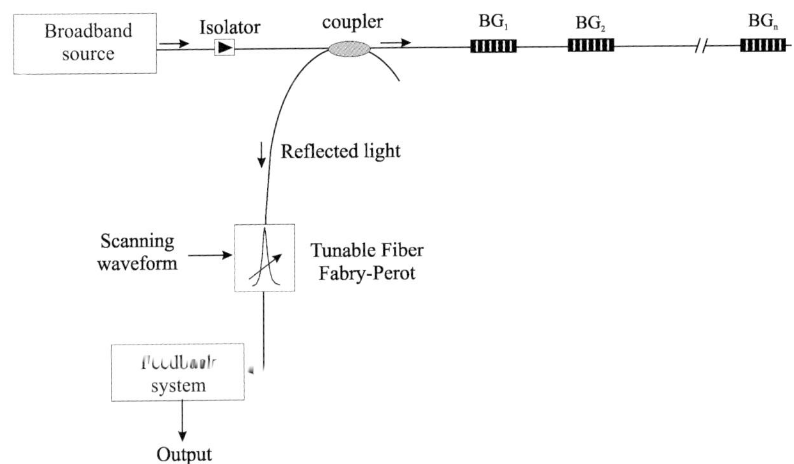

Fig. 5.45. Schematic of multiplexed fibre Bragg grating sensor array with scanning Fabry–Perot filter

Interferometric Interrogation

A sensitive technique for detecting the wavelength shifts of fibre Bragg grating sensors makes use of a fibre interferometer. The principle behind such system is shown in Fig. 5.46. Light from a broadband source is coupled along a fibre to the Bragg grating element. The wavelength component reflected back along the fibre toward the source is tapped off and fed

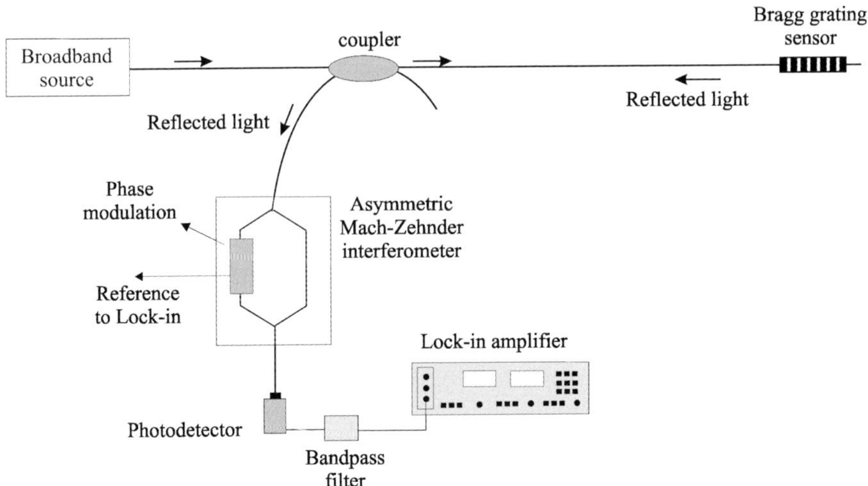

Fig. 5.46. Schematic of interferometer interrogation technique for fibre Bragg grating sensors

to unbalance the Mach–Zehnder interferometer. In effect, this light becomes the light source into the interferometer, and wavelength shifts induced by perturbation of the Bragg grating sensor resemble a wavelength-modulated source. The unbalanced interferometer behaves as a spectral filter with a raised cosine transfer function. The wavelength dependence at the interferometer output can be expressed as

$$I(\lambda_B) = A\left[1 + k\cos\left(\frac{2\pi n_{eff} d}{\lambda_B} + \phi\right)\right]$$

(5.33)

where A is proportional to the input intensity and system losses, d is the length imbalance between the fibre arms, n_{eff} is the effective index of the core, λ_B is the wavelength of the return light from the grating sensor, and ϕ is a bias phase offset of the Mach–Zehnder interferometer. Pseudoheterodyne phase modulation is used to generate two quadrature signals with a 90°-phase shift with respect to each other, thus providing directional information. Wavelength shifts are tracked using a phase demodulation system developed for interferometric fibre optic sensors. In practical applications, a reference wavelength source is used to provide low frequency drift compensation. Strain resolution as low as $0.6\,\text{n}\varepsilon/\text{Hz}^{-0.5}$ at 500 Hz have been reported [120].

Active Laser Interrogation

In active interrogation the fibre Bragg grating sensor is used as an optical feedback element of an optical laser cavity [121]. Compared to the passive broadband base system, forming a fibre Bragg laser sensor generally provides stronger optical signals and has thus the potential to provide improved signal to noise performance. The basic concept is shown in Fig. 5.47. The laser cavity is formed between the mirror and the fibre Bragg grating element, which may be located at some sensing point. A gain section within the cavity can be provided via a semiconductor or doped fibre (such as the erbium-doped fibre). Once the laser gain is greater than unity, the fibre laser will lase at the wavelength determined by the fibre Bragg grating wavelength. As the Bragg grating changes its periodicity due to strain or temperature, the lasing wavelength will also shift. Reading of the laser wavelength using filtering, tracking filters or interferometric techniques can then be used to determine induced shifts. This laser sensor configuration is, however, limited to a single fibre Bragg grating element. A means to increase the number of Bragg gratings that can be addressed is to incorporate an additional tuning element within the cavity,

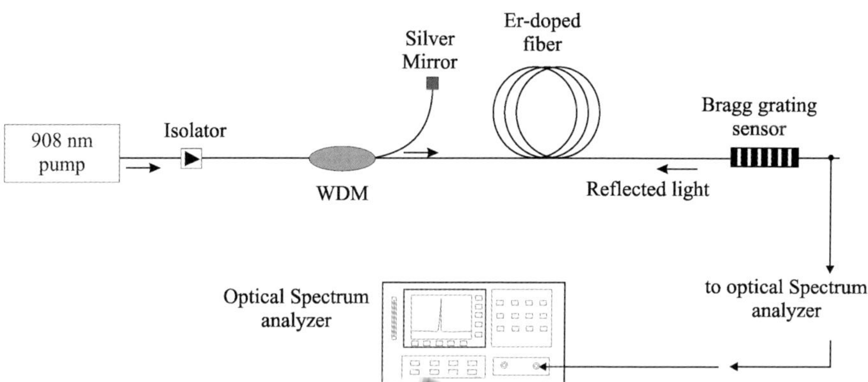

Fig. 5.47. Schematic of fibre laser sensor configuration with fibre Bragg grating elements

which selectively optimises the gain at certain wavelengths. In this way, a number of fibre Bragg gratings, each operating at a nominally different wavelength, can be addressed in a sequential manner to form a quasi-distributed fibre laser sensor. By tuning a wavelength selective filter located within the laser cavity over the gain bandwidth the laser selectively lases at each of the Bragg wavelengths of the sensors. Thus strain induced shifts in the Bragg wavelengths of the sensors are detected by the shift in the lasing wavelengths of the system.

An alternative multiplexed fibre laser sensor is based on a single element fibre laser sensor utilizing wavelength division multiplexing. Theoretically, since erbium is a homogeneously broadened medium it will support only one lasing line simultaneously. To produce several laser lines within a single length of optical fibre, a section of erbium-doped fibre is placed between the successive Bragg gratings. With sufficient pump power and enough separation between the Bragg grating centre wavelengths, a multiplexed fibre laser sensor is possible. The maximum number of sensors utilized would depend on the total pump power, the required dynamic range, and finally the gain profile of the active medium. A schematic configuration of the serially multiplexed Bragg grating fibre laser is shown in Fig. 5.48. One of the drawbacks in such a serial multiplexed configuration is that the cavities are coupled, so their respective gains are not independent. In fact, gain coupling is a common effect in such systems.

At the cost of adding more elements in a fibre laser sensor system, an alternative is to multiplex the fibre laser sensor in a parallel configuration. In essence, this system incorporates several single fibre lasers, one for each fibre Bragg grating.

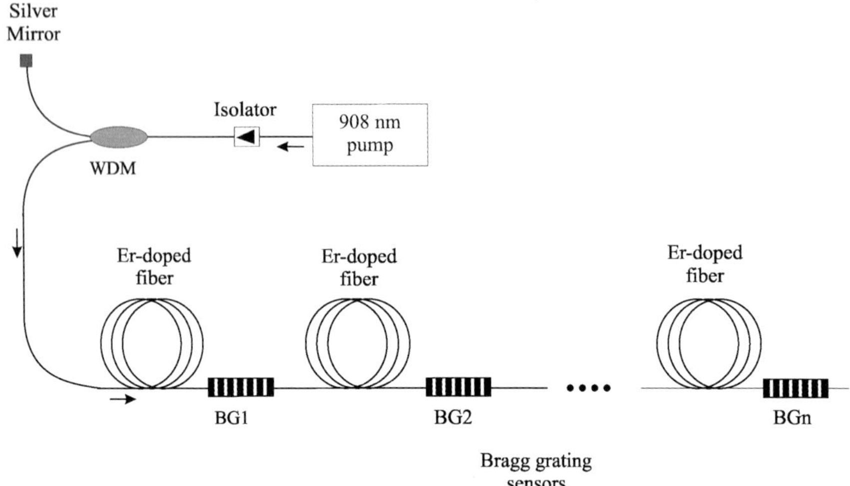

Fig. 5.48. Schematic configuration of serially multiplexed Bragg grating fibre laser

Simultaneous Measurements of Strain and Temperature

Although Bragg gratings are well suited for measuring strain and temperature in a structure, one of the drawbacks is the actual separation of the temperature from the strain component (Sect. 5.2.4). This complicates the Bragg grating application as a strain or a temperature gauge. In the case of a single measurement of the Bragg wavelength shift it is impossible to differentiate between the effects of changes in strain and temperature. Various schemes for discriminating between these effects have been developed. These include the use of a second grating element contained within a different material and placed in series with the first grating element [122] and the use of a pair of fibre gratings surface-mounted on opposite surfaces of a bent mechanical structure [123]. However, these methods have limitations when it is required to interrogate the wavelength of a large number of fibre gratings. Techniques such as measuring two different wavelengths, two different optical or grating modes have been employed [18]. In another scheme two superimposed fibre gratings with different Bragg wavelengths (850 nm and 1300 nm) have been used to simultaneously measure strain and temperature (Fig. 5.49). The change in the Bragg wavelength of the fibre grating due to a combination of strain and temperature can be expressed as

$$\Delta\lambda_B(\varepsilon,\lambda) = \Psi_\varepsilon \Delta\varepsilon + \Psi_T \Delta T \qquad (5.34)$$

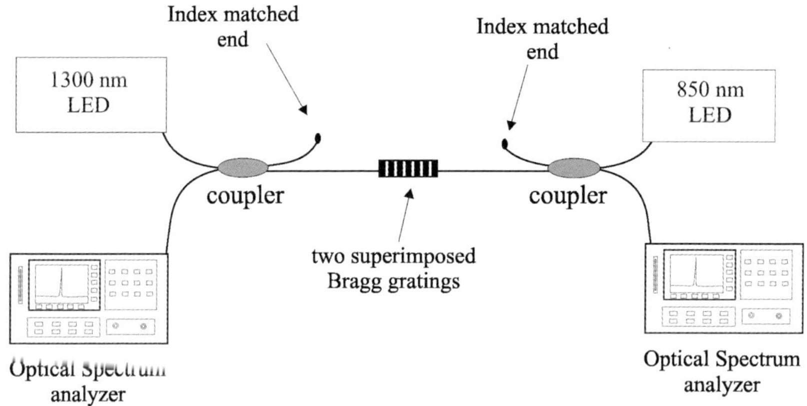

Fig. 5.49. Schematic diagram for simultaneous measurement of strain and temperature using two superimposed Bragg gratings at 1300 nm and 850 nm

In the case of two Bragg gratings with different wavelengths (referred to as 1 and 2) the following relation holds:

$$\begin{bmatrix} \Delta\lambda_{B_1} \\ \Delta\lambda_{B_2} \end{bmatrix} = \begin{bmatrix} \Psi_{\varepsilon 1} & \Psi_{T1} \\ \Psi_{\varepsilon 2} & \Psi_{T2} \end{bmatrix} \begin{bmatrix} \Delta\varepsilon \\ \Delta T \end{bmatrix} \tag{5.35}$$

The elements of the Ψ-matrix can be determined experimentally by separately measuring the Bragg wavelength changes with strain and temperature. Once Ψ is known, changes in both strain and temperature can be determined using the inverse of the above equation. The measured values of Ψ obtained in Ref. 124 (for the fibre used in that investigation) were

$$\begin{aligned} \Psi_{\varepsilon 1} &= 0.96 \pm 6.5 \; x10^{-3} \, \text{pm/}\mu\text{strain} \\ \Psi_{\varepsilon 2} &= 0.59 \pm 3.4 \; x10^{-3} \, \text{pm/}\mu\text{strain} \\ \Psi_{T1} &= 8.72 \pm 7.7 \; x10^{-2} \, \text{pm/}\,^{0}\text{C} \\ \Psi_{T2} &= 6.30 \pm 3.7 \; x10^{-2} \, \text{pm/}\,^{0}\text{C} \end{aligned} \tag{5.36}$$

where $\text{pm} = 1 \times 10^{-12}$ m. If the inverse matrix is used, strain and temperature may be obtained from the two wavelength shifts. However, a drawback of this approach is the disparate wavelengths that require the use of expensive multiplexing schemes. Furthermore, the actual temperature and strain coefficients are fixed by the wavelength difference.

Recent developments have led to ways of tailoring the temperature and strain coefficients of fibre Bragg gratings (Type I and IA) by influencing the photosensitivity pre-sensitisation of the host optical fibre. Controlling the level of hydrogen saturation via hot and cold hydrogenation, can produce gratings with tailored thermal and strain coefficients, resulting in a significant improvement in the matrix condition number, which impacts the ability to recover accurate temperature and strain data. Kalli et al. [24] designed a Type I-IA dual grating sensor that accurately decoupled temperature and strain. The key advantages of this scheme were the utilisation of two Bragg gratings having good wavelength proximity thereby avoiding costly multiplexing schemes, quick and efficient inscription using a single phase mask, common annealing cycles, and the precise placement of sensors located in a compact sensor head.

5.6.4 Aerospace Applications

The aerospace industry is a potentially important user of optical fibres, particularly for data links and fibre optic sensors. Although research projects have shown that optical fibre sensors can operate within tolerances prescribed for applications in aircraft, they are still considered an immature technology. To date efforts are directed towards the sensor development for harsh environments unsuitable for conventional electro-mechanical sensors, taking advantage of radiation resistance and EMI immunity. Increases in sensor reliability, ease of installation and maintenance with little training and without special handling are demanded, ideally leading to the so-called "fit and forget" systems.

Sensing strategies for aerospace applications broadly follow the same directions. The most important requirements are to have passive, low weight and ideally common sensors that may be multiplexed over optical links. By carefully defining sensor requirements it may be possible to specify a range of optical sensors, satisfying the majority of avionics applications, that are either interchangeable or use at least common interrogation instrumentation. Currently, many sensor types perform similar functions without being interchangeable. The Bragg grating sensor solves one of the major drawbacks of optical fibre sensors: the lack of a standard demodulation approach, while maintaining a completely passive network. The largest class of sensors measures the position of flight control elements such as landing gear status, flap and rudder position and so forth. When taking into account high levels of system redundancy well in excess of 100 sensors are employed, therefore size and weight savings become critical.

5.6.5 Applications to Civil Engineering

There is growing concern over the state of civil infrastructure in both the US and Europe. It is essential that mechanical loading be measured for maintaining bridges, dams, tunnels, buildings and sport stadiums. By measuring the distributed strain in buildings one can predict the nature and grade of local loads, for example after an earthquake, whereas the mechanical health of bridges is increasingly under scrutiny, as old structures are often excessively loaded leading to a real possibility of increased structural failure rates. In fact, a 1996 US Department of Transportation survey estimates that 40% of all bridges in the United States are seriously deteriorated. There is concern with 50-year old railroad bridges in the US as regulatory limits on railcar loads are relaxed.

The current inspection routine depends on periodic visual inspection. The use of modern optic-based sensors can lead to real time measurements, monitoring the formation and growth of defects and optical fibre sensors allow for data to be transmitted over long distances to a central monitoring location. The advantage of optical fibres is that they may either be attached to an existing structure or embedded into concrete decks and supports prior to pouring, thereby monitoring the curing cycle and the condition of the structure during its serviceable lifetime. One of the most important applications of Bragg gratings as sensors is for "smart structures" where the grating is embedded directly into the structure to monitor its strain distribution, however, for error-free, quasi-static strain measurement temperature compensation of thermal fluctuations is required. This could lead to structures that are self-monitoring or even self-scheduling their maintenance and repair through the combination of optical fibre sensors and artificial intelligence with material science and structural engineering. Several types of fibre optic sensor are capable of sensing structural strain, for example, the intrinsic and extrinsic fibre Fabry–Perot sensor. Lee et al. [125] have used a multiplexed array of 16 fibre Fabry–Perot sensors to monitor strain on the Union Pacific Bridge that crosses the Brazos River at Waco, Texas. The fibre sensors are located at fatigue critical points for measuring dynamic loads induced by trains crossing the bridge, and the recorded data correlate well with those recovered by resistive strain gauges. Nevertheless, the general consensus is that fibre Bragg gratings are presently the most promising and widely used candidates for smart structures. The instrumentation for multiplexing large grating sensor arrays can be the same, offering a potentially low cost solution for monitoring structural strain. As the wavelength shift with strain is linear and with zero offset, long-term measurements are possible and because the measurement can be interrupt-immune one can avoid perpetual monitoring of a structure, performing periodic measurements when necessary.

References

1. A. Othonos and K. Kalli: *Fiber Bragg Gratings: Fundamentals and Applications in Telecommunications and Sensing* (Artech House, Boston, London, 1999)
2. I. Bennion, J. A. R. Williams, L. Zhang, K. Sugden, and N. J. Doran: "UV-written in-fiber Bragg gratings," Opt. and Quantum Electron. **28**, 93–135 (1996)
3. A. Othonos: "Fiber Bragg gratings," Review of Scientific Instruments **68**, 4309–4341 (1997)
4. B. J. Eggleton, A. Ahuja, P. S. Westbrook, J. A. Rogers, P. Kuo, T. N. Nielsen, and B. Mikkelsen: "Integrated tunable fiber gratings for dispersion management in high-bit rate systems," J. Lightwave Technol. **18**, 1418–1432 (2000)
5. R. I. Laming, W. H. Loh, X. Gu, M. N. Zervas, M. J. Cole, and A. D. Ellis: "Dispersion compensation with chirped fiber Bragg grating to 400 km at 10 Gbit/s in nondispersion-shifted fiber," *Opt. Fiber Commun. Conf.* (OFC'96), Techn. Digest, (San Jose, CA, USA, 1996), Vol. **2**, 203–204 (1996)
6. J. F. Brennan III, M. R. Matthews, W. V. Dower, D. J. Treadwell, W. Wang, J. Porque, and X. Fan: "Dispersion correction with a robust fiber grating over the full C-band at 10-Gb/s rates with < 0.3-dB power penalties," IEEE Photon. Technol. Lett. **15**, 1722–1724 (2003)
7. X. Chen, X. Xu, M. Zhou, D. Jiang, X. Li, J. Feng, and S. Xie: "Tunable dispersion compensation in a 10 Gb/s optical transmission system by employing a novel tunable dispersion compensator," IEEE Photon. Technol. Lett. **16**, 188–190 (2004)
8. K.-M. Feng, J.-X. Cai, V. Grubsky, D. S. Starodubov, M. I. Hayee, S. Lee, X. Jiang, A. E. Willner, and J. Feinberg: "Dynamic dispersion compensation in a 10 Gbit/s optical system using a voltage controlled tuned nonlinearly chirped fiber Bragg grating," IEEE Photon. Technol. Lett. **11**, 373–375 (1999)
9. J. Lauzon, S. Thibault, J. Martin, and F. Ouellette: "Implementation and characterization of fiber Bragg gratings linearly chirped by temperature gradient," Opt. Lett. **19**, 2027–2029 (1994)
10. P. C. Hill and B. J. Eggleton: "Strain gradient chirp of fiber Bragg grating," Electron. Lett. **30**, 1172–1174 (1994)
11. M. Pacheco, A. Medez, L. A. Zenteni, and F. Mendoz–Santoyo: "Chirping optical fiber Bragg gratings using tapered-thickness piezo-electric ceramic," Electron. Lett. **34**, 2348–2350 (1998)
12. M. M. Ohn, A. T. Alavie, R. Maaskant, M. G. Xu, F. Bilodeau, and K. O. Hill: "Dispersion variable fiber grating using a piezoelectric stack," Electron. Lett. **32**, 2000–2001 (1996)
13. P. I. Reyes, N. Litchinitser, M. Sumetsky, and P. S. Westbrook: "160-Gb/s tunable dispersion slope compensator using a chirped fiber Bragg grating and a quadratic heater," IEEE Photon. Technol. Lett. **17**, 831–833 (2005)
14. D. K. W. Lam and B. K. Garside: "Characterization of single-mode optical fiber filters," Appl. Opt. **20**, 440–445 (1981)
15. P. St. J. Russell, J. L. Archambault, and L. Reekie: "Fiber gratings," Physics World, October 1993 issue, 41–46 (1993)

16. G. Meltz and W. W. Morey: "Bragg grating formation and germanosilicate fiber photosensitivity," International Workshop on Photoinduced Self-Organization Effects in Optical Fiber, Quebec City, Quebec, May 10–11, *Proc. SPIE* **1516**, 185–199 (1991)
17. K. O. Hill and G. Meltz: "Fiber Bragg grating technology fundamentals and overview," J. Lightwave Technol. **15**, 1263–1276 (1997)
18. G. P. Brady, K. Kalli, D. J. Webb, L. Reekie, J. L. Archambault, and D. A. Jackson: "Simultaneous measurement of strain and temperature using the first- and second-order diffraction wavelengths of Bragg gratings," IEE Proceed. Optoelectron. **144**, 156–161 (1997)
19. E. Delevaque, S. Boj, J. F. Bayon, H. Poignant, J. Lemellot, and M. Monerie: "Optical fibre design for strong gratings photoimprinting with radiation mode suppression," *Opt. Fiber Commun. Conf.* (OFC'95), Techn. Digest (San Diego, CA, USA), postdeadline paper PD5 (1995)
20. H. Patrick and S. L. Gilbert: "Growth of Bragg gratings produced by continuous-wave ultraviolet light in optical fiber," Opt. Lett. **18**, 1484–1486 (1993)
21. Y. Liu, J. A. R. Williams, L. Zhang, and I. Bennion: "Abnormal spectral evolution of fibre Bragg gratings in hydrogenated fibres," Opt. Lett. **27**, 586–588 (2002)
22. A. G. Simpson, K. Kalli, K. Zhou, L. Zhang, and I. Bennion: "Formation of type IA fibre Bragg gratings in germanosilicate optical fibre," Electron. Lett. **40**, 163–164 (2004)
23. A. G. Simpson, K. Kalli, L. Zhang, K. Zhou, and I. Bennion: "Abnormal photosensitivity effects and the formation of type IA FBGs," *Conf. Bragg Gratings, Photosensitivity and Poling in Glass Waveguides* (BGPP), Techn. Digest (Monterey, CA, USA) paper MD31 (2003)
24. K. Kalli, A. G. Simpson, K. Zhou, L. Zhang, and I. Bennion: "Tailoring the temperature and strain coefficients of type I and type IA dual grating sensors – the impact of hydrogenation conditions," Measurement Science and Technology **17**, 949–954 (2006)
25. A. G. Simpson, K. Kalli, K. Zhou, L. Zhang, and I. Bennion: "An idealised method for the fabrication of temperature invariant IA-I strain sensors," postdeadline session, OFS-16 Nara, Japan, PD4 (2003)
26. K. Kalli, H. Dobb, A. G. Simpson, M. Komodromos, D. J. Webb, and I. Bennion: "Annealing and temperature coefficient study of type IA fibre Bragg gratings inscribed under strain and no strain - implications to optical fibre component reliability," *Proc. SPIE* **6193**, Reliability of Optical Fiber Components, Devices, Systems, and Networks III, 119–130 (2006)
27. I. Riant and F. Haller: "Study of the photosensitivity at 193 nm and comparison with photosensitivity at 240 nm influence of fiber tension: type IIA aging," J. Lightwave Technol. **15**, 1464–1469 (1997)
28. J. L. Archambault, L. Reekie, and P. St. J. Russell: "High reflectivity and narrow bandwidth fibre gratings written by single excimer pulse," Electron. Lett. **29**, 28–29 (1993)
29. A. Yariv: "Coupled-mode theory for guided-wave optics," IEEE J. Quantum Electron. **QE-9**, 919–933 (1973)

30. M. Yamada and K. Sakuda: "Analysis of almost-periodic distributed feedback slab waveguide via a fundamental matrix approach," Appl. Opt. **26**, 3474–3478 (1987)

31. K. O. Hill, S. Theriault, B. Malo, F. Bilodeau, T. Kitagawa, D. C. Johnson, J. Albert, K. Takiguchi, T. Kataoka, and K. Hagimoto: "Chirped in-fiber Bragg grating dispersion compensators: Linearization of the dispersion characteristic and demonstration of dispersion compensation in a 100 km, 10 Gbit/s optical fiber link," Electron. Lett. **30**, 1755–1756 (1994)

32. V. Mizrahi and J. E. Sipe: "Optical properties of photosensitive fiber phase gratings," J. Lightwave Technol. **11**, 1513–1517 (1993)

33. J. Albert, K. O. Hill, B. Malo, S. Theriault, F. Bilodeau, D. C. Johnson, and L. E. Erickson: "Apodisation of the spectral response of fiber Bragg gratings using a phase mask with variable diffraction efficiency," Electron. Lett. **31**, 222–223 (1995)

34. B. Malo, S. Theriault, D. C. Johnson, F. Bilodeau, J. Albert, and K. O. Hill: "Apodised in-fiber Bragg grating reflectors photoimprinted using a phase mask," Electron. Lett. **31**, 223–225 (1995)

35. R. Kashyap, A. Swanton, and D. J. Armes: "Simple technique for apodising chirped and unchirped fiber Bragg gratings," Electron. Lett. **32**, 1226–1228 (1996)

36. K. O. Hill, F. Bilodeau, B. Malo, T. Kitagawa, S. Theriault, C. Johnson, J. Albert, and K. Takiguchi: "Aperiodic in-fiber Bragg gratings for optical fiber dispersion compensation," *Opt. Fiber Commun. Conf.*(OFC'94), Techn. Digest (San José, CA, USA), post deadline paper PF-77 (1994)

37. K. O. Hill, F. Bilodeau, B. Malo, and D. C. Johnson: "Birefringent photosensitivity in monomode optical fibre: application to external writing of rocking filters," Electron. Lett. **27**, 1548–1550 (1991)

38. K. O. Hill, Y. Fujii, D. C. Johnson, and B. S. Kawasaki: "Photosensitivity in optical fiber waveguides: Application to reflection filter fabrication," Appl. Phys. Lett. **32**, 647–649 (1978)

39. B. S. Kawasaki, K. O. Hill, D. C. Johnson, and Y. Fujii: "Narrow-band Bragg reflectors in optical fibers," Opt. Lett. **3**, 66–68 (1978)

40. G. Meltz, W. W. Morey, and W. H. Glenn: "Formation of Bragg gratings in optical fibers by a transverse holographic method," Opt. Lett. **14**, 823–825 (1989)

41. M. L. Dockney, J. W. James, and R. P. Tatam: "Fiber Bragg grating fabricated using a wavelength tuneable source and a phase-mask based interferometer," Meas. Sci. Technol. **7**, 445 (1996)

42. R. Kashyap, J. R. Armitage, R. Wyatt, S. T. Davey, and D. L. Williams: "All-fiber narrow band reflection grating at 1500 nm," Electron. Lett. **26**, 730–732 (1990)

43. B. J. Eggleton, P. A. Krug, and L. Poladian: "Experimental demonstration of compression of dispersed optical pulses by reflection from self-chirped optical fiber Bragg gratings," Opt. Lett. **19**, 877–880 (1994)

44. H. G. Limberger, P. Y. Fonjallaz, P. Lambelet, Ch. Zimmer, R. P. Salathe, and H. H. Gilgen: "Photosensitivity and self-organization in optical fibers and waveguides," *Proc. SPIE* **2044**, Photosensitivity and Self-Organization in Optical Fibers and Waveguides, 272–285 (1993)

45. A. Othonos and X. Lee: "Narrow linewidth excimer laser for inscribing Bragg gratings in optical fibers," Rev. Sci. Instr. **66**, 3112–3115 (1995)

46. J. Cannon and S. Lee: "Fiberoptic Product News," Laser Focus World **2**, 50–51 (1994)

47. K. O. Hill, B. Malo, F. Bilodeau, D. C. Johnson, and J. Albert: "Bragg gratings fabricated in monomode photosensitive optical fiber by UV exposure thorough a phase-mask," Appl. Phys. Lett. **62**, 1035–1037 (1993)

48. A. Othonos and X. Lee: "Novel and improved methods of writing Bragg gratings with phase-masks," IEEE Photon. Technol. Lett. **7**, 1183–1185 (1995)

49. P. E. Dyer, R. J. Farley, and R. Giedl: "Analysis and application of a 0/1 order Talbot interferometer for 193 nm laser grating formation," Optics Commun. **129**, 98–108 (1996)

50. B. Malo, K. O. Hill, F. Bilodeau, D. C. Johnson, and J. Albert: "Point-by-point fabrication of micro-Bragg gratings in photosensitive fiber using single excimer pulse refractive index modification techniques," Electron. Lett. **29**, 1668–1669 (1993)

51. K. O. Hill, B. Malo, K. A. Vineberg, F. Bilodeau, D. C. Johnson, and I. Skinner: "Efficient mode-conversion in telecommunication fiber using externally written gratings," Electron. Lett. **26**, 1270–1272 (1990)

52. www.stratosphere.com

53. S. J. Mihailov, C. W. Smelser, P. Lu, R. B. Walker, D. Grobnic, H. Ding, and J. Unruh: "Fiber Bragg gratings (FBG) made with a phase mask and 800 nm femtosecond radiation," *Opt. Fiber Commun. Conf.* (OFC'03), Techn. Digest (Atlanta, GA, USA, 2003), Vol. **3**, postdeadline paper PD30 (2003)

54. S. J. Mihailov, C. W. Smelser, D. Grobnic, R. B. Walker, P. Lu, H. Ding, and J. Unruh: "Bragg gratings written in all-SiO/sub 2/ and Ge-doped core fibers with 800-nm femtosecond radiation and a phase mask," J. Lightwave Technol. **22**, 94–100 (2004)

55. D. Grobnic, C. W. Smelser, S. J. Mihailov, R. B. Walker, and P. Lu: "Fiber Bragg gratings with suppressed cladding modes made in SMF-28 with a femtosecond IR laser and a phase mask," IEEE Photon. Technol. Lett. **16**, 1864–1866 (2004)

56. A. Martinez, M. Dubov, I. Khrushchev, and I. Bennion: "Direct writing of fibre Bragg gratings by femtosecond laser," Electron. Lett. **40**, 1170–1172 (2004)

57. A. Martinez, Y. Lai, M. Dubov, I. Khrushchev, and I. Bennion: "Vector bending sensors based on fibre Bragg gratings inscribed by infrared femtosecond laser," Electron. Lett. **41**, 472–474 (2005)

58. G. D. Marshall and M. J. Withford: "Rapid production of arbitrary fiber Bragg gratings using femtosecond laser radiation," *18th Ann. Meeting IEEE Lasers & Electro-Optics Soc.* (LEOS 2005), Techn. Digest (Sydney, Australia, 2005) 935–936 (2005)

59. D. P. Hand and P. St. J. Russell: "Single-mode fibre grating written into sagnac loop using photosensitive fibre: transmission filters," *7th Internat. Conf. Integr. Optics and Opt. Fiber Commun.* (IOOC'89), Techn. Digest (Kobe, Japan), 64 (1989)

60. K. O. Hill, D. C. Johnson, F. Bilodeau, and S. Faucher: "Narrow-bandwidth optical waveguide transmission filters: A new design concept and applications to optical fiber communications," Electron. Lett. **23**, 464–465 (1987)

61. F. Bilodeau, K. O. Hill, B. Malo, D. C. Johnson, and J. Albert: "High-return-loss narrowband all-fiber bandpass Bragg transmission filter," IEEE Photon. Technol. Lett. **6**, 80–82 (1994)

62. D. C. Johnson, K. O. Hill, F. Bilodeau, and S. Faucher: " New design concept for a narrowband wavelength-selective optical tap and combiner," Electron. Lett. **23**, 668–669 (1987)

63. A. Fielding, T. J. Cullen, and H. N. Rourke: "Compact all-fiber wavelength drop and insert filter," Electron. Lett. **30**, 2160–2161 (1994)

64. D. C. Reid, C. M. Ragdale, I. Bennion, D. J. Robbins, J. Buus, and W. J. Stewart: "Phase-shifted Moiré grating fiber resonators," Electron. Lett. **26**, 10–11 (1990)

65. S. Legoubin, E. Fertein, M. Douay, P. Bernage, P. Niay, F. Bayon, and T. Georges: "Formation of Moiré grating in core of germanosilicate fiber by transverse holographic double exposure method," Electron. Lett. **27**, 1945–1946 (1991)

66. L. Zhang, K. Sugden, I. Bennion, and A. Molony: "Wide-stopband chirped fiber moiré grating transmission filters," Electron. Lett. **31**, 477–479 (1995)

67. L. Brilland, D. Pureur, J. F. Bayon, and E. Delevaque: "Slanted gratings UV-written in photosensitive cladding fibre," Electron. Lett. **35**, 234–235 (1999)

68. K. Takahashi, M. Tamura, T. Sano, K. Saito, and H. Suganuma: "Reconfigurable optical add/drop multiplexer using passive temperature-compensated wavelength tunable fiber Bragg grating," *Opt. Fiber Commun. Conf.*(OFC'01), Techn. Digest (Anaheim, CA, USA) Vol. **3**, paper WDD93 (2001)

69. P. Yvernault, D. Méchin, E. Goyat, L. Brilland, and D. Pureur: "Fully functional optical add and drop multiplexer using twin-core fiber based Mach–Zehnder interferometer with photoimprinted fiber Bragg gratings," *Opt. Fiber Commun. Conf.* (OFC'01), Techn. Digest (Anaheim, CA, USA) Vol. **3**, paper WDD92 (2001)

70. Y.-L. Lo and C.-P. Kuo: "Packaging a fiber Bragg grating without preloading in a simple athermal bimateerial device," IEEE Trans. Adv. Packaging **25**, 50–53 (2002)

71. www.gouldfo.com

72. P. Yvernault, D. Durand, D. Méchin, M. Boitel, and D. Pureur: "Passive athermal Mach–Zehnder interferometer twin-core fiber optical add/drop multiplexer," *Proc. 27*th *Europ. Conf. Opt. Commun.* (ECOC'01), Amsterdam, The Netherlands, Vol. **6**, 88–89 (2001)

73. W. W. Morey: "Tuneable narrow-line bandpass filter using fiber gratings," *Opt. Fiber Commun. Conf.* (OFC'91), Techn. Digest (San Diego, CA, USA), PDP 20, 96 (1991)

74. G. P. Agrawal: *Nonlinear Fiber Optics, 3*rd *ed.* (Academic, New York, USA, 2001)

75. www.corning.com

76. J. A. R. Williams, I. Bennion, K. Sugden, and N. J. Doran: "Fiber dispersion compensation using a chirped in fiber Bragg grating," Electron. Lett. **30**, 985–987 (1994)

77. B. J. Eggleton, P. A. Krug, and L. Poladian: "Experimental demonstration of compression of dispersed optical pulses by reflection from self-chirped optical fiber Bragg gratings," Opt. Lett. **19**, 877–880 (1994)

78. W. Loh, M. Cole, M. Zervas, S. Barcelos, and R. Laming: "Complex grating structures with uniform phase masks based on the moving fiber-scanning beam technique," Opt. Lett. **20**, 2051– (1995)

79. L. Quetel, L. Rivoallan, M. Morvan, M. Monerie, E. Delevaque, J. Y. Guilloux, and J. F. Bayon: "Chromatic dispersion compensation by apodised Bragg gratings within controlled tapered fibers," Optical Fiber Technology **3**, 267–271 (1997)

80. M. Ibsen, M. K. Durkin, M. J. Cole, and R. I. Laming: "Sinc-sampled fiber Bragg gratings for identical multiple wavelength operation," IEEE Photon. Technol. Lett. **10**, 842–844 (1998)

81. A. V. Buryak, K. Y. Kolossovski, and D. Yu. Stepanov: "Optimization of refractive index sampling for multichannel fiber Bragg gratings," IEEE J. Quantum Electron. **39**, 91–98 (2003)

82. M. Guy, F. Trépanier, and Y. Painchaud: "Manufacturing of high-channel count dispersion compensators using complex phase mask technology," OSA Topical Meeting on Bragg Gratings, Photosensitivity and Poling in Glass Waveguides (BGPP), Monterey Bay, CA, USA, 269–271 (2003)

83. A. Othonos, X. Lee, and R. M. Measures: "Superimposed multiple Bragg gratings," Electron. Lett. **30**, 1972–1974 (1994)

84. J. Lauzon, S. Thibault, J. Martin, and F. Ouellette: "Implementation and characterization of fiber Bragg gratings linearly chirped by a temperature gradient," Opt. Lett. **19**, 2027–2029 (1994)

85. T. Komukai, T. Inui, and M. Nakazawa: "Very low group delay ripple characteristics of fibre Bragg grating with chirp induced by an S-curve bending technique," Electron. Lett. **37**, 449–451 (2001)

86. A. Mugnier, E. Goyat, P. Lesueur, and D. Pureur: "Wide tuning range and low insertion loss variation dispersion compensator," Electron. Lett. **40**, 1506–1508 (2004)

87. A. Mugnier, E. Goyat, D. Pureur, and P. Yvernault: "Tunable dispersion compensating fibre Bragg grating using pure bending of a simply supported beam," *Proc. 28*[th] *Europ. Conf. Opt. Commun.* (ECOC'02), Copenhagen, Denmark, paper 10.3.5 (2002)

88. www.teraxion.com

89. C. Scheerer, C. Glingener, G. Fischer, M. Bohn, and W. Rosenkranz: "Influence of filter group delay ripples on system performance," *Proc. 25*[th] *Europ. Conf. Opt. Commun.* (ECOC'99), Nice, France, Vol. **I**, 410–411 (1999)

90. K. Ennser, M. Ibsen, M. Durkin, M. N. Zervas, and R. Laming: "Influence of nonideal chirped fiber grating characteristics on dispersion cancellation," IEEE Photon. Technol. Lett. **10**, 1476–1478 (1998)

91. M. Derrien, D. Gauden, E. Goyat, A. Mugnier, P. Yvernault, and D. Pureur: "Wavelength-frequency analysis of dispersion compensator group delay ripples," *Opt. Fiber Commun. Conf.* (OFC'03), Techn. Digest (Atlanta, GA, USA), Vol. **1**, 34–35 (2003)

92. D. Gauden, A. Mugnier, M. Gay, L. Lablonde, F. Lahoreau, and D. Pureur: "Experimental measurement of 10 Gbit/s system power penalty spectrum created by group delay ripple of fiber Bragg grating chromatic dispersion compensator," *Proc. 29*[th] *Europ. Conf. Opt. Commun.* (ECOC'03), Rimini, Italy, 684–685 (2003)

93. M. Eiselt, C. Clausen, and R. Tkach: "Performance characterization of components with group delay fluctuations," IEEE Photon. Technol. Lett. **15**, 1076–1078 (2003)

94. S. James, X. Fan, and J. Brennan III: "Performance effect in optical communication systems caused by phase ripples of dispersive components," Appl. Opt. **43**, 5033–5036 (2004)

95. A. M. Vengsarkar, P. J. Lemaire, J. B. Judkins, V. Bhatia, T. Erdogan, and J. E. Sipe: "Long-period fiber gratings as band-rejection filters," J. Lightwave Technol. **14**, 58–65 (1996)

96. M. Guy and F. Trépanier: "Chirped fiber Bragg gratings equalize gain," *WDM Solutions*, Vol. **3**(3) 77–82 (2001)

97. H. Chotard, Y. Painchaud, A. Mailloux, M. Morin, F. Trépanier, and M. Guy: "Group delay ripple of cascaded Bragg grating gain flattening filters," IEEE Photon. Technol. Lett. **14**, 1130–1132 (2002)

98. M. Guy, F. Trépanier, A. Doyle, Y. Painchaud, and R. L. Lachance: "Novel applications of fiber Bragg grating components for next-generation WDM systems," Annales des Télécommunications **58**, 1275–1306 (2003)

99. H. Renner: "Effective-index increase, form birefringence and transition losses in UV-side-illuminated photosensitive fibers," Opt. Express **9**, 546–560 (2001)

100. K. Kalli, A. G. Simpson, K. Zhou, L. Zhang, D. Birkin, T. Ellingham, and I. Bennion: "Spectral modification of type IA fibre Bragg gratings by high power near infra-red lasers," Measurement Science and Technology **17**, 968–974 (2006)

101. A. D. Kersey, M. A. Davis, J. Patrick, M. LeBlanc, K. P. Koo, C. G. Askins, M. A. Putnam, and E. J. Friebele: "Fiber grating sensors," J. Lightwave Technol. **15**, 1442–1463 (1997)

102. D. M. Bird, J. R. Armitage, R. Kashyap, R. M. A. Fatah, and K. H. Cameron: "Narrow line semiconductor laser using fiber grating," Electron. Lett. **27**, 1115–1116 (1991)

103. P. A. Morton, V. Mizrahi, P. A. Andrekson, T. Tanbun-Ek, R. A. Logan, P. Lemaire, D. L. Coblentz, A. M. Sergent, K. W. Wecht, and P. F Sciortino, Jr.: "Mode-locked Hybrid soliton pulse source with extremely wide operating frequency range," IEEE Photon. Technol. Lett. **5**, 28–31 (1993)

104. L. Reekie, R. J. Mears, S. B. Poole, and D. N. Payne: "Tuneable single-mode fiber laser," J. Lightwave Technol. **LT-4**, 956–957 (1986)

105. G. A. Ball, W. W. Morey, and J. P. Waters: "Nd^{3+} fiber laser utilizing intra-core Bragg reflectors," Electron. Lett. **26**, 1829–1830 (1990)

106. G. A. Ball and W. H. Glenn: "Design of a single-mode linear-cavity erbium fiber laser utilizing Bragg reflector," J. Lightwave Technol. **10**, 1338–1343 (1992)

107. G. A. Ball, W. H. Glenn, W. W. Morey, and P. K. Cheo: "Modeling of short, single-frequency fiber laser in high-gain fiber," IEEE Photon. Technol. Lett. **5**, 649-651 (1993)

108. G. A. Ball, W. W. Morey, and P. K. Cheo: "Single- and multi-point fiber-laser sensors," IEEE Photon. Technol. Lett. **5**, 267–270 (1993)

109. V. Mizrahi, D. J. D. DiGiovanni, R. M. Atkins, S. G. Grubb, Y. K. Park, and J. M. P. Delavaux: "Stable single-mode erbium fiber-grating laser for digital communications," J. Lightwave Technol. **11**, 2021–2025 (1993)

110. A. Othonos, X. Lee, and D. P. Tsai: "Spectrally broadband Bragg grating mirror for an erbium-doped fiber laser," Opt. Eng. **35**, 1088–1092 (1996)

111. G. A. Ball, W. W. Morey, and W. H. Glenn: "Standing-wave monomode erbium fiber laser," IEEE Photon. Technol. Lett. **3**, 613–615 (1991)

112. J. L. Zyskind, V. Mizrahi, D. J. DiGiovanni, and J. W. Sulhoff: "Short single frequency erbium-doped fiber laser," Electron. Lett. **28**, 1385–1386 (1992)

113. J. L. Zyskind, J. W. Sulhoff, P. D. Magill, K. C. Reichmann, V. Mizrahi, and D. J. DiGiovanni: "Transmission at 2.5 Gbits/s over 654 km using an erbium-doped fiber grating laser source," Electron. Lett. **29**, 1105–1106 (1993)

114. J. T. Kringlebotn, J.-L. Archambault, L. Reekie, J. E. Townsend, G. G. Vienne, and D. N. Payne: "Highly efficient, low-noise grating-feedback Er3+:Yb3+ codoped fibre laser," Electron. Lett. **30**, 972–973 (1994)

115. G. S. Grubb: "High-power 1.48 µm cascaded Raman laser in germanosilicate fibers," *Opt. Fiber Commun. Conf.* (OFC'95), Technical Digest Series, Post-conference ed. Vol. **8**, 41–42 (1995)

116. S. M. Melle, K. Liu, and R. M. Measures: "A passive wavelength demodulation system for guided-wave Bragg grating sensors," IEEE Photon. Technol. Lett. **4**, 516–518 (1992)

117. A. D. Kersey, T. A. Berkoff, and W. W. Morey: "Multiplexed fiber Bragg grating strain-sensor system with a fiber Fabry–Perot wavelength filter," Opt. Lett. **18**, 1370–1372 (1993)

118. H. Geiger, M. G. Xu, N. C. Eaton, and J. P. Dakin: "Electronic tracking system for multiplexed fiber grating sensors," Electron. Lett. **31**, 1006–1007 (1995)

119. D. A. Jackson, A. B. Lobo Ribeiro, L. Reekie, and J. L. Archambault: "Simple multiplexing scheme for fiber-optic grating sensor network," Opt. Lett. **18**, 1192–1194 (1993)

120. A. D. Kersey, T. A. Berkoff, and W. W. Morey: "High-resolution fiber-grating based strain sensor with interferometric wavelength-shift detection," Electron. Lett. **28**, 236–238 (1992)

121. A. Othonos, A. T. Alavie, S. Melle, S. E. Karr, and R. M. Measures: "Fiber Bragg grating laser sensor," Opt. Eng. **32**, 2841–2846 (1993)

122. W. W. Morey, G. Meltz, and J. M. Weiss: "Evaluation of a fiber Bragg grating hydrostatic pressure sensor," *Proceed. Opt. Fiber Sensors Conf.* (OFS-8), Monterey, CA, USA, 1992, postdeadline paper PD-4.4 (1992)

123. M. G. Xu, J. L. Archambault, L. Reekie, and J. P. Dakin: "Thermally-compensated bending gauge using surface mounted fiber gratings," Int. J. Optoelectron. **9**, 281–283 (1994)

124. M. G. Xu, J. L. Archambault, L. Reekie, and J. P. Dakin: "Discrimination between strain and temperature effects using dual-wavelength fibre grating sensors," Electron. Lett. **30**, 1085–1087 (1994)

125. W. Lee, J. Lee, C. Henderson, H. F. Taylor, R. James, C. E. Lee, V. Swenson, W. N. Gibler, R. A. Atkins, and W. G. Gemeiner: "Railroad bridge instrumentation with fiber optic sensors," *Proceed. Opt. Fiber Sensors Conf.* (OFS-12), Williamsburg, VA, USA, 1997, 412–415 (1997)

6 Fabry–Perot Interferometer Filters

Ton Koonen

6.1 Operating Principles

6.1.1 Multi-beam Interference Process

A Fabry–Perot (FP) interferometer filter uses a multiple-beam interference process for obtaining wavelength selectivity. Usually, the filter has one input and one output port and employs two highly reflecting plates which together constitute the resonating cavity creating the multiple-beam interference process.

The basic concept of an FP filter is shown in Fig. 6.1. It was described first by Charles Fabry and Albert Perot in 1899 (Ann. Chim. Phys. Vol. 16). Two highly reflective planar plates are accurately positioned in parallel and thus form a cavity. A light beam entering the cavity is reflected multiple times between the plates. Each time when the beam hits a plate, a small part of its power escapes. When the two plates are aligned perfectly in parallel, the multiple beams escaping at each side of the FP cavity are exactly parallel. Each beam has a fixed phase difference with respect to the preceding one; this phase difference corresponds to the extra path length travelled in the cavity. The multiple parallel beams are brought into a common focus point with a lens, and in this point the actual multiple beam interference takes place. Hence, the amplitude of the transmitted electrical field E_t can be described by (see also [1])

$$E_t = at^2 E_i + at^2 a^2 r^2 e^{-i\delta} E_i + at^2 a^4 r^4 e^{-i\cdot2\delta} E_i + ...$$
$$...+ at^2 a^{2k} r^{2k} e^{-i\cdot k\delta} E_i \tag{6.1}$$
$$= E_i \cdot t^2 \cdot a \cdot \sum_{k=0}^{\infty} \left(a^2 r^2\right)^k e^{-i\cdot k\delta} = E_i \cdot \frac{a \cdot t^2}{1 - a^2 r^2 \cdot e^{-i\delta}}$$

where a is the amplitude attenuation factor when travelling once through the cavity, t is the amplitude transmission factor of a plate, r is its amplitude reflection factor.

Similarly, the amplitude of the reflected electrical field E_r is

$$E_r = rE_i + t^2 a^2 r e^{-i\delta'} E_i + t^2 a^4 r^3 e^{-i \cdot 2\delta'} E_i + ... + t^2 a^{2k} r^{2k-1} e^{-i \cdot k\delta'} E_i$$

$$= E_i \cdot \left(r + \frac{t^2}{r} \sum_{k=1}^{\infty} \left(a^2 r^2 e^{-i\delta'} \right)^k \right)$$

$$= E_i \cdot \left(r + \frac{t^2}{r} \cdot \frac{a^2 r^2 e^{-i\delta'}}{1 - a^2 r^2 e^{-i\delta'}} \right)$$

$$(6.2)$$

The phase shift δ (or δ') is the shift experienced between the directly transmitted (/reflected) beam and the beam which is transmitted (/reflected) after one roundtrip in the cavity.

With n being the refractive index of the medium between the plates, d the distance between the plates, θ the angle of incidence of the light, λ_0 its wavelength in vacuum, and v is its frequency, the optical path length difference ΔS between the directly transmitted beam and the one transmitted after one cavity roundtrip is (see Fig. 6.2)

$$\Delta S = n \cdot (BC + CD) - BF = \frac{2nd}{\cos\theta'} - \sin\theta \cdot 2d \tan\theta' = \frac{2nd\,(1 - \sin^2\theta')}{\cos\theta'} = 2nd\,\cos\theta'$$

$$(6.3)$$

where according to Snell's law $\sin\theta = n \sin\theta'$. The phase shift δ is therefore given by

$$\delta = \Delta S \cdot \frac{2\pi}{\lambda_0} = \frac{4\pi \cdot nd\,v}{c_0} \cdot \cos\theta'$$

$$(6.4)$$

Similarly, the optical path length difference $\Delta S'$ and the phase difference δ' between the directly reflected beam and the one reflected after one roundtrip in the cavity, taking into account an additional phase change of π when reflecting at an optically more dense medium, are

$$\Delta S' = n \cdot (AB + BC) - AE = \frac{2nd}{\cos\theta'} - \sin\theta \cdot 2d \tan\theta' = 2nd\,\cos\theta'$$

$$\delta' = \pi + \Delta S' \cdot \frac{2\pi}{\lambda_0} = \pi + \frac{4\pi \cdot nd\,v}{c_0} \cdot \cos\theta'$$

$$(6.5)$$

The sharpest filter characteristics are obtained when all the multiple beams are collected by the lenses, so by operating as close as possible to normal incidence, i.e. $\theta \approx 0$.

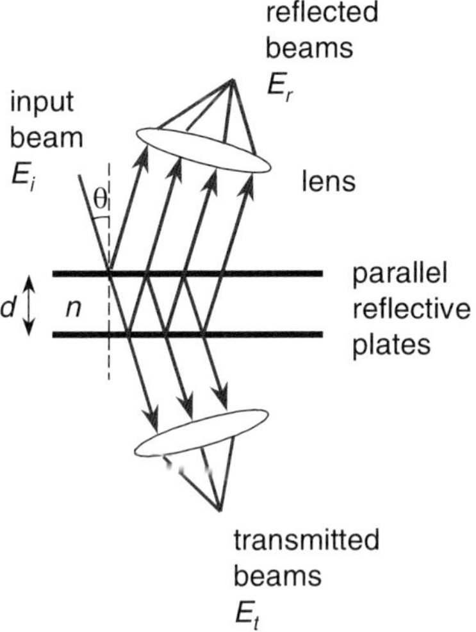

Fig. 6.1. Fabry–Perot Interferometer

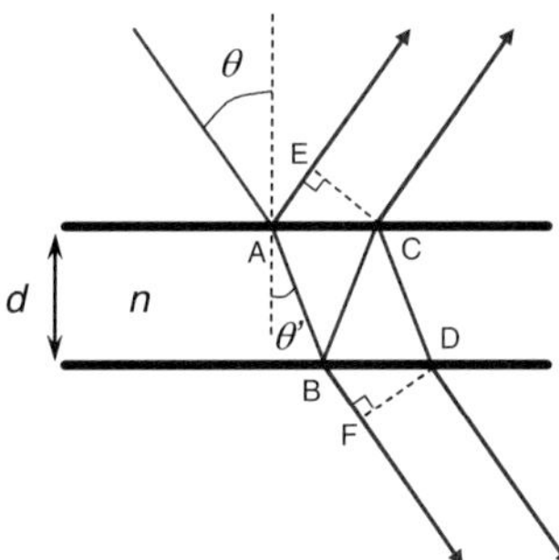

Fig. 6.2. Path length differences between transmitted and reflected beams

The intensity transmission factor T_{FP} of the Fabry–Perot filter, defined as the ratio of transmitted light intensity and the incident light intensity, follows the Airy function according to

$$T_{FP}(v) = \frac{I_t}{I_i} = \frac{E_t\, E_t^*}{E_i\, E_i^*} = \frac{T_0}{1 + \left\{\frac{2}{\pi} F_R \sin\left(\frac{\delta}{2}\right)\right\}^2} \tag{6.6}$$

where the reflectivity finesse F_R and the maximum transmission factor T_0 are respectively

$$F_R = \frac{\pi\sqrt{AR}}{1 - AR} \tag{6.7}$$

$$T_0 = \frac{AT^2}{(1 - AR)^2} = \frac{A\,(1-R)^2}{(1 - AR)^2}$$

where $R = |r|^2$ is the intensity reflection coefficient of the plates, $T = |t|^2$ the intensity transmission coefficient, the plates are assumed to be loss free (so $R + T = 1$), and $A = |a|^2$ is the attenuation factor of the light intensity when travelling from one plate to the other. The intensity transmission factor depends on the optical frequency v through the dependence of the phase shift δ on v.

Similarly, the intensity reflection factor R_{FP} of the Fabry–Perot filter, defined as the ratio of reflected light intensity and the incident light intensity, is

$$R_{FP}(v) = \frac{I_r}{I_i} = \frac{E_r E_r^*}{E_i E_i^*} = \frac{R_0 + \left\{ \frac{2}{\pi} F_R \sin\left(\frac{\delta}{2}\right) \right\}^2}{1 + \left\{ \frac{2}{\pi} F_R \sin\left(\frac{\delta}{2}\right) \right\}^2} \tag{6.8}$$

where the minimum reflection factor is

$$R_0 = R \cdot \left(\frac{1-A}{1-AR} \right)^2 \tag{6.9}$$

If the medium between the plates is lossless (so $A = 1$, which is a good approximation when air is used between the plates), then, as to be expected,

$$T_{FP}(v) + R_{FP}(v) = 1 \tag{6.10}$$

6.1.2 Frequency Characteristics (Bandwidth, finesse, contrast ratio)

The dependence of $T_{FP}(v)$ on the optical frequency v for a lossless FP is shown in Fig. 6.3. The characteristics $T_{FP}(v)$ and $R_{FP}(v)$ are periodic with a period called the Free Spectral Range FSR, given in the frequency domain by

$$FSR_v = \frac{c_0}{2nd} \quad [\text{Hz}] \tag{6.11}$$

and in the wavelength domain by

$$FSR_\lambda = \frac{\lambda_0^2}{2nd} \quad [\text{m}] \tag{6.12}$$

For example, when the plate distance $d = 120\,\mu\text{m}$, $\lambda_0 = 1.55\,\mu\text{m}$, and $n = 1$, $FSR_\lambda = 10\,\text{nm}$.

Another useful performance parameter is the contrast factor C which is defined as the ratio of the maximum and the minimum of the intensity transmission factor $T_{FP}(v)$:

$$C = \frac{T_{FP,\max}}{T_{FP,\min}} = \left(\frac{1+AR}{1-AR} \right)^2 \tag{6.13}$$

This factor determines the crosstalk attenuation which is achievable when using the FP for selecting a wavelength channel out of a set of channels. For example, for a lossless medium between the FP plates (i. e. $A = 1$) and a plate reflectivity $R = 0.9$, $C = 361$, which amounts to a crosstalk attenuation of 25.6 dB.

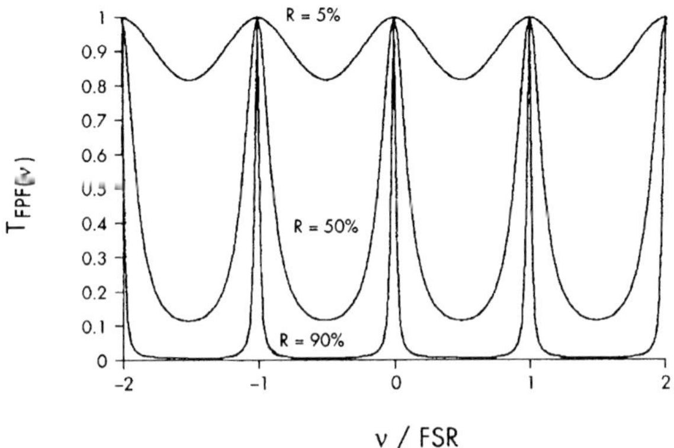

Fig. 6.3. Intensity transmission factor $T_{FP}(v)$ of a Fabry–Perot filter versus the light frequency v normalized with repect to the free spectral range FSR (where R is the reflectivity of the plates)

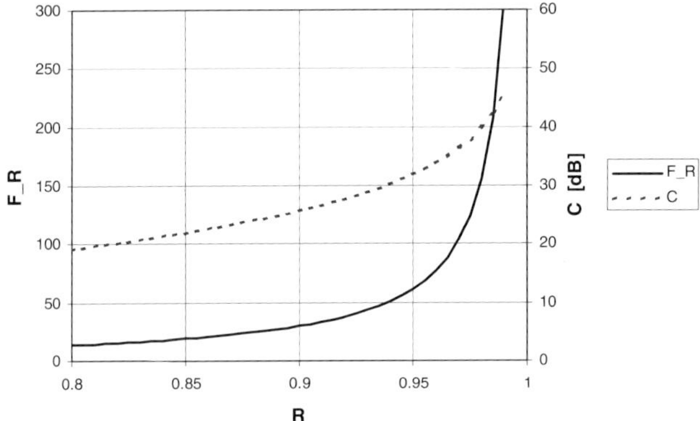

Fig. 6.4. Reflectivity finesse F_R and contrast factor C of a lossless Fabry–Perot filter versus the plate reflectivity R

The -3 dB bandwidth Δv_{FWHM} (FWHM, full-width at half-maximum) of the FP bandpass curves $T_{FP}(v)$ is

$$\Delta v_{FWHM} = \frac{c_0}{2\pi\, n\, d} \cdot \frac{1 - AR}{\sqrt{AR}} \quad [\text{Hz}] \tag{6.14}$$

and is related to the Free Spectral Range according to

$$\Delta v_{FWHM} = \frac{FSR_v}{F_R} \qquad (6.15)$$

Hence, the ratio of the −3 dB bandwidth Δv_{FWHM} and the FSR is only determined by the reflectivity R of the plates and the attenuation factor A of the medium between them.

6.2 FP Design Aspects

6.2.1 Instrument Finesse

The effective finesse of a Fabry–Perot filter may be lowered when the plates are not perfectly aligned: they may be not perfectly parallel, and/or not perfectly flat.

For instance, if the plates have a spherical curvature such that the distance between the plates measured in the centre and measured at the edges varies by λ_0 / M (where M is a positive real number), then the flatness finesse factor is to a good approximation given by [1]

$$F_F \approx M / 2 \qquad (6.16)$$

The resulting effective instrument finesse F_I of the FP is

$$F_I^{-2} = F_R^{-2} + F_F^{-2} \qquad (6.17)$$

Commercial Fabry–Perot filters are available with instrument finesses from less than 30 to more than 200. The instrument finesse F_I is a measure of the resolution for wavelength filtering. In a multi-wavelength system, it e.g. determines the maximum number of wavelength channels which can be resolved. For instance, when a crosstalk attenuation of better than 10 dB is required (which implies a crosstalk power penalty of less than 0.5 dB), this maximum number of wavelength channels N_{max} is

$$N_{max} \approx \frac{F}{3} \qquad (6.18)$$

Hence an instrument finesse of $F = 100$ would allow to resolve up to some 30 wavelength channels.

6.2.2 Tuning

A Fabry–Perot filter may be tuned by changing the optical path length between the plates which can be done by changing the refractive index and/or the physical distance.

The refractive index may be changed for instance by changing the pressure of the gas (air) in the cavity; this will allow slow tuning only.

Changing the spacing between the plates can be done manually (by micrometer screws for coarse adjustment) and/or electro-mechanically. The fastest tuning is achieved with piezo-electrical transducers.

The maximum transmission is obtained when the optical path length matches an integer number times half the wavelength of the light in the cavity, so when

$$nd = m \cdot \lambda_0 / 2 \tag{6.19}$$

where m is an integer. Hence the transmission maxima occur at the optical resonance frequencies

$$\nu_{res} = \frac{c_0}{\lambda_0} = c_0 \frac{m}{2nd} \tag{6.20}$$

which are spaced by the Free Spectral Range $FSR_\nu = c_0 / 2nd$.

6.3 Practical Implementations

6.3.1 Free-space Bulk Fabry–Perot Filter

Planar etalon

In laboratory setups for high-resolution wavelength measurements with free-space optics, relatively large, modular planar FP etalons are used. The plate spacing is usually adjustable manually over a wide range which allows very high resolutions. Large diameter mirrors with high flatness (diameter up to ca. 2 inches, flatness $\lambda/200$ or even better) are commercially available, with broadband multilayer dielectric coatings achieving reflectivities over 98%.

As shown in Fig. 6.5, the setup consists of two lenses which collimate the light beam from the light source into a parallel beam which traverses the FP cavity, and subsequently through a pinhole focus the light on a detector. The cavity is composed of two mirror plates of which the facing sides are coated with a highly reflective coating; these sides are aligned in accurate parallelism. The plates are slightly wedge-shaped in order to prevent the outer sides to form a secondary FP cavity. When using a monochromatic source, a concentric Airy disc pattern appears in the focus plane of the second lens. On the Airy rings, the condition of constructive multiple beam interference is met. When the distance d between the plates is varied by means of e. g. piezo-electric transducers, the rings of the Airy

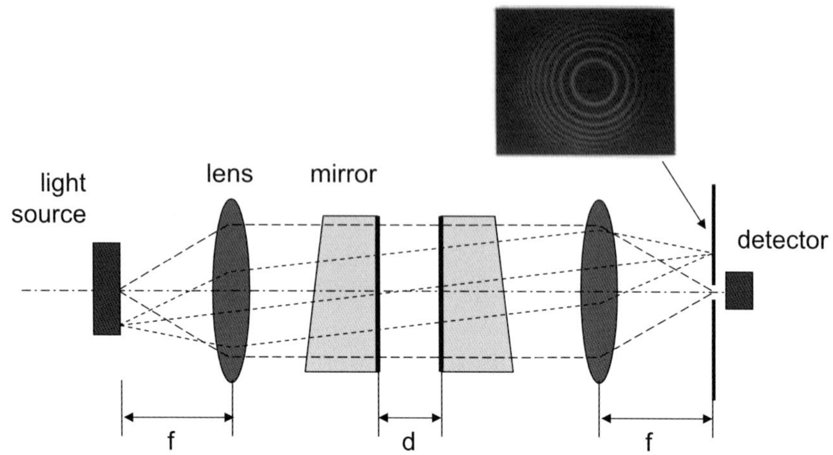

Fig. 6.5. Planar Fabry–Perot etalon

disc pattern are expanding or contracting. Through the pinhole, a detector is illuminated and will detect maximum signal when the cavity length d is an exact integer multiple of half the light source's wavelength (the resonance condition).

The FP mirror plates are mounted in a setup such as the one shown in Fig. 6.6. The distance between the plates can be set coarsely by sliding one of the mirror holders. The coarse adjustment of the mirrors to achieve parallelism is done with the three fine-threaded screws. One of the FP mirror plates is mounted on a piezo-electric transducer with three elements positioned under 120 degrees in order to enable both translation and tilting of the mirrors. The fine-tuning of the parallelism can be done by steering the elements individually in order to achieve the appropriate tilting. Sweeping of the plate distance is done by steering the three elements together. Putting

Fig. 6.6. Mounting setup for planar Fabry–Perot etalon [2]

ramp-shaped driving voltages on the piezo-s yields a linear sweep of the plate distance, and by applying the sweep voltage on the horizontal channel and the detector output on the vertical channel of an oscilloscope, the actual spectrum of the light source can be observed in nearly real time. The ramp slopes should be tuned to match the sensitivity of the piezo-electric transducers in order to maintain strict parallelism of the plates during scanning.

The planar FP etalon is very versatile, and the FSR can be changed over a wide range by changing the mirror separation. Mirror spacings may be set from a few µm up to more than 10 cm, yielding e. g. an $FSR_v = 1500$ GHz at $d = 100$ µm down to $FSR_v = 1.5$ GHz at $d = 10$ cm. However, the alignment of the mirrors needs careful tuning and is a delicate operation (asking extra caution when the mirror spacing is small, including careful removal of dust particles on the mirror surfaces). Instrument finesses typically can be around 150, yielding resolutions down to 10 MHz.

Confocal etalon

A confocal Fabry–Perot cavity uses two concave spherical mirrors which are separated by a distance equal to the radius of curvature of the mirrors [2]. As illustrated in Fig. 6.7, a complete roundtrip of a light beam through the cavity covers a path length of $L = 4 \cdot d = 4R$. The resonance condition $L = m \cdot \lambda_0/n$ (with integer m) thus implies that transmission maxima occur at the optical frequencies

$$v_{res} = \frac{c_0}{\lambda_0} = m \cdot \frac{c_0}{4nR} \qquad (6.21)$$

which are spaced by the Free Spectral Range $FSR_v = c_0 / 4nR$.

It is relatively easy to obtain a high finesse with a confocal FP interferometer because the narrow width of the incident beam reduces the finesse degradation due to mirror surface imperfections. Also the alignment procedure is simple as the mirrors just have to be positioned in a common

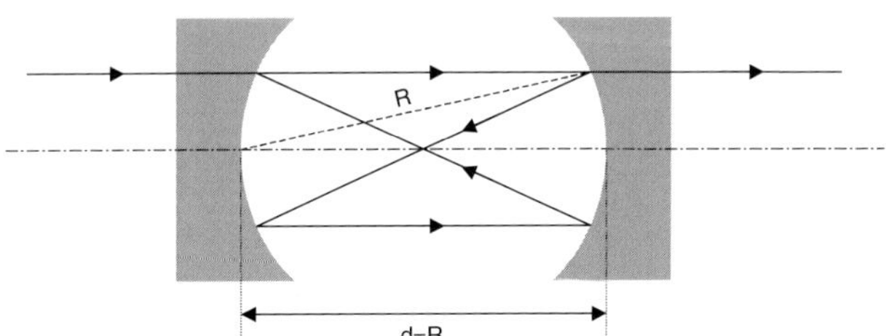

Fig. 6.7. Confocal Fabry–Perot interferometer

focus. However, the FSR is fixed by the mirror curvature radius; if the FSR needs to be changed, the mirrors must be replaced.

Commercial confocal etalons typically have FSR-s around 2 or 8 GHz, with instrument finesses from 200 to 300, yielding resolutions down to some 7 MHz. These etalons are well suited for detailed investigations of the spectra of narrowband lasers.

6.3.2 Fibre Fabry–Perot Filter

Fibre Fabry–Perot (FFP) filters fit readily into optical fibre communication links as they are equipped with single-mode fibre pigtails. No lenses or other collimating optics are used. The FP cavity is formed by two carefully aligned end faces of fibres on which a highly reflective coating has been deposited. Between the coated end faces the medium is air; in case of larger mirror spacings, a short piece of anti-reflection coated single-mode fibre may be inserted in the cavity which takes care of appropriate confined light guiding and of which the end face is anti-reflection coated in order to prevent secondary cavities. The basic structure of such a Fibre Fabry–Perot is shown in Fig. 6.8; a typical commercially available device is shown in Fig. 6.9 [3].

The spacing d between the highly reflective mirrors can be varied by means of piezo-electric transducers. Typically, maximum tuning voltages may be up to 70 V, and less than 12 V voltage swing is needed to traverse a Free Spectral Range. Depending on the coating design, the device may operate in the S-band (1480–1520 nm), the C-band (1520–1570 nm), the

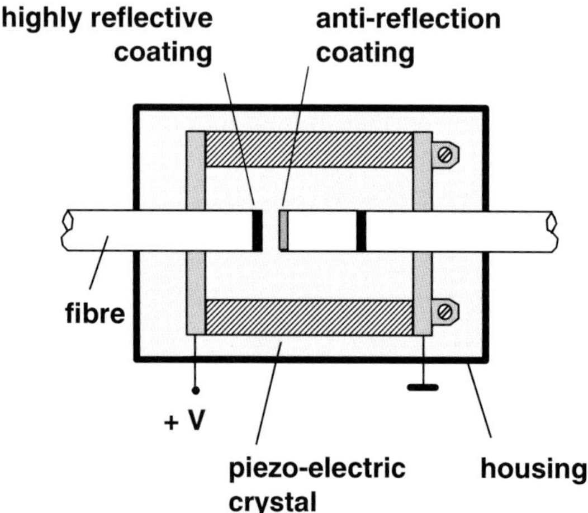

Fig. 6.8. Fibre Fabry–Perot filter

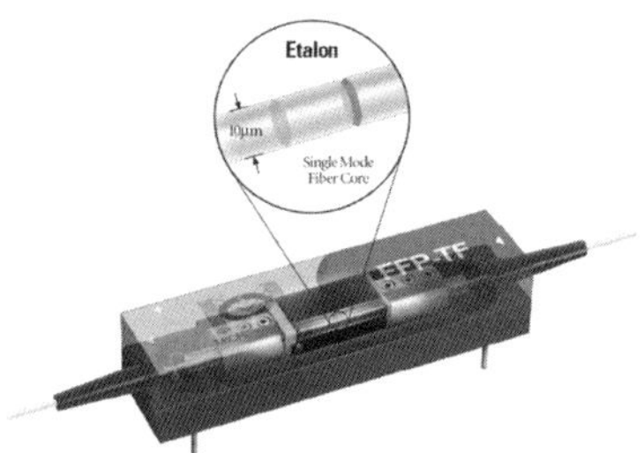

Fig. 6.9. Fibre Fabry–Perot module (courtesy of Micron Optics [3])

L-band (1520–1620 nm), or in the C- plus L-band. Instrument finesses may range from 10 up to 6000, and Free Spectral Ranges from 80 pm up to 250 nm (10 GHz to more than 31 THz). Due to the fibre-based design without extra optics, the fibre-to-fibre insertion losses are low (typically less than 2.5 dB). The circularly nearly symmetrical design makes the polarisation dependency low; typically less than 0.25 dB. The tuning speed is limited by the piezo-electric crystals; tuning over one FSR typically requires about 0.5 ms.

6.3.3 Gires–Tournois Filter

Similarly as a Fabry–Perot filter, a Gires–Tournois (GT) filter has a resonating cavity consisting of two parallel reflective plates. In a GT filter, however, one plate is fully reflective, whereas the other one is partly reflective. Hence, the GT interferometer can only work in the reflection mode, not in the transmission mode. The resonance condition is the same as with the FP, i. e. maximum reflection occurs when the plate spacing $d = m \cdot \lambda/2$, where m is an arbitrary integer. Equivalent to the Fabry–Perot, the Free Spectral Range is given by $FSR_v = c_0 / 2nd$. The GT's intensity reflection function basically has the same Airy function shape as the FP's one.

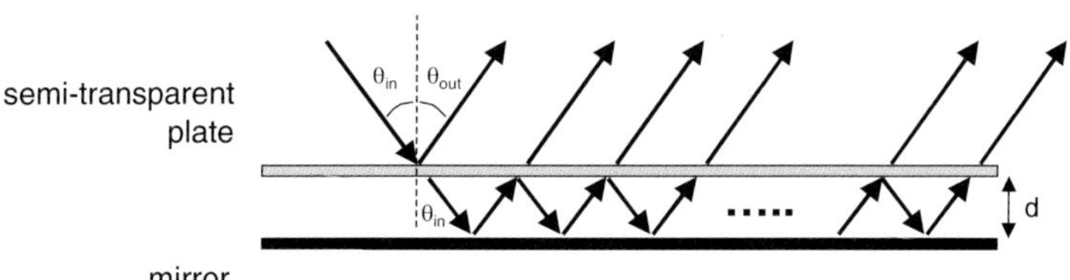

Fig. 6.10. Gires–Tournois interferometer

The Gires–Tournois filter is particularly suited for advanced spectral filter shaping, e. g. by inserting it in a Michelson interferometer setup, and by using e. g. micro-electromechanical systems (MEMS) – based micro-mirror arrays [4].

6.3.4 Interference Fabry–Perot Filter

A compact FP cavity can be realized by thin-film technologies. As shown in Fig. 6.11, a spacer region sandwiched between two reflecting layer stacks, deposited on a glass substrate, can constitute such an FP (see also [5]). The reflecting layer stacks usually are composed of $\lambda/4$-thickness layers, alternating of a high and a low refractive index. The reflection characteristics of these stacks become more square-top shaped with steeper edges when the number of layers is increased. Dielectric materials such as titanium dioxide ($n \approx 2.35$) and silicon dioxide ($n \approx 1.52$) are typically used in the 1.3–1.5 µm wavelength region. The dielectric spacer layer may be of high or low refractive index; the thickness of this layer is tuned such that the absolute frequency position and the Free Spectral Range requirements are met. The passband curves of the filter can be shifted to shorter wavelengths by tilting the incidence angle of the light. By this tilting, the phase shift δ between the transmitted beams would decrease when the wavelength is kept constant (see Sect. 6.1.1, equation (6.4)), and thus the wavelength needs to be reduced to get the same phase shift as with normal incidence, hence the filter curves shift to shorter wavelengths. However, the characteristics then also become dependent on the polarisation state; different curves shifted with respect to each other are experienced for TE- and TM-polarised light. This

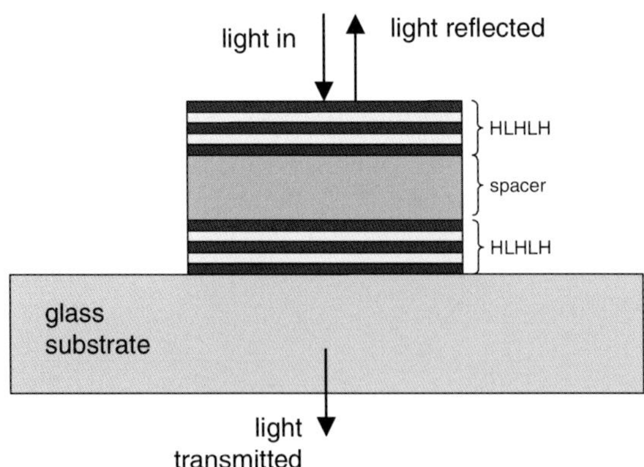

Fig. 6.11. Thin-fim Fabry–Perot interference filter

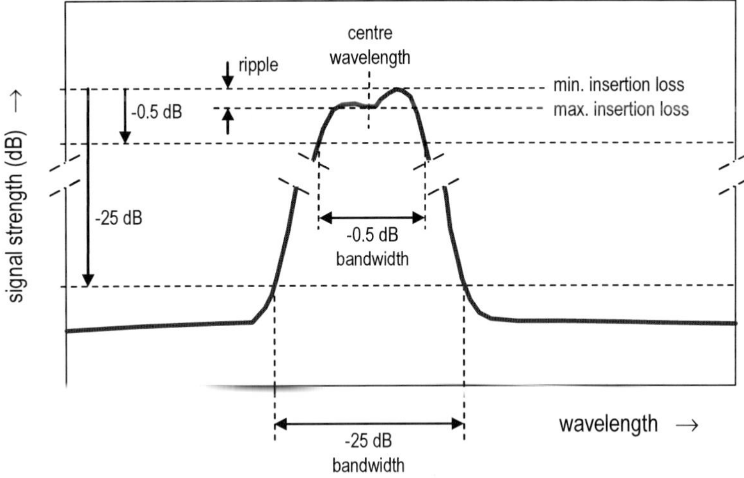

Fig. 6.12. Typical characteristics of a dielectric bandpass filter

polarisation-splitting deteriorates the filter performance for unpolarised light.

For smaller bandwidths, the peak transmission of the filter declines. DWDM dielectric bandpass filters are commercially available for wavelength channel selection according to the ITU grid (see e. g. [6]), having a −0.5 dB bandwidth of more than 0.7 nm and −25 dB bandwidth of less than 2.4 nm. The passband ripple can be less than 0.5 dB. The bandpass characteristics typically look as shown in Fig. 6.12.

6.4 Applications

6.4.1 Narrowband Single-channel Filtering

A tuneable Fibre Fabry–Perot can attractively be applied for channel selection in a multi-wavelength system, as illustrated in Fig. 6.13. The FSR should be larger than the range over which the wavelength channels are spread. Furthermore, the finesse should be high enough to resolve a single channel without too much crosstalk; as stated before, as a rule of thumb, the finesse should be at least three times the number of channels.

When an optical pre-amplifier is used (see Fig. 6.13), the tuneable FFP also cuts away the amplified spontaneous emission (ASE) noise spectrum which is not in the direct neighbourhood of the selected wavelength channel. This reduces the ASE-ASE beat noise and thus improves the receiver performance.

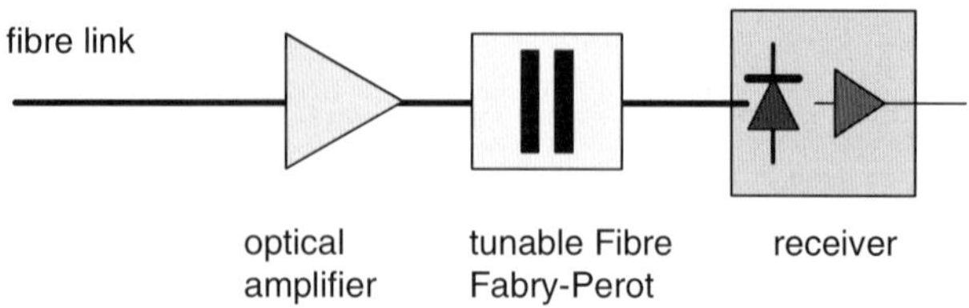

Fig. 6.13. Channel selection and pre-amplifier ASE filtering

The FFP can enable narrowband bandpass filtering, e. g. according to the ITU frequency grid definition for DWDM systems.

6.4.2 Optical Wavelength Channel Dropping

In multi-wavelength transmission networks, specific wavelength channels may be selected for dropping at intermediate nodes. For fixed-wavelength channel dropping, a Fibre Bragg Grating provides suitable characteristics. An FBG, however, is not easily tuneable over a wide wavelength range. A widely tuneable wavelength channel drop operation can be implemented with a tuneable FFP, as shown in Fig. 6.14 [9]. The FFP passes the selected wavelength channel λ_x for local dropping and reflects the other channels which then via the circulator and a coupler exit to the output fibre. At the coupler, another data signal at nominally the same wavelength λ'_x can be added. In its reflection path, the tuneable FFP needs to suppress adequately the selected wavelength channel and hence should have adequate notch-type characteristics (see also [3]).

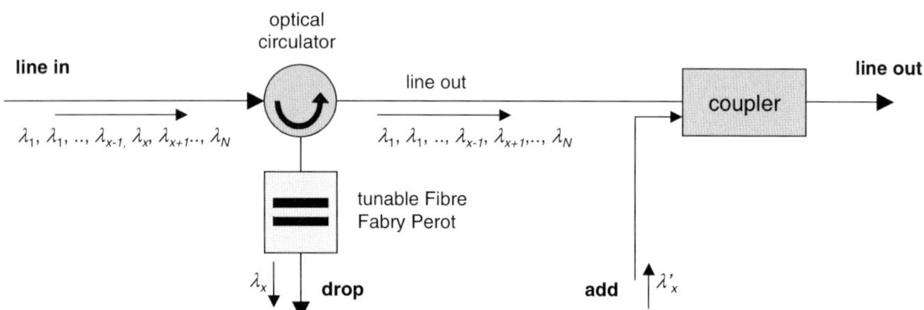

Fig. 6.14. Wavelength channel drop- and add-node, using a tuneable FFP

6.4.3 Multi-passband Filter

Multi-passband optical filters are useful in dense wavelength-division multiplexed transmission systems, e. g. when a number of wavelength channels need to be selected simultaneously from a noisy background (such as to cut ASE noise), or when a set of narrowly-spaced wavelength channels is to be split in two de-interleaved sets in order to ease channel processing. It is desirable to have flat-top square-shaped multi-passband characteristics in order to accommodate slight tolerances on the wavelength channel positions and to reduce any frequency-to-intensity modulation, such as may happen due to signal-induced chirp of the laser transmitters. These advanced multi-passband filter characteristics can be realized by e. g. including a Gires–Tournois interferometer in a Michelson interferometer, such as shown in Fig. 6.15. Such a filter can also be realized in all-fibre technology, as shown in Fig. 6.16, where the mirror function is implemented with a chirped Fibre Bragg Grating (FBG) and the Gires–Tournois filter with a composition of a weak fibre grating and a strong fibre grating, representing the beamsplitter and the mirror respectively in the GT architecture.

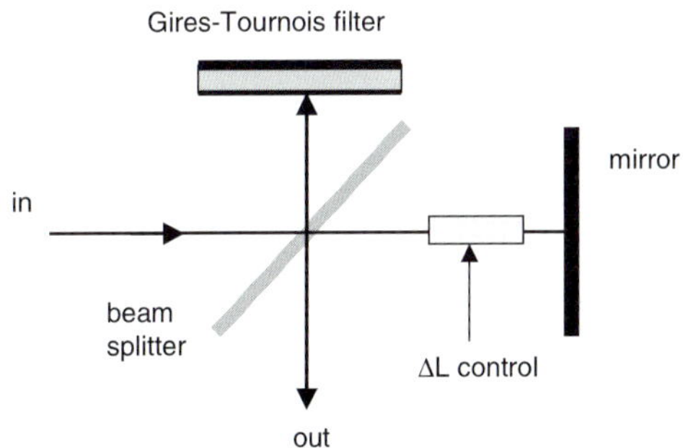

Fig. 6.15. Michelson–Gires–Tournois wavelength channel de-interleaver

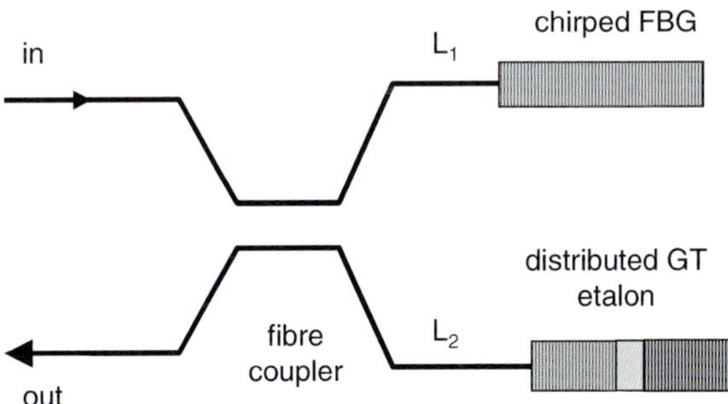

Fig. 6.16 All-fibre Michelson–Gires–Tournois filter

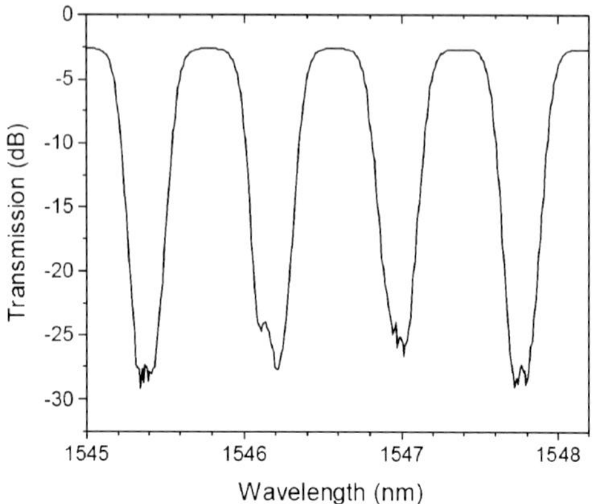

Fig. 6.17. Measured performance of all-fibre Michelson–Gires–Tournois filter

Measured spectra on this device show a nice wavelength de-interleaver performance (see Fig. 6.17) [7].

6.4.4 Wavelength Locking

In order to stabilize the wavelength of laser diodes which e. g. have to conform to the ITU frequency grid specifications of a high-density wavelength multiplexed system, Fabry–Perot filters can be applied in a so-called wavelength locker. The optical schematics are shown in Fig. 6.18 (see also [8]). Part of the output light of the laser to be stabilized is fed into the wavelength locker module where the total output power is monitored as well as the power at the desired wavelength position (to be set with the FP etalon). These two monitor output signals are then fed back to the laser transmitter unit for power- and wavelength-control. Commercially available wavelength lockers

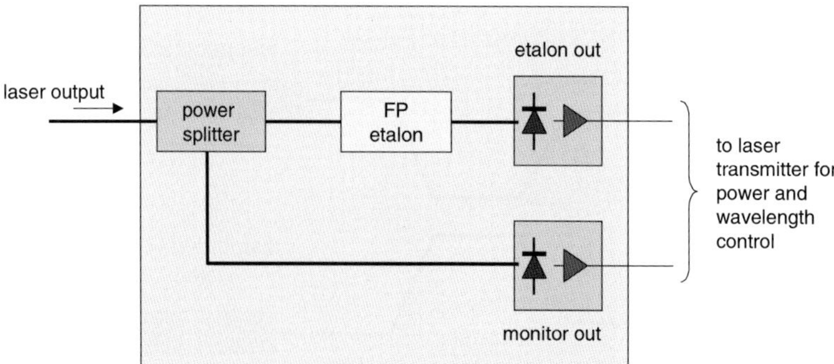

Fig. 6.18. Wavelength locker using a Fabry–Perot etalon

are designed for stabilization at 50 or 100 GHz wavelength channel spacing in the 1520–1620 nm range, conform the ITU grid. The centre channel accuracy can be within ±2.5 GHz; input powers up to 10 mW can typically be handled.

References

1. M. Born and E. Wolf: *Principles of Optics*, Chap. 7.6 (Pergamon Press, Oxford, 1980)
2. EXFO Burleigh, www.exfo.com
3. Micron Optics, www.micronoptics.com
4. O. Solgaard, D. Lee, K. Yu, U. Krishnamoorthy, K. Li, and J. P. Heritage: "Microoptical phased arrays for spatial and spectral switching," IEEE Commun. Mag. **41**, 96–102 (2003)
5. Melles Griot, www.mellesgriot.com
6. Edmund Optics, www.edmundoptics.com
7. X. Shu, K. Sugden, and I. Bennion: "Flattop multi-passband filter based on all-fiber Michelson-Gires-Tournois interferometer," *Proc. 30ᵗʰ Europ. Conf. Opt. Commun.* (ECOC'04), Stockholm, Sweden, 122–123 (2004), paper Tu1.3.1
8. JDS Uniphase, Broadband (Fabry-Perot) wavelength locker, www.jdsu.com
9. A. M. J. Koonen: "Multiwavelength Add/Drop Multiplexer," US Patent No. 5,751,456, issued on May 12, 1998

7 Dielectric Multilayer Filters

Markus K. Tilsch, Robert B. Sargent and Charles A. Hulse

7.1 Introduction

Dielectric multilayer or *thin-film optical interference* filters were the first filter type to be widely deployed in wavelength division multiplexing (WDM) systems in the 1990s. In addition to enabling channel separation in 200 GHz and later 100 GHz WDM systems, the technology has been applied to a number of vital optical network applications such as gain-flattening filters (GFFs) and pump WDMs for erbium-doped fibre amplifiers (EDFAs), wideband *band splitting* filters for separating bands of channels, and more recently for implementing low-cost modules for coarse WDM (CWDM) and access networks. Dielectric multilayer filters have become ubiquitous in optical communications systems.

Thin-film filter technology was useful in implementing WDM optical networks because it was relatively mature. Optical interference coatings have been fabricated and used since the 1930s. One early application was antireflection (AR) coatings [1], which found important military uses during World War II. Since that time the technology has evolved rapidly and expanded from largely military and scientific applications into diverse commercial uses. Demanding applications in fields such as aerospace and medical diagnostics motivated the development of very stable filters, and by the 1980s processes for depositing fully dense, bulk-like metal oxide coatings had been developed and were in production. The technology had reached a high level of maturity by the 1990s and was well positioned to meet the filtering requirements of optical fibre networks.

Dielectric multilayer filters enjoy wide acceptance in optical networks because of their availability, proven reliability, and long-term stability. The filters can be made to have excellent wavelength stability, which makes them ideal candidates for passive components. Indeed, dielectric filters are sometimes employed for the wavelength stabilization of active components. Thin-film dense wavelength division multiplexing (DWDM) filters exhibit low insertion loss (IL), low polarisation dependent losses (PDL), and exceptional isolation. The technology is very flexible; AR, narrow bandpass, wide bandpass, edge, gain flattening, dispersion compensation, and other filters can be designed and produced from similar materials. Thin-film filters may be employed in modular architectures that start with just a few channels but can later be expanded as the need for more bandwidth arises.

We begin this chapter with a brief review of the theory of thin-film filters, then describe the physical vapour deposition (PVD) processes used to fabricate the dense metal-oxide coatings most commonly used in telecommunications applications. Next we review the excellent temperature stability of filter elements and related properties, before moving on to a description of common methods for interfacing thin-film coatings to fibre-optic systems. We conclude with a description of the most common applications of thin-film filters and a short discussion of new developments.

For clarity, we use the terms *coating*, *stack*, or *dielectric multilayer* to indicate any multiple-layer thin-film structure, while *filter* is reserved for a coating having a spectrally-selective characteristic.

7.2 Theory of Interference Filters and Devices

7.2.1 Interference Filter Theory

Thin-film coatings consist of thin layers of controlled index and thickness. The thickness of each layer is typically a fraction of a wavelength; the reflections at every layer boundary interact coherently to generate the spectral performance. The theory of interaction of plane waves is well understood and is described in numerous references [2–6]. Depending upon the application, the filter may consist of a few layers to a few hundred layers of two or more different material types to generate almost any arbitrary spectral response. Common applications encountered in telecommunications include AR coatings, high reflectors, bandpass filters, and GFFs. Each of these coating types has a basic theory and character, which is discussed below.

Antireflection (AR) coatings
Whenever light crosses an interface, a fraction is reflected due to the mismatch in refractive indices. AR coatings quench this reflection through the insertion of additional layers to create destructive interference of the reflected wave. These filters are important for two reasons. First, they maximize the throughput of the system. Secondly, and perhaps more importantly

for telecommunication systems, they minimize unwanted back reflections. Usually termed return loss, insufficient reflection suppression can appear as crosstalk in a communication channel or as undesired feedback to a laser source.

Many design approaches to AR coatings exist; numerous references describe the design of simple to very complex ARs [7–10]. The complexity of an AR coating is increased by increasing the operational bandwidth (wavelength range), by increasing the required return loss (decreasing the allowed reflectance), and other factors such as increasing the range of angles that the AR coating must perform over. Telecommunication ARs usually operate only at a single wavelength or a narrow wavelength spectrum, but the return loss requirements can be quite challenging. Some applications require 40 dB to 50 dB return loss over extended wavelength regions [11].

High reflector (HR) coatings
HRs enhance the natural reflection due to an interface. The basic structure for a dielectric HR consists of alternating layers of two materials, one with a high and one a low refractive index. The physical thickness of each layer is $\lambda_0/(4n)$, where λ_0 is the desired centre wavelength and n is the refractive index of the respective material. Layers with this thickness are referred to as *quarterwave* layers. The HR design can be noted as:

$$\text{Substrate} \mid (\text{HL})^p \mid \text{Ambient}. \tag{7.1}$$

In design formulae, H and L represent quarterwave layers of high and low refractive indices at the desired wavelength. The power p determines how often the period is repeated; $(\text{HL})^2$ is the same as HLHL.

Figure 7.1 shows the spectral reflection of a HR design for 5, 10, and 15 periods.

The high reflecting region is called the stopband of the HR. The stopband width $2\Delta g$ does not depend on the number of periods, but instead on the ratio of the high and low refractive indices. Macleod [Ref. 2, p. 191] defines the normalized frequency parameter g:

$$g = \frac{\lambda_c}{\lambda} \tag{7.2}$$

The edges of the stopbands are at $g_{edge} = 1 \pm \Delta g$ with Δg being:

$$\Delta g = \frac{2}{\pi} \sin^{-1}\left(\frac{n_H - n_L}{n_H + n_L} \right) \tag{7.3}$$

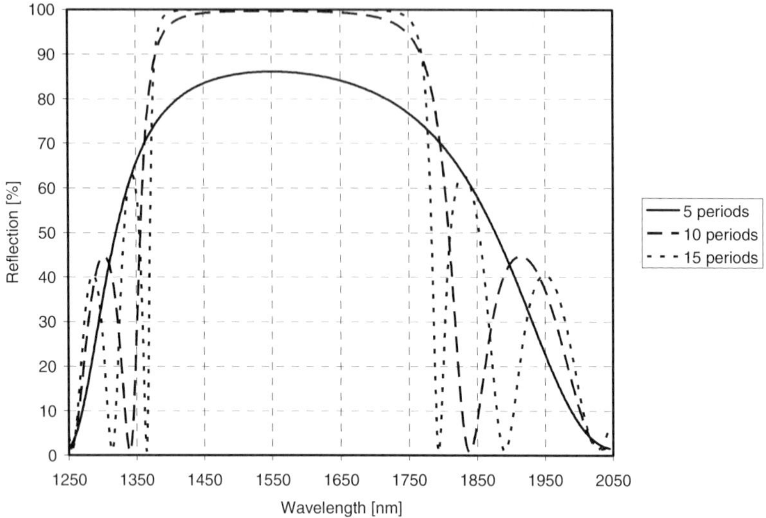

Fig. 7.1. Reflectance of HR designs with 5, 10, and 15 layer periods at centre wavelength 1550 nm. The refractive indices are 1.45 (L) and 2.1 (H)

The stopband width becomes wider for a higher refractive index ratio. In the example from Fig. 7.1, $n_L = 1.45$ and $n_H = 2.1$. Equation (7.2) yields a stopband from 1387 to 1756 nm for a filter centred at 1550 nm.

The strength of the reflection in the stopband depends both on the ratio of the refractive indices and the number of periods. A higher ratio and more periods lead to higher reflection [cf. Ref. 2, p. 186].

$$R \cong 1 - 4\left(\frac{n_L}{n_H}\right)^{2p} \frac{n_S}{n_A}, \tag{7.4}$$

where n_A is the refractive index of the ambient medium and n_S is the refractive index of the substrate.

Usually it is desirable to select two materials with a high refractive index ratio. Fewer layers are required to achieve a specified rejection level and the stopband width is wide. It is possible to cascade two or more mirror structures with different centre wavelengths to increase the stop band width. The HR design in its basic form can be found in coatings for some lasers and other applications where the absorption of a metal mirror would be objectionable.

A variation of the basic HR design can be used for band edge filters. For a shortpass filter the centre wavelength is shifted up to align the low-wavelength edge with the desired band edge. The layer thicknesses can be tuned to reduce the ripple in the passband. Similarly, a longpass filter can be generated by shifting the centre wavelength to a shorter wavelength. In band edge filters, the number of periods determines the edge steepness. In many telecommunication applications, wide bandpass filters, discussed in the next section, are preferred over edge filters for their higher achievable edge steepness.

Bandpass filter

A thin-film bandpass filter consists of one or more coupled thin-film Fabry–Perot filters. A thin-film Fabry–Perot bandpass filter consists of a thin-film etalon, called the *spacer*, surrounded by metal reflectors or all-dielectric thin-film HRs of the type described above. The basic design formula for the more commonly used all-dielectric version is:

$$\text{Substrate} \mid (\text{LH})^{\text{p}} \, (2\text{L})^{\text{s}} \, (\text{HL})^{\text{p}} \mid \text{Ambient}. \tag{7.5}$$

p describes the number of periods in the HRs and s determines the order (thickness) of the spacer. The optical spacer thickness at the centre wavelength for the first order ($s = 1$) is one halfwave, for the second order ($s = 2$) two halfwaves, etc. In a Fabry–Perot filter, only a small fraction of light normally penetrates the first reflector, but at certain resonant wavelengths, the light intensity builds up in the spacer layer until a significant fraction, nominally 100%, of the input light is transmitted. The transmission of the filter at non-resonant wavelengths, as well as the width of the passband, are determined by the number of periods in the HRs. As the reflectors are made stronger, the transmission at non-resonant wavelengths is suppressed, but the resonant transmission is preserved, narrowing the filter. The width of the passband can also be narrowed by increasing the spacer order, which decreases the free-spectral range (FSR) of the Fabry–Perot. Chapter 6 discusses the theory of Fabry–Perot filters in more detail.

In Fig. 7.2 the curve labelled *1 cav* shows the spectral response of a thin-film Fabry–Perot filter. The spectral response of a Fabry–Perot filter is very rounded and the rejection outside the passband increases only gradually. It is possible to stack multiple Fabry–Perot structures on top of each other to form a coupled multi-cavity filter. Figure 7.2 shows the spectral transmission for several filters from one to ten cavities. The spectral behaviours are of *square top* type, where the designs are tailored to approximate an ideal shape of 100% transmission in the pass band and full rejection outside the pass band.

The coupling of the cavities is critical to avoid excessive ripple in the passband, and is the subject of considerable literature. Thelen [3] and Macleod [2] provide introductions to the subject of optical bandpass filter design. Baumeister [4] adapted powerful techniques from microwave theory

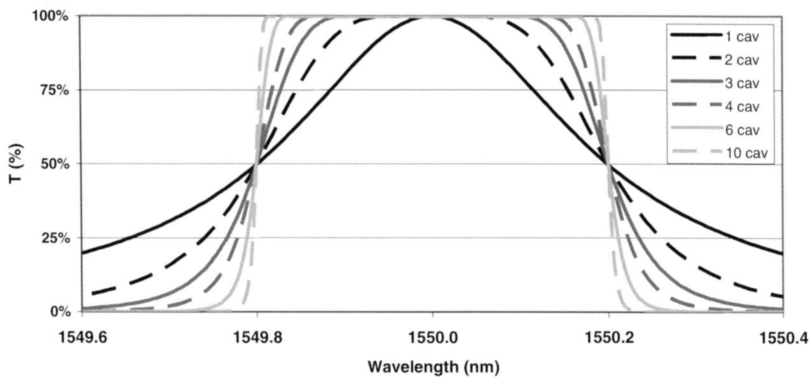

Fig. 7.2. Spectral response of bandpass filters with varying number of cavities

to the optical domain. Other recent publications present methods tailored specifically to DWDM bandpass filters [12, 13].

Figure 7.3 shows the transmission of 3 five-cavity filters over an extended wavelength spectrum. The refractive indices used in this example are 1.45 and 2.1. The solid line shows a design for a 100 GHz filter. Out-of-band blocking reaches a level that is in excess of 200 dB. The blocking stays well above 50 dB down to 1390 nm and up to 1750 nm. The refractive indices of the materials used in the HRs determine the rejection width as discussed above in the HR section.

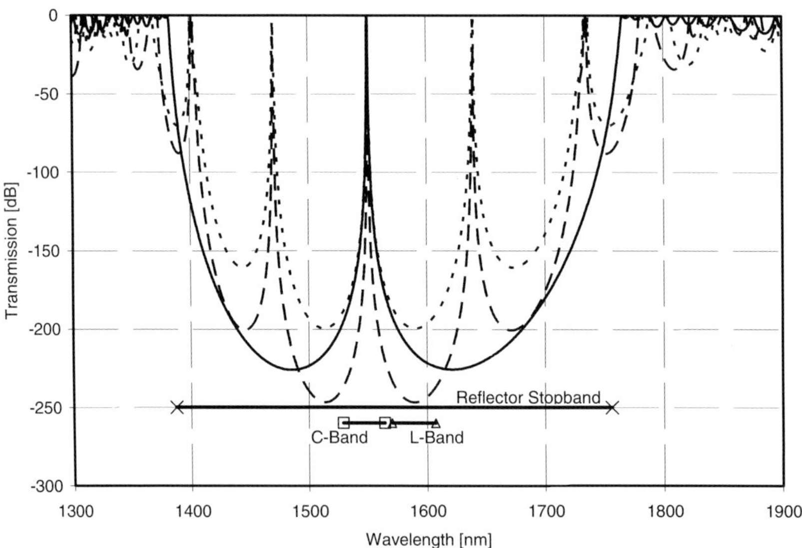

Fig. 7.3. Transmission spectra of 3 five-cavity filters. See text for details

The dashed curve in Fig. 7.3 shows the spectrum of the design where all spacer thicknesses are made four times larger. As mentioned above, this reduces the FSR of the filter, which reduces the width of the passband significantly. The reduction in FSR also leads to additional transmission peaks within the blocking range of the HRs. The free spectral range is still large compared to the C-band or L-band. A further increase of spacer thickness turns such filters into interleavers.

The dotted curve in Fig. 7.3 gives the spectrum of the design having four times larger spacer thicknesses where the reflector strengths (cf. (7.4)) were reduced to maintain the original passband width. The free spectral range is primarily determined by the spacer thicknesses and is unchanged from the previous case. The maximum rejection is reduced to 200 dB due to the reduction in reflector strengths. The material properties of the reflectors drive the limits of the rejection band.

Many options are available in the design of narrow bandpass thin-film filters. That gives high flexibility in tailoring the spectra to the customer's requirements in the passband, but also in the rejection area. Extended blocking or additional passbands for pump lasers are typical out of band requirements.

Gain-flattening filter (GFF)

Gain flattening is necessary to counteract the innate spectral dependence of an EDFA, ensuring that the gain is constant across the desired wavelength range. Thin-film filters can be designed with nearly arbitrary spectral performance [14, 15], and are a good choice for this application [16, 17]. Figure 7.4 illustrates this concept. The diamonds show the targeted reflectance

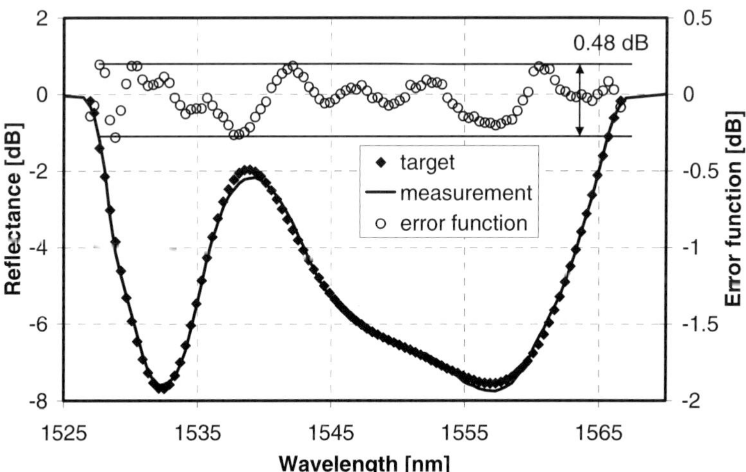

Fig. 7.4. GFF target spectrum, measurement of a filter and error when measured in cascade with the amplifier (after [17])

values for a GFF. The solid line shows the reflectance measurement from a fabricated filter. The error function was recorded when the GFF was measured in cascade with the amplifier. Numerous GFF-specific design approaches have been proposed, including [18] and [19]; all require extensive computer optimization. In many applications, a GFF can flatten the overall shape of the amplifier to less than 5% of the maximum excursion of the uncompensated amplifier.

7.2.2 Dispersion Effects

Thin-film filters can be designed and built to have nearly square passband characteristics, which makes them useful as DWDM bandpass filters. In this section, the implications of filter design on the delay characteristic are introduced. Possible tradeoffs in design to minimize the delay variation, and the notion of allpass filters as an alternative to solving the dispersion of thin-film filters are discussed.

This section assumes the reader is familiar with the relations between phase, group delay (GD), and chromatic dispersion (CD), as detailed in Chap. 2. The familiar square shape of a DWDM filter is achieved by inducing a very strong engineered resonance; the light is trapped within the filter for an appreciable time, especially near the band edges, to achieve constructive interference and eventual transmission through the multicavity structure. Lenz and co-workers have pointed out that the square passband edge and strong delay are linked through a Hilbert Transform [20]. The CD increases approximately linearly with the order of the filter, and quadratically with the bandwidth [21]. A typical transmission dispersion characteristic is superimposed on the IL characteristic for a 100 GHz filter in Fig. 7.5.

Strong CD is a limiting effect in a WDM transmission system. Differential delay to the frequencies constituting a pulse can lead to broadening and distortion. A general rule of thumb is that the GD variation of a channel must be kept to less than one-quarter of the pulse width to maintain decipherability [23].

Thin-film filters can be optimized to take into account the dispersion penalty and minimize overall system penalty. While the transmission of a square top filter is high over the whole pass band, the CD is low in the centre of the filter and rises rapidly towards the pass band edges. By centring the filter tightly, the dispersion penalty can be reduced. A filter with a wider passband pushes the onset of CD excursion further away from the centre, at the expense of adjacent channel isolation. To maintain isolation, a filter with higher cavity count may have steeper edges, increasing the CD

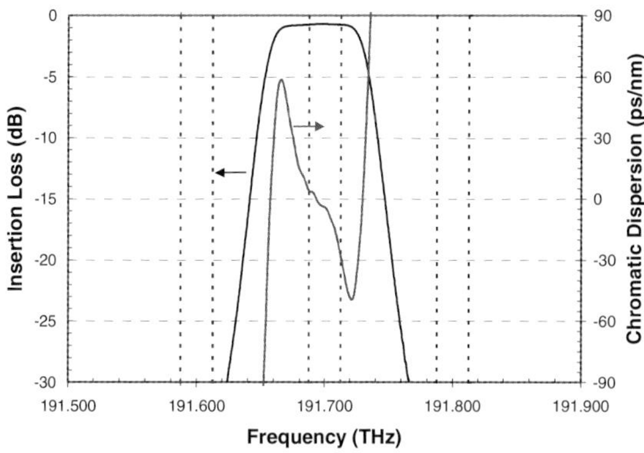

Fig. 7.5. Measured insertion loss and chromatic dispersion of a 100 GHz thin-film filter (after [22])

but still pushing it from the channel centre [24]. Another approach is to give up some of the squareness of the filter, trading transmission penalty for dispersion penalty. Introducing a controlled ripple across the passband reduces the magnitude of the CD [24]. This approach can be taken to the extreme. Instead of designing a filter for a square passband a filter can be designed for maximum CD flatness. This Bessel–Thomson shape imparts nearly constant delay to the transmitted pulses, at the expense of a nearly Gaussian passband [25]. The performance of such a filter is shown in Fig. 7.6.

All discussed approaches require excellent process control in the fabrication of the components. Exact centring and achieving precise transmission IL are equally critical. Tight tolerances need to be held under all operating conditions.

A more complex approach to correcting the dispersion penalty is to add an additional component to correct the CD imparted by the thin-film filter.

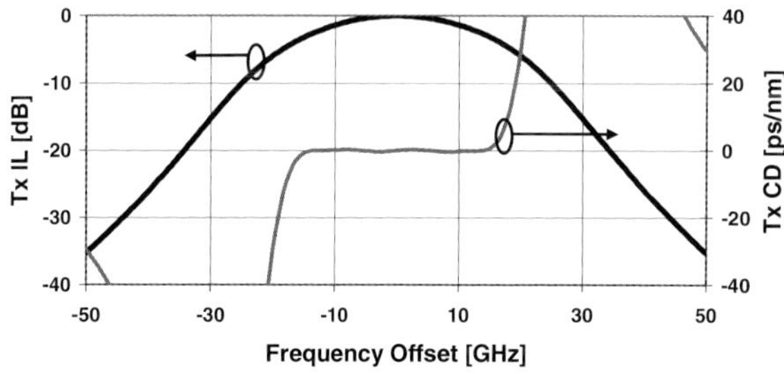

Fig. 7.6. Transmission and CD of a thin-film Bessel filter (after [25])

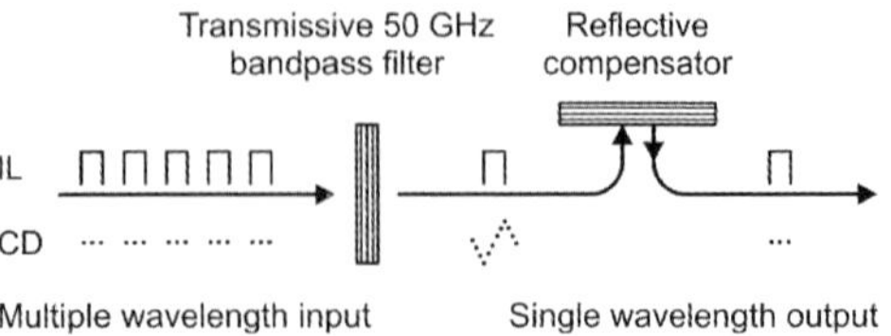

Fig. 7.7. Concept of dispersion compensation for a thin-film filter using a thin-film allpass filter (after [28])

This is possible since CD is a deterministic phenomenon. Concepts to compensate filter CD have been discussed [20, 26]. In particular, the CD of a thin-film filter in reflection does not have the same constraints as in transmission; a purely reflective *allpass* filter can be designed to modify the delay of a signal without imparting a transmission effect. Figure 7.7 shows the concept of a thin-film allpass filter solution for a 50 GHz network. The reflective compensator is designed to trap the light for a wavelength dependent amount of time, which allows correcting the CD of the 50 GHz filter [19, 27]. This technique has been experimentally demonstrated [28]. Centring tolerances between the bandpass filter and the compensator are extremely tight.

Thin-film allpass filters have also been used to compensate for CD in other elements in the telecommunication network [29, 30].

7.2.3 Angle-of-incidence Effects

Thin-film filters are usually operated at a small angle of incidence in telecommunications components. This angle separates the input and reflected beams, lessening backreflections and enabling multiport components without the need for a circulator. The use of thin-film filters at angle also introduces two angle-dependent effects: an angle-dependent wavelength shift and a polarisation splitting effect. This section discusses the origin and impact of these effects for small angles of incidence.

When light strikes a filter at angle, the layers that comprise the filter appear thinner than they do at normal incidence. For small angles, each layer appears to shift in thickness an amount proportional to the sine of the incident angle. This effect can be used to tune the operational wavelength of a filter at manufacture or even in operation (see Sect. 7.6.6). This effect is only useful for small angles; for angles of incidence of more than a few degrees, Snell's Law must be used to calculate the effective thickness of each layer in the thin film, and operation can vary markedly.

The second effect of angle, polarisation splitting, occurs because electromagnetic field boundary conditions are different for fields normal to an interface than they are for fields parallel to the interface. At a nonzero angle, any incident light may be split into an s field with an electric field component normal to the plane of incidence and a p field with the electric field purely parallel to the plane of incidence. The intralayer reflection for each of these fields is different, and becomes more pronounced as the angle of the incident light beam increases. For certain applications, such as MacNeill polarisation filters, the filters are designed to be used at large angle and this polarisation difference is exploited. For most telecommunications applications, however, this is a source of PDL, and components minimize the effect by using the filter as close to normal incidence as is practical.

7.2.4 Expanded Beam Optical Devices

Thin-film filters can be designed with very narrow, square passbands for multiplexing or demultiplexing, as well as nearly arbitrary transmission and reflection shapes for equalization purposes. These shapes are typically maintained only over a narrow range of angles without significant design effort. Furthermore, the random polarisation of typical fibre-optic applications requires polarisation-insensitive design, which becomes much more difficult further away from normal incidence. This section introduces expanded-beam devices as a means to interface single-mode fibres to thin-film filters, and how they overcome the limitations of angular spread, angle of incidence and beam intensity.

Standard SMF-28 fibre has a 9 micron core, and supports a near-Gaussian mode of roughly 10 microns in diameter. Using Gaussian beam optics [31], one can calculate that the beam exiting a fibre has a divergence angle of roughly 5.5°. An AR coating can work over these angles, but most DWDM filter designs cannot maintain their shape and wavelength centre. In addition, the high divergence of the fibre mode requires the filter and substrate to be very thin; even separated by 100 micrometers (a thin substrate), a calculation of fibre-to-fibre coupling of the mode finds approximately 1.5 dB of loss [32].

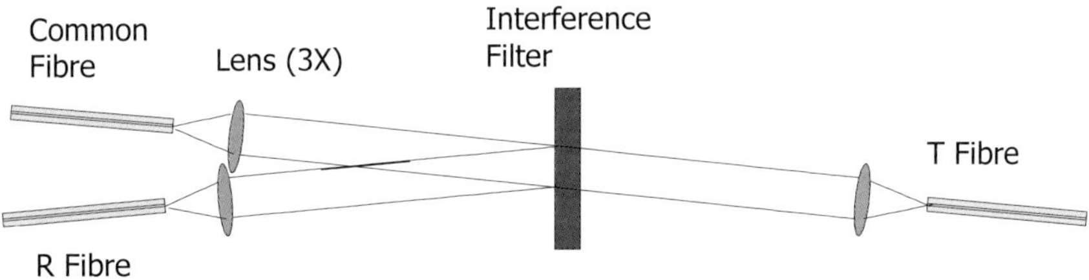

Fig. 7.8. Schematic of three-port thin-film filter component

To ameliorate divergence problems associated with fibre modes, single-mode fibres are usually interfaced to thin films using collimators [33] to expand the waist of the beam and reduce its divergence. This is shown schematically in Fig. 7.8.

A typical gradient-index (GRIN) collimator expands the fibre mode into a beam between 0.15 and 0.75 mm in diameter. This arrangement has three main benefits:

1. A collimator reduces the divergence of the interacting beams, making them more like plane waves. The reduced divergence both reduces the range of angles in the beam and increases the Rayleigh range. For a 0.15 mm diameter beam, the range of angles in the beam is reduced to less than $\pm 0.5°$, and the Rayleigh range increases to tens of mm, easily able to incorporate a filter on a substrate 1 mm thick.
2. A collimator reduces the angle of incidence required to separate the beams in a multi-port device [34]. Smaller angles of incidence reduce the polarisation sensitivity of the filter. Typical arrangements use angles of only a few degrees [35, 36].
3. A collimator reduces the electric field intensity in the device. Thin-film filters function by trapping light resonantly within the cavities, and total electric field intensity would be a concern for a 10 micron diameter beam. Collimators drop the electric field intensity by orders of magnitude; a 0.15 mm diameter beam has a field intensity 225 times less than a $10\,\mu m$ fibre mode.

Although Fig. 7.8 depicts the filter in the waist of the beam, this location is not necessary. The angular constituents of the beam are independent of location; the only advantage to placing the filter in the beam waist is the decreased spot size at that point.

Section 7.5 discusses the integration of thin-film filters in fibre-optic packages in detail.

7.2.5 Interaction of Beams with Filters – Beam Shift Effect

The primary effect of increasing the angle of incidence on the spectral performance of an interference filter was described in Sect. 7.2.3: the spectral performance is shifted to a shorter wavelength. In practice, filters are used in optical systems in which the light is not perfectly collimated. A common way to analyze this situation is to compute the transmission as a function of wavelength as an incoherent sum over the range of angles of incidence present in the beam [Ref. 2, pp. 288–292]. This type of calculation is valid over a wide range of conditions where the phase relationship between the various angular components of the beam is not important. The result is that the transmission of the filter is reduced and bandwidth becomes wider as compared with the filter performance in collimated light.

In situations involving filters having very narrow bandwidths used in conjunction with small beams, the assumption stated above may no longer be valid. In such cases, an analysis that is akin to the treatment of the Goos–Haenchen shift using an angular spectrum approach [37] is useful. In addition to predicting transmission and bandwidth changes, this approach predicts a lateral beam shift. This effect was first noted by McGuirk [37] and Hsue [38], but was of little practical importance until the advent of very narrowband filters for DWDM [39, 40]. A brief summary of the effect is given below.

Thin-film DWDM filters operate with extremely high Q-factors, so the transiting beam accumulates large amounts of phase as it goes through the filter. This phase may induce a lateral shift ΔL of the beam, proportional to the derivative of the accumulated phase ϕ with respect to its transverse angular frequency $k_\perp$:

$$\Delta L = \frac{d\phi}{dk_\perp},\tag{7.6}$$

where $k_\perp$ is the component of the propagation vector $\boldsymbol{k}$ parallel to the filter interface.

Beam shift increases as the filter bandwidth decreases, or as the angle of incidence increases. Typical beam shift is tens or hundreds of micrometers for a typical 100 GHz application. Table 7.1 summarizes some typical beam shift numbers. The magnitude of the beam shift is independent of the size of the beam, and the effect is therefore less significant for larger beams. Since the angular content of the beam does not change as the beam propagates, the beam shift does not depend upon the placement of the filter relative to the beam waist.

Table 7.1. Typical displacement vs. angle of incidence for 100 and 200 GHz thin-film filters (after [41])

Channel Spacing (GHz)	0° Incidence Shift (μm)	3° Incidence Shift (μm)	5° Incidence Shift (μm)
200	0	47	78
100	0	81	137

Klinger et al. [39] experimentally verified the presence of lateral beam shift, and developed a full three-dimensional model for calculating the propagation of a small beam through a thin-film DWDM filter. The program was based on expressing the beam as an angular spectrum of plane waves [42], then propagating each plane wave through the filter independently. The model predicted beam shift as well as other non-ideal effects, such as passband rounding and distortion, wavelength dependence of beam shift, and size-dependent shift [39, 41]. The work stressed the importance of the size of the beam in designing thin-film optical couplers.

Beam shifts can be accommodated in the optical design of thin-film components, for example by enlarging the beam to minimize loss or by shifting the output collimator [43]. Section 7.7 describes work to harness the beam shift as a means to demultiplex DWDM channels.

7.3 Filter Materials and Processes

7.3.1 Brief History of Thin-Film Technology

Thin-film filters can be created with a variety of technologies. The focus here is on PVD processes that are currently used for most telecommunication filters.

The advent of reliable vacuum generation in the 1930s enabled PVD development. One of the most important early processes for optical coatings was thermal evaporation from directly heated crucibles or *boats*. The application of electron beam guns to melt or sublime source materials expanded the range of materials that could be evaporated to include high-melting-point materials such as refractory oxides (cf. Sect. 7.3.3). Another innovation was reactive evaporation or the introduction of a low oxygen partial pressure during the evaporation of the oxide or sub-oxide material. Through the 1970s, conventional evaporation processes dominated the optical coating industry, though preliminary work in using other PVD processes such as sputtering was underway.

The evaporation of high-melting-point materials onto substrates at relatively low temperatures often yields films with poor properties. The microstructure of such films has been found to be significantly different from that of the bulk materials. Typically the films exhibit a porous microstructure of density lower than that of the bulk material. Moisture may be observed to penetrate such coatings, and the properties have been observed to vary as a function of temperature and humidity. In many cases a solution to this problem is to increase the substrate temperature, which usually results in films of greater density.

By the 1980s energetic PVD processes capable of depositing robust films of refractory materials were under development [44]. Such processes provided significant property improvements over previous methods, resulting in films of greater density, higher refractive index, reduced sensitivity in spectral performance to temperature and humidity, and greater durability. Three of the most commonly used energetic PVD processes are ion-assisted deposition (IAD), ion beam sputtering (IBS), and reactive magnetron sputtering; all of these processes have been applied to the deposition of thin-film optical communication filters. The processes will be described in more detail in Sect. 7.3.4 after the introduction of substrate and thin-film materials.

7.3.2 Substrates

Some coatings are directly applied to devices. A laser diode facet may be directly coated with an AR or HR design. A high-power-carrying fibre tip may need an AR. Partial reflectors on GRIN lenses can act as taps. WDM filters have been directly coated on GRIN lenses [45]. Challenges include the cleaning of such devices prior to coating and supporting them in the coating tool. Outgassing of the devices may impact the coating performance. The coating process needs to be tailored to comply with the device requirements. Restrictions in temperature and energetic bombardment are common.

The majority of coatings are deposited on designated substrates that are later inserted into the optical beam path. Often the coating is applied on a large substrate, which is diced and thinned in post-deposition steps (see Sect. 7.3.6). The required optical, mechanical, and environmental properties, as well as availability and cost drive the substrate selection.

Important attributes of a DWDM filter substrate are its coefficient of thermal expansion (CTE) and mechanical properties. Filters can be athermal, i.e. their spectral response does not shift to different wavelengths when the filter undergoes thermal changes. This allows the fabrication of DWDM filters without the need of active temperature stabilization. The theory for the temperature shift phenomena is discussed in Sect. 7.4.1. Glass companies [46–48] offer special substrates that are tailored to the DWDM requirements.

The substrate material must be clean to promote proper adhesion of the coating to the substrate. DWDM coatings require very smooth surfaces. In many coating processes defects grow larger throughout deposition of the filter [49, 50]. Defects tend to scatter light and can limit filter performance.

7.3.3 Filter Materials

Refractory materials, or materials having high melting temperatures, tend to be among the most durable of materials. Refractory oxides have been employed for thousands of years for such applications as the manufacture of glass and ceramics; the durability and transparency of many of these materials have made them attractive materials for optical thin-film coating applications. Representative refractory oxides that have been used as materials for optical thin films include silicon dioxide, aluminium oxide, magnesium oxide, a number of transition metal oxides, and even a few lanthanide and actinide oxides [2, 51]. Silicon dioxide is of particular interest, as it exhibits a relatively low refractive index of about 1.44 at 1550 nm. SiO_2 is commonly used with a refractory metal oxide such as TiO_2, ZrO_2, HfO_2, Nb_2O_5, or Ta_2O_5 as the high index material. Most of the thin film literature specific to WDM filter deposition cites the use of Ta_2O_5 and SiO_2 as the materials of choice.

When deposited using an appropriate energetic PVD process, these refractory oxides tend to remain amorphous over a wide range of temperatures. This property minimizes scattering within the coatings. These materials may also exhibit very low absorption, with an extinction coefficient $k < 10^{-6}$ [52].

7.3.4 Filter Deposition Processes

Three of the most commonly used energetic PVD processes are IAD, IBS, and reactive magnetron sputtering [53–55]. All of these processes have been applied to the deposition of thin-film optical communication filters, and are briefly considered in turn.

IAD is a modification of the evaporation process that employs traditional direct thermal sources or electron beam sources to deposit material. The modification is the addition of an ion source that serves to bombard the substrate with ions to densify the growing film. Various ions may be used; commonly oxygen and/or argon ions are used in the deposition of an oxide. The IAD process has been demonstrated to produce refractory oxide coatings with excellent properties [44, 56].

A notable example of an IAD process that has been applied to the deposition of WDM filters employs an advanced plasma source (APS) [57, 58]. Figure 7.9 shows a photograph of an APS 1100 deposition system from Leybold Optics [59]. The APS produces a high uniform flux of ions having energies of up to 200 eV. Relatively high deposition rates are reported; for example, a 3-cavity 100 GHz filter may be produced in 12 hours.

Challenges of evaporation based processes include the random variation in the spatial distribution of the evaporant plume over long time periods. The control of the evaporation process and control of the APS are critical.

Fig. 7.9. Interior of APS 1100 IAD deposition system (Courtesy of Leybold Optics, Alzenau, Germany)

Progress on a system that uses a quartz crystal monitoring system to adjust electron beam gun parameters has been reported [60].

IBS employs an ion source to bombard a metal or oxide target with high velocity ions. Material sputtered from the target condenses on the substrate. The target material can be of the film material, like SiO_2 and Ta_2O_5. Alternatively a metallic or substoichiometric target can be used and oxygen can be added to the process as a reactive gas. Keeping the target neutralized is one of the challenges in the process. Optionally, a second ion source may be directed toward the substrate to bombard the growing film with the reactive oxygen, or argon, etc.; this variation of the process is often called Dual Ion Beam Sputtering (DIBS). IBS is known to produce films with very low loss and the process has been shown to be very stable over a long period of time [44, 61]. The deposition rates associated with the IBS process have been considered to be rather slow, leading to long deposition times. For example, the time required to deposit a 3–4 cavity 100 GHz filter was on the order of 24–40 hours. A significant improvement in the rate of an IBS process for WDM filters has been reported [62]. Another disadvantage of IBS is that the resulting films usually exhibit a high degree of compressive stress (see Sect. 7.4.2). This is understood to result from the neutralization and reflection of a percentage of the primary sputtering ions (usually Ar^+), some of which bombard the growing film. Figure 7.10 shows a picture of Veeco Instruments Inc.'s DIBS chamber SPECTOR [63].

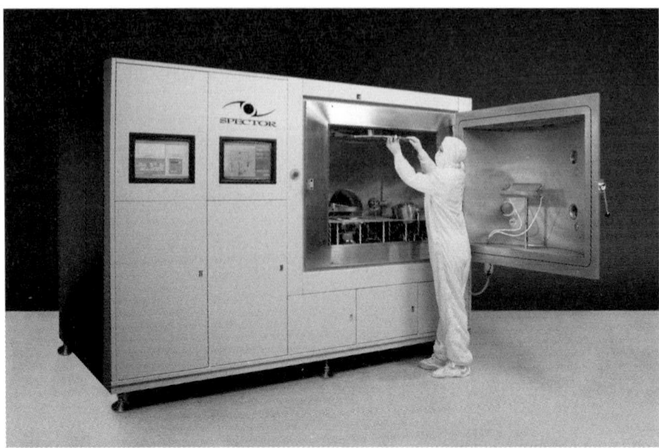

Fig. 7.10. SPECTOR DIBS deposition system (Courtesy of Veeco Instruments Inc., Woodbury, NY, US)

Reactive magnetron sputtering is an energetic PVD process that utilizes DC magnetron sputtering of a metal or semiconductor; the film is reacted to an oxide at the growing film interface at the substrate [64, 65]. Various modifications to this basic process have been proposed and demonstrated, such as the use of an auxiliary ion source to bombard the substrate with oxygen ions during film growth [66]. This last process has been employed in the manufacture of WDM filters [16]. The process is claimed to exhibit IBS-like film properties but with significantly higher deposition rates. Reported drawbacks of reactive magnetron sputtering include the potential for particulate contamination [67] and potential process instability which can arise from oxidation or *poisoning* of the sputter target surface [68].

The development of CVD processes for telecommunications has also been reported. Bauer et al. have applied plasma impulse chemical vapour deposition (PICVD) for the coating of telecommunication filters [69]. Domash reports on the use of PECVD coated films for tuneable devices [70]. Patent application publications describe using atomic layer deposition for the manufacture of DWDM filters [71, 72].

Common to any of the coating techniques above are very stringent requirements for low thickness variation across the substrates. The channel spacing for 200 GHz filters is about 1.6 nm. The layer thicknesses in filters for neighbouring channels differ by about 1.6 nm / 1550 nm $\approx$ 0.1%. For a high yielding and predictable process it is desirable to control the coating thickness distribution across the 'sweet' area of the substrate to that precision level or better. Most of the processes for the fabrication of DWDM filters use single rotation substrate holders. Many use masking to minimize the radial runoff. Rotational frequencies in excess of 1000 rpm are reported [63, 59]. Such a rotation system needs to be well balanced. The substrate size varies from a few millimetres to 300 mm in diameter. The substrate needs to be held perpendicular to the rotational axis. Otherwise parts of the substrate are systematically closer to the source than others and will pick up more coating material. Depending on the deposition process, height variation of only a few micrometers can be tolerated. Chamber geometry considerations are published [73–75]. Wavelength uniformity has been reported to be less than 0.05% with excellent filter shape on an area of sufficient size that yields thousands of good filters per coating run [52].

7.3.5 Layer Thickness Control

Knowledge of optical material properties and a repeatable process build the foundation for depositing multilayer thin-film filters. Controlling the layers to the correct thicknesses for the required coating design is crucial for achieving the desired spectral response.

Thickness control requirements depend on the design and the stability of the coating process. Simple AR coatings or HRs may be quite tolerant to layer thickness errors and can be coated by simply controlling the deposition time of each layer in a machine with a stable process.

Quartz crystal monitoring can be employed for more advanced control. A quartz crystal oscillates within the vacuum and is exposed to the coating flux. The added mass shifts the frequency which allows determination of the added thickness [76].

Other advancements in thickness monitoring make use of optical techniques. Light either passes through or is reflected off a part in the deposition chamber. The part can be a witness part or the actual filter. In a monochromatic monitoring system deposition is stopped when the signal reaches a predetermined photometric value or the trace reaches a predetermined phase. The OMS 4000 from Leybold Optics is a commercially available optical monitoring system of this type [77]. Alternatively broadband monitoring can be used where deposition is stopped when a merit function between a target and the in-situ measured spectrum reaches its minimum [78]. In a related published method 90% of the layer thickness is deposited by time. Then the substrate transfers to a broadband scanning station. The thickness of the layer is determined from reverse engineering and the remainder of the layer is deposited. Optionally the remainder of the coating design can be re-optimized to compensate for thickness errors of previous layers [79–81]. Thickness monitoring based on ellipsometric data is another option [82].

Coatings with complex spectral requirement like DWDM bandpass filters and GFFs require very elaborate control techniques. Companies usually incorporate some of the methods mentioned above to arrive at their proprietary systems or methods. Designing the coating and determining the thickness control become dependent tasks. Simulations of the process variations and thickness control process can give guidance which design to select for best manufacturability [19, 13].

The well-known composite turning point monitoring method can be employed for the deposition of DWDM bandpass multicavity filters [Ref. 2, pp. 500–509]. These filters require very tight tolerances on thickness control. Figure 7.11 shows the transmission spectrum of a four cavity 100 GHz filter. The solid line describes the performance as designed. The dotted line shows the performance where the thicknesses of all layers are randomly disturbed by 0.5%. This would be the expectation if the filter was monitored by quartz crystal monitoring, time, or some other indirect method. The filter is completely deteriorated and would be of no use for a 100 GHz WDM system. For that reason DWDM designs usually consist of quarterwave layers which can be controlled by composite turning point monitoring at the centre wavelength. This technique has the advantage of inherent error compensation. A cutpoint error in one layer is automatically compensated by the next layer with very little distortion to the filter shape [83–86]. The dashed curve in Fig. 7.11 shows a simulation for 1% thickness accuracy on each individual cutpoint. The final filter performance is almost indistinguishable from the design performance (solid line).

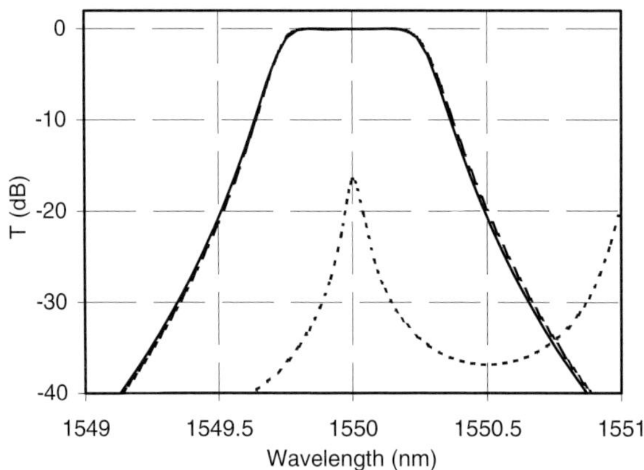

Fig. 7.11. Transmission spectra of a 100 GHz filter; filter as designed (*solid curve*); filter with 0.5% random thickness error on each layer, representing non-composite monitoring (*dotted curve*); filter with 1% thickness error for turning point monitoring at the centre wavelength (*dashed curve*, almost indistinguishable from design spectrum)

Two options exist for the hardware of a centre wavelength transmission monitoring system: a laser system, or a white light source coupled with a monochromator. The laser has an exceptionally high spectral irradiance at the desired wavelength. Excellent amplitude and frequency stability over a long period of time is critical. The long coherence length of a laser may pose a problem. The monitoring system performs cut points based on the change in transmission due to the growing film. But transmission changes as a consequence of optical thickness changes in the substrate due to temperature changes can be superimposed and disturb the cut points. Furthermore, lasers have a limited range of tunability. A white light system may be inherently more stable, can easily cover a wide spectral range but may give higher signal-noise problems specifically for very narrow bandpass filters. The spectral resolution of the monochromator needs to be kept under half the filter bandwidth [87]. Laser based and white light based systems are successfully used in the industry. Commercial instruments include systems from Veeco [63] and Intellemetrics [88].

The monitoring of an AR coating to match the WDM filter to air is discussed by R. Willey [89].

7.3.6 Post-deposition Fabrication and Filter Annealing

Thin-film filters for telecommunications are often deposited on a substrate that is thicker and larger than the final filter size. Manufacturers employ various optical fabrication methods to reduce the substrate thickness to the desired value and to dice the substrate into final-sized filter elements. The final filter dimensions are typically on the order of one to two millimetres square by one millimetre thick. Usually an AR coating is deposited on the rear surface of the filter element.

Thin-film filters are sometimes annealed to improve their stability; these processes are often proprietary to manufacturers and sometimes not even well understood [Ref. 2, pp. 417–418]. The procedure and function of annealing varies with the coating materials and processes and by application. For example, filters deposited using thermal sources or by electron beam evaporation typically exhibit a columnar microstructure; voids which may contain air or moisture are typically incorporated into these structures. Annealing such filters tends to collapse the voids, thereby densifying the structure and stabilizing the spectral performance of the filter. The centre wavelength of evaporation-deposited filters tends to decrease after annealing.

Thin-film coatings of the type used in telecommunication filters do not exhibit voids, rather they exhibit an amorphous and fully dense microstructure. The effect of annealing on these coatings is much more subtle.

Tilsch et al. [90] found that the absorption of ion-beam-sputtered silicon dioxide could be reduced by thermally annealing the coatings at temperatures of $200\,^{\circ}\mathrm{C}-300\,^{\circ}\mathrm{C}$. They also found that annealing the coatings slightly increased the film thickness, and reduced the coating stress and film density (though the density remained greater than the bulk density of silicon dioxide). Brown [91] reported on experiments in which ion-beam-sputtered single layer coatings of tantalum pentoxide and silicon dioxide were subjected to various annealing cycles. The optical thickness of such coatings increased slightly when annealed; associated observations were a small increase in the physical thickness and an even smaller decrease in the index of refraction. Prins et al. [92] baked ion-beam-sputtered WDM filters at temperatures above $300\,^{\circ}\mathrm{C}$ and observed positive shifts of the filter centre wavelength. For example, the centre wavelength of a filter baked at $340\,^{\circ}\mathrm{C}$ for 4 minutes shifted up 1 nanometre. A model for the observation was suggested that attributes the positive wavelength shift to a small increase in the thickness of the filter as a result of temperature-induced strain relaxation.

7.4 Properties of Filter Elements

One of the key advantages of thin-film filters in telecommunications applications is that they may be designed and constructed to have spectral performance that is stable with temperature. On the other hand, a potential limitation of thin-film filters results from the high compressive stress levels that are intrinsic to energetically-deposited fully-dense coatings. Both temperature stability and stress effects result from the optical and mechanical properties of the film-substrate system of the filter elements, and are considered in turn in this section.

7.4.1 Passive Temperature Stability of Filter Elements

Thin-film filters are ordinarily deposited on a substrate of some kind. Considering for a moment the simpler case of an unsupported thin-film filter, the primary effect of a temperature change is a change in the centre wavelength of the filter. A first step toward understanding the effect is to model the filter as a single *effective material* having an optical path length (*OPL*) given as the product of an effective film index of refraction n_{eff} and the physical film thickness d:

$$OPL = n_{eff} \cdot d. \tag{7.7}$$

For most materials n_{eff} increases with increasing temperature, and d also increases as a result of thermal expansion. The OPL therefore increases, and a corresponding positive shift in the filter centre wavelength is observed. For silica (SiO_2), the rate of shift of centre wavelength with temperature is about 0.015 nanometres per °C (where the centre wavelength is in the region of 1550 nm).

The key to the temperature stabilization of thin-film filters is through the interaction of the filter with the supporting substrate. Takahashi [93] demonstrated in a pioneering paper that a substrate with a large enough CTE can laterally stretch the filter and cause a sufficient reduction in the filter thickness d to counteract the increase in n_{eff}, thereby maintaining constant OPL over a useful temperature range. Figure 7.12a tabulates measurements of the thermal shift for filters deposited on substrates having a variety of CTEs, as reported in Takahashi's paper and in a subsequent paper by Kim et al. [94]. All of the filters listed employed tantalum pentoxide and silica as the coating materials, and were single cavity bandpass filters with low index spacers. Figure 7.12b displays a plot of the same data as a function of substrate CTE. The centre wavelengths of filters deposited on substrates with at CTE of about 10 to 11 parts per million (ppm) per °C are seen to exhibit a very low temperature dependence.

In the same paper, Takahashi introduced an elastic strain model that quantitatively describes this phenomenon. The model assumes that the substrate is perfectly rigid (unchanged by the presence of the filter), and

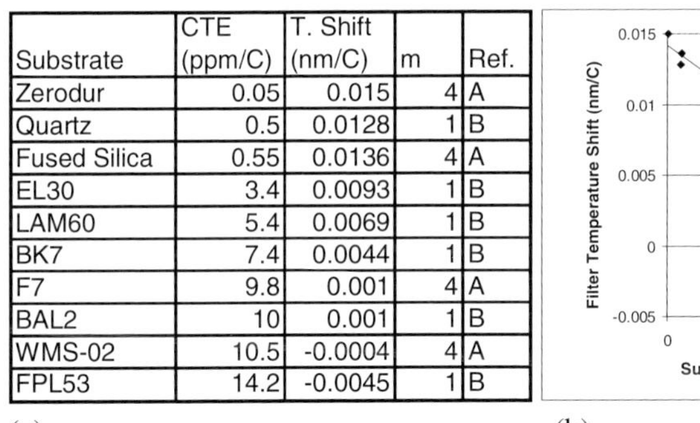

Substrate	CTE (ppm/C)	T. Shift (nm/C)	m	Ref.
Zerodur	0.05	0.015	4	A
Quartz	0.5	0.0128	1	B
Fused Silica	0.55	0.0136	4	A
EL30	3.4	0.0093	1	B
LAM60	5.4	0.0069	1	B
BK7	7.4	0.0044	1	B
F7	9.8	0.001	4	A
BAL2	10	0.001	1	B
WMS-02	10.5	-0.0004	4	A
FPL53	14.2	-0.0045	1	B

(a)

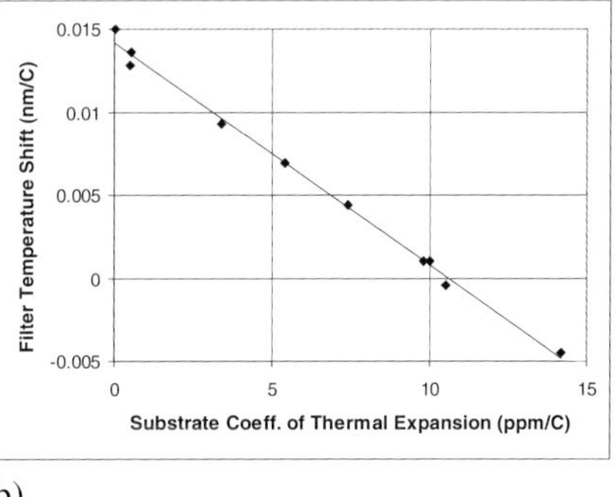

(b)

Fig. 7.12. (a) Tabulation of measured temperature shifts for various substrate types from two literature sources: A = [94], B = [93]. The filter centre wavelengths were approximately 1550 nm; m refers to the order number of the spacer of the filter. Zerodur, BK7, F7, are Schott glass types; BAL2, LAM60, FPL53, and WMS-02 are Ohara glass types; EL30 is a Hoya glass type; all three companies produce specially-formulated glasses for use as WDM filter substrates. **(b)** Plot of temperature shift as a function of the CTE of the glasses tabulated in **(a)**

therefore may be characterized by only the CTE. The filter thickness change is determined through the film mechanical properties (CTE and Poisson ratio). The film index change depends on the temperature-induced change of the refractive index (the bulk material property) as well as a strain-induced index change that is determined from the film material mechanical properties. The model agrees well with the experimental data.

Several recent workers have introduced models that remove some of the simplifying assumptions in the Takahashi model. Kim and Hwangbo [94] developed a model in which the elastic strain model of Takahashi is extended by including the effect of each individual thin-film layer in a filter (rather than using an effective index). A formula is derived which yields the optimum substrate CTE for a given film formulation. In addition to the temperature dependence of the index of refraction of the materials, the order number of the cavities (i. e. the number of half waves that constitute the spacer layers) is found to have a modest impact on temperature stability. Jiang et al. [95] have introduced a simulation in which the thermal stress and strain of each thin-film layer are determined. The model is used to predict the substrate curvatures and the correct filter centre wavelength as the substrate thickness is changed. The results agree (at least qualitatively) with measurements. In the representative examples considered, as the substrate becomes thinner, the substrate curvature increases, the centre wavelength decreases, and the centre wavelength shift with increasing temperature becomes more positive.

In addition to the basic method of employing an appropriate high CTE substrate, other methods have been proposed to passively athermalize thin-film interference filters. The methods typically employ external stress-inducing devices such as a clamp [96], a washer [97], a superstrate [98], or a ring or other encasement surrounding the periphery of the filter [99].

7.4.2 Stress Effects

The mechanical properties of materials in thin-film form are often very different from the corresponding bulk properties. One thin-film material property that has attracted significant attention is stress [Ref. 2, pp. 436–440, 100, 101]. Stress in evaporated coatings is often tensile in nature, and has been a limitation for example in limiting the adhesion of the coating to the substrate. Fully-dense metal-oxide coatings of the type used in telecommunications filters tend to have significant compressive stress, which can lead to a number of problems. Probably the most significant of these problems, and the topic of several recent papers in the literature, is degradation of the optical filtering performance and distortion of the transmitted and reflected wave fronts. The practical impact of the latter in a fibre link is increased optical loss.

The effects of stress become more pronounced as the film thickness increases. A typical manifestation of stress is a variation in centre passband wavelength across the filter, which leads to degraded filter performance. Ockenfuss et al. [102, 103] mapped the optical performance of 100 GHz filters across the filter element using a small beam; results are illustrated in Fig. 7.13a. The centre wavelength (CWL) is seen to change by about 1.5 nm from the centre to the edge of the 1.4 mm-square filter element. This corresponds to more than one 100 GHz channel! Such filters are still usable for the 100 GHz filtering application because the beam passes through the centre of the filter where the optical performance is not significantly changing, but the effect becomes limiting as filtering performance becomes even more stringent.

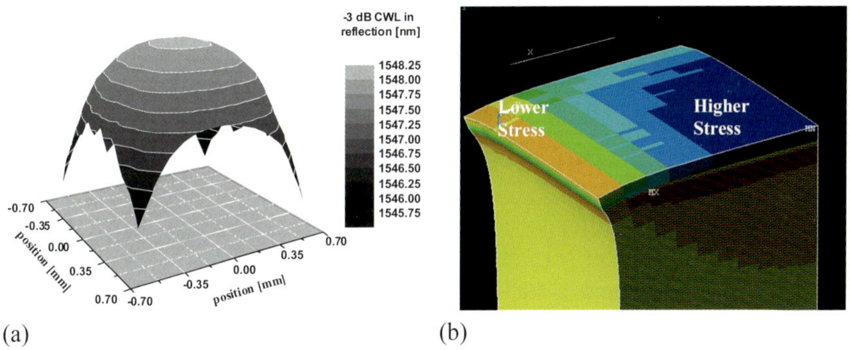

(a)   (b)

Fig. 7.13. (a) Variation in filter centre wavelength (CWL) across a 1.4 mm-square filter element; the centre wavelength variation is about 1.5 nm between the centre and the edge of the part (and about 2.5 nm between the centre and the corner of the part). (b) FEA of in-plane stress in the x-direction (roughly right to left) of a 40 micrometer thick coating on a 1.4 mm-square × 1 mm thick substrate; only ¼ of filter shown (reproduced from Ockenfuss et al. [103])

Ockenfuss et al. attributed the observed variation in centre wavelength to a non-uniform stress field in the filter element. A Finite Element Analysis (FEA) of the coating-substrate system was conducted; the results are shown in Fig. 7.13b. The film stress at the edge of the filter is relieved through a distortion of the substrate, and the optical thickness of the film stack becomes slightly smaller in this region. The film stress at the centre of the filter remains high, and the optical thickness of the film stack is slightly larger. The optical thickness difference in the filter at the centre compared to the edge is the reason for the centre wavelength difference.

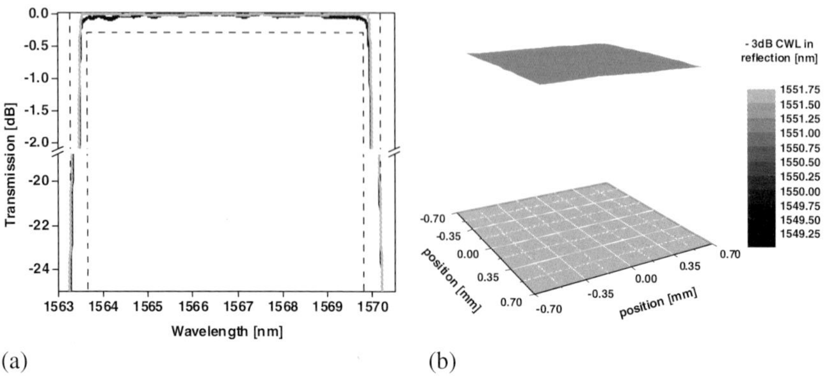

Fig. 7.14. (a) Design (grey) and measurement (black) of 8-skip-0 filter implemented with low-stress production process. **(b)** Centre wavelength variation of low-stress-process 100 GHz filter plotted as a function of position, shown with same vertical scale as Fig. 7.13a

The effect of stress is especially deleterious in bandsplitting coatings owing to the large number of cavities used to achieve the required square spectral response profile. It was demonstrated that film stress reduction can yield significant filter performance benefits. The process involved depositing the filter, removing the filter from the substrate it was deposited on, and re-attaching the filter to another substrate (but in a low stress condition). An 8-skip-0 filter (cf. Sect. 7.6.2) for 100 GHz channel spacing that required 17 cavities and a total physical thickness of 94 micrometers was demonstrated, as shown in Fig. 7.14a. The variation in centre wavelength of a 100 GHz filter made using this process is shown in Fig. 7.14b. The vertical scales in Fig. 7.13b and Fig. 7.14b are the same.

Prins et al. [104] reported a similar stress-induced position-dependent centre wavelength behavior in 50 GHz and 100 GHz filters. In addition, these authors characterized the optical wavefront transmitted and reflected by such filters. The transmitted wavefront distortion of a 100 GHz filter having dimensions of 1.4 mm × 1.4 mm × 1 mm (thick) is about 1 wave over the beam size of their coupler when illuminated at the centre wavelength. Filters of larger lateral dimensions have the edges further away from the beam and thus show less transmitted wavefront distortion. Filters of smaller thickness also have less transmitted wavefront distortion. The thinner substrate is less stiff which reduces the stress relaxation gradients from the centre to the edge of the filter. The reflected wavefront distortion of a 100 GHz filter having dimensions 1.4 mm × 1.4 mm × 0.55 mm (thickness) is about 1 wave over the size of the filter. Filters of greater substrate thicknesses were observed to have less reflected wavefront distortion.

Thus larger lateral dimensions improved the transmitted wavefront distortion, while increasing substrate thickness improved the reflected wavefront while distorting the transmitted wavefront.

Here we have given a brief review of studies on the effects of coating stress on thin-film filter performance, but it should be clear to the reader that this represents work in progress. For the time being, filter manufacturers are meeting the stringent requirements of telecommunications applications by utilizing lower stress deposition processes and through careful design and optimization of the filter-substrate system.

7.5 Device Packaging

To be used in an optical fibre communications system, thin-film interference filters must be interfaced with fibres. These interfaces are known variously as *packages, couplers, bulk optic packages, micro-optic devices,* etc. Thin-film filter packaging methods have been reviewed in several articles over the past few decades [34, 105, 106]. In this section the most common packaging techniques are reviewed and a few recent developments briefly described.

7.5.1 Filters Deposited on Fibre Ends

Perhaps the most straightforward way to interface thin-film filters with optical fibres is by depositing a filter directly on the end of a fibre; representative examples of couplers employing this method are shown in Fig. 7.15. Filters deposited on fibre ends suffer from several serious problems, including spectral degradation of the filter performance as a result of the large range of angles incident from the fibre and the absence of the temperature compensation mechanism described in Sect. 7.4.1. Couplers based on direct-coated fibres were demonstrated in early multimode fibre WDM system configurations where the very wide channel spacing reduced the impact of such effects [107, 108].

Most filter functions of interest cannot be realized using a direct coating-to-fibre interface. However, filter functions that have a constant transmittance and/or reflectance over a wide spectral region can be satisfactorily realized since the angle-averaging and temperature shift effects do not change the performance. Hence HR coatings and AR coatings on the ends of fibres are readily implemented and are widely used. The design, fabrication, and optical characterization of antireflection coatings for optical telecommunications has been recently reviewed by Stevenson [11]. A useful benchmark on the performance of such coatings are the results of Uehara et al. [109], who have reported reflectances of well under 50 dB for a four layer AR coating over a 70 nm bandwidth from 1550 nm to 1620 nm. It is noted that AR and HR interference coatings find many applications in telecommunications beyond coatings on fibres; for example, as the high reflection and output coupler mirrors in semiconductor lasers.

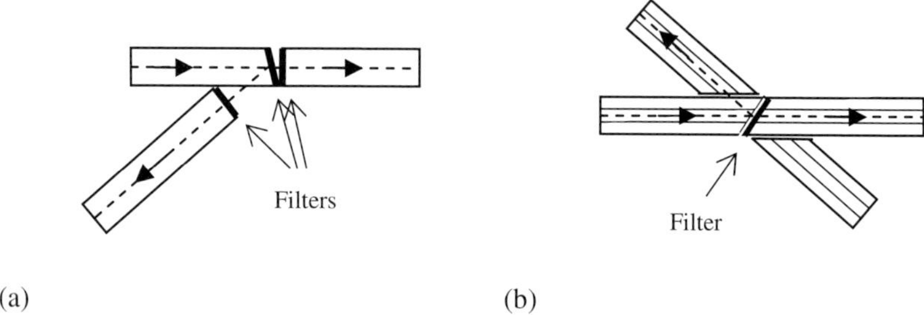

Fig. 7.15. Configurations employing thin films deposited on fibre ends: (**a**) after [107]; (**b**) after [108]

7.5.2 Three-port Couplers and Related Lens-based Couplers

The divergence of light from a fibre usually necessitates the use of lenses to collimate the light in devices that use thin-film filters. A convenient arrangement for performing this function is shown in Fig. 7.8 (in Sect. 7.2.4). Light shown emerging from the common fibre is collimated by a first lens and interacts with the thin-film filter. Light transmitted by the filter is then coupled into the T fibre by another lens, while light reflected by the filter is coupled into the R fibre. As shown, the device acts to separate two incoming spectral bands, that is, it functions as a demultiplexer (DEMUX). The device is reciprocal in that the direction of the arrows may be reversed. In the reversed arrangement the device acts to combine two spectral bands, and the device behaves as a multiplexer (MUX).

When the angle of incidence is small, a single lens usually replaces the lenses for the common and reflection ports. A device of this type may be implemented using GRIN lenses, as shown in Fig. 7.16. An early realization of such a device was reported by Sugimoto et al. [110]; these authors deposited the interference coating directly onto one of the GRIN lenses, and the second GRIN lens was cemented onto the other side of the coating. In modern implementations the filter is frequently deposited on a separate substrate and positioned between the lenses. The device may be tuned in

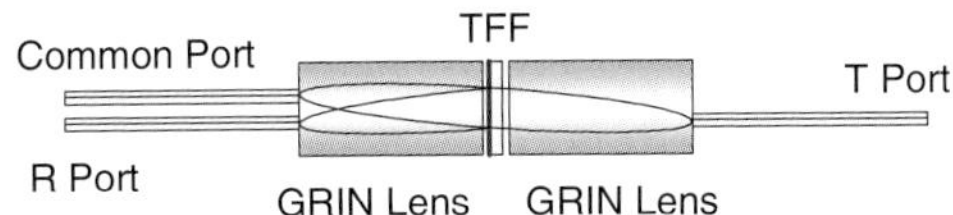

Fig. 7.16. GRIN-lens-based three-port coupler

wavelength by changing the separation of the Common Port and R port fibres and moving the T port fibre appropriately. This has the effect of changing the angle of incidence of light on the filter, thereby changing the centre wavelength [35].

GRIN lenses are convenient for the construction of interference-filter-based devices, but other lens types have been employed for this application as well. Nicia [111] compared GRIN-rod, ball, and plano-convex rod lenses and found that the expected losses in micro-optic applications are similar. Ball-lenses have been used in the design and demonstration of 200 GHz and 100 GHz add-drop MUXs [112]. Whichever lens type is employed, the device has high symmetry about the central axis of the coupler, and is sometimes referred to as a cylindrically-symmetric coupler. Since the most common version employs three ports, the device is frequently called a three-port coupler.

The approach outlined above and shown in Fig. 7.16 is the dominant method for packaging thin-film filters today. A significant reason for this dominance is that the approach has been proven reliable, and is readily available from a number of manufacturers. Devices which multiplex or demultiplex a large number of channels may be constructed through the concatenation of three-port cylindrically-symmetric couplers, as described in Sect. 7.6.

Variations of the basic device include the introduction of additional ports and the incorporation of multiple filters in the structure. For example, Wagner [113] describes a single GRIN lens device with up to 4 filters and 5 ports, as shown in Fig. 7.17.

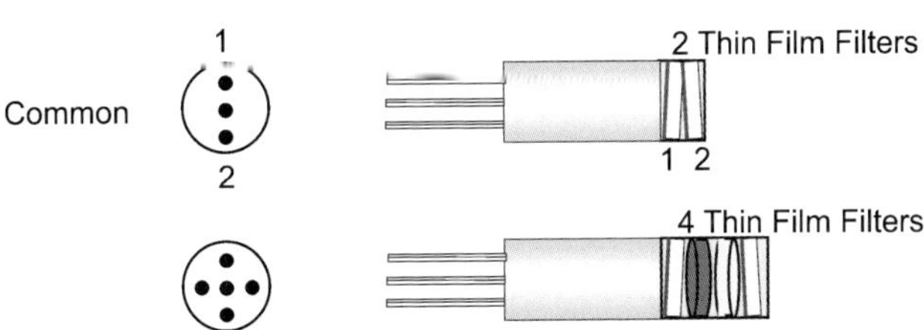

Fig. 7.17. Device using one GRIN lens incorporating up to 4 filters and 5 ports, as described by Wagner [113]. (**a**) three-port version; (**b**) five-port version

7.5.3 Zigzag Devices

The devices described in the preceding paragraphs operate by surrounding a filter with lenses, so that for every reflection or transmission operation the beam is coupled out of and back into a fibre. This construction suffers from the inherent loss associated with the many coupling operations, as well as the cost incurred because of the numerous lenses employed. A device that reduces the effect of coupling losses was proposed by Nosu et al. [114], and is shown in Fig. 7.18. The device was used to demultiplex 6 channels spaced about 40 nm apart in the 725 nm – 920 nm wavelength region. These devices are sometimes referred to as *block*-type devices as the interference filters are mounted to an optical block, or alternately as *zigzag* devices because of the path of the beam within the block. The zigzag geometry was implemented in the 1550 nm band in the mid-1990s [16], and the construction of a 16 channel device of this type has been reported [115].

Zigzag type devices have also been demonstrated for the implementation of CWDM in local area network (LAN) data links. A device implemented in a polymer waveguide was used to realize a 4 channel LAN DEMUX in the 850 nm wavelength band [116]. Such a device was subsequently implemented in injection-molded plastic to realize a 4 channel DEMUX in the 1310 nm band. Filtering was achieved using four interference filters which were deposited on 400 micrometer × 500 micrometer glass substrates and glued into the device [117, 118]. Recently a 250 Gbit/s parallel/CWDM optical interconnect has been demonstrated that employs MUXs and DEMUXs in an optically-transparent semiconductor material; the system employs 48 channels operating over 12 fibres with 4 wavelengths per fibre [119].

A disadvantage of the zigzag device architecture is that the beam is continually expanding as it propagates within the block. Thus, unless couplers

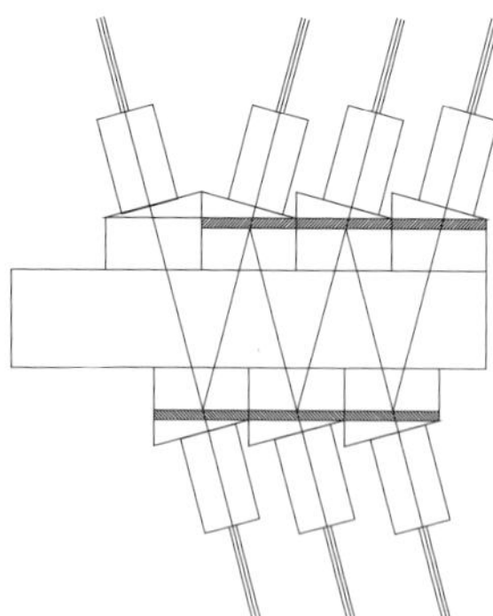

Fig. 7.18. Block or zigzag device, after Nosu et al. [114]

optimized for slightly different beam parameters are used at each point in the device, channel losses depend on the drop position. A method for alleviating this problem was introduced by Honda et al. [120]. In the proposed arrangement, filters on a slightly curved surface are employed to continuously slightly refocus the beam as it propagates within the block so that it maintains more or less a constant beam diameter. As described in Sect. 7.4.2, stress induces curvature in thin-film filters anyway, and the amount and direction of curvature may be engineered to make this device realizable.

7.5.4 Other Devices

Most commonly-used devices that incorporate interference filters fit into the two categories described above. A device that does not is based on a stacked assembly of interference filters and an array of diffractive optical elements (DOEs), as shown in Fig. 7.19 [121, 122]. The DOEs are lenses which serve to couple light in and out of the fibre ports. After collimation, light from the common port is directed to the correct port by one of the three interference filters or the mirror. The device described was used to multiplex four wavelengths (1280, 1300, 1320, 1340 nm) into an output fibre. The performance of the prototype device was not very good in that the IL from the common port to each of the other ports was high – in the

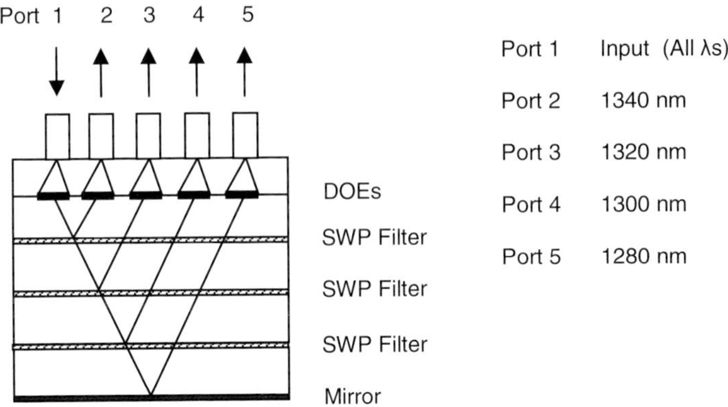

Fig. 7.19. Structure of device reported by Okabe and Sasaki [121, 122]. Figure is schematic, SWP: short wave pass; the actual angle of incidence on the filters was reported to be 5°

range of 4 to 5 dB – and the isolation between some of the channels was poor – approximately 10 dB. The poor performance appeared to be attributable to the IL of the DOE lenses in combination with the imperfect parallelism of the interference filter substrates.

7.5.5 Device Reliability

A variety of device packaging technologies have been applied to interference filters. These packaging technologies are also used for the manufacture of isolators, circulators, and other bulk optic devices. In such devices epoxies are widely used to secure components to each other. Over the years manufacturers have optimized processes that generally reduce the variation in such processes, for example, by increased precision of epoxy bond line dimensions. Epoxy is now rarely used in the optical path of a micro-optic device. Solder and laser welding technologies are often employed to thin-film device packaging.

Thin-film interference filters hold up well in adverse environments, and hermetically-sealed packages are not a requirement (though they are occasionally employed). Robust fibre optic devices incorporating interference filters are widely available. A number of manufacturers supply devices that meet or exceed the Telcordia 1209-GR-CORE [123] and 1221-CORE [124] requirements. For example, JDSU supplies a laser-welded package that has been demonstrated to exceed the requirements of Telcordia GR-1221-CORE. The design of components which meet reliability requirements has been the subject of several papers in the past few years [125, 126].

7.6 Applications

Thin-film filters are used for a wide range of applications. It is beyond the scope of this chapter to cover all of them. The focus in this section is on the most important applications in which thin-film filters have played a significant and enabling role. Noteworthy examples that are not included are pump WDMs for optical amplifiers and telemetry / supervisory add/drop filters. Specifications for these devices may be found on the web sites of JDSU and other component suppliers.

This section contains tables of specifications of commercially available products. The tables were reduced to *key* specifications for briefness and to omit redundant information.

7.6.1 DWDM Single Channel Couplers

The single channel coupler is a fundamental building block of many WDM systems. The usual implementation uses the three-port coupler as described in Sect. 7.5.2. Used as a DEMUX, a spectrum of channels enters the input port. The internal thin-film filter directs one channel to the transmitted output port and all other channels to the reflected output port. The device is reciprocal, and may be used as a MUX by reversing the direction of the light. Figure 7.20 shows a three-port coupler. Typical dimensions for a coupler are 5 to 6 mm in diameter and 30 to 40 mm length.

A straight forward way to create a multiple channel DEMUX is to concatenate a series of couplers as shown in Fig. 7.21. The architecture can be used in reverse to implement a MUX module.

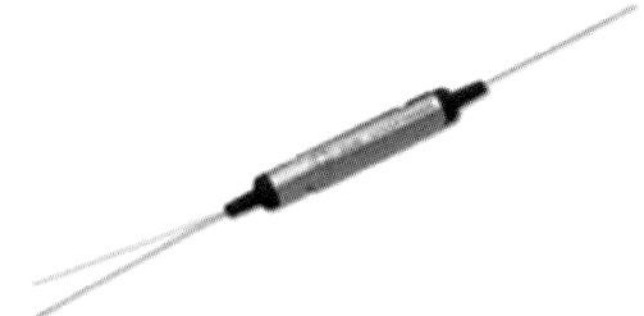

Fig. 7.20. DWDM single channel three-port coupler (courtesy of JDSU)

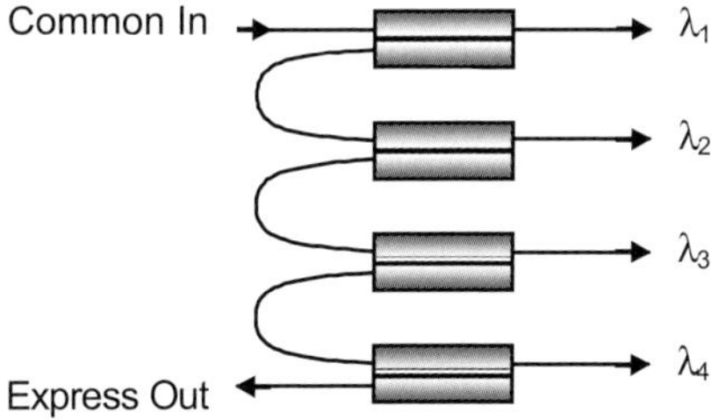

Fig. 7.21. Four channel DEMUX based on cascaded thin-film filter couplers. The *Express Out* port contains any portion of the input spectrum that has not been dropped by preceding couplers

By far the most widely deployed thin-film based single channel couplers have been for 200 and 100 GHz channel spacing. Manufacturers usually offer a standard product, but can tailor the performance to the specific customer requirement e. g. for higher isolation or lower CD. Table 7.2 lists specifications for a commercially available standard 100 GHz coupler and the low CD version.

Table 7.2. Key specifications of standard and low CD 100 GHz coupler (courtesy of JDSU)

Parameter		Standard	Low CD
Wavelength range		C and L band ITU channels	
Passband width (@-0.5 dB)	min	± 12.5 GHz	± 14.1 GHz
Transmission passband ripple	max	0.4 dB	0.3 dB
Reflection passband ripple	max	0.2 dB	0.2 dB
Transmission IL	max	1.1 dB	1.1 dB
Reflection IL	max	0.4 dB	0.4 dB
Transmission isolation (adjacent channel)	min	25 dB	25 dB
Transmission isolation (non-adjacent channel)	min	50 dB	50 dB
Reflection isolation	min	12 dB	16 dB
Transmission CD	max	464 ps/nm	17 ps/nm
Reflection CD	max	17.5 ps/nm	17 ps/nm
Directivity	min	45 dB	45 dB
Return loss	min	45 dB	50 dB
PDL	max	0.2 dB	0.1 dB
PMD	max	0.2 ps	0.1 ps
Optical power handling	max	1 W	1 W
Operating temperature range		-5 to $75\,^{\circ}$C	-5 to $75\,^{\circ}$C

7.6.2 Band Splitting Couplers

Couplers that separate a band of two or more adjacent channels from a fibre optic stream are often referred to as *band splitting*, *wave band*, or *wideband* couplers. They can be used in the same way as single channel couplers, but all operations are performed for a band of channels. Two examples of band splitting filters are illustrated in Fig. 7.22. A 4-skip-1 filter as shown in Fig. 7.22a transmits 4 channels and achieves isolation between the transmitted and reflected channels by skipping one channel. The 4-skip-0 filter in Fig. 7.22b also transmits 4 channels, but does not skip any channel. The latter filter has higher performance, but is more difficult to manufacture due to the extreme edge steepness requirements.

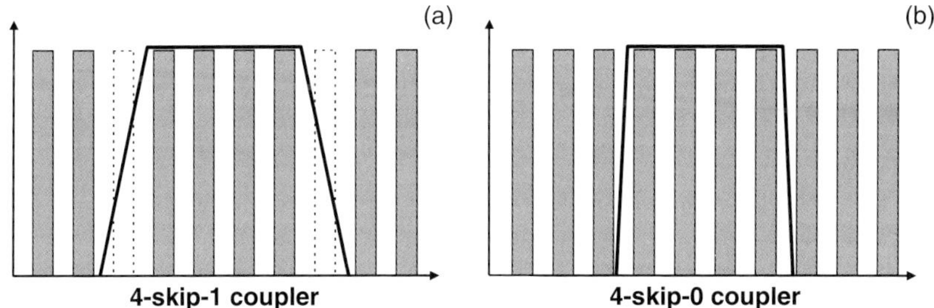

Fig. 7.22. Conceptual filter performance of (**a**) 4-skip-1 and (**b**) 4-skip-0 coupler

Band splitting couplers are often combined with single channel couplers in the architecture of MUX or DEMUX modules for higher channel count systems. Figure 7.23 shows two 40 channel DEMUX architectures. Figure 7.23a is a cascade of 40 single channel bandpass couplers. The first channel is transmitted without any reflection, the last channel after 39 reflections. That could lead to a *tilt* in IL from the first to the last channel due to losses incurred in each reflection operation. Figure 7.23b employs five band splitting 8-skip-0 filters. That reduces the maximum number of reflections and thus minimizes the variation in IL between the 40 channels.

Skipping channels is undesirable, as it may have the effect of reducing the available communication bandwidth. This has driven the development of filters with steeper slopes to skip only one or in the best case no channel. Table 7.3 lists three specifications for commercially available band splitting couplers on the 100 GHz ITU grid.

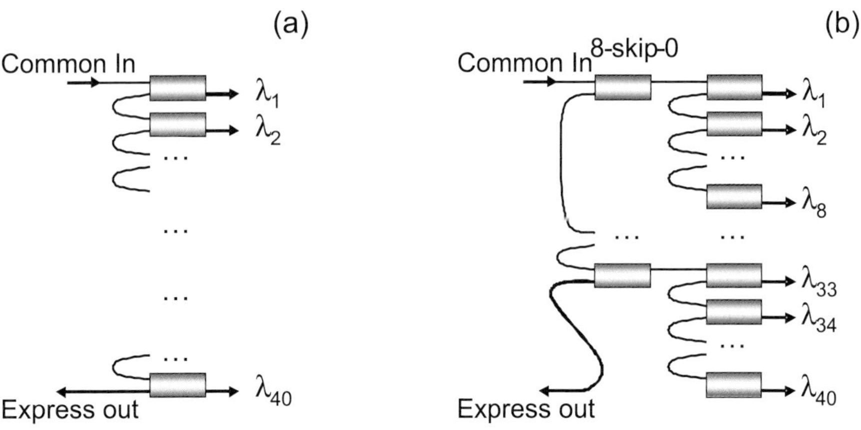

Fig. 7.23. 40 channel DEMUX based on (**a**) cascaded thin-film filter couplers and (**b**) employing five 8-skip-0 bandsplitting filters together with single channel filters

Table 7.3. Key specifications for three commercially available 100 GHz band splitting couplers (courtesy of JDSU)

Parameter		4-Skip-0	8-Skip-2	20-Skip-4
Passband width	min	± 162.5 GHz	± 387.5 GHz	± 987.5 GHz
Passband ripple	max	0.3 dB	0.25 dB	0.2 dB
IL (transmission)	max	1.2 dB	0.8 dB	1.0 dB
IL (reflection)	max	0.6 dB	0.4 dB	0.6 dB
Isolation (adjacent band)	min	25 dB	20 dB	25 dB
Isolation (reflection)	min	12 dB	15 dB	15 dB
Directivity	min	50 dB	50 dB	50 dB
Return loss	min	45 dB	45 dB	45 dB
PDL	max	0.2 dB	0.1 dB	0.15 dB
Polarisation mode dispersion	max	0.7 ps	0.1 ps	0.15 ps
CD (Tx)	max	± 50 ps/nm	± 5 ps/nm	± 3 ps/nm
CD (Rx)	max	± 30 ps/nm	± 5 ps/nm	± 2 ps/nm

For 100 GHz channel spacing DEMUX modules are typically offered for 4, 8, 16, and 40 channels. The specifications for individual channels of MUX/DEMUX devices are similar to the single coupler specification (Table 7.2), but maximum IL typically drops to 3 to 7 dB depending on channel count. An example for a 40 channel DEMUX spectrum is shown in Fig. 7.24.

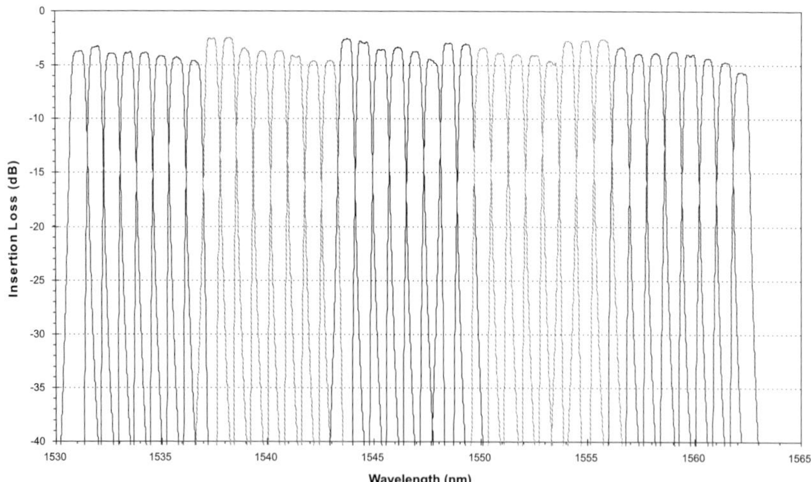

Fig. 7.24. Sample spectrum of a 40-channel DEMUX based on the 100 GHz ITU grid, circa year 2000

Most modules provide an express port that contains the remaining spectrum of channels that have not been dropped. That allows for modular future upgrades. This *pay as you grow* option is an advantage in many cases allowing service providers to initiate service with modest capacity at initial modest capital cost, then increasing capacity by adding additional channels when needed.

7.6.3 DWDM Add/Drop Modules

Add/drop modules drop one or multiple channels at a node and add new signals on these channels. All other channels pass through unchanged. Modules commonly integrate 2 three-port couplers as shown in Fig. 7.25

Integrated modules are available from the major component manufacturers. A specification for a single channel 100 GHz add/drop module is given in Table 7.4.

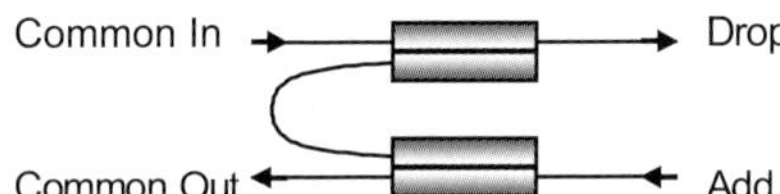

Fig. 7.25. Add/drop module built from 2 three-port couplers

Table 7.4. Key specifications for single channel 100 GHz add/drop (courtesy of JDSU)

Parameter		Specification
Wavelength range		C or L band ITU channels 186.6 to 196.1 THz
Passband	min	± 12.5 GHz ($\approx \pm 0.10$ nm)
Passband ripple	max	0.35 dB
Drop IL[1]	max	1.1 dB
Express IL[1]	max	0.9 dB
Add IL[1]	max	1.1 dB
Drop isolation (adjacent channel)	min	25 dB
Drop Isolation (non-adjacent channel)	min	50 dB
Express isolation	min	30 dB
Directivity	min	50 dB
Return loss	min	45 dB
PDL	max	0.2 dB
Polarisation mode dispersion	max	0.15 ps

[1] Losses include one connector. If no connector option is selected, the maximum loss should be 0.2 dB lower.

7.6.4 CWDM and FTTX Modules

CWDM is used for short distance transmission in the Metro/Access market, where cost is a key consideration. A relatively wide channel spacing is employed so that low-cost uncooled lasers may be used. The channel spacing is typically 20 nm with 15 nm wide passbands (cf. Appendix, A.1.2). Modules for 4 to 8 channels are standard from telecommunication companies. Table 7.5 lists a typical CWDM module specification.

Table 7.5. Key specifications for four and eight channel CWDM modules (courtesy of JDSU)

Parameter		4 Channel	8 Channel
Operating wavelength range		1300 to 1620 nm	
Wavelength centres (ITU wavelengths)		1470, 1490, 1510, 1530, 1550, 1570, 1590, 1610 nm	
Wavelength centres (ITU+1 wavelengths)		1471, 1491, 1511, 1531, 1551, 1571, 1591, 1611 nm	
Passband	min	$\lambda_c \pm 6.5$ nm	
Passband ripple	max	0.35 dB	
IL[1]			
DEMUX module	max	2.0 dB	3.2 dB
MUX module	max	1.7 dB	2.9 dB
Isolation (adjacent channel)			
DEMUX module	min	30 dB	
MUX module	min	15 dB	
Isolation (non-adjacent channel)			
DEMUX module	min	50 dB	
MUX module	min	15 dB	
Directivity	min	50 dB	
Return loss	min	45 dB	
PDL	max	0.2 dB	
Polarisation mode dispersion	max	0.2 ps	
Operating temperature range		−40 to 75°C	
Fibre type		SMF-28 or equivalent	

[1] Losses include one connector. If no connector option is selected, the maximum loss should be 0.2 dB lower.

Fibre to the *something* (such as home, premises, building, business, curb, or node) is commonly abbreviated as FTTX. FTTX systems utilize two or three bands, centred at 1310, 1490, and 1550 nm [127, see also

Appendix, A.2]. The 1310 nm band is used for data from the user to the network. The other bands are for data or video to the user. MUX/DEMUX components can be employed as discussed in section 7.6.2. However, specific low cost FTTX device architectures are also in use which integrate filters, receivers, and transmitters.

7.6.5 Gain-flattening Couplers

Thin-film filter technology dominates the important application of gain-equalization in optical amplifiers. First employed in the mid 1990s, a common implementation used two or three filters to provide the correct equalization function for EDFAs [16]. By 2000 single element GFFs were available [17] and being deployed. An example of a GFF spectrum was shown in Fig. 7.4. For a given requirement, the error function achievable depends largely on the maximum excursion in the loss profile, but also on the precise shape profile requirement of the filter. Minimizing spectral changes with temperature in GFFs is very important. In many cases commercially available GFFs reduce the gain excursion to 5% of the uncompensated gain spectrum under all operating conditions. E. g. a filter with 6 dB excursion may be equalized to below 0.3 dB excursion. Table 7.6 shows a typical specification for a GFF component.

Table 7.6. Key specifications for GFF coupler (courtesy of JDSU)

Parameter	Performance
Wavelength	C, L, or S band or customer specified
IL error function (peak to peak)	0.2 to 0.8 dB typical
PDL	0.1 dB maximum
PMD	0.1 ps maximum
Optical power handling	0.5 W standard 2 W high power GFF
Operating temperature range	-5 to 75°C

7.6.6 Tuneable Graded Filters

Tuneable filters are needed in DWDM systems for wavelength monitoring, re-configurable add/drops, variable attenuation, and other applications. Several approaches have been implemented to tune a conventional thin-film filter, typically through angle-of-incidence or thermal effects. In addition,

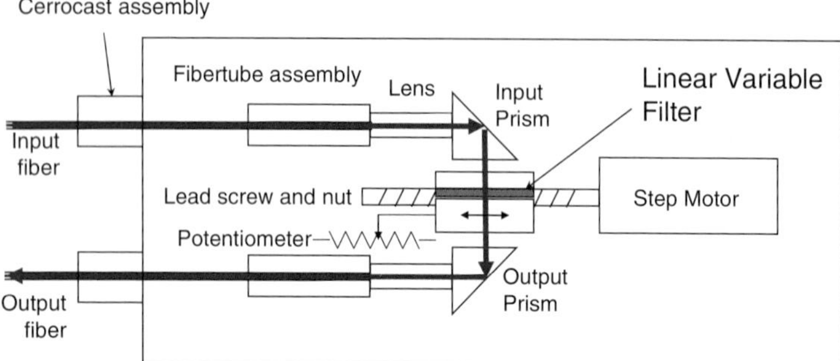

Fig. 7.26. Tuneable filter assembly (courtesy of JDSU)

graded thin-film filters have been developed, which are only used for tuneable filtering. This section briefly discusses approaches to tuning a conventional thin-film filter and then focuses on the tuneable graded filter.

Prior sections have detailed engineering efforts to make thin-film filters stable over their lifetime. For a tuneable device, the filter can be designed to enhance a sensitivity to environmental conditions, notably temperature or angle. For example, changing the cavity material in a multicavity filter can lead to temperature tuning coefficients of more than 100 pm/°C [128]; this approach has been used to build a filter than can tune over the entire C-band. Alternatively, a filter with low-index cavities will be very sensitive to angle of incidence variations; a single-cavity air-gap etalon can tune over the entire C-band with a change in angle of around 15°. Each of these approaches requires considerable optimization; the thermally-tuned filter must operate reliably over tens or hundreds of degrees of temperature, while the angle-tuned filter must operate without significant polarisation splitting or degradation.

Tuneable graded filters are employed in a different approach to tuneability. A tuneable graded filter employs a thin-film filter which possesses a varying spectral response with position. These filters are deposited with controlled gradients in thickness or optical properties across their surface. A mechanism moves the filter mechanically relative to the beam path thus tuning the response of the device. Figure 7.26 shows an example of such a device where the filter is moved by an internal stepper motor.

If the filter response varies smoothly with location, a gradient in response is present in the expanded beam path. The smaller the physical dimension of the filter and the wider the spectral range, the larger is the gradient. The beam size, the angle of incidence, the gradient, and the actual filter design all influence the spectral performance of the device.

Table 7.7. Key specifications for tunable 50 GHz module (courtesy of JDSU)

Parameter		Performance
Optimized wavelength range		1525 to 1570 or 1565 to 1610 nm
Bandwidth (@-0.5 dB)	typical	15 GHz
Bandwidth (@-1 dB)	min	20 GHz
Bandwidth (@-3 dB)	typical	35 GHz
Bandwidth (@-20 dB)	max	84 GHz
Out of band rejection	min	−40 dB
IL (over all conditions)	max	3.0 dB
PDL	max	0.3 dB
PMD	max	0.2 ps
CD		±50 ps/nm
Frequency setting accuracy	max	±4 GHz
	typical	±3 GHz
Max power handling	max	23 dBm (all channels)
Return loss	min	40 dB
Tuning speed		5000 ms
Operating temperature range		−5 to 70°C

Devices in the field today include single and multi-cavity 50, 100, and 200 GHz filters that are tuneable over the C or the L bands. A device specification for a multi-cavity 50 GHz filter is shown in Table 7.7.

A similar device is offered in which the filter is a graded neutral density filter. The device functions as an optical attenuator.

7.6.7 Filters for Wavelength-locking of Lasers

Thin-film filters are used to lock the wavelengths of lasers. The DFB lasers used as the light sources in most DWDM systems exhibit a long term shift in wavelength with time of about 0.01 nm/year [129, 130]. DWDM systems with channel spacings of 100 GHz or 0.8 nm cannot tolerate this magnitude of drift if the lifetime of the system is to be 20 years or more. Thus, a means of stabilizing the wavelength is required. Devices that provide the feedback are usually called *laser wavelength lockers* or *laser wavelength controllers*. Several means for controlling the centre wavelength of the laser can be chosen: current control or temperature control. Temperature control is normally chosen as it affects other laser properties the least. A means of providing feedback for the control loop is needed, and a number of filtering technologies have been applied to this application including Fabry–Perot etalons (Chapter 6). Two methods that employ interference filters are described below.

Villeneuve et al. [131, 132] describe a wavelength locker that may be employed in a separate package or within a laser package to stabilize the wavelength of a laser. A single interference filter is employed at angle in the device, and detectors behind the filter capture light which has been transmitted through the filter at different angles. As a result, the peak transmission for the two detectors occurs at different wavelengths. By comparing the signal levels of the two detectors, an error signal can be generated and used to control the temperature and wavelength of the laser. The scheme is shown in Fig. 7.27.

Munks et al. [133] describe another arrangement which employs two interference filters to perform a laser wavelength-locking function. A signal tapped from the laser to be controlled is split into two beams, and before being detected each of the two beams is filtered by filters having different spectral responses. The filters may be narrow bandpass filters, chosen so that the bandpass of one is centred at a wavelength lower than the desired laser wavelength set-point, while the other is centred at a higher wavelength. The power captured by the corresponding detectors is processed to provide an error signal. One example is to divide one of the detector outputs by the sum of the detector outputs.

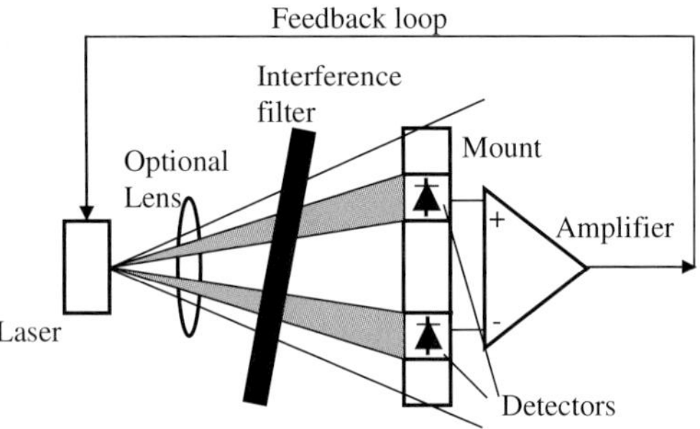

Fig. 7.27. Wavelength-locking scheme employing a single interference filter and two detectors that receive light that is transmitted through the filter at different angles of incidence (after [132])

The most critical specification for a laser wavelength-locking device is its own stability of the centre wavelength over the operating temperature, polarisation, and over the lifetime of the device. A common requirement is a centre wavelength stability of ±0.02 nm or better over 20 years [134, 135].

7.7 Recent Developments in Thin Films

The previous sections of this chapter have focused on telecommunications applications where thin-film-based devices have evolved to a high level of maturity and near-universal availability. This section discusses more recent developments that may prove bellwethers of future trends. In particular, two exciting developments in recent years have been the demonstration of actively tuned filters and the use of photonic crystal theory as a driver of thin-film applications.

Until recently, tuned or switched filters in telecommunications networks have been tuned mechanically, as in Sect. 7.6.6. Filters which are tuned without moving parts have recently been demonstrated. As an example, DWDM filters with switchable or tuneable passbands were recently fabricated by depositing a multicavity filter with strong temperature sensitivity directly atop a thin-film heater [128]. For a comprehensive review of this and other so-called *active filters*, see the review by Lequime [136].

A second development of note has been the influence of photonic crystal theory on thin-film design. Thin-film filters can be understood as a one-dimensional photonic crystal, and this yields a number of useful insights, including analogies between beam steering [137] and prism effects [138] and between thin-film reflectors [139, 140] and bandgap metamaterials [141, 142]. Prism-like phenomena, coupled with the walkoff effects of Sect. 7.2.5, have led to demonstrations of wavelength DEMUXs which work by spatially separating multiple channels simultaneously [143–145]. Bandgap metamaterials have led to the development of hollow waveguides which carry light at exotic wavelengths [146, 147].

7.8 Conclusions and Outlook

Dielectric multilayer filters are widely deployed in modern fibre optic networks. AR coatings, HR coatings, output couplers, and wavelength stabilization filters are integral parts of laser sources. GFFs are key components in most optical amplifiers. WDM bandpass and edge filters enable MUX, DEMUX, and add/drop modules. Besides these main applications of thin-film filters, many more exist in the modern network, including filters for supervisory channel monitoring, dispersion management, and switching.

The success of thin-film filters is due in large part to the maturity and versatility of the design and fabrication technologies. Processes exist to manufacture dense, robust coatings which do not require thermal compensation or hermetic packaging. Effects which limit thin-film performance, including film stress, beam shift, and CD are well-understood; trade-offs are available and part of the design process. Thin-film technology has been adapted to meet a wide range of design and application environments, from high isolation bandpass couplers in a submarine environment to press-fit filters in injection-molded FTTx modules. Filters can be produced in large volumes, at low cost, or customized for a specific application and prototyped in a matter of days.

Given the flexibility in design, the ability for rapid prototyping, the ease of scaling to high volume, and continued improvements in fabrication processes, thin-film filters are expected to remain one of the key technologies used in fibre optic networks.

Acknowledgements

The authors would like to thank their colleagues at JDSU for many contributions to this work; special thanks go to Karen Hendrix, Paul Colbourne, Georg Ockenfuss, Fred Van Milligen, and Andrew Clark who carefully reviewed the manuscript. Alfons Zoeller of Leybold Optics contributed useful materials for the section on WDM filter deposition processes, and Veeco Instruments made contributions to this section as well. We also thank Herbert Venghaus, the editor of this volume, who gave us valuable and detailed feedback at several stages during the writing of this chapter.

References

1.   J. Strong: "On a method of decreasing the reflection from nonmetallic substances," J. Opt. Soc. Am. **26**, 73–74 (1936)

2.   H. A. Macleod: *Thin-Film Optical Filters* (Institute of Physics Publishing, Dirac House, Temple Back, Bristol BS1 6BE, UK, 2001)

3.   A. Thelen: *Design of Interference Coatings* (McGraw-Hill Book Company, New York, 1989)

4.   P. Baumeister: *Optical Coating Technology* (SPIE Optical Engineering Press, Bellingham, WA, 2004)

5.   J. D. Rancourt: *Optical Thin Films: User Handbook* (SPIE Optical Engineering Press, Bellingham, WA, 1996)

6.   P. Yeh: *Optical Waves in Layered Media* (John Wiley and Sons, Hoboken, NJ, 2005)

7.   A. Thelen: "Antireflection Coatings," in *Design of Interference Coatings,* Chap. 4 (McGraw-Hill Book Company, New York, 1989)

8.   J. A. Dobrowolski, A. V. Tikhonravov, M. K. Trubetskov, B. T. Sullivan, and P. G. Verly: "Optimal single-band normal-incidence antireflection coatings," Appl. Opt. **35**, 644–658 (1996)

9.   P. Baumeister: "Reflection reducing coatings," in *Optical Coating Technology,* Chap. 4 (SPIE Optical Engineering Press, Bellingham, WA, 2004)

10. U. Schallenberg, U. Schulz, and N. Kaiser: "Multicycle AR coatings: a theoretical approach," *Proc. SPIE* **5250**, 357–366 (2004)

11. Stevenson: "High performance antireflection coatings for telecommunications," *Proc. SPIE* **5527**, 79–92 (2004)

12. P. G. Verly: "Simple technique for the accurate design of square bandpass WDM interference filters," *Proc. SPIE* **5250**, 378–383 (2004)

13. A. V. Tikhonravov and M. K. Trubetskov: "Automated design and sensitivity analysis of wavelength-division multiplexing filters," Appl. Opt. **41**, 3176–3182 (2002)

14. B. T. Sullivan and J. A. Dobrowolski: "Implementation of a numerical needle method for thin-film design," Appl. Opt. **35**, 5484–5492 (1996)

15. J. A. Dobrowolski: "Numerical methods for optical thin films," Optics and Photonics News **8**, 25–33 (1997)

16. M. Scobey, D. E. Spock, M. E. Grasis, and J. A. Beattie: "EDFA gain equalization using interference films," NFOEC 1996, Technical Proceedings, Denver, CO, pp. 969–972 (1996)

17. M. Tilsch, C. A. Hulse, K. D. Hendrix, and R. B. Sargent: "Design and demonstration of a thin-film based gain equalization filter for C-band EDFAs," NFOEC 2000, Technical Proceedings, 390–395 (2000)

18. P. G. Verly: "Design of a robust thin-film interference filter for erbium-doped fiber amplifier gain equalization," Appl. Opt. **41**, 3092–3096 (2002)

19. A. Thelen, M. Tilsch, A. V. Tikhonravov, M. K. Trubestskov, and U. Brauneck: "Topical Meeting on Optical Interference Coatings (OIC'2001): design contest results," Appl. Opt. **41**, 3022–3038 (2002)

20. G. Lenz, B. J. Eggleton, C. R. Giles, C. K. Madsen, and R. E. Slusher: "Dispersive properties of optical filters for WDM systems," IEEE J. Quantum Electron. **34**, 1390–1403 (1998)

21. G. Lenz and C. K. Madsen: "General optical all-pass filter structures for dispersion control in WDM systems," J. Lightwave Technol. **17**, 1248–1254 (1998)

22. R. B. Sargent: "Recent advances in thin film filters," Optical Fiber Communication Conference, 2004 (OFC 2004), TuD6.

23. J. A. Buck: *Fundamentals of Optical Fibers* (John Wiley and Sons, New York, NY, 1995)

24. R. M. Fortenberry, M. E. Wescott, L. P. Ghislain, and M. A. Scobey: "Low chromatic dispersion thin film DWDM filters for 40 Gb/s transmission systems," Optical Fiber Communication Conference and Exhibit, 2002 (OFC 2002), pp. 319–320

25. C. A. Hulse, K. D. Hendrix, F. K. Zernik, M. Tilsch, G. Ockenfuss, R. B. Sargent, A. Zhao, H. Pinkney, and S. Moffat: "Demonstration of a novel low-dispersion thin-film DWDM filter for high data rate applications," *Optical Interference Coatings (OIC'04),* OSA Tech. Digest (Opt. Soc. America, Washington, DC, 2004), ThD7

26. C. Madsen and G. Lenz: "Optical all-pass filters for phase design with applications for dispersion control," IEEE Photon. Technol. Lett. **10**, 994–996 (1998)

27. T. D. Noe: "Design of reflective phase compensator filters for telecommunications," Appl. Opt. **41**, 3183–3186 (2002)

28. M. Tilsch, C. A. Hulse, F. K. Zernik, R. A. Modavis, C. J. Addiego, R. B. Sargent, N. A. O'Brien, H. Pinkney, and A. V. Turukhin: "Experimental demonstration of thin-film dispersion compensation for 50-GHz filters," IEEE Photon. Technol. Lett. **15**, 66–68 (2003)

29. M. Jablonski, Y. Takushima, and K. Kikuchi: "The realization of all-pass filters for third-order dispersion compensation in ultrafast optical fiber transmission systems," J. Lightwave Technol. **19**, 1194–1205 (2001)

30. M. Jablonski, K. Sato, D. Tanaka, H. Yaguchi, S. Y. Yet, K. Furuki, K. Yamada, B. Buchholz, and K. Kikuchi: "A compact thin-film-based all-pass device for the compensation of the in-band dispersion in FBG filters," IEEE Photon. Technol. Lett. **15**, 1725 (2003).

31. B. E. A. Saleh and M. C. Teich: *Fundamentals of Photonics* (John Wiley and Sons, New York, NY, 1991)

32. D. Marcuse: "Loss analysis of single-mode fiber splices," Bell Syst. Tech. J. **56**, 703–718 (1977)

33. W. J. Tomlinson: "Wavelength multiplexing in multimode optical fibers," Appl. Opt. **16**, 2182–2194 (1977)

34. W. J. Tomlinson: "Applications of GRIN-rod lenses in optical fiber communication systems," Appl. Opt. **19**, 1127–1139 (1980)

35. Y. C. Si, G. S. Duck, J. Ip, and N. Teitelbaum: "Narrow band filter and method of making same," U.S. Patent 5,612,824 (March 18, 1997)

36. M. A. Scobey and D. E. Spock: "Passive DWDM components using microplasma optical interference filters," Optical Fiber Communication Conference, 1996 (OFC'96), pp. 242–243

37. M. McGuirk and C. K. Carniglia: "An angular spectrum representation to the Goos–Hanchen shift," J. Opt. Soc. Am. **67**, 103–107 (1977)

38. C. W. Hsue and T. Tamir: "Lateral displacement and distortion of beams incident upon a transmitting-layer configuration," J. Opt. Soc. Am. A **2**, 978–988 (1985)

39. R. E. Klinger, C. A. Hulse, and R. B. Sargent: "Beam displacement and distortion effects in narrowband optical thin film filters," *Optical Interference Coatings (OIC'04)*, OSA Tech. Digest (Opt. Soc. America, Washington, DC, 2004), ThE5

40. C. K. Carniglia, D. G. Jensen, and A. J. Fielding: "Lateral shift and internal electric fields in multi-cavity narrow-band-pass filters," *Optical Interference Coatings (OIC'04)*, OSA Tech. Digest (Opt. Soc. America, Washington, DC, 2004), ThE4

41. C. A. Hulse, R. E. Klinger, and R. B. Sargent: "Wavelength-dependent beam displacement in narrowband optical thin film filters," 10th Microoptics Conference (MOC'04), Jena, Germany, C10 (2004)

42. J. W. Goodman: *Introduction to Fourier Optics* (McGraw-Hill, Columbus, OH, 1996)

43. C. A. Hulse, R. E. Klinger, and R. B. Sargent: "Optical coupler device for dense wavelength division multiplexing," U.S. Patent 6,215,924 (April 10, 2001)

44. P. J. Martin and R. P. Netterfield: "Optical Films Produced By Ion-Based Techniques," in *Progress in Optics Vol. 23* (E. Wolf, ed.), (Elsevier, New York, 1986)

45. http://www.nsgamerica.com/press/dsp_view.cfm?specialid=140

46. http://www.us.schott.com/optics_devices/english/products/dwdm.html

47. http://www.hoyaoptics.com/specialty_glass/wdm_substrate.htm

48. http://www.oharacorp.com/swf/ps.html

49. D.J. Smith: "Modeling of modular defects in thin films for various deposition techniques," *Proc. SPIE* **821**, 120–128 (1987)

50. R.J. Trench, R. Chow, and M.R. Kozlowski: "Characterization of defect geometries in multilayer optical coatings," J. Vac. Sci. Technol. A **12**, 2808–2813 (1994)

51. E. Ritter: "Dielectric film materials for optical applications," in *Physics of Thin Films 8* (G. Hass, M.H. Franscombe, and R.W. Hoffman, eds.), 1–49 (Academic Press, New York, 1975).

52. N.A. O'Brien, M.J. Cumbo, K.D. Hendrix, R.B. Sargent and M.K. Tilsch: "Recent advances in thin film interference filters for telecommunications," *44th Annual Technical Conference Proceedings of the Society of Vacuum Coaters,* (Philadelphia, PA 2001) pp. 255–261

53. J.S. Colligon: "Energetic condensation: Processes, properties, and products," J. Vac. Sci. Technol. A **13**, 1649–1657 (1995)

54. W.D. Westwood: *Sputter Deposition* (AVS, New York, 2003)

55. J.L. Vossen and W. Kern: *Thin Film Processes II* (Academic Press, New York, 1991)

56. F. Bovard: "Ion-Assisted Deposition," in *Thin Films For Optical Systems* (F. Flory, ed.), p. 117 (Marcel Dekker, New York, 1995)

57. R. Faber, K. Zhang, and A. Zöller: "Design and manufacturing of WDM narrow band interference filters," *Proc. SPIE* **4094**, 58–64 (2000)

58. F. Zöller, S. Beißwenger, R. Gotzelmann, and K. Matl: "Plasma ion assisted deposition: A novel technique for the production of optical coatings," *Proc. SPIE* **2253**, 394–402 (1994)

59. Leybold Optics GmbH, Siemensstrasse 88, 63755 Alzenau, Germany

60. D. Gibson: "Fast Coatings for DWDM Filters," Photonics Spectra **35** 114 (2001)

61. D.T. Wei, H.R. Kaufman, and C.-C. Lee: "Ion beam sputtering," in *Thin Films For Optical Systems* (F.R. Flory, ed.), Chap. 6 (Marcel Dekker, New York, 1995)

62. C. Montcalm, S.M. Lee, D. Burtner, A. Dummer, D. Siegfried, I. Wagner, and M. Watanabe: "High-rate dual ion beam sputtering deposition technology for Optical telecommunication filters," *45th Annual Technical Conference Proceedings of the Society of Vacuum Coaters,* (Buena Vista, FL, 2002) pp. 245–249

63. Veeco Instruments Inc., 100 Sunnyside Blvd. Ste. B, Woodbury, NY 11797, US

64. W.T. Pawlewicz, P.M. Martin, R.W. Knoll, and I.B. Mann: "Multilayer optical coating fabrication by dc magnetron reactive sputtering," *Proc. SPIE* **678**, *Optical Thin Films II: New Developments,* 134–140 (1986)

65. S.M. Edlou, A. Smajkiewicz, and G.A. Al-Jumaily: "Optical properties and environmental stability of oxide coatings deposited by reactive sputtering," Appl. Opt. **32**, 5601–5605 (1993)

66. M.A. Scobey: "Low pressure reactive magnetron sputtering apparatus and method," U.S. Patent 5,851,365 (December 22, 1998)

67. G.S. Selwyn and C.A. Weiss: "Particle contamination formation in magnetron sputtering processes," J. Vac. Sci. Technol. A **15**, 2023–2028 (1997)

68. W.D. Sproul and B.E. Sylvia: "Multi-level control for reactive sputtering," *45th Annual Technical Conference Proceedings of the Society of Vacuum Coaters,* (Buena Vista, FL, 2002) pp. 11–15

69. S. Bauer, L. Klippe, U. Rothhaar, and M. Kuhr: "Optical multilayers for ultranarrow bandpass filters fabricated by PICVD," Thin Solid Films **442**, 189–193 (2003)

70. L. H. Domash: "Thermo-optically tunable thin film devices," *Proc. SPIE* **5225**, 1–6 (2003)

71. G. T. Mearini and L. Takacs: "Optical filter constructed by atomic layer deposition for next generation dense wavelength division multiplexers," U.S. Patent Application, Pub. No. US2002/0003664 A1 (2002)

72. G. T. Mearini and L. Takacs: "Atomic layer controlled optical filter design for next generation dense wavelength division multiplexers," U.S. Patent Application, Pub. No. US2002/0003665 A1 (2002)

73. A. Musset and I. C. Stevenson: "Thickness distribution of evaporated films," *Proc. SPIE* **1270**, 287–291 (1990)

74. M. Yang, J. Liu, Q. Chen, and B. Zhang: "Uniformity analysis and design optimization of multi-layer thin film filter used in fiber optics communication system," *Proc. SPIE* **5250**, 691–696 (2004)

75. C. Lee, K. Chuang, and J. Wu: "Thickness distribution of thin films deposited by ion beam sputtering," *Optical Interference Coatings (OIC'01)*, OSA Tech. Digest (Opt. Soc. America, Washington DC, 2001), MB4-2

76. R. P. Riegert: "Optimum usage of quartz crystal monitor based devices," IVth International Vacuum Congress, Bristol: Institute of Physics and the Physical Society, pp. 527–530, (1968)

77. A. Zoeller, M. Boos, H. Hagedorn, W. Klug, and C. Schmitt: "High accurate in-situ optical thickness monitoring for multilayer coatings," *47th Annual Technical Conference Proceedings of the Society of Vacuum Coaters*, (Dallas, TX, 2004), pp. 72–78

78. D. Ristau: "Characterisation and Monitoring," in *Optical Interference Coatings* (N. Kaiser and H. K. Pulker, eds.), 181–205 (Springer, Berlin, Heidelberg, 2003)

79. B. T. Sullivan and J. A. Dobrowolski: "Deposition error compensation for optical multilayer coatings. I. Theoretical description," Appl. Opt. **31**, 3821–3835 (1992)

80. B. T. Sullivan and J. A. Dobrowolski: "Deposition error compensation for optical multilayer coatings. II. Experimental results - sputtering system," Appl. Opt. **32**, 2351–2360 (1993)

81. B. T. Sullivan, G. A. Clarke, T. Akiyama, N. Osborne, M. Ranger, J. A. Dobrowolski, L. Howe, A. Matsumoto, Y. Song, and K. Kikuchi: "High-rate automated deposition system for the manufacture of complex multilayer coatings," Appl. Opt. **39**, 157–167 (2000)

82. D. E. Morton: "Optical monitoring of thin films using spectroscopic ellipsometry," Vacuum Technology & Coating, August (2003)

83. H. A. Macleod: "Turning value monitoring of narrow-band all-dielectric thin-film optical filters," Optica Acta **19**, 1–28 (1972)

84. P. Bousquet, A. Fornier, R. Kowalczyk, E. Pelletier, and P. Roche: "Optical Filters: monitoring process allowing the auto-correction of thickness errors," Thin Solid Films **13**, 285–290 (1972)

85. K. Postava and J. Pistora: "Thickness monitoring of optical filters for DWDM applications," Optics Express **11**, 610–616 (2003)

86. R. R. Willey: "Simulation of errors in monitoring of narrow bandpass filters," Appl. Opt. **41**, 3193–3195 (2002)

87. C. Kuo, S. Chen, C. Lee, D. Lu, and C. Wei: "Influence of monitor passband width to layer thickness determination during depositing a DWDM filter," *Optical Interference Coatings (OIC'04)*, OSA Tech. Digest (Opt. Soc. America, Washington, DC, 2004), ThD9

88. http://www.intellemetrics.com/dwdmspec.html

89. R. R. Willey: "Monitoring the last two (AR) layers in narrow bandpass filters," *Proc. SPIE* **5250**, 400–405 (2004)

90. M. Tilsch, V. Scheuer, and T. Tschudi: "Effects of thermal annealing on ion beam sputtered SiO_2 and TiO_2 optical thin films," *Proc. SPIE* **3133**, 163–175 (1997)

91. J. T. Brown: "Center wavelength shift dependence on substrate coefficient of thermal expansion for optical thin-film interference filters deposited by ion-beam sputtering," Appl. Opt. **43**, 4506–4511 (2004)

92. S. L. Prins, A. C. Barron, W. C. Herrmann, and J. R. McNeil: "Effect of stress on performance of dense wavelength division multiplexing filters: thermal properties," Appl. Opt. **43**, 633–637 (2004)

93. H. Takahashi: "Temperature stability of thin-film narrow-bandpass filters produced by ion assisted deposition," Appl. Opt. **34**, 667–675 (1995). Note that the author's last name is misspelled as "Takashashi" in the article; the correct spelling is "Takahashi."

94. S. Kim and C. K. Hwangbo: "Derivation of the center-wavelength shift of narrow-bandpass filters under temperature change," Optics Express **12**, 5634–5639 (2004)

95. J. Jiang, J. J. Pan, J. Guo, and G. Keiser: "Model for analyzing manufacturing-induced internal stresses in 50-GHz DWDM multilayer thin-film filters and evaluation of their effects on optical performances," J. Lightwave Technol. **23**, 495–503 (2005)

96. H. S. Lee, M. C. Lo, and B. Chiang: "Highly temperature stable filter for fiberoptic applications and system for altering the wavelength or other characteristics of optical devices," U.S. Patent 6,269,202 (July 31, 2001)

97. W. Fan, Z. Yan, and C. H. Hsia: "Temperature compensated optical filter," U.S. Patent 6,469,847 (October 22, 2002)

98. M. Tilsch: "Sandwiched thin film optical filter," U.S. Patent 6,721,100 (April 13, 2004)

99. R. J. Ryall, C. A. Hulse, A. T. Taylor, and R. I. Seddon: "Extrinsically athermalized optical filter devices," U.S. Patent 6,707,609 (March 16, 2004)

100. C. C. Fang, F. Jones, and V. Prasad: "Effect of gas impurity and ion-bombardment on stresses in sputter-deposited thin films: A molecular-dynamics approach," J. Appl. Phys. **74**, 4472–4482 (1993)

101. M. Ohring: *Materials Science of Thin Films* (Academic Press, San Diego, CA, 2002)

102. G. J. Ockenfuss, N. A. O'Brien, and E. Williams: "Ultra-low stress coating process: an enabling technology for extreme performance thin film interference filters," Optical Fiber Communications Conference, 2002 (OFC 2002), post-deadline paper FA8

103. G. J. Ockenfuss and R. E. Klinger: "Ultra-low stress thin film interference filters," *Optical Interference Coatings (OIC'04)*, OSA Tech. Digest (Opt. Soc. America, Washington DC, 2004), ThE2

104. S. L. Prins, A. C. Barron, W. C. Herrmann, and J. R. McNeil: "Effect of stress on performance of dense wavelength division multiplexing filters: optical properties," Appl. Opt. **43**, 626–632 (2004)

105. J. Straus and B. Kawasaki: "Passive optical components," in *Optical-Fiber Transmission* (E. E. Basch, ed.), pp. 241–264 (Howard Sams & Co., Indianapolis, IN, 1987)

106. Y. C. Si and Y. Cheng: "Optical multiplexer/demultiplexer: discrete," in *WDM Technologies: Passive Optical Components, Volume* I (A. K. Dutta, N. K. Dutta, and M. Fujiwara, eds.), pp. 39–78 (Academic Press, San Diego, CA, 2003)

107. E. Miyauchi, T. Iwama, H. Nakajima, N. Tokoyo, and K. Terai: "Compact wavelength multiplexer using optical-fiber pieces," Opt. Lett. **5**, 321–322 (1980)

108. G. Winzer, H. F. Mahlein, and A. Reichelt: "Single-mode and multimode all-fiber directional couplers for WDM," Appl. Opt. **20**, 3128–3135 (1981)

109. N. Uehara, R. Okuda, and T. Shidara: "Super antireflection coating at 1.5 µm," *Optical Interference Coatings (OIC'04)*, OSA Tech. Digest (Opt. Soc. America, Washington DC, 2004), WA5

110. S. Sugimoto, K. Minemura, K. Kobayashi, M. Shikada, H. Nomura, K. Kaede, A. Ueki, and S. Matsushita: "Wavelength division two-way fibre-optic transmission experiments using micro-optic duplexers," Electron. Lett. **15**, 15–17 (1978)

111. A. Nicia: "Lens coupling in fiber-optic devices: efficiency limits," Appl. Opt. **20**, 3136–3145 (1981)

112. W. Jiang, Y. Sun, R. T. Chen, B. Guo, J. Horwitz, and W. Morey: "Ball-lens based optical add-drop multiplexers: design and implementation," IEEE Photon. Technol. Lett. **14**, 825–827 (2002)

113. R. E. Wagner: "Optical multi/demultiplexer using interference filters," U.S. Patent 4,474,424 (October 2, 1984)

114. K. Nosu, H. Ishio, and K. Hashimoto: "Multireflection optical multi/demultiplexer using interference filters," Electron. Lett. **15**, 414–415 (1979)

115. M. A. Scobey, W. J. Lekki , and T. W. Geyer: "Filters create thermally stable, passive multiplexers," Laser Focus World **33**, 111–116 (1997)

116. B. E. Lemoff, L. B. Aronson, and L. A. Buckman: "Zigzag waveguide demultiplexer for multimode WDM LAN," Electron. Lett. **34**, 1014–1016 (1998)

117. L. A. Buckman, B. E. Lemoff, A. J. Schmit, R. P. Tella, and W. Gong: "Demonstration of a small-form-factor WWDM transceiver module for 10-Gb/s Local Area Networks," IEEE Photon. Technol. Lett. **14**, 702–704 (2002)

118. B. E. Lemoff: "Coarse WDM transceivers," Optics and Photonics News **13**, S8–S14 (2002)

119. B. E. Lemoff, M. E. Ali, G. Panotopoulos, E. de Groot, G. M. Flower, G. H. Rankin, A. J. Schmit, K. D. Djordjev, M. R. T. Tan, A. Tandon, W. Gong, R. P. Tella, B. Law, L. Chia, and D. W. Dolfi: "Demonstration of a compact low-power 250-Gb/s parallel-WDM optical interconnect," IEEE Photon. Technol. Lett. **17**, 220–222 (2005)

120. T. Honda, A. C. Liu, J. Valera, J. Colvin, K. Sawyer, and R. R. McLeod: "Diffraction-compensated free-space WDM add-drop module with thin-film filters," IEEE Photon. Technol. Lett. **15**, 69–71 (2003)

121. Y. Okabe and H. Sasaki: "A simple wide wavelength division multi/demultiplexer consisting of optical elements," Optical Fiber Communications Conference, 2002 (OFC 2002), pp. 322–323

122. Y. Okabe and H. Sasaki: "Compact multi/demultiplexer system consisting of stacked dielectric interference filters and aspheric lenses," *Proc. SPIE* **4652**, 197–203 (2002)

123. GR-1209-CORE: "Generic Requirements for Passive Optical Components," Issue 3 (Telcordia Technologies, 2001).

124. GR-1221-CORE: "Generic reliability assurance requirements for passive optical components," Issue 2 (Telcordia Technologies, 1999).

125. R. R. McLeod, M. Wolkin, V. Morozov, and K. A. Sawyer: "Packaging of micro-optic components to meet Telcordia standards," Optical Fiber Communications Conference, 2002 (OFC 2002), paper WS7

126. W. Moore and T. Kiktyeva: "Optical damage in fiber optic components," Optical Fiber Communications Conference, 2003 (OFC 2003), pp. 525–527 vol. 2.

127. http://www.ftthcouncil.org

128. L. H. Domash, M. Wu, N. Nemchuk, and E. Ma: "Tunable and switchable multiple-cavity thin film filters," J. Lightwave Technol. **22**, 126–134 (2004)

129. R. S. Vodhanel, M. Krain, R. E. Wagner, and W. B. Sessa: "Long-term wavelength drift of the order of -0.01 nm/yr for 15 free-running DFB laser modules," Optical Fiber Communications Conference, 1994 (OFC'94), paper WG5

130. Y. C. Chung and J. Jeong: "Aging-induced wavelength shifts in 1.5-um DFP lasers," Optical Fiber Communications Conference, 1994 (OFC'94), paper WG6

131. B. Villeneuve, H. B. Kim, M. Cyr, and D. Gariepy: "A compact wavelength stabilization scheme for telecommunications transmitters," 1997 Digest of the LEOS Summer Topical Meetings (IEEE, Piscataway, NJ, 1997), Paper WD2

132. B. Villeneuve and H. B. Kim: "Wavelength monitoring and control assembly for WDM optical transmission systems," U.S. Patent 5,825,792 (October 20, 1998)

133. T. C. Munks, P. E. Dunn, and D. J. Allie: "Method and apparatus for monitoring and control of laser emission wavelength," U.S. Patent 6,134,253 (October 17, 2000)

134. http://www.jdsu.com/site/images/products/pdf/Wavelength_Locker_Narrowband_012704.pdf

135. http://www.santec.com/pdf/components/OWL-1020.pdf

136. M. Lequime: "Tunable thin film filters : review and perspectives," *Proc. SPIE* **5250** 302–311 (2004)

137. R. Zengerle: "Light propagation in singly and doubly periodic planar waveguides," J. Mod. Opt. **34**, 1589–1617 (1987)

138. H. Kosaka, T. Kawashima, A. Tomita, M. Notomi, T. Tamamura, T. Sato, and S. Kawakami: "Superprism phenomena in photonic crystals," Phys. Rev. B **58**, R10096–R10099 (1998)

139. P. Yeh, A. Yariv, and E. Marom: "Theory of Bragg fiber," J. Opt. Soc. Am. **68**, 1196–1201 (1978)

140. C. K. Carniglia: "Perfect mirrors from a coating designer's point of view," *Proc. SPIE* **3902**, 68–84 (1999)

141. Y. Fink, J. N. Winn, S. Fan, C. Chen, J. Michel, J. D. Joannopoulos, and E. L. Thomas: "A dielectric omnidirectional reflector," Science **282**, 1679–1682 (1998)

142. S. Kim and C. K. Hwangbo: "Design of omnidirectional high reflectors with quarter-wave dielectric stacks for optical telecommunication bands," Appl. Opt. **41**, 3187–3192 (2002)

143. M. Gerken: "Wavelength multiplexing by spatial beam shifting in multilayer thin-film structures," Electrical Engineering Ph.D. Dissertation, Stanford University, CA, March 2003

144. M. Gerken and D. A. B. Miller: "Multilayer thin-film stacks with steplike spatial beam shifting," J. Lightwave Technol. **22**, 612–618 (2004)

145. M. Gerken and D. A. B. Miller: "Photonic nanostructures for wavelength division multiplexing," *Proc. SPIE* **5597**, 82–96 (2004)

146. Y. Fink, D. J. Ripin, S. Fan, C. Chen, J. D. Joannopoulos, and E. L. Thomas: "Guiding optical light in air using an all-dielectric structure," J. Lightwave Technol. **17**, 2039–2041 (1999)

147. M. Ibanescu, Y. Fink, S. Fan, E. L. Thomas, and J. D. Joannopoulos: "An all-dielectric coaxial waveguide," Science **289**, 415–418 (2000)

8 Ring-Resonator-Based Wavelength Filters

Douwe H. Geuzebroek and Alfred Driessen

8.1 Introduction

Microring resonators (MR) represent a class of filters with characteristics very similar to those of Fabry–Perot filters. However, they offer the advantage that the injected and reflected signals are separated in individual waveguides, and in addition, their design does not require any facets or gratings and is thus particularly simple. MRs evolved from the fields of fibre optic ring resonators and micron scale droplets [1–4]. Their inherently small size (with typical diameters in the range between several to tens of micrometres), their filter characteristics, and their potential for being used in complex and flexible configurations make these devices particularly attractive for integrated optics or VLSI photonics applications [5–10].

MRs for filter applications, delay lines, as add/drop multiplexers, and modulators will be covered in detail in this chapter, while other applications such as in optical sensing, in spectroscopy or for coherent light generation (MR lasers) are outside the scope of this chapter.

This chapter focuses primarily on 4-port microrings, while 2-port devices will play a minor role here and are covered in more detail in Chap. 9. The present chapter starts with design considerations, the functional behaviour, and key characteristics of a single microring resonator and continues with the design of cascaded MRs allowing the implementation of higher order filters. Finally, complex devices like add-drop filters, tuneable dispersion compensators, all-optical wavelength converters, and tuneable cross-connects are treated.

8.2 Fundamentals of Microring Resonators

8.2.1 General Considerations

The fundamental building blocks of microring-based devices are a microring plus one or two waveguides. In the former case this leads to two-port devices which act as all-pass filters and introduce a wavelength-dependent phase shift only (lossless case). This property is exploited for the realization of dispersion compensators and will be illustrated in Sect. 8.6.5.

On the other hand, a microring resonator consisting of a ring plus two straight waveguides represents a 4-port structure, as illustrated in Fig. 8.1.

The ring (radius R) and the port waveguides are evanescently coupled and a fraction κ_1 of the incoming field is transferred to the ring. When the optical path-length of a roundtrip is a multiple of the effective wavelength, constructive interference occurs and light is 'built up' inside the ring: the MR is ON resonance. As a consequence, periodic fringes appear in the wavelength response at the output ports as shown in Fig. 8.2. At resonance the drop port shows maximum transmission since a fraction κ_2 of the built-up field inside the ring is coupled to this port. In the through port the ring exhibits a minimum at resonance. In the ideal case with equal coupling constants at resonance all the power is directed to the drop port. Light coupled back to the through port after a roundtrip experiences an additional $180°$ phase shift with respect to the light coming directly from the in port, and as a consequence no light exits from the through port at resonance.

Before coming to a more detailed description of MRs it is worthwhile to introduce a number of key parameters of MRs.

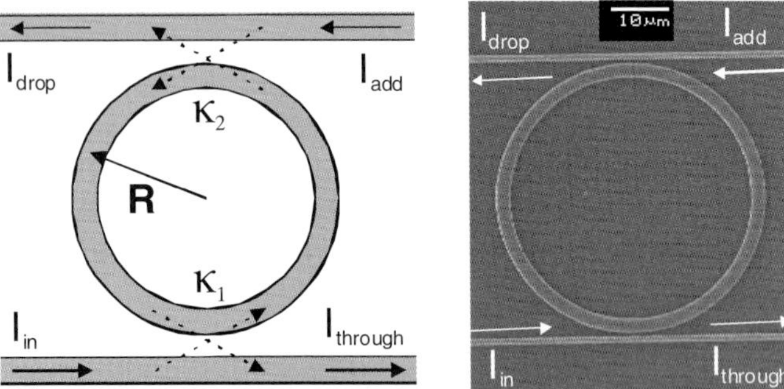

Fig. 8.1. Schematic drawing of a 4-port microring resonator (*left*) and SEM picture (*right*) with in- and output port waveguides

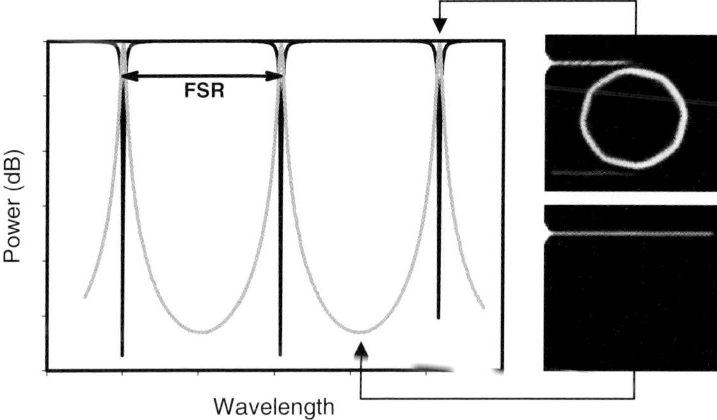

Fig. 8.2. Simulated response of a MR for the through- (*black*) and drop port (*grey*). Also shown is the Free Spectral Range, the distance between two consecutive fringes. At the right simulated fields of the MR are given; top: ON resonance, Bottom: OFF resonance

The difference in position between two consecutive resonant peaks, see Fig. 8.2, is called the Free Spectral Range and can be defined either in the frequency or wavelength domain (FSR_f or FSR_λ, respectively):

$$FSR_f = \Delta f = \frac{c}{n_g\, 2\pi R} \quad , \quad FSR_\lambda = \Delta\lambda \approx \frac{\lambda^2}{n_g\, 2\pi R} \qquad (8.1)$$

The group index n_g is defined as [9]

$$n_g = n_{eff}(f_0) + f_0 \left.\frac{dn_{eff}}{df}\right|_{f_0} = n_{eff}(\lambda_0) + \lambda_0 \left.\frac{dn_{eff}}{d\lambda}\right|_{\lambda_0} \qquad (8.2)$$

Other parameters are: n_{eff} is the effective refractive index, R the radius of the ring, c the speed of light in vacuum, and f_0 is defined by $f_{1,2} = f_0 \pm \Delta f / 2$, with f_1 and f_2 two consecutive resonance frequencies.

Another important parameter of the MR is the Full-Width at Half-Maximum (FWHM) or 3-dB bandwidth, which is a measure for the bandwidth of the device. A quality measure of the microring is the Finesse F which is the ratio between the FSR and the 3-dB bandwidth [11]

$$F = \frac{FSR}{FWHM} = \frac{\pi\left(X_1 X_2\right)^{1/4}}{1-\left(X_1 X_2\right)^{1/2}} \qquad (8.3)$$

$$X_i = \sqrt{(1-\kappa_i^2)}\, e^{L_r \alpha}$$

with α being the losses per length inside the ring and L_r the length of the optical path in the ring. High finesse devices have a small FWHM and a strong intensity build-up in the ring when in resonance. Therefore high F devices are suitable for applications where high intensities in the cavity are necessary like lasers or non-linear optical (NLO) devices. Instead of the finesse, the quality factor Q can also be used as an absolute measure for the wavelength selectivity of the microring resonator according to:

$$Q = \omega / \Delta\omega_{FWHM} = \lambda / \Delta\lambda_{FWHM} \tag{8.4}$$

where $\Delta\omega_{FWHM}$ and $\Delta\lambda_{FWHM}$ are the 3-dB bandwidths in the frequency and wavelength domain, respectively.

The speed of high bit rate communication is limited by the bandwidth of the optical filters. As this bandwidth in high finesse MR devices is becoming increasingly small, only medium finesse devices ($F \sim 10-20$) with relative large coupling constants ($\kappa \sim 0.4-0.6$) are employed. The FWHM is determined by the coupling constants and the loss inside the ring according to [12, 13]:

$$FWHM_f = \frac{c}{2\pi R n_g}\left(\frac{\sqrt{X_1 X_2}}{\pi \sqrt[4]{X_1 X_2}} \right) \tag{8.5}$$

8.2.2 Bent Waveguides

Bent waveguides exhibit radiation losses which -for a given confinement of the guided wave- increase as the bend radius decreases. On the other hand, for a given bend radius, the bending losses decrease as the confinement of the guided wave gets stronger, or alternatively spoken, the bending losses decrease as the (effective) index contrast between the ring and its surrounding/substrate increases. In the following we will focus on high index contrast rings ($\Delta n > 0.1$) since this is a prerequisite for small footprint devices offering the required integration potential. Δn is the difference of the effective indices of the ring waveguide and its surrounding, respectively.

In relation to the index contrast the term 'minimum bend radius' is used to define the lowest acceptable (technology-dependent!) radius keeping the bend losses sufficiently low. 'Sufficiently low' is dependent of the application and also related to the coupling fractions as this chapter will show. Typically one designs for a bending loss below 1 dB per $360°$ roundtrip. The minimum bend radius as a function of effective index contrast is illustrated in Fig. 8.3. As can be seen, a high index contrast enables extremely

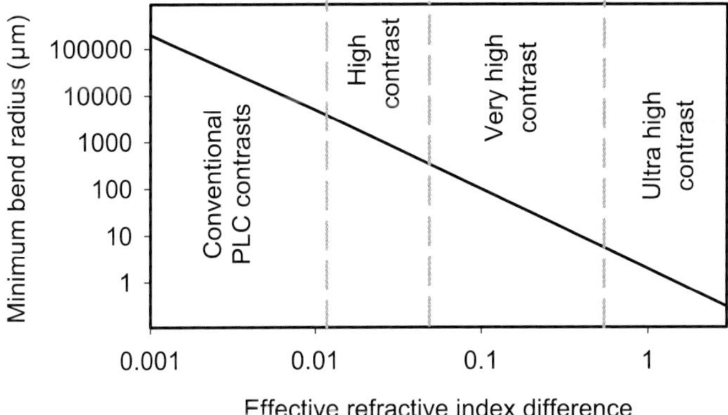

Fig. 8.3. Minimum bending radius (< 1 dB/360°) as a function of effective index difference

small dimensions and thus compact structures, but at the same time puts severe challenges to the fabrication technology. In addition, the stronger the confinement and the smaller the ring dimensions the more polarisation-related issues become important as will be discussed in the next section.

8.2.3 Polarisation Dependence

Polarisation dependence is primarily determined by the waveguide characteristics (refractive index contrast, geometry [14]). However, even if a chosen geometry would exhibit low birefringence in a straight waveguide (i. e. the propagation constants for TE- and TM-polarised light are approximately the same), this is normally no longer true in bends or rings with small radii. In that case, the modes are no longer pure TE- or TM-like, the resonance wavelengths become polarisation dependent, and even polarisation conversion can occur [15–18].

If pure TE- and TM-like modes with equal coupling constants are assumed, the difference in resonance wavelength of the modes propagating inside the ring is

$$\Delta\lambda_c = \frac{\lambda(n_{eff,TE} - n_{eff,TM})}{n_{eff}} \tag{8.6}$$

with $n_{eff,TE}$ and $n_{eff,TM}$ being the effective indices of the TE- and TM-mode inside the ring.

In general, in addition to the propagation constants, the coupling constants between the ring and waveguides are also different for TE- and TM-polarisation. Nevertheless, polarisation independence can in principle be achieved by carefully balancing material- and geometry-dependent effects [19]. In reality this is restricted to material with low index contrast [20, 21], and it requires the potential to keep technological tolerances extremely tight. An alternative, although more complex, solution is polarisation diversity [22].

8.3 Design of Single Microrings

8.3.1 Evaluation of Parameters

One frequently chosen way of modelling the response of a single MR is the use of a scattering matrix [23, 24] as illustrated in Fig. 8.4.

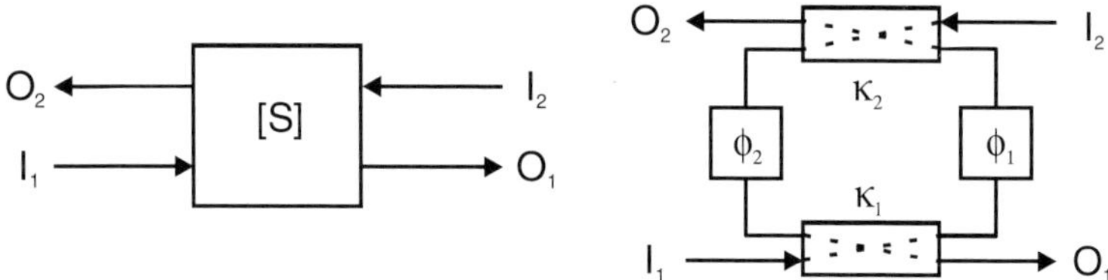

Fig. 8.4. Scattering matrix model of a microring resonator

This approach does not include polarisation effects, but it is sufficient in many cases for the derivation of fundamental design parameters for different material systems such as III-V semiconductors or silicon oxi-nitride (SiON) for example. We will present this approach in more detail in the following as it illustrates relevant device characteristics and fundamental dependencies. As long as the MRs under consideration exhibit polarisation dependent characteristics without TE-TM mode conversion (cf. Sect. 8.2.), one may design MRs for polarisation diversity, while a significantly more sophisticated treatment is needed if TE- and TM-modes are no longer good eigenstates.

In the scattering matrix model the MR is modelled as two couplers which couple a fraction κ_1 or κ_2 over to the cross port, and two delays ϕ_1 and ϕ_2 in-between as shown in Fig. 8.4. The optical fields in the inputs and outputs of the ring are related as follows:

$$\begin{bmatrix} O_1 \\ O_2 \end{bmatrix} = \begin{bmatrix} S_{11} & S_{12} \\ S_{21} & S_{22} \end{bmatrix} \begin{bmatrix} I_1 \\ I_2 \end{bmatrix} \tag{8.7}$$

with

$$S_{11} = \frac{\mu_1 - \mu_2 \alpha_r^2 e^{-j\Delta\omega}}{1 - \mu_1 \mu_2 \alpha_r^2 e^{-j\Delta\omega}}$$

$$S_{21} = -S_{12} = \frac{\kappa_1 \kappa_2 \alpha_r e^{-j\Delta\omega/2}}{1 - \mu_1 \mu_2 \alpha_r^2 e^{-j\Delta\omega}} \tag{8.8}$$

$$S_{22} = \frac{\mu_2 - \mu_1 \alpha_r^2 e^{-j\Delta\omega}}{1 - \mu_1 \mu_2 \alpha_r^2 e^{-j\Delta\omega}}$$

where $\mu_{1,2} = \sqrt{1 - \kappa_{1,2}^2}$, α_r is the loss per roundtrip, $\Delta\omega = \omega_0 - \omega$, with $\omega_0 = \dfrac{c}{2\pi R n_g} m$ being the resonance frequency, while m is the resonance number. The delays ϕ_1 and ϕ_2 are related to the frequency and the optical path-lengths $L_{opt1,2}$ by $\phi_{1,2} = \dfrac{\omega L_{opt1,2}}{c}$, and they are not necessarily identical to each other. The place of the couplers does not matter as long as the total roundtrip phase is the same.

By use of this model one can extract values for the coupling constants from the desired functional behaviour and a given loss and radius. For example, in order to have as much power in the drop port as possible the two coupling constants must be the same [12]. In this case the device is said to be symmetric and has the lowest possible insertion loss in the drop port. Different parameters are found when optimizing the extinction ratio between the drop and through port. In that case the MR must be critically coupled [25], that means $\mu_1 = \mu_2 \alpha_r$. Under this condition all power is extracted from the through-port leading to the highest possible extinction ratio.

With the scattering matrix model, the influence of the loss parameter on the MR response can be determined as well. Parameters such as filtering bandwidth, insertion loss, crosstalk, and channel separation can be determined in this way. Three corresponding examples are given in Figs. 8.5 to 8.7.

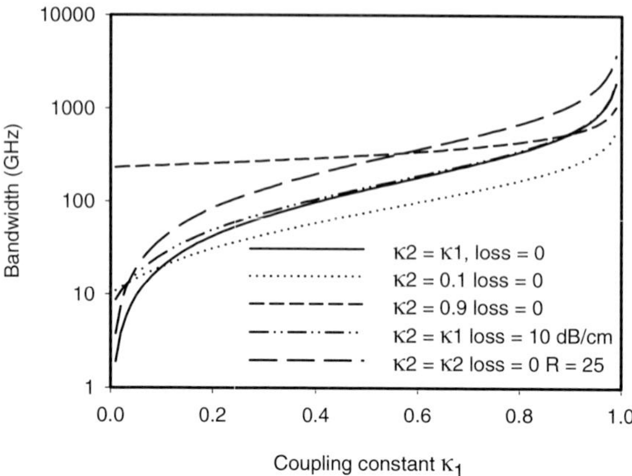

Fig. 8.5. Bandwidth of a MR as a function of coupling constant κ_1

Figure 8.5 shows the 3-dB bandwidth of a microring-based filter with 50 μm radius, as given by (8.5), as a function of the coupling constant κ_1. The bandwidth increases as the coupling ratio gets higher. The effect of asymmetry in coupling is also shown in the figure by the lines with a fixed κ_2. Losses do not have a large impact on the bandwidth. Only for the lower coupling constants a difference can be seen. This figure focuses on bandwidth only, and one might conclude that having κ_2 fixed to 0.9 is favourable since it assures high bandwidths over a wide range of κ_1. But when

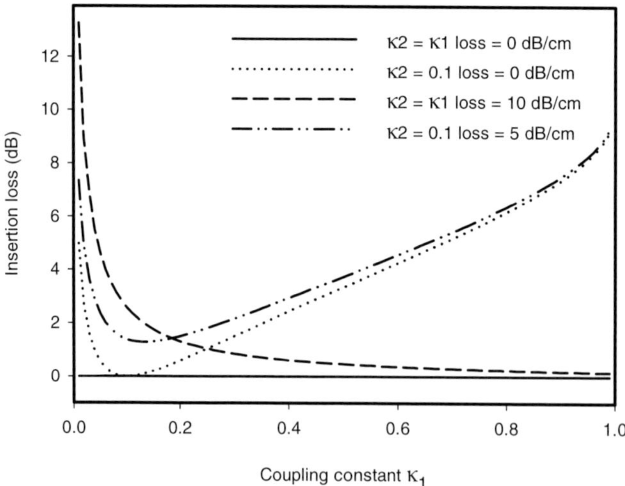

Fig. 8.6. Insertion loss for signals sent to the drop port as function of the coupling constant κ_1 for a MR on resonance

looking at the insertion loss, which is another relevant parameter, this conclusion turns out to be premature as can be seen in Fig. 8.6.

Figure 8.6 shows the insertion loss in the drop port of a microring resonator as a function of the coupling constant κ_i for the MR on resonance. The insertion loss (IL) can be calculated according to [11, 12]:

$$IL = -\log\left(\frac{\kappa_1\kappa_2 e^{-2\pi R}}{\left(1-\sqrt{X_1 X_2}\right)^2}\right) \qquad (8.9)$$

A symmetric lossless device has no insertion loss as can be seen from the solid line in Fig. 8.6. When the ring exhibits losses, the insertion loss increases, especially at lower coupling constants. The insertion loss does also increase with rising asymmetry of the coupling constants. Thus in order to get low insertion loss devices with bandwidths exceeding 10 GHz, the ring must be designed symmetrically with coupling constants above 0.4.

In order to get maximum field inside the cavity, as required for example for all-optical switching, totally different specifications are derived. In this case the coupling constants must be in the order of 0.1 as is described in [26] and as can be seen in Fig. 8.7. The intra-cavity power (P_{cav}) has a maximum when the coupling constant is symmetric and relatively small (around 0.1).

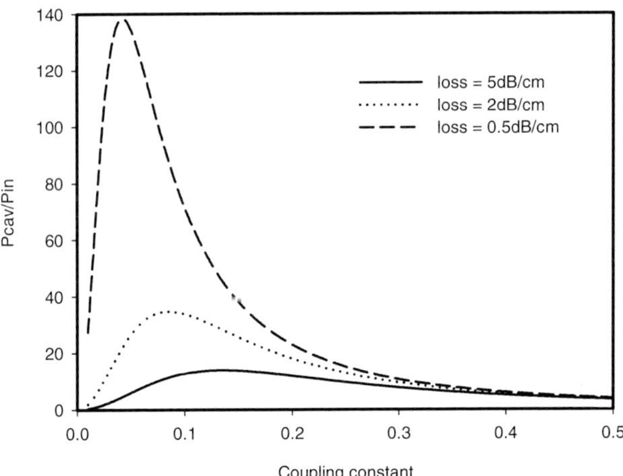

Fig. 8.7. Power inside cavity as a function of coupled fraction κ assuming symmetric coupling for a ring with radius $R = 50\,\mu m$

The scattering matrix model has proven to be a good tool to design single microring resonators, whatever the application-specific demands are. Furthermore, this model allows for extracting 'experimental' parameters by fitting the model to measured data, and the scattering matrix model is also applicable to more complex devices. In particular, cascading more than one ring can simply be expressed by matrix multiplication (cf. Sect. 8.6).

A design parameter of high relevance is the microring radius R. Via the round trip path length the FSR (8.1) and the finesse (8.3) are radius dependent, and R affects the finesse by the round trip losses as well. For telecom applications in the 1.5 µm wavelength range the FSR should be >35 nm for the C-band for example. This implies very small rings with $R < 6$ µm, and this requires a high index contrast technology in order to avoid excessive bending losses. The trade-off between large FSR and low resonator loss can be circumvented by using more than one ring. When two rings of different radii are used, the Vernier effect [27–29] causes the total FSR to be a multiple of the respective single ring FSRs according to:

$$FSR_{tot} = N \cdot FSR_1 = M \cdot FSR_2 \tag{8.10}$$

with N, M being integers without common divisor (cf. Sect. 8.6.4).

8.3.2 Geometry of Single Microrings

Microrings can be designed and fabricated in two generic coupling configurations as shown in Fig. 8.8. When the ring and waveguide are structured in the same waveguiding layer, the configuration is called 'laterally coupled'. When the ring and the waveguides are in different layers, the configuration is called 'vertically coupled'. The vertical coupling configuration has the advantage that the coupling depends mainly on the thickness of the layer in between which can be controlled very accurately during deposition, however, at the expense of an additional processing step for the ring layer. The lateral configuration uses a single layer only, but requires very accurate lithography and etching processes to open the gap between the straight waveguide parts and the ring with high precision. Another advantage of the vertical configuration is that the ring and waveguide layers do not have to be the same thickness which enhances the design freedom.

The fabrication of microring resonators is limited by the tolerances of the lithography and the etching processes. For the realization of the gap between the ring and the straight waveguide in a lateral configuration and using high index contrast, nanometre precision is required which can only be obtained by direct e-beam writing, focused ion beam milling, or high precision wafersteppers. In the vertical configuration this problem is circumvented as the coupling is now determined by the deposition process where nanometre precision can be obtained more easily. In this case, however, alignment is an issue as ring- and port waveguides are structured in two separate lithographic steps. This is especially important in the case of symmetrically coupled devices.

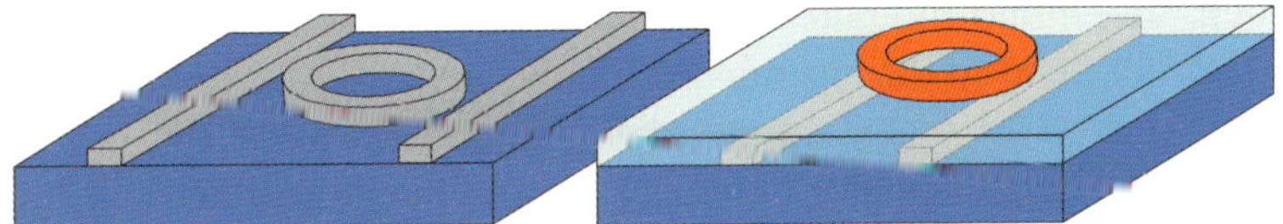

Fig. 8.8. 3D drawing of ring laterally coupled (*left*) and vertically coupled (*right*) to the straight waveguides

The optical modes of the straight waveguide cross-section and the bend cross-section can be calculated by the use of (bend-) mode solvers. Since the two modes are different in behaviour, they can not be calculated in a single step. In addition, different simulation tools may be needed. The most relevant parameters obtained by these calculations are i) the effective index, ii) the propagation and bending loss of the mode, and iii) the number of higher order modes besides the desired fundamental one. Figure 8.9 shows an example of a bend mode and a straight waveguide mode calculated by a commercial mode solver [30]. The deformed mode profile with a shift to the outer rim of the curved waveguide (i. e. the right side in Fig. 8.9) is clearly visible.

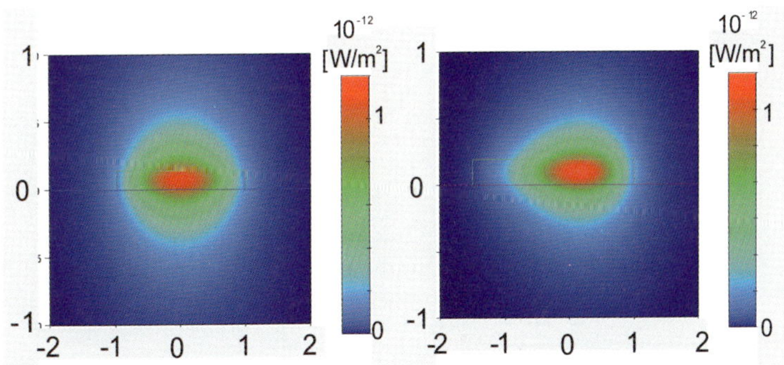

Fig. 8.9. Optical mode of straight waveguide (*left*) and bend (*right*)

An important design consideration is the phase-matching between the mode that propagates within the ring resonator and the mode of the port waveguides. Ideally the effective indices of these modes should be the same. In general, however, this is not easily achieved because the index of the mode in the micro-resonator needs to be relatively high to keep the radiation-induced bending losses at an acceptable level. The effective index of the bend mode is therefore generally higher than the index of the straight waveguide which is bound by the condition of mono-modality. Although the mono-modality condition is also important for the ring, a higher order bend mode has rapidly increasing losses.

The mismatch in phase causes the coupling between the port waveguides and the ring resonator to be less efficient, as illustrated in Fig. 8.10, which shows calculations for a coupler which consists of a straight waveguide vertically coupled to a bent waveguide as schematically given in the right part of the figure. The vertical distance between the two is $1\,\mu m$, the lateral distance is the gap. The calculation assumes waveguide indices of 1.97, an index of 1.45 for the surrounding, and a radius of $50\,\mu m$. According to Fig. 8.10 the maximum achievable coupling constant gets the lower the larger the phase mismatch. In the modelling described above losses inside the coupler region were not taken into account. However, these can be included by extending the scattering matrix as described e. g. in [31].

The dependence of bending losses on index contrast is illustrated in Fig. 8.11. Parameters for the calculation are $n_{sep} = n_{buff} = 1.45$ and $n_{wg} = 1.97$ (corresponding to a nitride waveguide). As can be seen the bending losses

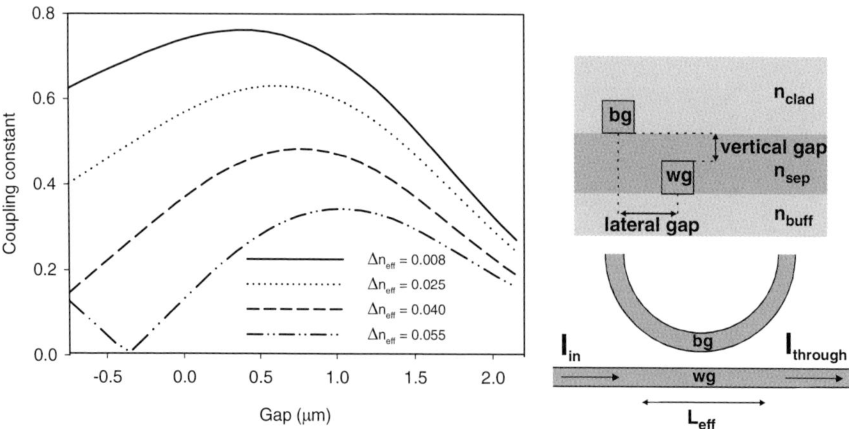

Fig. 8.10. Effect of phase-mismatch (Δn_{eff}) on the coupling constants for various lateral gaps

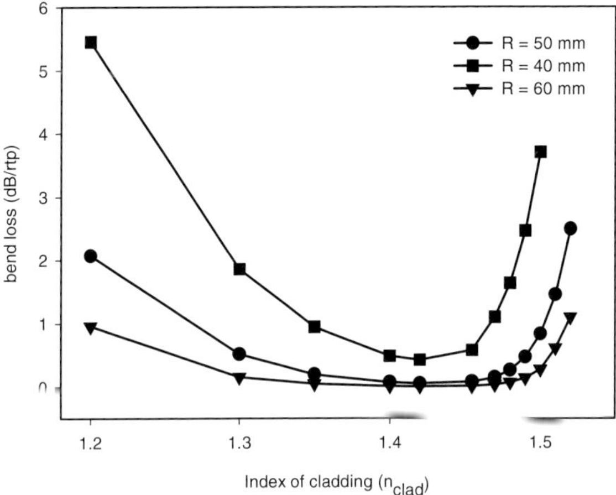

Fig. 8.11. Bending losses of SiON microrings of different radii for different cladding indices

have a minimum around $n_{clad} = 1.45$. Furthermore, when the contrast decreases the bending losses increase. So higher index contrasts allow smaller radii which implies larger FSR. However, when the index of the cladding layer is increased over that of the buffer and separation layers, the losses increase again, counter-intuitively. But as explained in [32] the lateral contrast decreases in this case which causes the losses to increase.

8.3.3 Group Delay and Dispersion

For applications in telecommunication the time behaviour of the microring or alternatively the frequency response to a time-varying signal is of great importance. Since the filter is a resonant filter, the delay depends on the frequency (wavelength) position with respect to the resonance. This dependence can conveniently be described by the structural or quadratic dispersion D [9]. This dispersion is the second derivative of the transmission phase-response $\varphi(\omega)$ with respect to frequency. The normalized group-delay τ_n is the negative derivative of the phase-response with T being the inverse of the FSR:

$$\tau_n = -\frac{d}{d\omega}\varphi(\omega)$$

$$D = -c\frac{T}{\lambda^2}\frac{d}{d\omega}\tau_g(\omega)$$

(8.11)

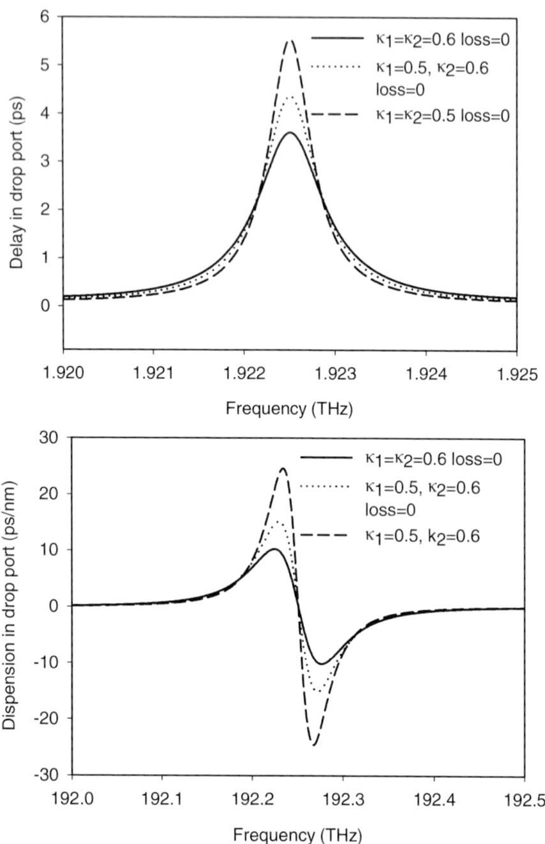

Fig. 8.12. Simulated absolute group delay (*top*) and dispersion (*bottom*) in microrings for various coupling constants

The absolute group delay is given by $\tau_g = T\tau_n$. Figure 8.12 shows a typical simulation of the group-delay and the dispersion response of the drop port of a single ring with radius 50 µm. In the symmetric case ($\kappa_1 = \kappa_2$) the through port exhibits the same response as the drop port.

According to Fig. 8.12 the absolute delay increases when the coupling constants decrease (in that case F and Q increase) since more time is needed to 'build up' the signal. Also non-symmetric coupling constants increase the delay as can be seen comparing the dotted line with the solid one. This can be intuitively explained since raising one of the coupling constants effectively decreases the finesse. At these higher coupling constants loss does not have much influence on the drop phase response. Larger ring radii cause higher delays as it takes longer to get into resonance.

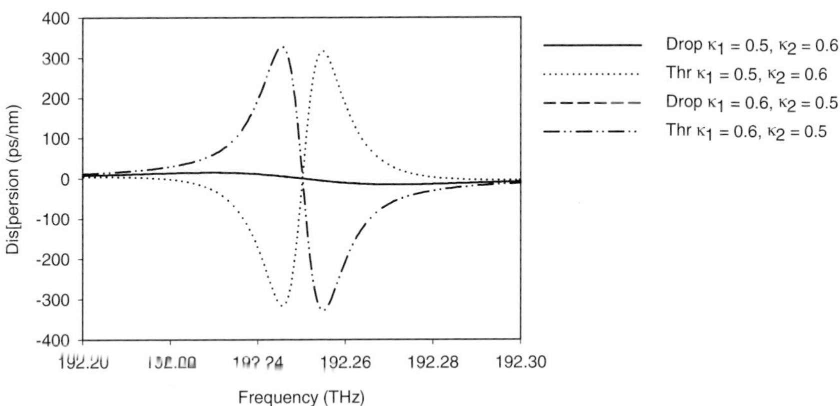

Fig. 8.13. Dispersion in drop and through port for two different kinds of asymmetry

Figure 8.12 also shows the simulated dispersion for various coupling constants and the effect of asymmetry. In the case of symmetric coupling ($\kappa_1 = \kappa_2$) the total dispersion experienced by the signal leaving the MR from the drop and the through port is moderate, and it is identical for both ports. However, in the case of asymmetric coupling ($\kappa_1 \neq \kappa_2$) the dispersion observed from the through port gets significantly larger (note the different scales of Figs. 8.12 and 8.13!), and the shape of the dispersion curve is inverted if κ_1 and κ_2 are interchanged. This is illustrated in Fig. 8.13. The minimum phase filter becomes a maximum phase filter [9]. This effect can be used in filters used to optimize the time domain parameters of a signal i. e. dispersion compensators.

8.3.4 Other Micro-resonator Geometries

The shape of MRs is evidently not restricted to a circle. Nearly any geometrical path that provides optical feedback will act like the microring. A frequently used geometry is the racetrack [33] which is shown in Fig. 8.14. In this geometry the couplers of the resonator are straight waveguides that allow accurate control of the coupling constants at the expense of being somewhat larger and consequently having a reduced FSR.

Another common shape is the disk [34] instead of a ring as shown in Fig. 8.14. A disk is more difficult to make single mode since the lateral width of the cross section is large. But since the lateral contrast of the disk is higher than the contrast for a ring, potentially lower losses can be obtained.

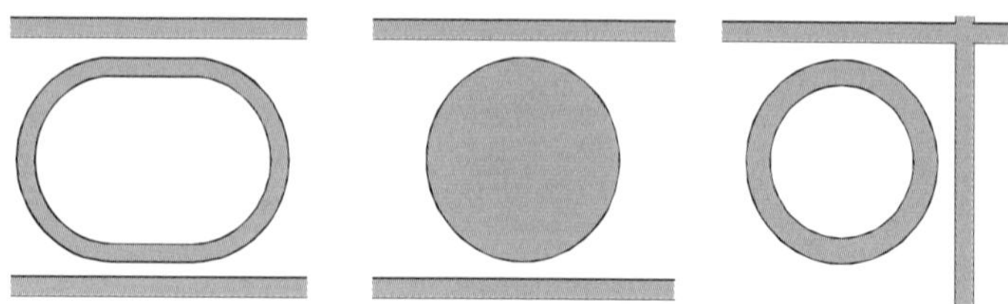

Fig. 8.14. Other microring resonator shapes: Racetrack (*left*), disk (*middle*), and Manhattan configuration (*right*)

The place of the adjacent waveguides is not important. As can be seen from (8.8), only the total roundtrip length influences the response, but not the place of the coupling regions. An additional geometry is the so-called Manhattan configuration as shown in Fig. 8.14. This structure allows for optimum use of area and is scalable to a large extent.

8.4 Tuning and Modulation of Microrings

Until now the ring was assumed to be totally passive, i.e. all geometrical and materials parameters being constant in time. To add functionality or to overcome fabrication errors, the MR can be made active by varying (tuning) some of the parameters and consequently its wavelength response. Furthermore, additional functionality can be added by externally tuning the roundtrip loss or coupling constants of the ring.

8.4.1 Resonance Wavelength Tuning

The resonant wavelength of a MR can be tuned in several ways. The most straightforward approach is to change the effective index of the ring by any of the following means:

- Thermo-optic effect: by applying heat to the ring the refractive index of the material changes [35]
- Electro-optic effect: an electrical field causes a change in refractive index [36]
- Carrier injection: optical pumping creates free carriers (single photon- or two-photon absorption) which change the loss parameter and the refractive index of the material [37, 38]
- Changing the material (sensor) [39, 40]
- Opto-optical effect: the light itself causes a change in index via nonlinear effects [11]

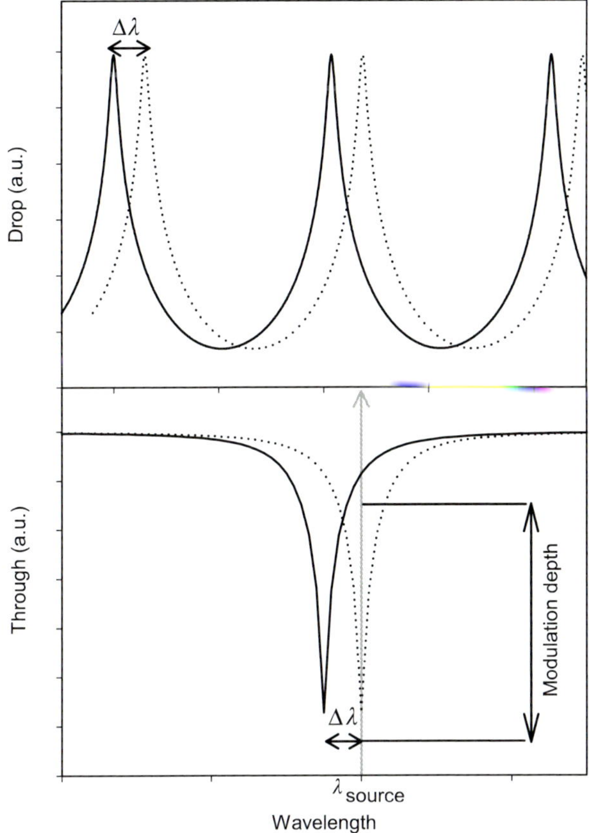

Fig. 8.15. Tuning the passband of a MR filter to different wavelengths (*top*) and modulation by tuning the MR wavelength response for fixed operation wavelength (*bottom*)

A change in effective index shifts the total wavelength response by an amount of $\Delta\lambda$. This shift can be used in filter applications to tune the passband to a desired wavelength. The principle of tuning is shown in Fig. 8.15.

The same phenomenon leads to a different functionality when the operation wavelength λ_{source} is kept constant. A small change in effective index around the steep slope of the MR response induces a modulation at λ_{source} with high extinction ratio.

The thermo-optic effect can be used by applying a thin film heater on top of the ring. The heater must be placed as close as possible to the ring to minimize the driving power. On the other hand, the heater can not be placed too close since the metal from the heater will induce large additional losses inside the ring.

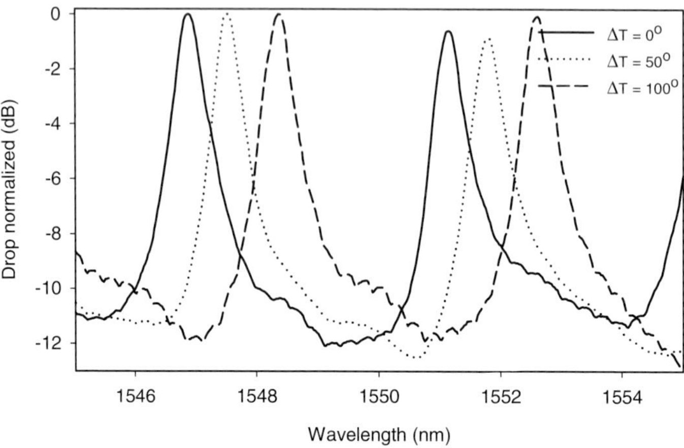

Fig. 8.16. Shift of drop response as function of heater temperature of a SiO_2/Si_3N_4 MR with radius 50 μm

The shift in centre wavelength of the ring $\Delta\lambda_c$ is a function of the difference in effective index induced by heating the device, according to:

$$\Delta\lambda_c = \frac{\lambda\Delta n_{eff}}{n_{eff}} \tag{8.12}$$

The change in effective refractive index induced by the heater depends largely on the materials used and the distance between the heater and the ring. Glass-like materials have a temperature-dependence of the refractive index (dn/dT) of around 10^{-5}, whereas polymers have a coefficient being an order of magnitude higher. Thermal shifts of 20 pm/mW have been shown for glass-based MRs [41], see Fig. 8.16. For polymer MRs this shift is about 1 nm/mW [42], i. e. around two orders of magnitudes higher which can be explained by the larger thermo-optic coefficient and a better conductance of the heat as described in [43]. In conclusion, thermal tuning is simple to implement, but the thermo-optic effect is slow with typical response times in the order of hundreds of microseconds. This can be sufficient for switching applications, but other effects must be used for modulation applications.

The electro-optic effect changes the i-th component of the refractive index n_i as function of the applied electric field E_k according to:

$$\Delta\left(\frac{1}{n_i^2}\right) = \sum_k r_{ik} E_k \tag{8.13}$$

with r_{ik} being the eo tensor [44, 45]. The direction of change of refractive index is determined by the material and can be positive as well as negative. Applying a reverse voltage will lead to the reversal of the effect. The electro-optic effect is inherently much faster than the thermo-optic effect. In practice now the limiting factor will be the 'build-up' time of the resonance given by

$$\tau_{cav} = \frac{F \cdot R \cdot n_g}{c} \tag{8.14}$$

For well designed ring resonators electro-optic modulation speeds of several tens of GHz can be expected. In recent experiments [45] modulation up to 1 GHz has been demonstrated in a Mach–Zehnder interferometer loaded on one arm with a microring. The device has been fabricated in PMMA-DR1 polymer, the relevant electro-optic coefficient is $r_{33} = 10\,\text{pm/V}$ [44, 45].

In semiconductor materials, with highly confining waveguides, which have a bandgap energy higher than the photon energy, it is possible to change the refractive index by injecting free carriers [46]. An empirical relation between the change in refractive index and the carrier density [47] leads to an index change of around −0.002.

For all-optical or opto-optical tuning mostly the Kerr non-linearity of a material is used [11] i. e. the total effective refractive index of the material is dependent on the intensity of the light (I_{eff}) :

$$n_{eff} = n_{eff,0} + n_{eff,2} I_{eff} \tag{8.15}$$

where $n_{eff,2}$ is a parameter dependent on the Kerr non-linearity. Recently, GHz modulation in silicon-on-insulator (SOI) MRs based on this effect has been shown [38].

8.4.2 Resonance Wavelength Trimming

Statistical errors in the radius during fabrication can be cancelled by post-deposition trimming. In the methods described above an active function is introduced to change the resonant wavelength. When permanent changes are needed in order to compensate for fabrication errors, other methods are required. The effective index of the ring can e. g. be altered by locally injecting high optical power by a laser. This post-fabrication process can be used to trim every MR to the proper position [48]. For low-cost mass production, however, the need of trimming should be completely avoided.

8.4.3 Loss Parameter Tuning

In the previous section methods of tuning were discussed where the optical roundtrip path length was changed which resulted in a shift of the resonant wavelength. Alternatively, the roundtrip loss can be changed which affects the Q-factor and the shape of the resonance curve. In this way a MR can be used as space-switch: at the resonance in high Q, weakly coupled MRs practically all power is directed to the drop port, but at low Q to the through port. There are several methods of controlling the Q of the cavity dependent on the materials in the system. In semiconductor devices electro-absorption or free carrier injection can be used to control the loss parameter [46]. Other materials like polymers and glasses allow for doping with rare-earth ions like Erbium [49]. In these materials the losses can be largely reduced by optical pumping. In this way switching of light by light is feasible, i. e. opto-optical switching.

8.4.4 Coupling Constant Tuning

Finally, the individual coupling constants can also be tuned. This has been done mainly in relatively large racetrack structures where enough coupling length is present to allow for thermo-optic tuning restricted to the coupling sections [50]. Furthermore, some more exotic methods have been demonstrated making use of MEMS (Micro Electro-Mechanical Systems) technology. Fleming et al. changed the coupling between port waveguides and ring by moving the ring, which was attached to a membrane, in a controlled way by electro-static forces to the evanescent field of the waveguides underneath. In this way light at the resonance wavelength is coupled from one waveguide to the other [51].

8.5 Characterisation Methods

The full characterisation of microring-based devices requires the determination of their amplitude and phase characteristics (as outlined in detail in Chap. 2). Wavelength-dependent measurements of transmission, add and drop characteristics, and of crosstalk properties as well can be done using broadband or tuneable narrow-band light sources. Essential points have been outlined in Chap. 4, Sect. 4.7 for the characterisation of AWGs and hold for microrings as well. Particular requirements for the investigation of high index-contrast devices are sub-μm precision with negligible drift over tens of minutes.

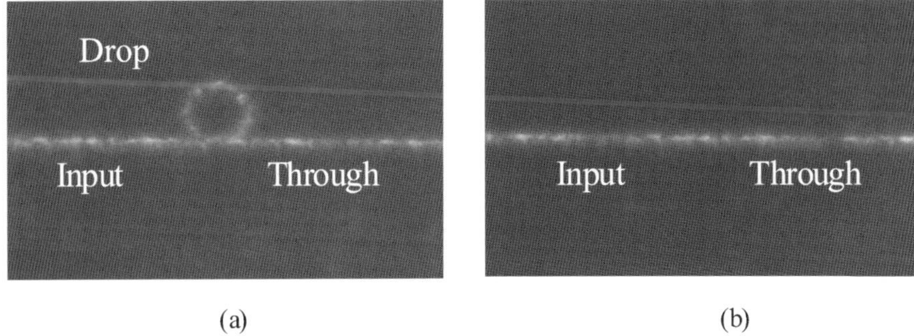

Fig. 8.17. IR Camera image of MR on-resonance (*left*) and off-resonance (*right*)

Another characterisation method is a lens system and an IR camera placed on top of the device under test. In this way scattered light can be captured, and by image processing also wavelength scans of particular parts of the image can be made [52]. The technique provides the possibility to measure every point on the device, and this is particularly useful in more complex systems where not all inputs and outputs of the individual rings are directly connected to chip input/output ports. With the camera technique one can look 'in between' rings, and on-chip insertion loss can be assessed rather easily as scattered intensity inside the waveguides decreases along the propagation direction. The principle is limited to devices or channels that exhibit sufficient scattering, thus it is primarily suited for device characterisation in an early development stage. Figure 8.17 shows an example of a SiO_2/Si_3N_4 MR with a radius of 25 µm on resonance (left picture) and off resonance captured by an IR camera setup.

The phase characteristics of microring-based devices can be measured by techniques such as the phase-shift method [53, 54] or by optical low-coherence reflectometry (OLCR) which are discussed in detail in Chap. 2, Sects. 2.3.1 and 2.3.2.

OLCR measurements can be performed in reflection and in transmission as well. The distance between consecutive interference peaks is a measure for the FSR, while the exponential envelope of the interference peaks e. g. provides information on the coupling constants and the losses [17, 50, 55].

8.6 Multiple Ring Resonator Devices

For many applications, in particular in fibre optic communication, the characteristics of single microrings are not well suited. However, much more favourable device performance can be achieved if a number of microrings are combined, and this can be done in a variety of architectures. This will be outlined in more detail in the following section.

8.6.1 Higher Order Filters

Higher order MR filters are devices where several feedback paths contribute to the response of the filter. Figure 8.18 shows two different setups, normally designated as serial cascade [9,56–60] and parallel cascade [52,61–66] in the literature. In both cases, the additional feedback paths result in an improved filter response, i. e. a more flat-top response in the pass-band and a steeper roll-off at the pass-band edges. The design of a serial cascade using the z-transform approach is described in detail e. g. in [9].

Concerning the parallel cascade a number of different design concepts have been reported in the literature. Examples are considering the cascade as a generalized DFB grating [63,66], describing the filter in terms of a prototype filter as known in the microwave domain [61] or using the scattering matrix description [64,65]. Recently, the Z-transform description of these filters has also been applied [67].

In the serially cascaded structure it is necessary to control the resonance wavelength of each ring in the filter accurately. In the parallel cascade the distance between the rings must be well controlled with sub-wavelength precision. In both cases the filter responses can be further improved by special designs of all the coupling constants in the cascade [66]. By apodizing the coupling constant for each individual ring along the array, even better filter shapes can be obtained.

Higher order filters in a serial cascade of 3 to 11 rings have been made by Little Optics in their Hydex^{TM} technology which enables the fabrication of lightwave circuits including microring structures with appealing characteristics. Hydex^{TM} is claimed to overcome all the limitations of other high-Δn systems, the details of which are not disclosed so far. However, it is deposited through a conventional chemical vapour deposition process, its

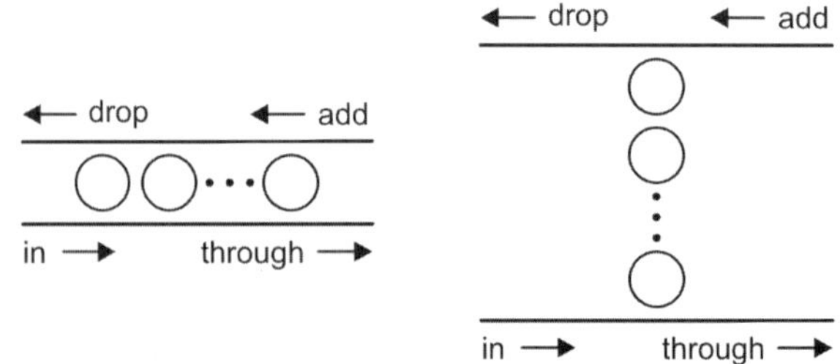

Fig. 8.18. Parallel- (*left*) and serially cascaded ring filters (*right*)

refractive index contrast is adjustable from 0% to over 20%, the fabrication does not require any annealing step, and the material absorption is low all over the S-, C-, and L band [68].

Figure 8.19 shows a serial cascade of three rings and the transmission spectra of 1st-, 3rd-, and 11th-order serial filters.

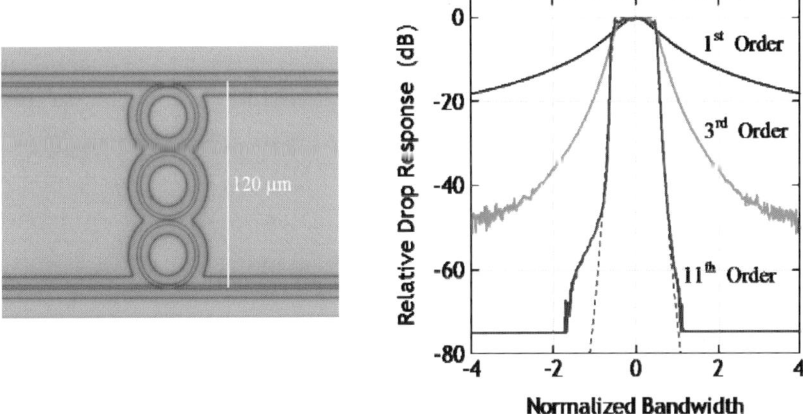

Fig. 8.19. (a) 3rd-order filter and **(b)** Transmission spectra of different order ring filters (Little Optics Inc.)

As can be seen, the roll-off of the filter characteristics can be freely adjusted by choosing an appropriate number of cascaded rings. Typical characteristics of these commercially available tuneable filters based on cascaded rings are compiled in Tab. 8.1.

Tab 8.1. Characteristics of tuneable ring filters (Little Optics Inc.)

Channel spacing	25	50	100	GHz
Tuning range		192.1 – 196.1		THz
3-dB pass-band	> 17	> 35	> 42	GHz
Insertion loss		< 5		dB
PDL		< 0.5		dB

8.6.2 Wavelength-selective MR Switch

A schematic drawing of a MR-based switch is shown in Fig. 8.20. This structure is based on two MRs connected through a straight waveguide and acts as a wavelength selective switch. The first MR selects a certain wavelength,

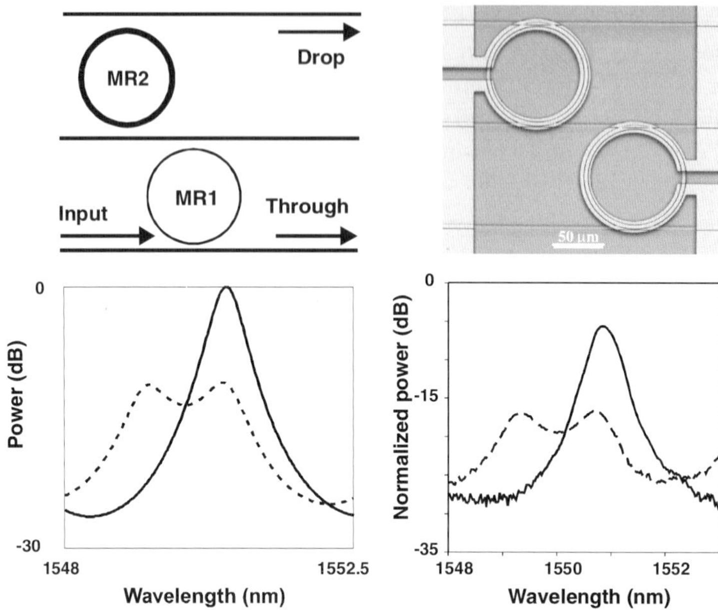

Fig. 8.20. Wavelength-selective switch based on two MRs: schematic lay-out and microscope picture of device (*top*) and wavelength response of the drop port (*bottom*). *Lower left*: simulated response of single ring; solid line: switch in ON-state, dotted line: switch in OFF-state. *Lower right*: experimental results

and the second ring, which is tuneable, thereafter switches this wavelength to the drop port or to a non-used channel leading to an absorber [54].

8.6.3 Reconfigurable Optical Add/drop Multiplexer

Another example of complex structures based on MRs is the $N \times M$ matrix switch based on the Manhattan configuration [69] as shown schematically in Fig. 8.21. This compact configuration allows adding and dropping any of the M wavelengths to any of the N input/output ports. In order to improve the performance each building block consisting of a single MR can be replaced by a higher order filter [69, 70].

The SiON technology looks promising as a platform for these complex MR structures. With a large index range between $1.45\,(SiO_2)$ and $2\,(Si_3N_4)$ SiON allows for relatively high index contrast, low-loss devices with small bending radius [19, 71]. This is a prerequisite for complex structures since multi-MR devices with large FSR can only meet the specifications when they are small in size and have low losses. It is therefore not surprising that quite some interest has been shown for MRs in the SiON technology, see

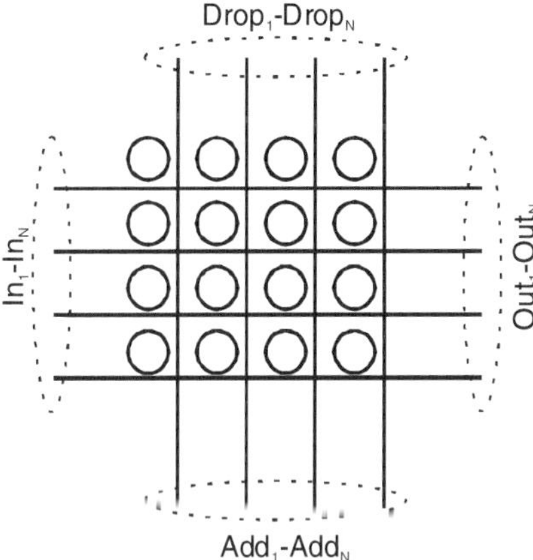

Fig. 8.21. $N \times M$ Manhattan microring configuration with N input/output channels and add/drop ports for each of the M wavelengths

for example [20, 62, 69, 72]. Some of the previous figures in this chapter (Fig. 8.1, Fig. 8.17 and Fig. 8.20) are MR structures in SiON technology as well. As another example in Fig. 8.22 a single row of a cross-grid structure already discussed above is shown. This structure, known as a Reconfigurable Optical Add/Drop Multiplexer (ROADM), is based on Si_3N_4 channels embedded in SiO_2. The individual rings are thermally tuneable across the full FSR by means of a chromium thin-film heater. In this way the ROADM can, for example, be switched performing as a 4 channel demultiplexer where every ring filters a particular wavelength from the in-port to the respective drop-ports. Alternatively, when the microrings are tuned to the same wavelength, the structure can be used as a single wavelength broadcasting device, directing one input wavelength to more than one drop port. This is shown in the upper right picture in Fig. 8.22, where the through port of the device is shown when the four rings are set to different wavelengths (4 single channels) or all at the same wavelength (single combined channel). The lower right picture displays the response for the four individual add ports when light enters one of the add ports and is directed to the through port. This demonstrates that every channel can add power to the through port.

The device was pigtailed and wire-bonded and can be controlled by a computer. It has an inter-channel crosstalk of 12 dB and can tune to any wavelength within the FSR of 4.2 nm at a maximum driving power of 380 mW per channel. The device shows 12 dB extinction per channel in the through port and 17 dB in the drop port [73], and it has been tested at 40 Gbit/s without showing reduction of performance [74].

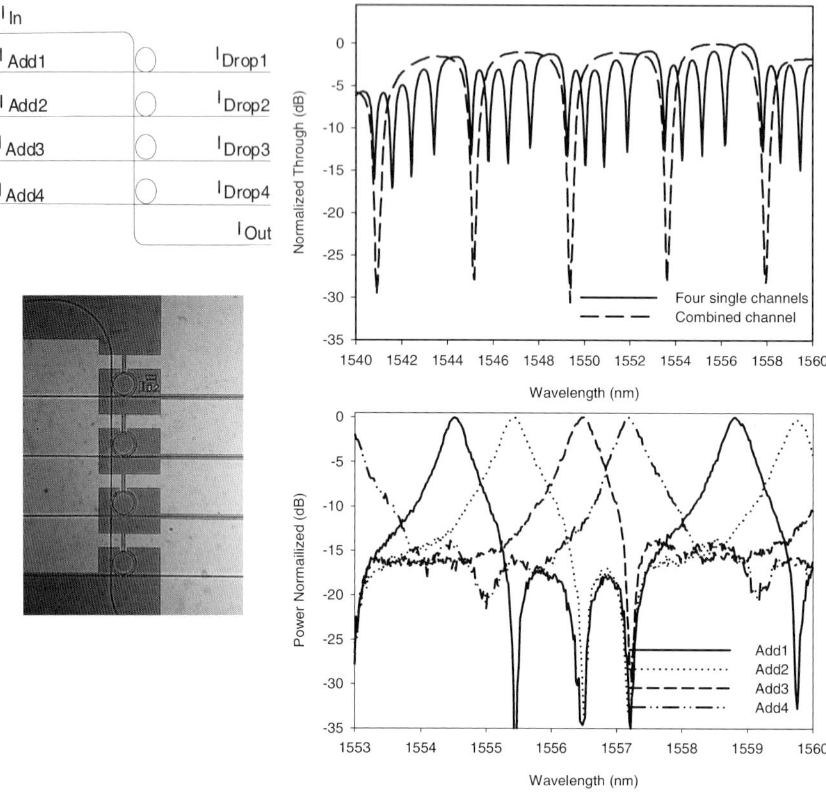

Fig. 8.22. Schematic (*top left*) and photograph (*bottom left*) of a SiO_2/Si_3N_4 MR based ROADM and the measured through (*top right*) and add (*bottom right*) response

Currently, there are several commercial suppliers of MR-based products in SiON technology. Two of them, Little Optics (recently acquired by Nomadics [68]) and Lambda Crossing [75], deliver products based on MR technology, whereas another one, LioniX [71], offers foundry services and MR technology for use by others.

8.6.4 Microring-based Filters with Extended FSR

For many applications in fibre optic systems the FSR of filters should be comparable to the width of the C- or the L-band, for example, i. e. several 10 nm. According to (8.1) large FSR values require small bend radii (where the actual numbers vary widely depending to the material system chosen), and achieving several 10 nm FSR is demanding due to technological restrictions.

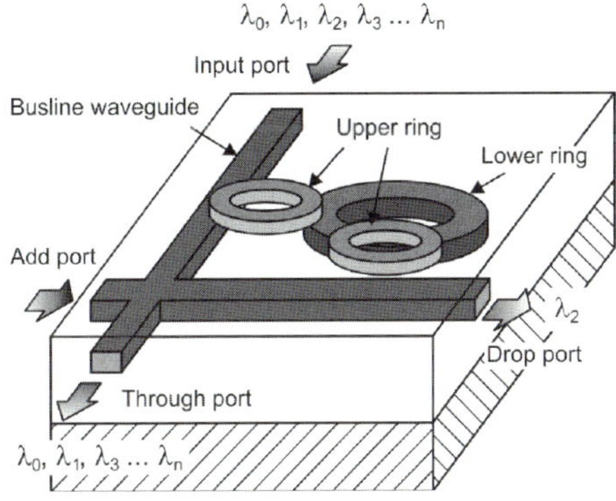

Fig. 8.23. Triply-coupled microring resonator filter in stacked configuration (after [77])

One means to increase the FSR without resorting to correspondingly small bend radii relies on the Vernier effect, i. e. microrings of different radius are combined, and the resulting device exhibits resonances only where the individual resonances of all microrings involved match. (The approach is similar to one concept to design widely tuneable lasers by using gratings with slightly different pitch which provide super modes with rather large wavelength separation [76], cf. also (8.10)). A corresponding microring-based device is illustrated in Fig. 8.23 [77].

The device has been fabricated in SiO_2/Si and the ring radii are 28.5 µm for the two smaller (upper) rings and 39.3 µm for the larger (lower) ring corresponding to a FSR of the individual rings of 8 and 6 nm, respectively, while the integrated filter exhibited about 20 nm FSR.

8.6.5 Microring-based Dispersion Compensators

Dispersion compensation is a functionality particularly needed in high-bit rate optical communication systems since the adverse effects of dispersion increase proportional to the square of the data rate. Dispersion compensation can be accomplished by many approaches, including bulk optics and in particular fibre Bragg gratings (cf. Chap. 5, Sect. 5.5), but planar waveguide technology as well. Examples of the latter are finite impulse response filters implemented as resonant couplers [78] or infinite impulse response (IIR) filters which exhibit periodic frequency characteristics and are thus particularly suited for dispersion compensation of multiple wavelength channels. Dispersion compensators using microrings rely on the all-pass filter characteristics of MR-based two-port structures, and one example of a corresponding dispersion-compensating filter is illustrated in Fig. 8.24 [79, 80].

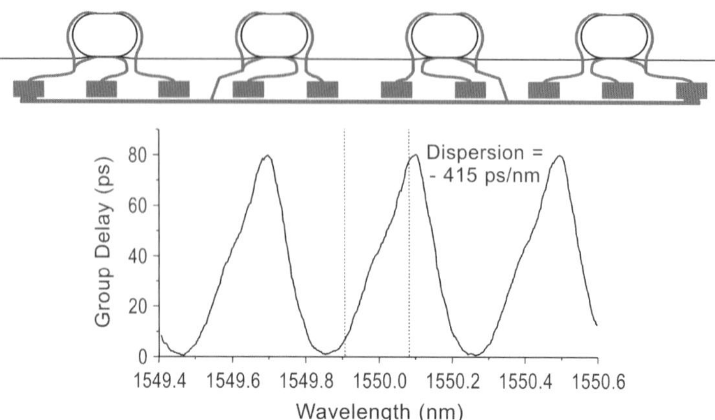

Fig. 8.24. Ring-resonator-based dispersion compensator. Upper part: device structure, lower part: device characteristics, after [80]

The tuneable dispersion compensator illustrated in Fig. 8.24 consists of four cascaded resonator loops and has been fabricated in high-refractive-index contrast SiON technology. An absolute index contrast of 0.05 or $\Delta = 3.3\%$ between waveguide core and cladding allows minimum bend radii of 0.8 mm with barely measurable radiation loss and 0.55 mm radii with <0.1 dB loss per $90°$ bend [80], and small ring radii are required in order to get large FSR values. The device illustrated in Fig. 8.24 is made adjustable by the implementation of two heaters per microring where one heater is used to vary the optical path length and the other to vary the strength of the coupler. In this way tuneable filters with linear delay slopes of $+100$, $+50$, -50, and -100 ps/nm were generated, but filters with quadratic dispersion-slope values of $+800$ and -800 ps/nm^2 as well. The technology lends itself to mass production and is expected to find applications in fibre optic systems operating at 10, 40, or even higher data rates.

8.7 Semiconductor-based Microrings

Microring- (including microdisk-) based devices have been fabricated in GaAs/AlGaAs [81], and in the GaInAsP/InP material system [82–85] as well. MRs in III-V semiconductors (SC) are considered attractive because they can be monolithically integrated with other (active) functionalities. They can be used as channel-dropping filters, WDM demultiplexers, as notch filters, for optical switching [86], or for all-optical wavelength conversion [81, 87].

Microrings and microdisks plus their straight coupling waveguides (WG) have on the one hand been fabricated as planar structures [81, 84], but in vertical coupler geometry as well [85]. The latter design is particularly favourable for the control of the coupling between the straight waveguides and the ring, however, the fabrication process is significantly more demanding (requires wafer bonding for example). The design of SC-based microrings has mostly been based on ridge waveguides (RWs) and to a much lower extent on buried waveguide structures [88] which are considerably more difficult to realize.

Layer composition, thicknesses, and WG geometry are determined on the basis of various, partly conflicting requirements: the straight input/-output WGs should enable efficient and alignment-tolerant coupling to optical fibres, coupling between the straight and the bent WGs should be efficient, guiding in the bent WGs (i.e. in the MR) should be strong enough to enable small bend radii. The geometry of the bent WGs may either be symmetric or asymmetric [82, 84] where a steeper etch at the outside wall of the RW-based MR structure assures a particularly large difference of the effective refractive indices and consequently the strong confinement of the guided mode which is required for low radiation losses in small bend radius microrings. An asymmetric, deeply etched bent WG is illustrated in Fig. 8.25.

As discussed in Sect. 8.2.3, microrings tend to be polarisation dependent, and this is particularly true for RW-based semiconductor microrings. In applications where the state of polarisation of a lightwave entering a microring is random, polarisation diversity could solve the problem, although at the expense of extra complexity. On the other hand, if the polarisation needs alignment anyway, for example in all-optical wavelength conversion, the polarisation dependence of the microring characteristics is less of an issue.

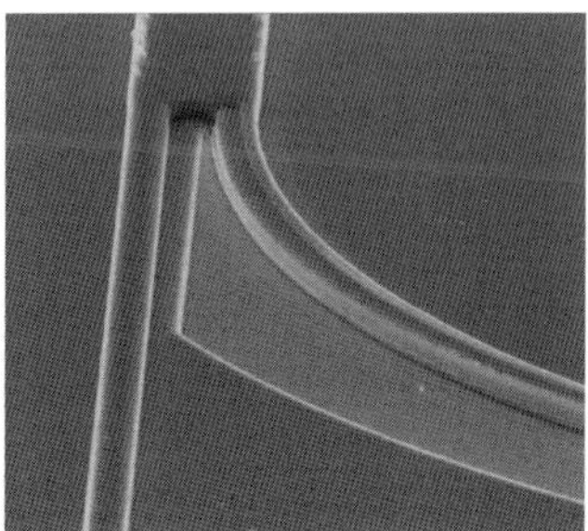

Fig. 8.25. SEM picture of bent WG region illustrating deep etch at the outside microring wall, after [84]

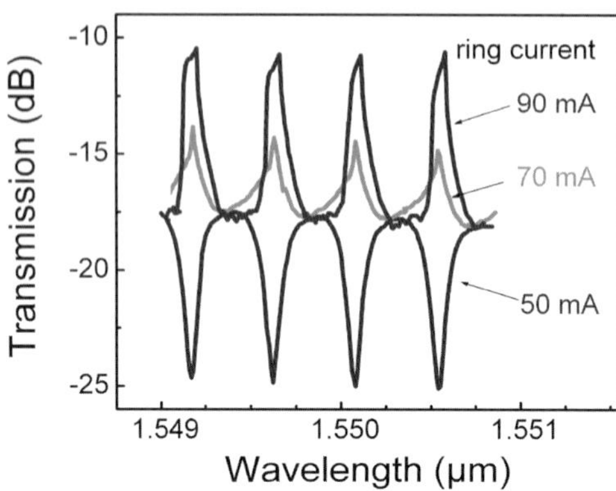

Fig. 8.26. Transmission characteristics of an active ring resonator (with SOAs) under different driving conditions, after [90]

Radiation losses as well as scattering losses due to sidewall roughness do essentially determine the residual loss in bent WGs [89], and they cannot be reduced easily, so that the losses may add up to an intolerable amount if a larger number of rings is cascaded in order to get flat-top and steep roll-off filter transmission characteristics. One means to overcome these problems is the incorporation of active parts in the microrings, i. e. adding semiconductor optical amplifier (SOA) sections which compensate WG losses. The effect of such SOAs is illustrated in Fig. 8.26. At low drive currents of the SOAs the microring acts as a wavelength-dependent filter with adjustable loss. If the current is increased so as to completely compensate for the losses within the ring, the structure gets an all-pass filter, and for even higher SOA currents the device becomes a ring laser.

Another application area of microrings, which is of general interest for fibre optic systems, is all-optical wavelength conversion since it can be made particularly efficient by an appropriately strong confinement of the optical waves within the microring [81, 87]. Figure 8.27 shows corresponding results.

At present it is difficult to predict the future role of semiconductor-based microrings in fibre optic communication systems. Although microrings offer a number of attractive properties, they will always be in competition with alternative solutions, and competitiveness will in the end essentially be determined by the ratio of performance to cost including aspects like device-, subsystem-, operation-, and surveillance cost.

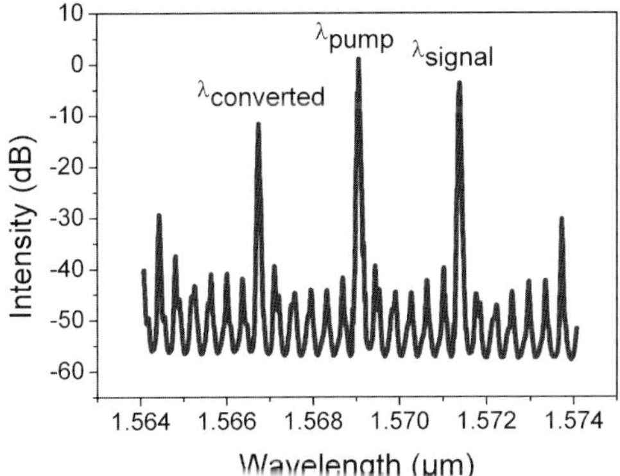

Fig. 8.27. All-optical wavelength conversion observed at the through-port of a micro–ring resonator, after [87]

On the other hand, microrings are also finding increasing interest in areas beyond telecom, in particular for sensor applications, which includes semiconductor-based devices, but microrings in other material systems as well.

8.8 All-optical Switching

All-optical switching has been performed in GaAs/AlGaAs [91] and more recently in SOI micro rings [38] as well. The device structure under investigation has been a single micro ring coupled to a straight waveguide using either horizontal [91] or vertical [38] coupling. According to the considerations in Sect. 8.4 one may either switch from on-resonance to off-resonance or the other way round by using any of the effects mentioned.

In the recently reported all-optical switching in SOI microrings [38] detuning is achieved by optical means, i. e. optically created free carriers modify the effective refractive index and provide a detuning of the MR resonances which is equivalent to switching. These experiments have aroused a lot of speculation about the prospects and the potential of semiconductor/silicon-based all-optical signal processing. However, before devices of the type described here might find their way into real and widespread applications, a number of topics need additional investigations and significant improvements: The wavelength to be switched and the wavelength inducing the detuning must both be accurately adjusted to a ring resonance. High finesse of the resonances makes the switching efficient, but is demanding with respect to wavelength accuracy. On the other hand, relaxed wavelength tolerances (lower finesse) raise the required switching power significantly. The requirements become even more demanding due to polarisation effects. In addition, the free carriers needed for changing the refractive index have been created so far by two-photon absorption, and this process requires high photon densities which in turn implies significant technological effort and cost. In the case of the SOI-ring-based switching 10 ps long pulses with 25 pJ energy were used, generated by a Ti:sapphire picosecond laser in combination with an optical parametric oscillator. Finally, the observed relatively long exponential relaxation ($\tau_{rel} = 450$ ps) of the SOI-device in combination with the short switching window of a few picoseconds is unfavourable for many application conditions. A better match between excitation and recovery time has been observed for the GaAs/AlGaAs microrings with about 50 ps recovery time which is attributed to strong surface-state recombination at the microring sidewalls, but even in this case the switching-window to recovery-time ratio is far from optimum.

Thus the all-optical switching experiments are encouraging, but there is still a very long way to go before all-optical signal processing might find its way into practical applications, which is true for III-V semiconductors and even more so for silicon-based optical switching, on the other hand the latter would have a tremendous economic impact.

8.9 Future Trends

Future developments of microring resonators are expected to go into different directions. One trend will be towards multi-stage filters with application-specific tailored characteristics. Devices will be mass-produced in non-expensive material systems, and consequently such filters are likely to find widespread applications.

Another trend will be towards MR-based devices which combine active and passive microrings and which can thus perform more complex functionalities. This will include active-passive functionalities on a single chip, but different kinds of MRs may equally well be part of even more complex hybrid (sub-)systems. One corresponding example is shown in Fig. 8.28 which illustrates a hybrid optical transceiver, comprising active, polymer and inorganic electro-optical materials, and passive, thermally tuneable microrings as well [92, 93].

Point-to-point links play a predominant role in current optical networks, but wavelength routing will become more significant in future optical networks, and under these circumstances reconfigurable wavelength-routers

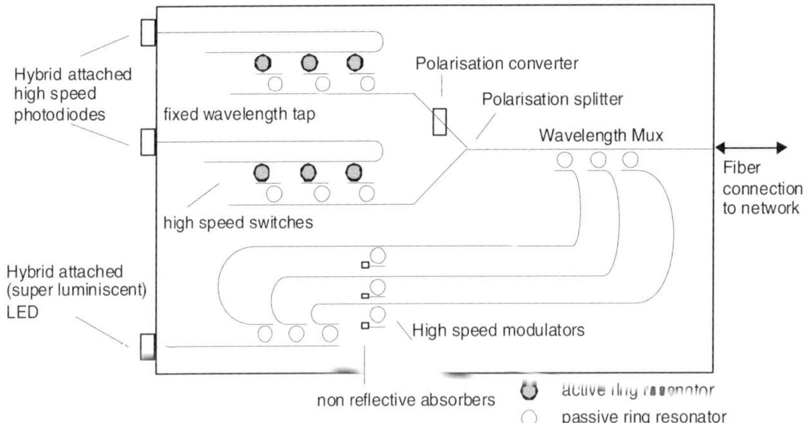

Fig. 8.28. Schematic overview of NAIS transceiver where active (high speed modulators) and passive MRs are combined to integrate different functions well (after [94])

will become particularly relevant. Key features of such routers are wavelength splitting (and combining) plus wavelength routing, they will be used in wavelength-selective and broadcast applications as well, and they should be wavelength agile to a high degree. Figure 8.29 shows a corresponding router architecture based on MR integrated in the so-called Manhattan architecture. Corresponding devices are currently under investigation in the Netherlands within the project Broadband Photonics (BBPhotonics), funded by the Freeband communications program [95].

There is no doubt that MRs offer a large potential for a variety of applications, however, the ultimate success will depend essentially on the extent to what technical performance and price targets can be met simultaneously which is true for all components in fibre optics.

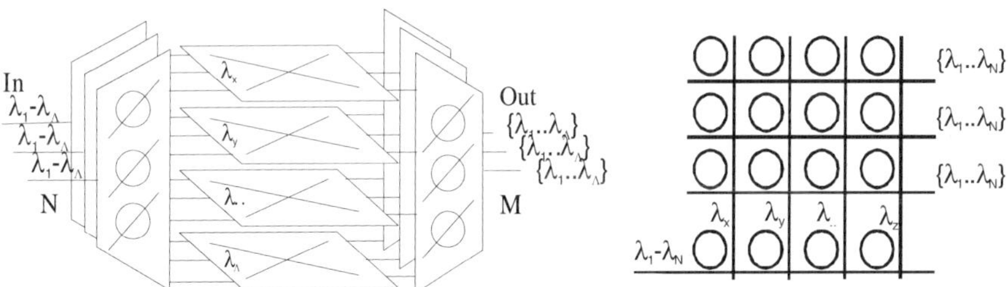

Fig. 8.29. Functional description of possible MR implementations (*left*) and schematic drawing of a MR based wavelength-router.

Acknowledgement

The authors would like to thank the following persons for their contribution to this chapter: Edwin Klein, Dion Klunder, Freddy Tan, Manfred Hammer, Arne Leinse, and Mart Diemeer, all from the University of Twente, The Netherlands, Dominik Rabus (Forschungszentrum Karlsruhe, Germany), and Brent E. Little (Little Optics Inc., Annapolis Junction, MD, USA).

References

1. A. Ashkin and J. M. Dziedzic: "Observation of optical resonances of dielectric spheres by light scattering," Appl. Opt. **20**, 1803–1814 (1981).
2. H. B. Lin, A. L. Huston, B. L. Justus, and A. J. Campillo: "Some characteristics of a droplet whispering-gallery-mode laser," Opt. Lett. **11**, 614–616 (1986)
3. S. C. Hill, D. H. Leach, and R. K. Chang: "Third-order sum-frequency generation in droplets: Model with numerical results for third- harmonic generation," J. Opt. Soc. Am. B **10**, 16–33 (1993)
4. M. M. Mazumder, S. C. Hill, D. Q. Chowdhury, and R. K. Chang: "Dispersive optical bistability in a dielectric sphere," J. Opt. Soc. Am. B **12**, 297–310 (1995)
5. J. Haavisto and G. A. Pajer: "Resonance effects in low-loss ring waveguides," Opt. Lett. **5**, 510–512 (1980)
6. W. Weiershausen and R. Zengerle: "Photonic highway switches based on ring resonators used as frequency-selective components," Appl. Opt. **35**, 5967–5978 (1996)
7. B. E. Little, S. T. Chu, W. Pan, and Y. Kokubun: "Microring resonator arrays for VLSI photonics," IEEE Photon. Technol. Lett. **12**, 323–325 (2000)
8. S. Suzuki, K. Shuto, and Y. Hibino: "Integrated-optic ring resonators with two stacked layers of silica waveguide on Si," IEEE Photon. Technol. Lett. **4**, 1256–1258 (1992)
9. J. H. Zhao and C. K. Madsen: *Optical Filter Design and Analysis* (Wiley, New York, 1999).
10. A. Driessen, D. H. Geuzebroek, H. J. W. M. Hoekstra, H. Kelderman, E. J. Klein, D. J. W. Klunder, C. G. H. Roeloffzen, F. Susanto, E. Krioukov, C. Otto, H. Gersen, N. F. van Hulst, and L. Kuipers: "Microresonators as building blocks for VLSI photonics," *AIP Conf. Proc.* **709** (F. Michelotti, A. Driessen, and M. Bertolotti, eds.), 1–18 (2003)
11. F. C. Blom, D. R. van Dijk, H. J. W. M. Hoekstra, A. Driessen, and T. J. A. Popma: "Experimental study of integrated-optics microcavity resonators: Toward an all-optical switching device," Appl. Phys. Lett. **71**, 747–749 (1997)
12. E. J. Klein, D. H. Geuzebroek, H. Kelderman, and A. Driessen: "Wavelength-selective switch using thermally tunable microring resonators," *Proc. IEEE Laser&Electro-Optics Soc. Annual Meeting* (LEOS 2003) (Tucson, AZ, USA, 2003) paper MM1 (2003)
13. J. T. Verdeyen: *Laser Electronics* (Prentice Hall, New Jersey, 1995)

14. H. Kogelnik: "Theory of optical waveguides" in *Guided-Wave Optoelectronics* (T. Tamir, ed.), Chap. 2 (Springer, Heidelberg, Berlin, 1988)

15. B. E. Little and S. T. Chu: "Theory of polarization rotation and conversion in vertically coupled microresonators," IEEE Photon. Technol. Lett. **12**, 401–403 (2000)

16. M. C. Larciprete, E. J. Klein, A. Belardini, D. H. Geuzebroek, A. Driessen, and F. Michelotti: "Polarization conversion in vertically coupled Si_3N_4/SiO_2 micro-ring resonators," *AIP Conf. Proc.* **709** (F. Michelotti, A. Driessen, and M. Bertolotti, eds., 415–416 (2003)

17. A. Melloni, F. Morichetti, and M. Martinelli: "Polarization conversion in ring resonator phase shifters," Optics Lett. **29**, 2785–2787 (2004)

18. S. S. A. Obayya, B. M. A. Rahman, K. T. V. Grattan, and H. A. El-Mikati: "Beam propagation modeling of polarization rotation in deeply etched semiconductor bent waveguides," IEEE Photon. Technol. Lett. **13**, 681–683 (2001)

19. K. Wörhoff, L. T. H. Hilderink, A. Driessen, and P. V. Lambeck: "Silicon oxinitride – a versatile material for integrated optics applications," J. Electrochem. Soc. **149**, F85– F91 (2002)

20. B. E. Little, S. T. Chu, P. P. Absil, J. V. Hryniewicz, F. G. Johnson, F. Seiferth, D. Gill, V. Van, O. King, and M. Trakalo: "Very high order microring resonator filters for WDM application," IEEE Photon. Technol. Lett. **16**, 2263–2265 (2004)

21. Y. Kokubun, S. Kubota, and S. T. Chu: "Polarisation-independent vertically coupled microring resonator filter," Electron. Lett. **37**, 90–92 (2001)

22. D. J. W. Klunder, C. G. H. Roeloffzen, and A. Driessen: "A novel polarization-independent wavelength-division-multiplexing filter based on cylindrical microresonators," IEEE J. Select. Topics Quantum Electron. **8**, 1294–1299 (2002)

23. J. Capmany and M. A. Muriel: "A new transfer-matrix formalism for the analysis of fiber ring resonators – compound coupled structures for FDMA demultiplexing," J. Lightwave Technol. **8**, 1904–1919 (1990).

24. O. Schwelb: "Transmission, group delay, and dispersion in single-ring optical resonators and add/drop filters – a tutorial overview," J. Lightwave Technol. **22**, 1380–1394 (2004)

25. A. Vorckel, M. Mönster, W. Henschel, P. Haring Bolivar, and H. Kurz: "Asymmetrically coupled silicon-on-insulator microring resonators for compact add-drop multiplexers," IEEE Photon. Technol. Lett. **15**, 921–923 (2003)

26. D. J. W. Klunder, E. Krioukov, F. S. Tan, T. van der Veen, H. F. Bulthuis, G. Sengo, C. Otto, H. J. W. M. Hoekstra, and A. Driessen: "Vertically and laterally waveguide-coupled cylindrical microresonators in Si_3N_4 on SiO_2 technology," Appl. Phys. B **73**, 603–608 (2001)

27. K. Oda, N. Takato, and H. Toba: "A wide-FSR wave-guide double-ring resonator for optical FDM transmission-systems," J. Lightwave Technol. **9**, 728–736 (1991)

28. S. Suzuki, K. Oda, and Y. Hibino: "Integrated-optic double-ring resonators with a wide free spectral range of 100 GHz," J. Lightwave Technol. **13**, 1766–1771 (1995)

29. O. Schwelb: "A design for a high finesse parallel-coupled microring resonator filter," Microwave Opt. Technol. Lett. **38**, 125–129 (2003)

30. C2V, Enschede, The Netherlands. www.c2v.nl

31. G. Cusmai, F. Morichetti, P. Rosotti, R. Costa, and A. Melloni: "Circuit-oriented modelling of ring-resonators," Optics and Quantum Electron. **37**, 343–358 (2005)

32. D. J. W. Klunder: *Photon physics in integrated optics microresonators*, (PhD Thesis, University of Twente, The Netherlands, 2002)

33. R. Grover, T. A. Ibrahim, T. N. Ding, Y. Leng, L.-C. Kuo, S. Kanakaraju, K. Amarnath, L. C. Calhoun, and P.-T. Ho: "Laterally coupled InP-based single-mode microracetrack notch filter," IEEE Photon. Technol. Lett. **15**, 1082–1084 (2003)

34. S. C. Hagness, D. Rafizadeh, S. T. Ho, and A. Taflove: "FDTD microcavity simulations: Design and experimental realization of waveguide-coupled single-mode ring and whispering-gallery-mode disk resonators," J. Lightwave Technol. **15**, 2154–2165 (1997)

35. M. B. J. Diemeer: "Polymeric thermo-optic space switches for optical communications," Opt. Materials **9**, 192–200 (1998)

36. I. L. Gheorma and R. M. Osgood, Jr.: "Fundamental limitations of optical resonator based high-speed EO modulators," IEEE Photon. Technol. Lett. **14**, 795–797 (2002)

37. T. A. Ibrahim, W. Cao, Y. Kim, J. Li, J. Goldhar, P.-T. Ho, and C. H. Lee: "All-optical switching in a laterally coupled microring resonator by carier injection," IEEE Photon. Technol. Lett. **15**, 36–38 (2003)

38. V. R. Almeida, C. A. Barrios, R. R. Panepucci, and M. Lipson: "All-optical control of light on a silicon chip," Nature **431**, 1081–1084 (2004)

39. E. Krioukov, D. J. W. Klunder, A. Driessen, J. Greve and C. Otto: "Sensor based on an integrated optical microcavity," Opt. Lett, **27**, 512–514 (2002)

40. E. Krioukov, D. J. W. Klunder, A. Driessen, J. Greve, and C. Otto: "Integrated optical microcavities for enhanced evanescent-wave spectroscopy," Opt. Lett. **27**, 1504–1506 (2002)

41. D. H. Geuzebroek, E. J. Klein, H. Kelderman, and A Driessen: "Wavelength tuning and switching of a thermo-optic microring resonator," *Proc. 11th Europ. Conf Integr. Optics* (ECIO'03), Prague, Czech Republic, 395–398 (2003)

42. P. Rabiei and W. H. Steier: "Tunable polymer double micro-ring filters," IEEE Photon. Technol. Lett. **15**, 1255–1258 (2003)

43. M. B. J. Diemeer: "Organic and inorganic glasses for microring resonators," *AIP Conf. Proc.* **709** (F. Michelotti, A. Driessen, and M. Bertolotti, eds.), 252–267 (2003)

44. P. Gunter: *Nonlinear Optical Effects and Materials* (Springer, Berlin, Heidelberg, 2000)

45. A. Leinse, M. B. J. Diemeer, A. Rousseau, and A. Driessen: "A novel high-speed polymeric eo modulator based on a combination of a microring resonator and an MZI," IEEE Photon. Technol. Lett. **17**, 2074–2076 (2005)

46. K. Djordjev, S. J. Choi, and P. D. Dapkus: "Active semiconductor microdisk devices," J. Lightwave Technol. **20**, 105–113 (2002)

47. L. A. Coldren and S. W. Corzine: *Diode lasers and photonic integrated circuits* (Wiley Series in Microwave and Optical Engineering, Wiley, New York, 1995)

48. H. Haeiwa, T. Naganawa, and Y. Kokubun: "Wide range center wavelength trimming of vertically coupled microring resonator filter by direct UV irradiation to SiN ring core," IEEE Photon. Technol. Lett. **16**, 135–137 (2004)

49. R. Dekker, D. J. W. Klunder, A. Borreman, M. B. J. Diemeer, K. Wörhoff, A. Driessen, J. W. Stouwdam, and F. C. J. M. van Veggel: "Stimulated emission and optical gain in LaF$_3$: Nd nanoparticle-doped polymer based waveguides," Appl. Phys. Lett. **85**, 6104–6106 (2004)

50. D. G. Rabus: Realization of optical filters using ring resonators with integrated semiconductor optical amplifiers in GaInAsP/InP (PhD Thesis, Berlin University of Technology, Germany, 2002)

51. M. C. Flemings, MR with MEMS, Sandia National Laboratory, www.sandia.gov

52. F. S. Tan: *Integrated Optical Filters based on Microring Resonators*, (PhD Thesis, University of Twente, The Netherlands, 2004)

53. D. Marcuse: *Principles of Optical Fiber Measurements* (Academic Press, New York, 1981)

54. D. H. Geuzebroek, E. J. Klein, H Kelderman, N. Baker, and A. Driessen: "Compact wavelength-selective switch for gigabit filtering in access networks," IEEE Photon. Technol. Lett. **17**, 336–338 (2005)

55. A. Küng, J. Budin, L. Thévenaz, and Ph. A. Robert: "Optical fiber ring resonator characterization by optical time-domain reflectometry," Opt. Lett. **22**, 90–92 (1997)

56. J. V. Hryniewicz, P. P. Absil, B. E. Little, R. A. Wilson, and P.-T. Ho: "Higher order filter response in coupled microring resonators," IEEE Photon. Technol. Lett. **12**, 320–322 (2000)

57. R. Orta, P. Savi, R. Tascone, and D. Trinchero: "Synthesis of multiple-ring-resonator filters for optical systems," IEEE Photon. Technol. Lett. **7**, 1447–1449 (1995)

58. D. G. Rabus, M. Hamacher, and H. Heidrich: "Resonance frequency tuning of a double ring resonator in GaInAsP/InP: Experiment and simulation," Jpn. J. Appl. Phys. **41**, 1186–1189 (2002)

59. C. K. Madsen and J. H. Zhao: "A general planar waveguide autoregressive optical filter," J. Lightwave Technol. **14**, 437–447 (1996)

60. B. E. Little, S. T. Chu, P. P. Absil, J. V. Hryniewicz, F. G. Johnson, F. Seiferth, D. Gill, V. Van, O. King, and M. Trakalo: "Very high-order microring resonator filters for WDM applications," IEEE Photon. Technol. Lett. **16**, 2263–2265 (2004)

61. A. Melloni: "Synthesis of a parallel-coupled ring-resonator filter," Opt. Lett. **26**, 917–919 (2001)

62. S. T. Chu, B. E. Little, W. G. Pan, T. Kaneko, and Y. Kokubun: "Second-order filter response from parallel coupled glass microring resonators," IEEE Photon. Technol. Lett. **11**, 1426–1428 (1999)

63. G. Griffel: "Vernier effect in asymmetrical ring resonator arrays," IEEE Photon. Technol. Lett. **12**, 1642–1644 (2000)

64. G. Griffel: "Synthesis of optical filters using ring resonator arrays," IEEE Photon. Technol. Lett. **12**, 810–812 (2000)

65. R. Grover, V. Van, T. A. Ibrahim, P. P. Absil, L. C. Calhoun, F. G. Johnson, J. V. Hryniewicz, and P.-T. Ho: "Parallel-cascaded semiconductor microring resonators for high-order and wide-FSR filters," J. Lightwave Technol. **20**, 900–905 (2002)

66. B. E. Little, S. T. Chu, J. V. Hryniewicz, and P. P. Absil: "Filter synthesis for periodically coupled microring resonators," Opt. Lett. **25**, 344–346 (2000)

67. C. J. Kaalund and G. Peng: "Pole-zero diagram approach to the design of ring resonator-based filters for photonic applications," J. Lightwave Technol. **22**, 1548–1558 (2004)

68. Little Optics, USA, www.littleoptics.com / www.nomadics.com/

69. S. T. Chu, B. E. Little, W. G. Pan, T. Kaneko, S. Sato, and Y. Kokubun: "An eight-channel add-drop filter using vertically coupled microring resonators over a cross grid," IEEE Photon. Technol. Lett. **11**, 691–693 (1999)

70. Z. Wang, W. Chen, and Y. J. Chen: "Unit cell design of crossbar switch matrix using micro-ring resonators," *Proc. 30th Europ. Conf. Opt. Commun.* (ECOC'04), Stockholm, Sweden, Vol. 3, 462–463 (2004)

71. LioniX BV, Enschede, the Netherlands; www.lionixbv.nl

72. S. Suzuki, K. Shuto, and Y. Hibino: "Integrated-optic ring resonators with two stacked layers of silica waveguide on Si," IEEE Photon. Technol. Lett. **4**, 1256–1259 (1992)

73. E. J. Klein, D.H Geuzebroek, H. Kelderman, G. Sengo, N. Baker, and A. Driessen: "Reconfigurable optical add-drop multiplexer using microring resonators," *Proc. 12th Europ. Conf. Integr. Optics* (ECIO'05), Grenoble, France, 180–183 (2005)

74. D. H. Geuzebroek, E. J. Klein, H. Kelderman, C. Bornholdt, and A. Driessen: "40 Gbit/s reconfigurable optical add-drop multiplexer based on microring resonators," *Proc. 31st Europ. Conf. Opt. Commun.* (ECOC'05), Glasgow, UK, 983–984 (2005)

75. Lambda Crossing, Israel, www.lambdax.com

76. J. Buus, D. J. Blumenthal, and M.-C. Amann: *Tunable laser diodes and related optical sources, 2nd ed.* (Wiley, New York, 2004)

77. Y. Yanagase, S. Suzuki, Y. Kokubun, and S. T. Chu: "Box-like filter response by vertically series coupled microring resonator filter," *Proc. 27th Europ. Conf. Opt. Commun.* (ECOC'01), Amsterdam, NL, Vol. 4, 634–635 (2001)

78. K. Jinguji and M. Kawachi: "Synthesis of coherent two-port lattice-form optical delay-line circuit," J. Lightwave Technol. **13**, 73–82 (1995)

79. G.-L. Bona, R. Germann, and B. J. Offrein: "SiON high-refractive-index waveguide and planar lightwave circuits," IBM J. Res. & Dev. **47**, 239–249 (2003)

80. G. L. Bona, F. Horst, R. Germann, B. J. Offrein, and D. Wiesmann: "Tunable dispersion compensator realized in high-refractive-index-contrast SiON technology," *Proc. 28th Europ. Conf. Opt. Commun.* (ECOC'02), Copenhagen, Denmark, Vol. 2, paper 4.2.1 (2002)

81. P. P. Absil, J. V. Hryniewicz, B. E. Little, P. S. Cho, R. A. Wilson, L. G. Joneckis, and P.-T. Ho: "Wavelength conversion in GaAs micro-ring resonators," Opt. Lett. **25**, 554–556 (2000)

82. G. Griffel, J. H. Abeles, R. J. Menna, A. M. Braun, J. C. Connolly, and M. King: "Low-threshold InGaAsP ring lasers fabricated using bi-level dry etching," IEEE Photon. Technol. Lett. **12**, 146–148 (2000)

83. R. Grover, P. Absil, V. Van, J. Hryniewicz, B. Little, O. King, L. Calhoun, F. Johnson, and P. Ho: "Vertically coupled GaInAsP InP microring resonators," Opt. Lett. **26**, 506–508 (2001)

84. D. G. Rabus and M. Hamacher: "MMI coupled ring resonators in GaInAsP/InP," IEEE Photon. Technol. Lett. **13**, 812–814 (2001)

85. K. Djordjev, Seung-J. Choi, Sang-J. Choi, and P. D. Dapkus: "High-Q vertically coupled InP microdisk resonators," IEEE Photon. Technol. Lett. **14**, 331–333 (2002)

86. K. Djordjev, Seung-J. Choi, Sang-J. Choi, and P. D. Dapkus: "Novel active switching components based on semiconductor microdisk resonators," *Proc. 28*th *Europ. Conf. Opt. Commun.* (ECOC'02), Copenhagen, Denmark, Vol. **1**, paper 2.3.5 (2002)

87. M. Hamacher, U. Troppenz, H. Heidrich, and D. G. Rabus: "Active ring resonators based on InGaAsP/InP," *Proc. SPIE Conf. Photonic Fabrication Europe,* vol. **4947**, 212–222 (2003)

88. Seung-J. Choi, Q. Yang, Z. Peng, Sang-J. Choi, and P. D. Dapkus: "High-Q buried heterostructure microring resonators," OSA Conf. Lasers and Electro Optics (CLEO 2004), San Francisco, CA, USA, paper CThF1 (2004)

89. Z. Bian, B. Liu, and A. Shakouri: "InP-based passive ring-resonator-coupled lasers," IEEE J. Quantum Electron. **39**, 859–865 (2003)

90. U. Troppenz, M. Hamacher, D. G. Rabus, and H. Heidrich: "All-active In-GaAsP/InP ring cavities for widespread functionalities in the wavelength domain," *Proc. 14*th *Internat. Conf. Indium Phosphide and Related Materials* (IPRM'02), Stockholm, Sweden, 475–478 (2002)

91. V. Van, T. A. Ibrahim, K. Ritter, P. P. Absil, F. G. Johnson, R. Grover, J. Goldhar, and P.-T. Ho: "All-optical nonlinear switching in GaAs-AlGaAs microring resonators," IEEE Photon. Technol. Lett. **14**, 74–76 (2002)

92. P. Rabiei, W. H. Steier, C. Zhang, and L. R. Dalton: "Polymer micro-ring filters and modulators," J. Lightwave Technol. **20**, 1968–1975 (2002)

93. P. Rabiei and W. H. Steier: "Tunable polymer double micro-ring filter," IEEE Photon. Technol. Lett. **15**, 1255–1257 (2003)

94. NAIS, Next-generation Active Integrated-optic Subsystems IST-2000–28018

95. Freeband Communications: "Broadband Photonics" www.freeband.nl

9 Interleavers

René M. de Ridder and Chris G. H. Roeloffzen

9.1 Introduction

In an optical communication system using wavelength division multiplexing (WDM), information is transmitted over several "channels", each at a different optical wavelength λ_i (or optical carrier frequency f_i). An *interleaver*, also known as a *slicer*, is an optical filter having at least one input and two complementary outputs with an optical transfer function that is periodic in frequency. In this way, for example, even-numbered optical channels can be routed to one output port while the odd-numbered channels will emerge from the other output port, as illustrated in Fig. 9.1. Such an interleaver can also be used the other way round, i. e. for combining two "combs" of optical channels, one shifted by half a channel spacing with respect to the other, into a single comb with half the channel spacing.

Besides the basic 1×2 interleaver function illustrated in Fig. 9.1, more complicated configurations may be used, as shown in Fig. 9.2.

In principle, any optical (de)multiplexer having a periodic response with frequency may be used as an interleaver, for example, arrayed waveguide gratings (AWG, cf. Chap. 4), Fabry–Perot resonators (Chap. 6), or ring resonators (Chap. 8). An overview of interleaver technology with an emphasis on bulk crystal optics is given by Cao et al. [1]. Interleavers that are applied in optical telecommunications should have a frequency-periodic

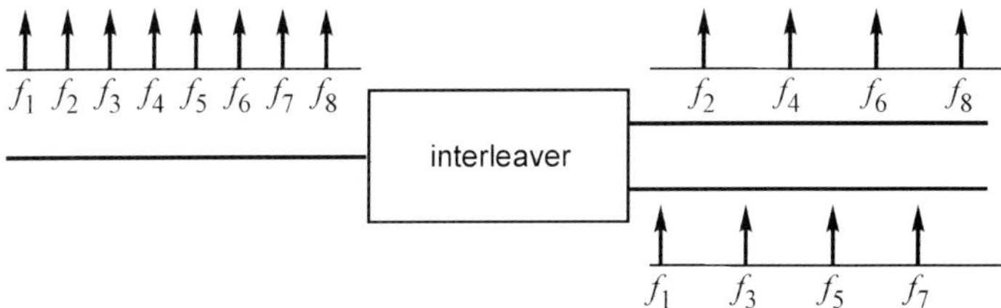

Fig. 9.1. Basic operation of an 1×2 interleaver as a frequency demultiplexer

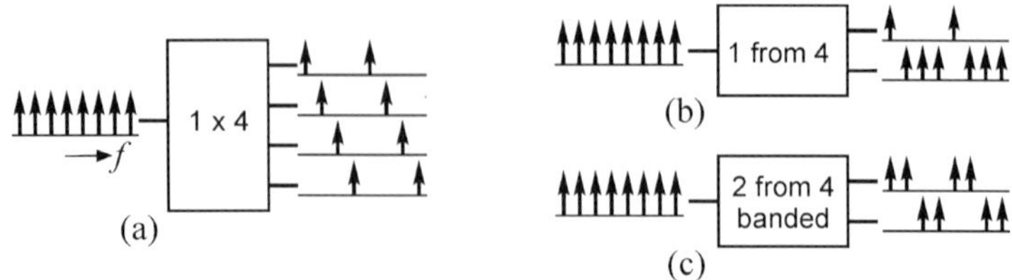

Fig. 9.2. Some examples of other interleaver configurations

response because the ITU grid specifying the optical channels defines a set of equidistant optical frequencies (cf. Appendix, Sect. A.1). Despite this, these devices are sometimes referred to as wavelength slicers or wavelength interleavers. (If an interleaver is periodic in frequency, it will not be strictly periodic in wavelength since $\lambda = c/f$, with c the vacuum speed of light.) If needed, however, strictly wavelength periodic interleavers can be designed [2].

The main applications of interleavers are in wavelength routing, (de)multiplexing, and pre-filtering. In the latter application, the interleaver is used as a first filtering stage which should have a transfer function approximating a rectangular shape as much as possible. The optical channels at its output ports still have the same bandwidth as before, but they are separated farther apart, thus making further routing or filtering operations less demanding. As an example, Fig. 9.3 shows the application of an interleaver for upgrading existing network nodes with 100 GHz channel spacing to ones with 50 GHz channel spacing. An additional advantage of integrating an interleaver with an AWG is the possibility of improving the AWG passband shape [3].

Figure 9.4 shows how a full demultiplexer can be built by cascading interleavers, in a binary tree structure, each stage having twice the free spectral

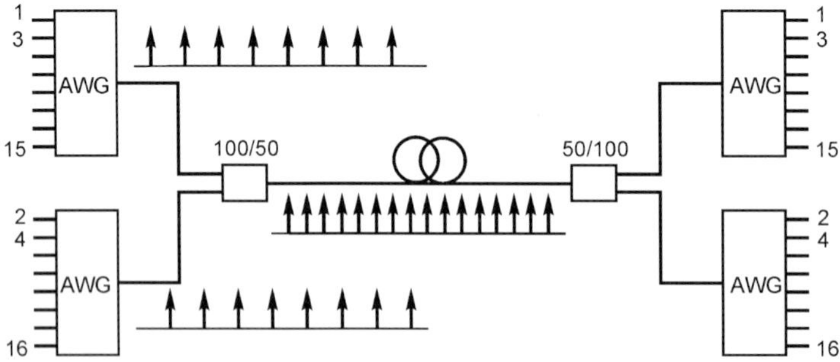

Fig. 9.3. 50/100 GHz interleaver as last-stage multiplexer and first-stage demultiplexer, combined with 100 GHz arrayed waveguide grating (AWG) (de)multiplexers to form a 50 GHz channel spacing system

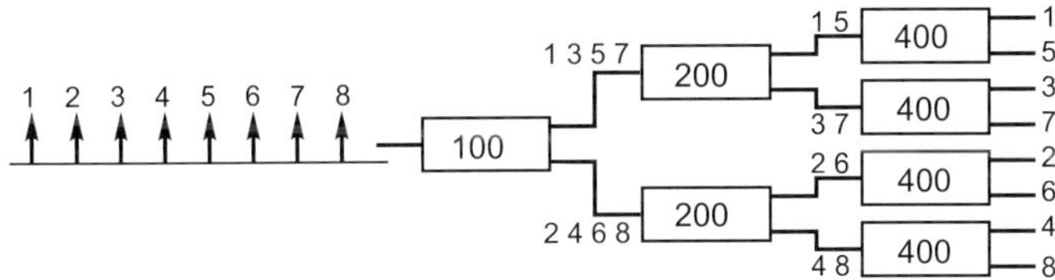

Fig. 9.4. Full binary tree 8-channel demultiplexer based on cascaded 1×2 interleavers. Eight channels with 50 GHz optical carrier frequency spacing are assumed, numbered 1–8. The numbers 100, 200, 400, written in the interleaver symbols of this example indicate the free spectral range (the frequency period in GHz) of these components, resulting in a 50 GHz input channel spacing

range (the period of the frequency response) as the preceding one, e. g. [4]. Finally, Fig. 9.5 illustrates the principle of an add-drop multiplexer built from two partial binary trees [5].

The obvious choice of a physical principle for implementing interleavers is interference since this phenomenon has an inherently frequency-periodic nature. The interleaver operation can be demonstrated using an asymmetric Mach–Zehnder interferometer (MZI).

The MZI is one of the most important building blocks in (integrated) optics. It is widely applied for such functions as switching, routing, and modulating optical signals, and it is a basic structure for many optical sensors. A schematic drawing of an MZI is shown in Fig. 9.6. Although MZI's can be implemented in several ways (e. g. with beam splitters and mirrors in free-space; using crystal optics; or in optical fibres), here we consider a structure based on planar optical waveguides. It consists of two 3-dB couplers connected by two single-mode optical channel waveguides where the signal in one path experiences a phase delay of $\Delta\varphi$ with respect to the other.

The couplers, which can be for example directional couplers, split the optical input power entering one of its input channels into two equal power output signals (each 50% or −3 dB of the input power), having 90° optical phase difference. For symmetry reasons, a signal in the other input would lead to two equal output signals having −90° phase difference. Since the

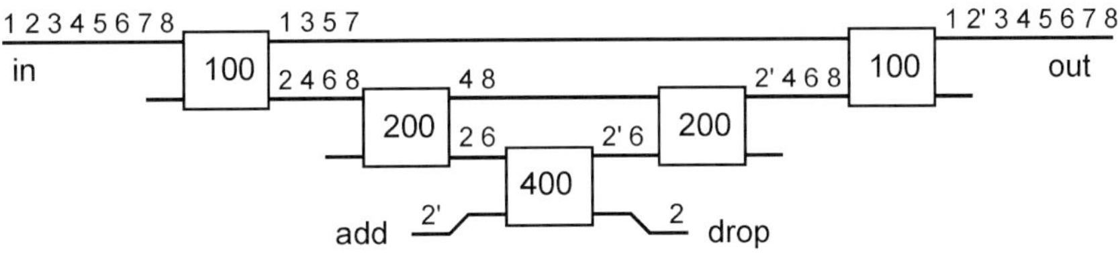

Fig. 9.5. Binary tree cascaded 2×2 interleavers organised to form a 1 from 8 add-drop multiplexer

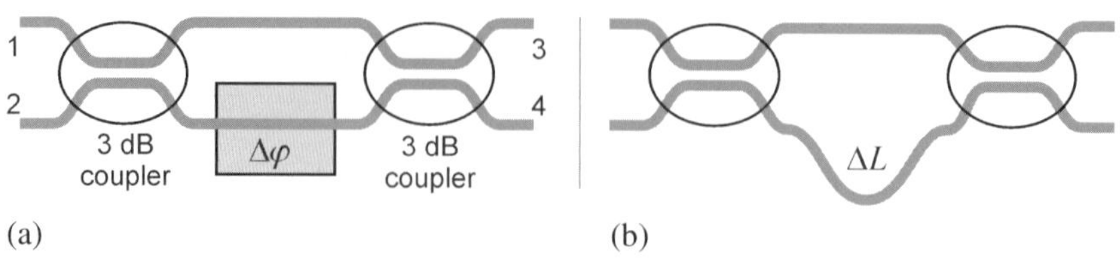

Fig. 9.6. Mach–Zehnder interferometer (MZI). (**a**) Basic structure. (**b**) Asymmetric MZI

coupler is reciprocal, two equal power coherent input signals with 90°
phase difference would combine into a single output signal at one output
channel, while −90° phase difference would direct all power to the other
output channel. With two in-phase coherent input signals, each output
channel would carry half of the total available power. Hence, it is obvious
that the phase difference determines the power distribution over the output
channels. It can be shown that the signals at the output ports of the MZI
are related to those at the input ports as

$$P_3 = P_1 \sin^2 \frac{\Delta\varphi}{2} + P_2 \cos^2 \frac{\Delta\varphi}{2}$$
$$P_4 = P_1 \cos^2 \frac{\Delta\varphi}{2} + P_2 \sin^2 \frac{\Delta\varphi}{2}$$

(9.1)

where P_i is the signal power in port i, as indicated in Fig. 9.6.

The phase difference $\Delta\varphi$ can be caused by a differential propagation delay in the interferometer arms due to a length difference ΔL between the
arms, as illustrated in Fig. 9.6b,

$$\Delta\varphi = \beta \Delta L = \frac{2\pi}{c} f\, n_{eff}\, \Delta L$$

(9.2)

where β is the propagation constant of the mode in the delay section, f is the
frequency of the wave, n_{eff} is the effective refractive index of the waveguide
mode, and c is the speed of light in vacuum. If the dispersion is neglected
(n_{eff} is assumed to be constant), the phase difference is proportional to both
the frequency f and the delay length ΔL. Combining (9.1) and (9.2), the
power transfer from port 1 to ports 3 and 4 as a function of f is depicted in
Fig. 9.7. The free spectral range (FSR), which is the period Δf_{FSR} in the frequency response, is given by

$$\Delta f_{FSR} = \frac{c}{n_{eff}\, \Delta L}$$

(9.3)

Figure 9.7 shows that a comb of optical signal frequencies with a spacing of $\Delta f = \Delta f_{FSR}/2$ may be split into two combs, each having the double
frequency spacing.

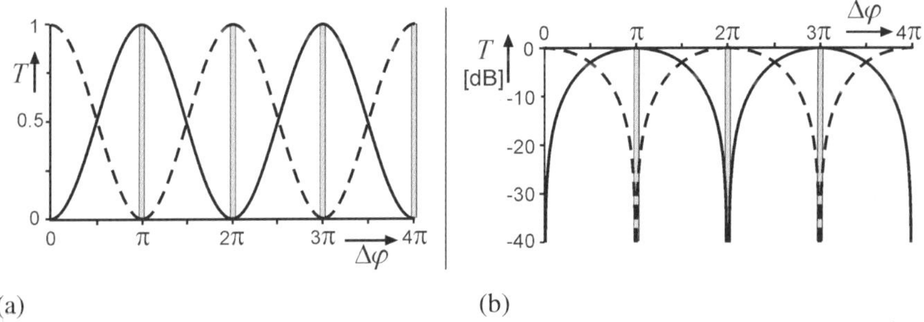

(a)   (b)

Fig. 9.7. Normalised power transfer functions of an asymmetric MZI; the solid curve indicates P_3/P_1, while the dashed curve indicates P_4/P_1. (a) Linear power scale; (b) logarithmic scale. The grey vertical bars indicate the optical signal channels which are routed from port 1 alternately to ports 3 and 4 (de-interleaving)

As may be obvious from Fig. 9.7, an MZI does not have an ideal rectangular-shaped transfer function. It does not provide sufficient tolerance for wavelength deviations, laser linewidth, and modulation bandwidth which should typically be about 40% of the grid spacing. Especially the suppression of unwanted channels is critical (25 dB is a common requirement, leading to a useful bandwidth of only 8% of the channel spacing for an MZI, see the narrow channel bars in Fig. 9.7).

One method for improving the transfer function comes from the observation that a periodic function with "arbitrary" shape can be composed, like a Fourier series, from a set of harmonically related sinusoidal functions which are added together with appropriate amplitude and phase. Figure 9.8a shows the conceptual picture where multiple delayed copies of the input signal are combined into a weighted sum. For a periodic filter response, the delays τ_i should be integer multiples of the unit delay τ_1 defining the FSR.

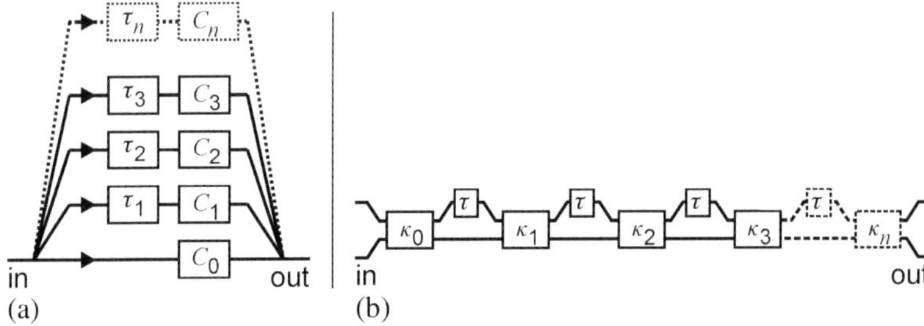

(a)   (b)

Fig. 9.8. Composite filters. (a) Multiple parallel weighted delay-line paths; for a Fourier-type synthesis, all τ_i should be an integer multiple of the smallest delay, τ_1. (b) Equivalent lattice filter with $\tau = \tau_1$

For approximating a "rectangular" filter transfer function only odd multiples of τ_1 are needed. The weighting constants C_i are the Fourier coefficients. Since it turns out to be rather difficult to realise low loss multiport optical splitters and combiners, the equivalent approach shown in Fig. 9.8b, requiring only 2×2-type couplers, is preferred. This type of structure is known as a lattice filter or resonant coupler. The equivalence with the parallel delay line filter may be intuitively understood by considering all possible paths that an optical signal can take through the lattice. The output signal can then be written as the sum of a number of terms, each involving an integer $(0, 1, .. n)$ multiple of the unit delay τ. A possible implementation, illustrated in Fig. 9.9, is by concatenation of MZI-like elements, using couplers with specific (non 50-50) power coupling ratios κ_i. In addition to the fixed path length differences ΔL_j generating the frequency-periodic response, in general also tuneable phase-shifters $\Delta \varphi_k$ are implemented in each MZI-section for fine-tuning purposes. Several design methods have been discussed in the literature, e. g. [2, 6–14]. This type of interleaver is analysed in detail in Sect. 9.3.1.

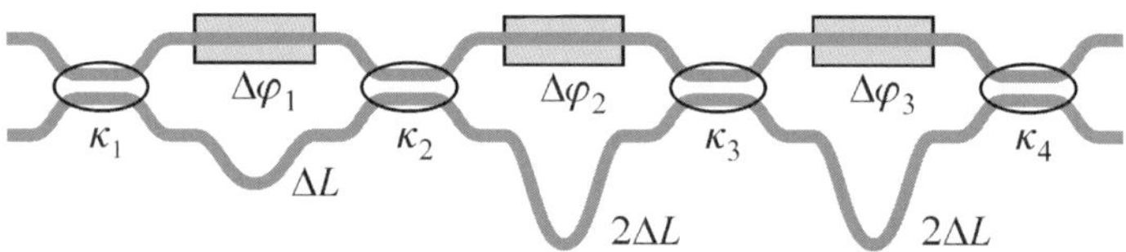

Fig. 9.9. Band flattening: lattice-type filter of concatenated MZI-like structures. The double delay length sections ($2\Delta L$) arise from combining two sections where the coupling ratio κ turned out to be zero in this example (see fifth-order interleaver design example at the end of Sect. 9.3.1)

Another method is based on manipulating the phase response in one of the MZI branches in a periodic and nonlinear way, so that the phase change due to the length difference ΔL is approximately compensated near the passband and stopband centres (flattening the frequency response there) while the phase will change quickly near the band edges, sharpening the transfer function there. An example is shown in Fig. 9.10 where a ring resonator coupled to one of the branches acts as an allpass filter having a strongly nonlinear phase response with a frequency period equal to half the FSR of the MZI. Again, some fine-tuning facilities $\Delta \varphi_k$ are generally needed. An analysis is given in Sect. 9.3.2. Several architectures and design methods for interleavers incorporating both lattice-type structures and ring resonators or other feedback elements have been described in the literature, e. g. general design strategies and overviews [15–19], a generic design algorithm [20], and various more specific design topics [21–25].

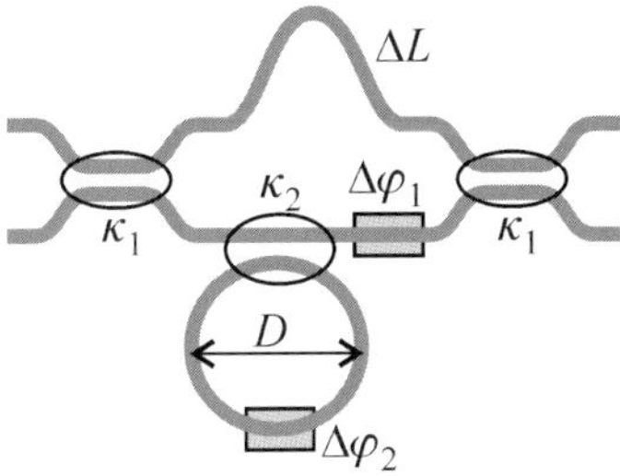

Fig. 9.10. Band flattening: MZI with internal phase response modified by a ring resonator

Signal transmission through an optical filter may be affected by dispersion which tends to distort the pulse shape and hence may cause an increased bit error rate. In some cases it is possible to build an interleaver from two complementary sections having inverse phase responses, thus realising an overall response with negligible dispersion. A general discussion of dispersion of different interleaver types is given in [26]. The dispersion characteristics of both lattice-type and MZI + ring-type interleaving filters, as well as possible compensation techniques, will be discussed in Sect. 9.4.

As indicated before, interleavers can be cascaded in order to build full (de)multiplexers and add-drop multiplexers. Here, the application of planar optical waveguide technology pays off because it enables the integration of many stages into a single planar optical circuit. The different requirements for the consecutive stages will be discussed in Sect. 9.5 where the design and realisation of a tuneable 1 from 16 add-drop multiplexer will be explained as an example.

Finally, in Sect. 9.6, an overview will be given of technologies that have been used for fabricating interleaver-circuits, and examples will be given, including bulk optics, fibre-based implementations, and planar guided wave optics. Also, the characteristics of some commercially available interleavers, based on bulk optics, fibre optics, and integrated optics will be shown.

9.2 Basic Mach–Zehnder Interferometer Interleavers

The basic lay-out of a Mach–Zehnder interferometer- (MZI-) based interleaver is shown in Fig. 9.6b, and its power transfer function is given in Fig. 9.7. The MZI can be considered as a concatenation of three building blocks, the input directional coupler, a differential delay section, and the output coupler. Although the MZI is a relatively simple device that does not require a specific approach for analysis, we will use it for demonstrating a general approach that is applicable to a large class of periodic frequency response filters that would be difficult to analyse otherwise.

Following the approach of Madsen and Zhao [27], this section will start with a short summary of the main analytical tool for analysing filters that have a periodic frequency response, the Z-transform [28] and its relation with the traditional transfer function. The MZI is a four-port device, and so are the lattice-type filters that will be discussed in later sections. Therefore, a brief review will be given of the transfer matrix, describing a four-port network. Next, the optical properties of the basic building blocks of an MZI, waveguides and directional couplers, will be summarised. The properties of an MZI interleaver then follow from putting all the pieces together. Although a simple MZI does not have the most desirable interleaver properties, it serves well as a building block for composite interleaver filters, and the mathematical analysis of this relatively simple device is easily extended to the more interesting higher-order interleavers to be discussed in following sections.

9.2.1 The Transfer Function and the Z-Transform

The type of wavelength slicers to be analysed have a periodic transfer function in frequency which is caused by the periodic phase response of a time delay. A lossless straight waveguide section of length L has a transfer function

$$H_{\text{straight}} = e^{-ik_0 n_{eff} L} \tag{9.4}$$

where $k_0 = 2\pi f/c$. The propagation delay through the waveguide is $\tau = L n_{eff}/c$, so that the transfer function (9.4), which just corresponds to a time delay, can be written explicitly as a function of frequency

$$H_{\text{delay}}(f) = e^{-i2\pi f \tau} \tag{9.5}$$

which clearly shows the frequency-periodic phase response. In an MZI there is only a single effective differential time delay τ, corresponding to a frequency period $\Delta f = 1/\tau$. In the lattice filters and MZI + ring structures to be discussed later, there are multiple delays which, however, are all integer multiples of an elementary delay τ_0. It then makes sense to introduce a normalised frequency f'

$$f' = f\tau_0 \tag{9.6}$$

which allows us to write (9.5) as

$$H_{\text{delay}'}(f') = e^{-i2\pi f'}$$

(9.7)

The Z-Transform

If we now introduce a new complex variable z

$$z = e^{i2\pi f'}$$

(9.8)

(9.7) can be written as

$$H_{\text{delay}}(z) = z^{-1}$$

(9.9)

The usefulness of this new variable is given by the property that the transfer function of n times the unit delay is simply z^{-n}, which greatly simplifies the transfer functions of more complicated periodic filters which can be expressed as rational functions in z^{-1}.

The mapping (9.8) of f (or f') to the complex z-plane is known as the Z-transform which is widely used in digital filter design. It can be considered as an analytic extension of the discrete-time Fourier transform (DTFT) for discrete signals [27]. The complex quantity z takes values on the unit circle, starting at $z=1$ and making a full turn around the origin as f' varies from 0 to 1.

From a given transfer function $H(z)$ in the z-domain – also known as the system function – the frequency response is found by evaluating $H(z)$ on the unit circle in the z-plane, i. e. $z = e^{i2\pi f'}$.

The transfer function can be written as a ratio of Mth- and N^{th}-order polynomials, or in an equivalent product form, e. g. [29], explicitly showing the zeros z_m and poles p_n of $H(z)$,

$$H(z) = \frac{\sum_{m=0}^{M} b_m z^{-m}}{1 + \sum_{n=1}^{N} b_n z^{-n}} = \Gamma z^{N-M} \frac{\prod_{m=1}^{M}(z - z_m)}{\prod_{n=1}^{N}(z - p_n)}$$

(9.10)

where Γ is the gain. The transfer function of a passive filter can never be greater than one, implying a maximum value of Γ that is determined by $\max(|H(z)|_{z=e^{2\pi f'}}) = 1$. Since the real frequency transfer function is found for $z = e^{i2\pi f'}$, only zeros that occur on the unit circle will correspond to zero transmission at the frequency corresponding to the argument of that zero.

A convenient graphical way to represent the transfer function is the pole-zero diagram. It shows the locations of each pole and zero in the complex plane. All zeros are designated by 'o' and a pole is marked by 'x'. An example of a pole-zero diagram is depicted in Fig. 9.13 in Sect. 9.2.4.

A filter that has only zeros in its transfer function has only feed-forward paths and is classified as a finite impulse response (FIR) or moving average (MA) filter. Filters containing feed-back paths will have at least one pole in their transfer function and are classified as infinite impulse response (IIR) filters. Sub-types of IIR filters contain either only poles (autoregressive (AR) filters), or both, poles and zeros (autoregressive moving average (ARMA) filters).

The Frequency Response

The frequency response function $H(f)$ is a complex function of f. It is usually expressed in terms of its magnitude $|H(f)|$ and phase $\varphi(f)$

$$H(f) = |H(f)| e^{i\varphi(f)} \tag{9.11}$$

Of course, the overall transfer function of a series connection of M transfer functions H_i is given by the product ΠH_i, implying a multiplication of magnitude responses $|H_i(f)|$ and summation of phase responses $\varphi_i(f)$.

Since $H(f)$ is obtained by evaluating $H(z^{-1})$ on the unit circle, the square of the magnitude response can be found as follows – if the coefficients of the transfer function are real [30]

$$|H(f)|^2 = H(f)H^*(f) = H(f)H(-f) = H(z)H(z^{-1})\Big|_{z=e^{i2\pi f}}. \tag{9.12}$$

where H^* denotes the complex conjugate of H. Equation (9.12) implies that reciprocal zeros, which are mirror images of each other about the unit circle, have identical magnitude characteristics. Based on the pole-zero representation (9.10) of $H(z)$, only the distance of each pole and zero from the unit circle, i.e. $|e^{i2\pi f'} - p_n|$ or $|e^{i2\pi f'} - z_m|$, affects the magnitude response. It can be shown that zeros z_m, z_m^*, $1/z_m$ and $1/z_m^*$ all have the same magnitude response. However, their phase characteristics will be different, depending on z_m being inside or outside the unit circle in the z-plane. A system with all zeros – and all poles as well – inside the unit circle ($|z_m| < 1$) has a so-called *minimum-phase* response (cf. Chap. 2) while all other systems (e.g. with one ore more zeros outside the unit circle) have a non-minimum phase response.

Four-port Networks

A basic 2×2 four-port device without reflection is schematically shown in Fig. 9.11. Its behaviour is conveniently described in terms of a transfer

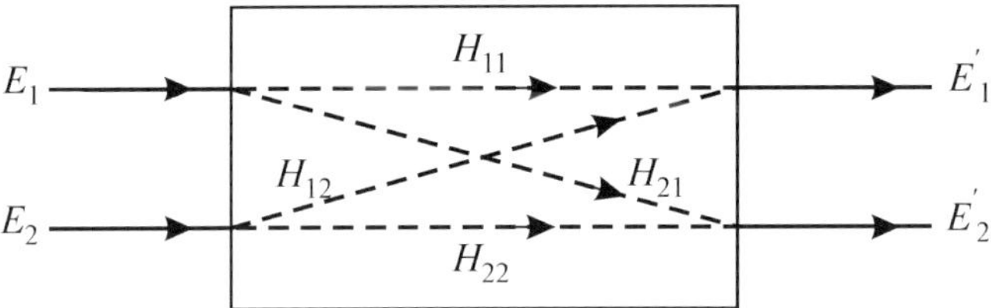

Fig. 9.11. Schematic drawing of the transfer functions in a 2×2 port without reflection

matrix relating the signals at its two output ports to those at the input ports, as given in (9.13),

$$\begin{bmatrix} E'_1 \\ E'_2 \end{bmatrix} = \mathbf{H} \begin{bmatrix} E_1 \\ E_2 \end{bmatrix} = \begin{bmatrix} H_{11} & H_{12} \\ H_{21} & H_{22} \end{bmatrix} \begin{bmatrix} E_1 \\ E_2 \end{bmatrix} \tag{9.13}$$

where the complex transfer matrix $\mathbf{H}$ contains two so-called bar transfer functions (H_{11} and H_{22}) and two cross transfer functions (H_{12} and H_{21}).

If several such four-port devices, having transfer matrices $\mathbf{H}_1, \mathbf{H}_2, ..., \mathbf{H}_{n-1}, \mathbf{H}_n$, are concatenated, the transfer matrix $\mathbf{H}_{tot}$ of the composite device is simply found by matrix multiplication as

$$\mathbf{H}_{tot} = \mathbf{H}_n \mathbf{H}_{n-1} \cdots \mathbf{H}_2 \mathbf{H}_1 \tag{9.14}$$

9.2.2 Differential Delay Section

The delay section of an MZI is formed by two independent waveguides having different lengths L_1 and L_2, respectively (we assume $L_1 > L_2$). Although we assume almost identical branches (regarding phase constant $\beta(f) = k_0 n_{eff}$ and attenuation coefficient α of the – single – guided mode), in order to model e. g. thermo-optic tuning, we do allow for a small deviation from the average effective index n_{eff}, leading to an additional tuning phase delay φ_t (indicated as $\Delta\varphi_k$ in Fig. 9.9) in branch 1 with respect to branch 2. The transfer matrix of the delay section is then given by

$$\mathbf{H}_{delay} = \begin{bmatrix} e^{-\alpha L_1} e^{-i k_0 n_{eff}(f) L_1} e^{-i\varphi_t} & 0 \\ 0 & e^{-\alpha L_2} e^{-i k_0 n_{eff}(f) L_2} \end{bmatrix} \tag{9.15}$$

The differential delay τ is given as

$$\tau = \frac{1}{\Delta f_{FSR}} = \frac{(L_1 - L_2) n_g}{c} = \frac{\Delta L \, n_g}{c} \tag{9.16}$$

where Δf_{FSR} is the FSR, defined in (9.3), ΔL is the path length difference, and n_g is the group index defined as

$$n_g = n_{eff} + f\frac{\partial n_{eff}}{\partial f} \tag{9.17}$$

Taking branch 2 as a reference, the transfer matrix can be written in terms of τ,

$$\mathbf{H}_{delay} = \eta_{L_2} e^{-i\beta(f)L_2}\begin{bmatrix} \eta_{\Delta L}e^{-i2\pi f\tau}e^{-i\varphi_t} & 0 \\ 0 & 1 \end{bmatrix} \tag{9.18}$$

where $\eta_{L2}=e^{-\alpha L2}$ is the loss along the path L_2 and $\eta_{\Delta L}=e^{-\alpha\Delta L}$ is the differential loss due to the path length difference ΔL. Applying the frequency normalisation with respect to the free spectral range as mentioned in the previous section, and applying the Z-transform, we arrive at

$$\mathbf{H}_{delay} = \eta_{L_2} e^{-i\beta(f)L_2}\begin{bmatrix} \eta_{\Delta L}z^{-1}e^{-i\varphi_t} & 0 \\ 0 & 1 \end{bmatrix} \tag{9.19}$$

In the remainder of this chapter we will omit the factor $\eta_{L_2}e^{-i\beta L_2}$ since it indicates just a constant loss and a linear phase. Furthermore, we introduce a new variable Z, defined by

$$Z^{-1} = \eta_{\Delta L}e^{-i\varphi_t}z^{-1} \tag{9.20}$$

so that $Z=z$ in the lossless case and without additional tuning phase shift. In general, for f' varying from 0 to 1, Z will take values on a circle centred in the complex plane with a radius $\eta_{\Delta L}\leq 1$ and starting with an argument offset φ_t at $f'=0$. The simplified transfer matrix for a differential delay becomes

$$\mathbf{H}_{delay} \cong \begin{bmatrix} Z^{-1} & 0 \\ 0 & 1 \end{bmatrix}. \tag{9.21}$$

9.2.3 Directional Coupler

Couplers are among the most elementary building blocks in planar lightwave circuits, performing the functions of splitting and combining guided optical waves. (Another variant are fibre couplers which are widely used in fibre optics.)

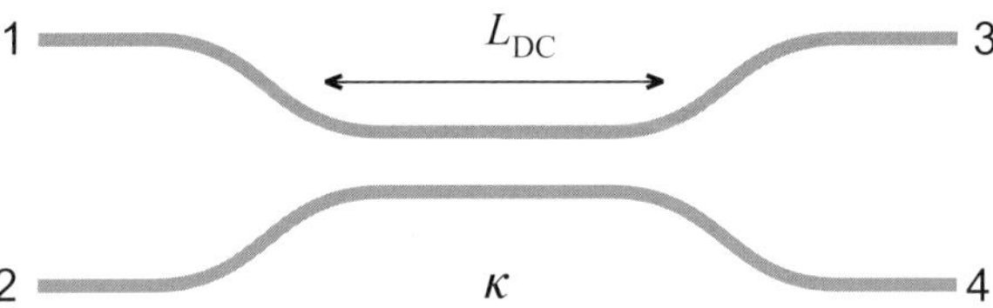

Fig. 9.12. Schematic top view of the directional coupler

Directional couplers (DC's) consist of two optical waveguides in close proximity to each other, see Fig. 9.12. Light will be coupled by the evanescent field of the mode, and power will be exchanged between the waveguides. The energy transfer process is similar to two mechanical pendulums that are weakly coupled by a spring. The two single-mode waveguides are usually chosen to be very close to each other in order to obtain a short coupling length.

One of the most convenient ways to understand the operation of a DC is by considering the two parallel waveguides as a single system which supports two system modes. The modes of the input and output channels are decomposed into these system modes, and the power exchange between the separate waveguides can be simply described as an interference phenomenon of the system modes. It can easily be shown that the field amplitudes at the output channels, E_3 and E_4, are related to those at the input channels, E_1 and E_2, as

$$\begin{bmatrix} E_3 \\ E_4 \end{bmatrix} = \begin{bmatrix} \cos\psi & -i\sin\psi \\ -i\sin\psi & \cos\psi \end{bmatrix} \begin{bmatrix} E_1 \\ E_2 \end{bmatrix} \tag{9.22}$$

where the parameter ψ is given as

$$\psi = \frac{\pi f \Delta n_{eff}}{c} L_{DC} \tag{9.23}$$

where Δn_{eff} is the difference in effective index of the two system modes, and L_{DC} is the effective length of the coupling secton, as indicated in Fig. 9.12.

The coupling length L_π is the propagation distance giving π phase difference between the system modes

$$L_\pi = \frac{c}{2f \Delta n_{eff}} \tag{9.24}$$

Arbitrary power splitting ratios can be obtained by simply choosing a length L_{DC} between 0 and L_π. The power is split equally over the output channels if the length of the coupler is half the coupling length. Equation

(9.25) gives a shorthand notation for the transfer matrix of a DC (as defined in (9.22)),

$$H_{DC} = \begin{bmatrix} C & -i\,S \\ -i\,S & C \end{bmatrix}$$

(9.25)

with

$$C = \cos\psi = \sqrt{1-\kappa}$$

$$-i\,S = -i\sin\psi = -i\sqrt{\kappa}$$

(9.26)

where C is the bar transfer function, and $-i\,S$ is the cross transfer function; ψ is equal to the coupling strength integrated over the length, and κ is the power coupling ratio. Coupling does not only occur in the straight waveguides. It occurs already in the leads, and the total phase is the sum of the phase in the straight waveguides and the leads.

9.2.4 Mach–Zehnder Interferometer Interleaver

The transfer matrix of the MZI can be calculated by simple multiplication of the transfer matrices of the type (9.19) and (9.25) of its component devices: input coupler, delay section, and output coupler.

$$\mathbf{H}_{MZI} = \mathbf{H}_{DC_2}\,\mathbf{H}_{delay}\,\mathbf{H}_{DC_1}$$

(9.27)

Using the simplified delay transfer matrix (9.21), the MZI transfer matrix is found to be:

$$\mathbf{H}_{MZI} = \begin{bmatrix} H_{11}(z) & H_{12}(z) \\ H_{21}(z) & H_{22}(z) \end{bmatrix} = \begin{bmatrix} A(z) & B^R(z) \\ B(z) & A^R(z) \end{bmatrix} =$$

$$= \begin{bmatrix} -S_1S_2 + C_1C_2Z^{-1} & -i(C_1S_2 + S_1C_2Z^{-1}) \\ -i(S_1C_2 + C_1S_2Z^{-1}) & C_1C_2 - S_1S_2Z^{-1} \end{bmatrix}$$

(9.28)

The two polynomials in the left-hand column, called the forward polynomials, are labelled $A(z)$ for the bar transfer and $B(z)$ for the cross transfer, respectively. The two polynomials in the second column, called the reverse polynomials, are labelled $B^R(z)$ and $A^R(z)$, respectively. These reverse polynomials appear in the Z-transform description of many optical filters.

Note that the coefficients of the polynomial $A^R = H_{22}$ are in reverse order compared to those of $A = H_{11}$. The same holds for B^R and B (H_{12} and H_{21}).

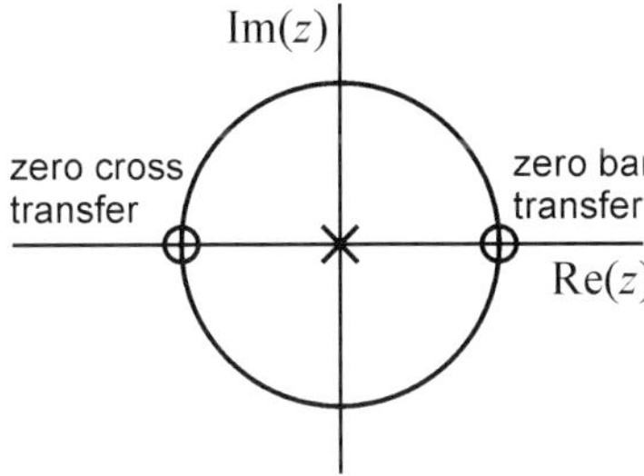

Fig. 9.13. Pole-zero diagram showing the zeros of the bar and cross transfer of the ideal MZI (lossless; perfect 3-dB couplers; no additional phase shift φ_t)

This symmetry property allows calculating H_{22} and H_{21} if H_{11} and H_{12} are known. The transfer matrix can also be written in terms of the roots of the polynomials as follows:

$$\mathbf{H}_{\mathrm{MZI}} = \begin{bmatrix} -S_1 S_2 Z^{-1}\left(Z - \dfrac{C_1 C_2}{S_1 S_2}\right) & -\mathrm{i}\, C_1 S_2 Z^{-1}\left(Z - \left(\dfrac{-S_1 C_2}{C_1 S_2}\right)\right) \\[2ex] -\mathrm{i}\, S_1 C_2 Z^{-1}\left(Z - \left(\dfrac{-C_1 S_2}{S_1 C_2}\right)\right) & -C_1 C_2 Z^{-1}\left(Z - \dfrac{S_1 S_2}{C_1 C_2}\right) \end{bmatrix} \quad (9.29)$$

The bar transfer $A(z)$, for example, has a zero for $Z = C_1 C_2/(S_1 S_2)$, or, using (9.20), $z = \eta_{\Delta L}\mathrm{e}^{-\mathrm{i}\varphi_t}C_1 C_2/(S_1 S_2)$ and a pole at the origin ($z = 0$). A way to get insight into the polynomials is to plot all the poles and zeros in the complex z-plane, see Fig. 9.13. Their position in the z-plane depends on the coupling ratios and the tuning phase φ_t. The zeros always lie on the real axis when $\varphi_t = 0$. The transfer is zero if z is equal to a zero point and would be infinite if equal to a pole.

Since passive devices never have an infinite transfer, possible poles will never occur on the unit circle $z = \mathrm{e}^{\mathrm{i}2\pi f'}$. An MZI transfer function, having a single pole at the centre, clearly satisfies this condition. The behaviour of a filter over its free spectral range can be investigated by evaluating its transfer matrix for all values of z encountered by travelling once around the unit circle. The transfer goes to zero if z crosses zero on the unit circle. However, a zero can also lie inside or outside the circle, see for example Fig. 9.17a. The closer z (on the unit circle) gets to a zero the lower the transfer is. Two zeros at mirrored positions with respect to the unit circle (z_m and $1/z_m^*$) will give the same amplitude transfer but a different phase transfer. The bar transfer (H_{11} and H_{22}) will have a zero on the unit circle, only if $\kappa_1 = 1 - \kappa_2$, and the cross transfer (H_{12} and H_{21}) needs $\kappa_1 = \kappa_2$ to have a zero on the unit circle. With $\kappa_1 = \kappa_2 = 0.5$, both the bar and the cross transfer functions have a zero on the unit circle as shown in Fig. 9.13. When the two couplers are identical the matrix reduces to

$$\mathbf{H}_{\text{MZI}} = \begin{bmatrix} -S^2 + C^2 Z^{-1} & -i\,S\,C(1 + Z^{-1}) \\ -i\,S\,C(1 + Z^{-1}) & C^2 - S^2 Z^{-1} \end{bmatrix} =$$

$$= \begin{bmatrix} -S^2 Z^{-1}(Z - \dfrac{C^2}{S^2}) & -i\,S\,C\,Z^{-1}(Z + 1) \\ -i\,S\,C\,Z^{-1}(Z + 1) & -C^2 Z^{-1}(Z - \dfrac{S^2}{C^2}) \end{bmatrix} \tag{9.30}$$

This matrix shows that $H_{12}(z)$ and $H_{21}(z)$ will always have a zero on the unit circle, but $H_{11}(z)$ and $H_{22}(z)$ will also have a zero on the unit circle only if $\kappa_1 = \kappa_2 = 0.5$. This means that it is much easier to have complete isolation at the cross port than at the bar port.

For this case of two perfect 3 dB couplers where $C = S = \sqrt{1/2}$ (cf. (9.26)), the complete frequency response (optical power transfer) is found from (9.30) by substituting (9.20), (9.8), and (9.6) to be

$$|H_{11}(f)|^2 = |H_{22}(f)|^2 = \eta_{\Delta L} \sin^2\left(\frac{2\pi f \tau}{2} + \frac{\varphi_t}{2}\right) + \frac{1}{4}(1 - \eta_{\Delta L})^2 \tag{9.31}$$

for the bar transfer and

$$|H_{12}(f)|^2 = |H_{21}(f)|^2 = \eta_{\Delta L} \cos^2\left(\frac{2\pi f \tau}{2} + \frac{\varphi_t}{2}\right) + \frac{1}{4}(1 - \eta_{\Delta L})^2 \tag{9.32}$$

for the cross transfer. The lossless filter ($\eta_{\Delta L} = 1$) satisfies the simple condition $|H_{11}(z)|^2 + |H_{21}(z)|^2 = 1$ which is obvious from power conservation.

Figure 9.14 shows the frequency response of the MZI filter for several values of the differential loss. Note that the filter curve has a very narrow stopband. The width of the stopband at -25 dB is only 4% of the FSR.

The phase response of the bar transfer of the non-ideal MZI with identical directional couplers, from (9.30), is

$$\varphi(f') = \tan^{-1}\left(\frac{\dfrac{C^2}{S^2}\eta_{\Delta L}\sin\left(2\pi f' + \varphi_t\right)}{1 - \dfrac{C^2}{S^2}\eta_{\Delta L}\cos\left(2\pi f' + \varphi_t\right)}\right) \tag{9.33}$$

The non-linear frequency dependence of the phase response leads to a frequency-dependent group delay and dispersion as will be explained in Sect. 9.4.

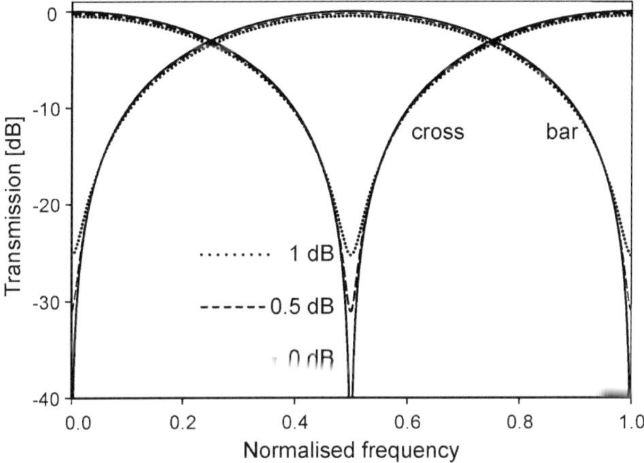

Fig. 9.14. Magnitude response for the Mach–Zehnder interferometer filter with differential loss of 0, 0.5, and 1 dB, respectively; tuning phase $\varphi_t = 0$, and coupling constants $\kappa_1 = \kappa_2 = 0.5$

9.3 Higher Order Interleavers

9.3.1 Finite Impulse Response Filters: The Resonant Coupler Approach

The disadvantage of the MZI filter is that its transfer function is sine-shaped. This results in a very narrow stopband. For example, if 25 dB isolation is required, the stopband width is only 8% of the channel spacing, so that 92% of the available spectrum must remain unused. Although the ideal rectangular-shaped filter transfer function, which would allow 100% spectrum use, cannot be realized for reasons of causality, several approaches are known from the literature, e. g. [31], for improving the simple MZI filter. One of them involves resonant couplers (RC, also called multi-stage moving average filters or lattice filters) [32, 33]. These filters can be implemented by cascading single MZI's, as shown in Fig. 9.15. Here a 2-stage filter is shown, consisting of 2 delay lines and 3 couplers. This concept can be extended to more stages. An N-stage filter has N delay lines and $N + 1$ couplers. The filter has 2 inputs and 2 outputs.

For simplicity the filters are assumed to have no loss. This means that the outputs are power complementary (the sum of the output powers is 100%). The best way to design such a filter is by using the *Z-transform* description and the accompanying zero diagram as described above. One can find a synthesis algorithm in the literature which calculates the power

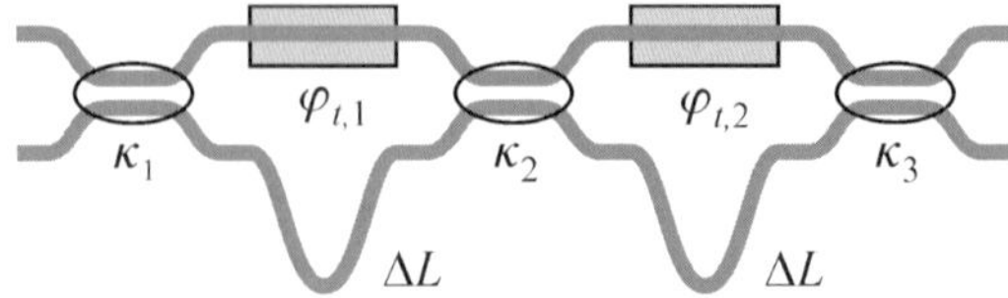

Fig. 9.15. Two stage resonant coupler filter, consisting of three couplers and two delay lines. Tuning phase shifts $\varphi_{t,1}$ and $\varphi_{t,2}$ might be provided by thermo-optic actuation

coupling ratios of each DC and the phase of the delay line from these polynomials [34, 27]. This is a very important algorithm since it opens the way for using all the design tools for digital filters in order to design a desired filter that can then be mapped to a real optical filter layout. Due to chip space restrictions and optical losses, it is not possible to make an optical filter with a large number of delay lines. For example, a polynomial filter of order one hundred, which is very common in digital filters, is not (yet) possible. Also, every additional delay line needs an independent tuning element. Therefore it is important to design a filter using as few delay lines as possible.

Filter Demands and Design Strategy

In order to be useful as interleavers, for example cascaded in a binary tree arrangement, the filters must satisfy certain requirements that can be summarised as follows:

- For interleaver operation, the bar transfer must be equal to the cross transfer shifted by half the FSR. This implies that the zero transfer frequencies are shifted by half the FSR compared to the frequencies of maximum transfer. The number of local maxima is equal to the number of local minima.
- To use the bandwidth as efficiently as possible, interleavers having broad passbands and stopbands are necessary.
- Low passband loss.
- Good isolation. It is difficult to fabricate filters having better isolation than 25 dB.

Figure 9.16 summarises the filter synthesis process. There are four general steps in the process as described below.

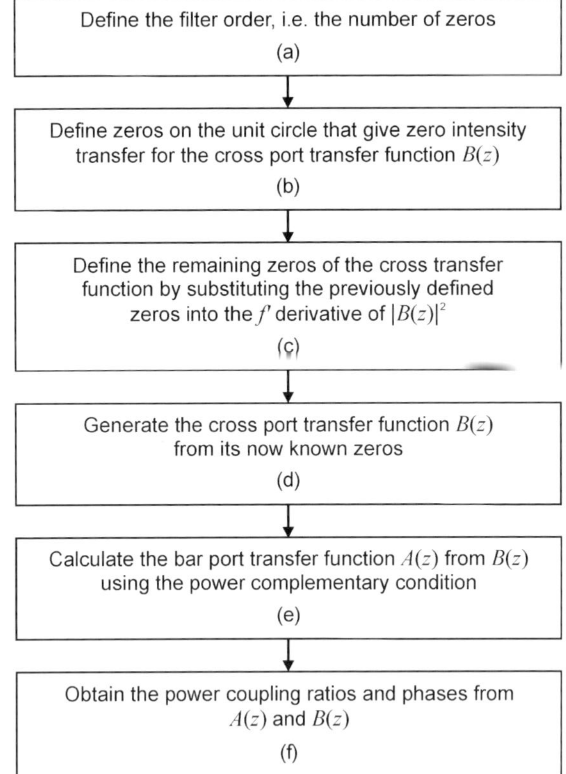

Fig. 9.16. Passband-flattened filter design synthesis flowchart

(1) Definition of the filter order

This first step, (a) in Fig. 9.16, defines the number of zeros which is equal to the order of the filter. The maximum number of zeros will be limited by the fabrication accuracy and available space. Generally, the larger the order, the better a desired transfer function (e. g. a rectangular shape) can be approximated. The filter shape is adjusted by carefully locating the zeros in the complex z-plane. A zero placed on the unit circle will create a zero transfer (absolute minimum) and thus contributes to a stopband. Local maxima will arise between two adjacent zeros in the stopband. A zero inside or outside the unit circle will create a local minimum and, in the cases of practical interest, two surrounding local maxima. The overall effect can be a flattening of the passband if the zero is not positioned too close to the circle, so that the local minimum is shallow and only a small ripple remains. Due to the required symmetry of the passband and stopband of an interleaver, the number of passband maxima should equal the number of stopband zeros. This leads to the requirement of having one fewer zero in the passband than in the stopband, implying that the total filter order should be odd.

(2) Generation of the cross port transfer function

This step is represented by (b)-(d) of Fig. 9.16. To generate the cross transfer function, the zeros on the unit circle are positioned first. These zeros give zero intensity transfer at their normalised frequencies and thus define the stopband width. There are side-lobes between each pair of zeros in the stopband. Increasing the distance between the zeros leads to a broader stopband but also to a higher side-lobe level. Hence, for a given number of zeros on the unit circle, there is a trade-off between isolation and stopband width. A common requirement is to have the maximum side lobe level at $\leq -25\,\mathrm{dB}$.

Next the passband shaping zeros are positioned inside or outside the circle. These zeros will create local minima at their corresponding positions and two surrounding local maxima. They should be positioned so that the passband becomes a half FSR shifted mirror image of the stopband with respect to the 50% power transmission level.

Adopting the formalism with A and B polynomials as introduced in (9.28), and applying (9.12), the magnitude squared of a cross transfer function, $B(z)$, with real coefficients is

$$\left|B(z)\right|^2 = \left|B(z)B(z^{-1})\right|_{z=e^{i2\pi f'}} \tag{9.34}$$

while the transfer function can be written in terms of its roots as

$$B(z) = \Gamma \prod_{m=1}^{M}(1 - z_m z^{-1}) \tag{9.35}$$

Substituting the unity zeros into (9.35) and applying (9.34), the magnitude squared function can be represented in terms of the unknown "passband" zeros and gain Γ. The locations of the maxima are given by the zeros of the derivative of (9.34) with respect to the normalised angular frequency f'. Since these zeros of the derivative are independent of the unknown constant Γ, the locations of the "passband zeros" can be determined from (9.34) and (9.35)

$$\frac{\mathrm{d}\left|B(z)\right|^2_{z=e^{i2\pi f'}}}{\mathrm{d}f'} = \Gamma^2 \prod_{m=1}^{M} \frac{\mathrm{d}}{\mathrm{d}f'}\left[(1 - z_m z^{-1})(1 - z_m z)\Big|_{z=e^{i2\pi f'}}\right] = 0$$

or

$$\prod_{m=1}^{M} \frac{\mathrm{d}}{\mathrm{d}f'}\left[(1 - z_m z^{-1})(1 - z_m z)\Big|_{z=e^{i2\pi f'}}\right] = 0 \tag{9.36}$$

Finally, the transfer function (9.35) should be normalised to one because the transmission of a passive device such as this Mach–Zehnder interferometer cannot exceed one. This procedure, which fixes Γ, is given by (9.37),

$$\max_{|z|=1} \left|B(z)\right| = \max_{\omega} \left|B(\omega)\right| = 1 \qquad (9.37)$$

(3) Calculation of the bar port transfer function

Once the cross transfer function has been obtained, the bar transfer function $A(z)$ can be calculated using the power complementary condition, (e) in Fig. 9.16. This condition, implying that the sum of the bar and the cross power transfer should be one, and assuming real coefficients of the transfer functions (allowing the use of (9.12)), provides a relationship between the A and B polynomials:

$$\left|A(f)\right|^2 = A(z)A(z^{-1})\Big|_{z=e^{i2\pi f}} = 1 - B(z)B(z^{-1})\Big|_{z=e^{i2\pi f}} \qquad (9.38)$$

The bar transfer function $A(z)$ is obtained by calculating its N zeros from (9.38). The $2N$ zeros of $1 - B(z)B(z^{-1})$ appear as pairs of $(a_k, 1/a_k^*)$ for $k = 1 \cdots N$. Using spectral factorization, each zero of $A(z)$ is determined by selecting one from each pair of zeros of $1 - B(z)B(z^{-1})$. There are 2^N selections that can be made to obtain the zeros of $A(z)$. Thus, 2^N different versions of $A(z)$ can be obtained from one known $B(z)$. They have the same amplitude response but different phase characteristics.

(4) Obtaining the optical parameters

The last step, (f) in Fig. 9.16, is the generation of power coupling ratios and phases of each directional coupler and delay line of the filter from the bar and cross port transfer functions. A simulation tool based on an algorithm derived by Jinguji and Kawachi [34] that can map the coefficients of the filter transfer function in a z polynomial to the optical parameters is used. The algorithm uses recursion equations to calculate the power coupling ratios of each directional coupler and the phase of each delay line.

Design Examples

Third order interleaver

The first logical extension of the (first order) MZI would be a second order device with a second zero on the unit circle. As explained before, this extra zero can be positioned so as to effect stopband broadening. In order to also flatten the passband, an additional zero (z_3) is needed which is located symmetrically with respect to the first two zeros, but at the opposite side of the imaginary axis, i. e. on the real axis.

The zero diagram of this third-order filter is shown in Fig. 9.17a. The power transfer is shown in Fig. 9.17c. Here the distance between the two zeros z_1 and z_2 has been chosen so that the maximum of the side lobe is -25 dB. Increasing the distance results in a higher side lobe and broader stopband.

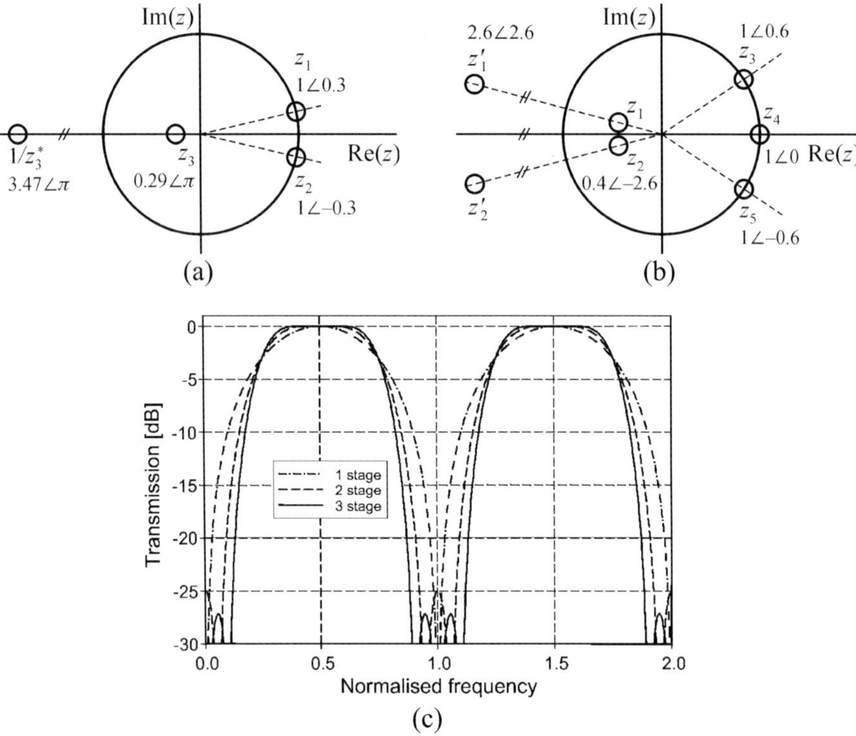

Fig. 9.17. (a) Zero diagram of the bar transfer $A(z)$ and the intensity transfer for the third order lattice filter. It has three zeros, two are on the unit circle (z_1, z_2) and give a zero transfer; the third zero is on the real axis and can be chosen inside (z_3) or outside ($1/z_3^*$) the unit circle. Both give the same amplitude transfer. (The notation $r \angle \varphi$ gives the modulus r and the argument φ [in radians] of complex z) (b) Zero diagram of the bar transfer $A(z)$ and intensity transfer for the fifth order lattice filter. It has five zeros, three are on the unit circle (z_3, z_4, z_5) and give a zero transfer, two zeros are on the opposite side of the imaginary axis (z_1, z_2) and can be chosen independently inside or outside the unit circle. Both choices give the same amplitude transfer. This graph must be mirrored about the origin to get the bar transfer. (c) Magnitude response for the 1, 2, and 3 stage slicer

Proper slicer operation requires identical cross and bar amplitude transfer functions, shifted over half the FSR. This condition is satisfied by correctly positioning the third zero z_3. There is still a degree of freedom left since the amplitude transfer does not change if z_3 is replaced by $1/z_3^*$. For both, the bar and the cross transfer, this zero can then be chosen to lie either inside (minimum phase) or outside (non-minimum phase) the unit circle, giving in total four possible solutions for this third order filter. These four different optical filter implementations have equal amplitude transfer but different phase transfer.

One of the couplers turns out to have $\kappa = 0$, which means that this coupler is removed and the two neighbouring delay lines are combined into one having the double delay. The three-stage filter is reduced to one having two stages. Since the number of tuning elements is equal to the number of delay lines, this implementation has also one tuning element less. The stopband width at -25 dB is 14% of the FSR or 28% of the channel spacing; 72% of the band is unavailable for data transmission. Since the filter is power complementary, -25 dB at the stopband can be calculated to correspond to a 0.014 dB ripple in the passband.

When looking at the four possible filter implementations, it can be seen that these four solutions can be split up into two groups where one contains mirror implementations of the other. This again shows that such a filter can be used both as filter and combiner. The last coupler can have a power coupling of either 0.923 or 0.077 $(= 1 - 0.923)$. The implementation with power coupling 0.077 is preferable since it is the shortest coupler and therefore less sensitive to wavelength.

Fifth order interleaver
Further improvement of the filter curve can be obtained by adding two more zeros to the diagram, as shown in Fig. 9.17b. Three zeros are on the unit circle (z_3, z_4, z_5) giving a broader stopband width (24% of the FSR at -25 dB). The distance between these zeros is chosen so that the side lobes in the stopband are -27 dB. The other two zeros are placed at the opposite side of the imaginary axis and not on the unit circle to obtain passband flattening (three local maxima). Due to the lower stopband sidelobes, this filter is also better passband flattened, and of course the cross and bar transfer shapes are equal, but shifted by half the FSR. The two "passband zeros" can be chosen individually to be inside or outside the unit circle. This results in four possible configurations for these two zeros, giving sixteen solutions for the bar and the cross functions. Some of the solutions have one coupler with $\kappa = 0$ allowing two neighbouring delay lines to be combined into a single one having the double delay. The best solution has two zero length couplers and chooses the shortest possible option for the remaining couplers.

9.3.2 An Infinite Impulse Response (IIR) Filter: MZI + Ring

There is a different way to design a passband-flattened interleaver, by combining a ring resonator inside an asymmetric MZI [35-37] as shown in Fig. 9.18. More complex structures can be found in [38]. The ring introduces a frequency-dependent nonlinear phase shift in one arm, while maintaining a unity amplitude response. The frequency response of the ring will be discussed before going into the details of this interleaver.

Figure 9.19 shows the ring resonator and a waveguide coupled to that ring. Part of the light that propagates through the channel is coupled into the ring waveguide and travels through the ring. After one roundtrip, part of that light is coupled back into the straight waveguide, and the remainder continues for a second roundtrip.

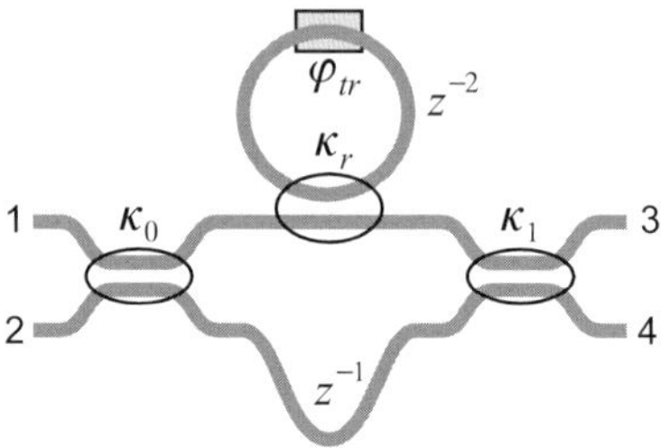

Fig. 9.18. Mach–Zehnder interferometer plus ring filter

This process continues until there is a stable solution. Two interrelated interference phenomena can be distinguished. First, there is the ring, where the total field distribution arises from waves that have made a number of roundtrips in the ring. Second, there is the straight waveguide where the direct light from the input interferes with waves that couple back from the ring. The output intensity is equal to the input if there is no loss. This interference clearly depends on the wavelength of the light. The roundtrip time of light propagating in the ring is τ_r, but we allow a small deviation in the average effective index n_{eff} of the ring, providing a phase adjustment φ_{tr}. The ring is in resonance, and intensity increases in the ring if the total roundtrip phase $\varphi_{ring} = 2\pi f \tau_r + \varphi_{tr} = 2\pi m$, where m is an integer.

The light that couples back into the channel is in anti-phase with the light from the input and will interfere destructively. The ring is in antiresonance when $\varphi_{ring} = \pi(2m + 1)$.

Introducing $z_r = \mathrm{e}^{\mathrm{i}2\pi f \tau_r}$, the transfer of the ring can be described by the following *Z-transform* polynomial [Ref. 27, pp. 306]

$$H_r(z_r) = \frac{\mathrm{e}^{-\mathrm{i}\varphi_{tr}}\left(c_r \mathrm{e}^{+\mathrm{i}\varphi_{tr}} - \eta_r z_r^{-1}\right)}{1 - c_r \mathrm{e}^{-\mathrm{i}\varphi_{tr}} \eta_r z_r^{-1}} = \frac{c_r\left(z_r - \dfrac{\eta_r}{c_r}\mathrm{e}^{-\mathrm{i}\varphi_{tr}}\right)}{z_r - \eta_r c_r \mathrm{e}^{-\mathrm{i}\varphi_{tr}}} \tag{9.39}$$

where $c_r = \sqrt{1 - \kappa_r}$, $\eta_r = \mathrm{e}^{-2\pi\alpha_r r}$, α_r is the ring waveguide attenuation coefficient and r is the ring radius.

We consider the lossless case, where the ring resonator acts as an allpass filter producing a non-linearly frequency-dependent phase shift which is given by

$$\varphi_r(f) = \tan^{-1}\left[\frac{\left(1 - c_r^2\right)\sin\left(2\pi f \tau_r + \varphi_{tr}\right)}{2c_r - \left(1 + c_r^2\right)\cos\left(2\pi f \tau_r + \varphi_{tr}\right)}\right] \tag{9.40}$$

and is shown in Fig. 9.19b for different values of the modulus of the pole location, $|z_p|$. In this lossless case $|z_p| = c_r$ is found from (9.39). The extreme case $|z_p| = c_r = 0$ corresponds to a power coupling constant $\kappa = 1$, meaning that all the light couples from the input into the ring, makes exactly one roundtrip, and then couples back completely to the straight waveguide. This is equivalent to a single waveguide which is lengthened by an amount equal to the circumference of the ring. As expected, its phase response is linear. For $|z_p| = 0.9$ only a small part of the power is coupled into the ring. Near

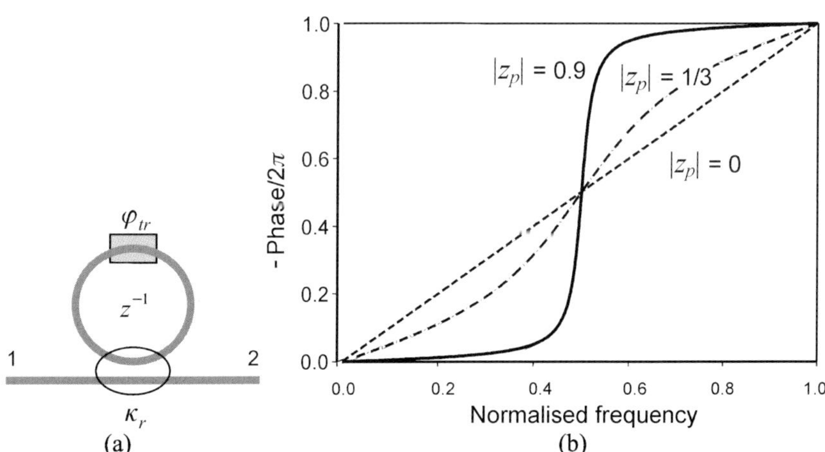

Fig. 9.19. (a) 1×1 port ring filter; (b) phase response for a lossless ring resonator with three different pole locations $|z_p|$, assuming that $\varphi_{tr} = \pi$

resonance a high intensity builds up, and the phase changes rapidly in a nonlinear fashion.

This non-linear phase shift can now be used inside the MZI (Fig. 9.18) where the ring is connected to the short channel. The time delay (or round-trip length) of the ring should be exactly twice the differential time delay of the MZI (τ_i) since the periodic nonlinear phase change should occur synchronously with the periodic MZI response curve in order for the ring to effect passband flattening and stopband broadening. The intensity transfer from port 1 to 3 (bar) and from 1 to 4 (cross) for the MZI + ring is given by (9.41) and (9.42), respectively,

$$\left| H_{11}(f) \right|^2 = \sin^2\left(\frac{\Delta\varphi(f)}{2} \right) \tag{9.41}$$

$$\left| H_{12}(f) \right|^2 = \cos^2\left(\frac{\Delta\varphi(f)}{2} \right) \tag{9.42}$$

where the DC's of the MZI have both $\kappa = 0.5$, and $\Delta\varphi(f) = \varphi_r(f) - 2\pi f \tau_i$, the difference between the phase of the ring-path and the phase of the through-path arm of the MZI. The bar transfer $\left| H_{11}(f) \right|$ is zero when $\Delta\varphi(f) = 2\,m\pi$, where m is an integer; it is 1 for $\Delta\varphi(f) = (2m + 1)\pi$. Passband flattening can be obtained by tuning the ring to be in anti-phase ($\varphi_{tr} = \pi$) at maximum transfer of the MZI.

Figure 9.20 shows the result obtained by adding the ring, with $\varphi_{tr} = \pi$. The intensity transfer is passband flattened and stopband broadened. The second graph shows the frequency-dependent phase of the two arms of the MZI with respect to the short arm of the same MZI without ring. The alternating long-short-dashed line represents the phase of the long arm (the one without ring). It has a phase change of 2π in one FSR. The dashed line gives the phase of the short branch in the case of 100% coupling to the connected ring. It has a phase change of 4π in one period ($\tau_r = 2\tau$). The solid line shows the phase for $\kappa_r = 0.82$ power coupling to the ring. The phase oscillates around the dashed line. There are exactly two periods of oscillations. The intensity transfer of the filter is now determined by the phase difference between the two channels as shown in the last graph. The centres of the passband and stopband occur at a phase difference of $m\pi$. The coupling coefficient has been calculated in such a way that near the centre of the pass- and stopbands the phase slope of the short branch + ring is equal to that of the long branch, resulting in a constant zero (or π) phase difference between the branches over a large fraction of these bands. As a result there is almost no change in the transfer. It is important that this stability occurs at a maximal or minimal transfer which is obtained by careful tuning of the phase of the ring relative to the MZI. The local maxima in the stop-band occur at the frequency where the slope of the phase difference is zero. There is a rapid transition in the transfer from passband to stopband because the ring is in resonance, which results in a fast phase change.

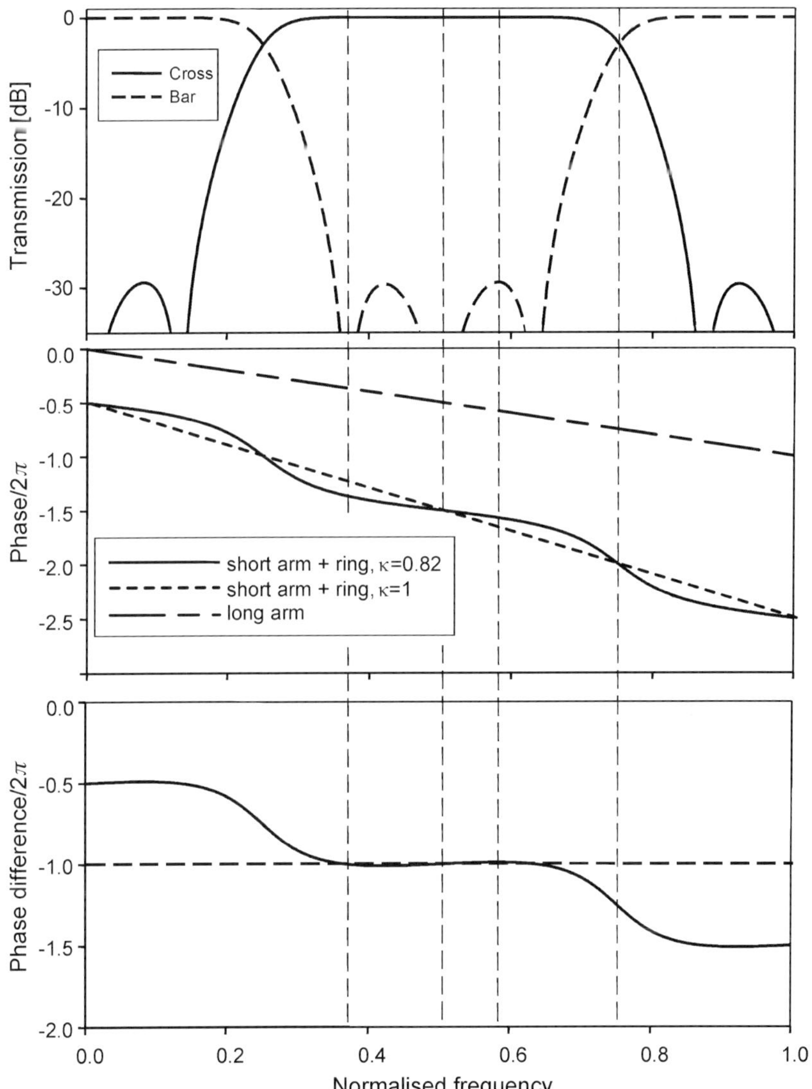

Fig. 9.20. Intensity transfer, phase of each arm (with respect to the short arm without ring), and the phase difference of a lossless, correctly tuned MZI + ring filter with perfect 3 dB couplers, and $\varphi_{tr} = \pi$

9.4 Group Delay and Dispersion

Dispersion is a measure of linearity of the phase response with respect to the frequency. The group delay is the local slope of the phase response curve, i.e., the slope of the phase at the frequency being evaluated. A filter's group delay or envelope delay is defined as the negative derivative of the phase response with respect to angular frequency ω as follows [30]:

$$\tau_g(\omega) = -\frac{\mathrm{d}\varphi(\omega)}{\mathrm{d}\omega} \tag{9.43}$$

If φ is in radians and ω is in radians per second, then the absolute group delay is given in seconds. For a sequence of discrete signals, each stage has a delay that is an integer multiple of a unit delay, τ_0. If the angular frequency is normalised to τ_0 such that $\omega' = \omega\tau_0$, then the normalised group delay τ_g' is given in number of unit delays τ_0, leading [27] to:

$$\tau_g = \tau_0\,\tau_g' \tag{9.44}$$

with a normalised group delay

$$\tau_g'(\omega') = -\frac{\mathrm{d}\varphi(\omega')}{\mathrm{d}\omega'} = -\frac{\mathrm{d}}{\mathrm{d}\omega'}\arg(H(z))\Big|_{z=\mathrm{e}^{\mathrm{i}\omega'}} \tag{9.45}$$

The filter *dispersion* is the derivative of the group delay. For normalised frequency $f' = f\tau$, the normalised dispersion D' is [27]

$$D' \equiv \frac{\mathrm{d}\tau_g'}{\mathrm{d}f'} = 2\pi\frac{\mathrm{d}\tau_g'}{\mathrm{d}\omega'} \tag{9.46}$$

and the filter dispersion D in absolute units [s/m] is [27]

$$D = \frac{\mathrm{d}\tau_g}{\mathrm{d}\lambda} = -c\left(\frac{\tau_0}{\lambda}\right)^2 D' \tag{9.47}$$

In comparison, for optical fibres the dispersion D_f is typically defined as the derivative of the group delay with respect to wavelength λ and normalised with respect to length L [27]

$$D_f = \frac{1}{L}\frac{\mathrm{d}\tau_g}{\mathrm{d}\lambda} \tag{9.48}$$

Practical units for D_f are [ps/(nm·km)].

9.4.1 MZI Group Delay and Dispersion

The normalised group delay of the non-ideal MZI with identical directional couplers can be calculated from (9.33) and (9.45)

$$\tau'_g(\omega') = \frac{\dfrac{C^2}{S^2}\left[\dfrac{C^2}{S^2} - \cos\omega'\right]}{1 - 2\dfrac{C^2}{S^2}\cos\omega' + \dfrac{C^4}{S^4}} \tag{9.49}$$

Figure 9.21a shows the normalised group delay of the MZI for different coupling constants. Note that the ideal MZI ($\kappa = 0.5$, $C = S$) has a constant group delay and thus no dispersion. Figure 9.21b shows the normalised dispersion of the MZI. Note that the dispersion sweep is in the stopband region and that dispersion is low in the passband region.

The group delay and dispersion go to infinity as κ goes to 0.5. This is possible since this is in the stopband region and the intensity transfer goes to zero.

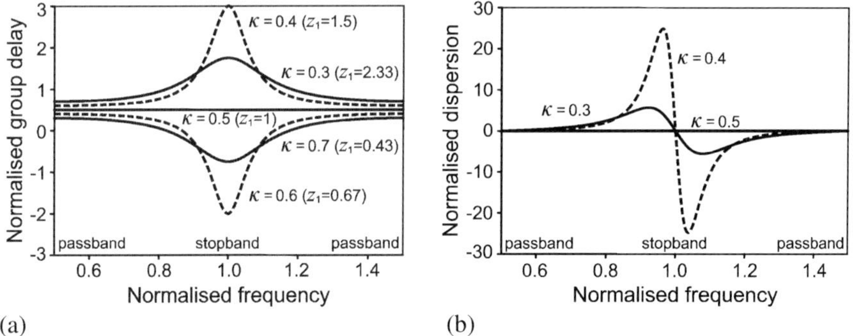

Fig. 9.21. Bar transmission of the MZI for various coupling constants. (**a**) Group delay. (**b**) Dispersion

9.4.2 Third- and Fifth-order Lattice Filter Group Delay and Dispersion

The dispersion for the 3rd order interleaver is non-zero as shown in Fig. 9.22a, in contrast to the ideal MZI. There is a frequency-dependent group delay having a minimum at the centre of the passband, resulting in a zero of the dispersion. The minimum normalised dispersion is -2.6 which is equivalent to 1.9 km of standard single-mode fibre for a 100 GHz FSR filter. For the 5th order interleaver, Fig. 9.22b, the dispersion goes to

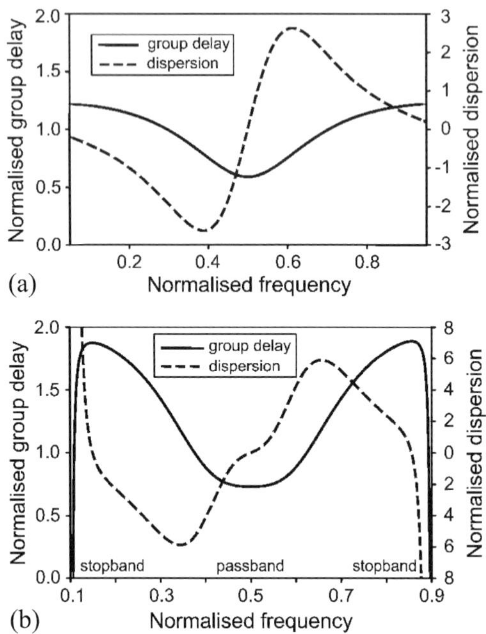

Fig. 9.22. Normalised group delay and normalised dispersion of bar response of minimum phase 3rd and 5th order filters. (**a**) 3rd order 2-stage filter; the minimum and maximum dispersion are –2.6 and 2.6, respectively. (**b**) 5th order 3-stage interleaver; the minimum and maximum dispersion are –5.9 and 5.9, respectively

infinity when reaching the stopband, which is not interesting since the intensity is low. The minimum normalised dispersion is –5.9 which is equivalent to 4.3 km of standard single-mode fibre for a 100 GHz FSR filter.

9.4.3 MZI + Ring Group Delay and Dispersion

Figure 9.23 shows the overall phase, delay, and dispersion of the bar transfer $H_{11}(z)$ of the MZI + ring filter. The normalised dispersion is zero at the centre of the passband and goes from negative to positive in the passband region. The extremes are –22 and +22 at a normalised frequency of 0.29 and 0.71, respectively. The transfer is –0.5 dB at these points. The dispersion of the filter does not depend on the chosen input and output ports. It will always give the same dispersion curve. So the dispersion will always be doubled when two MZI + ring interleavers are cascaded.

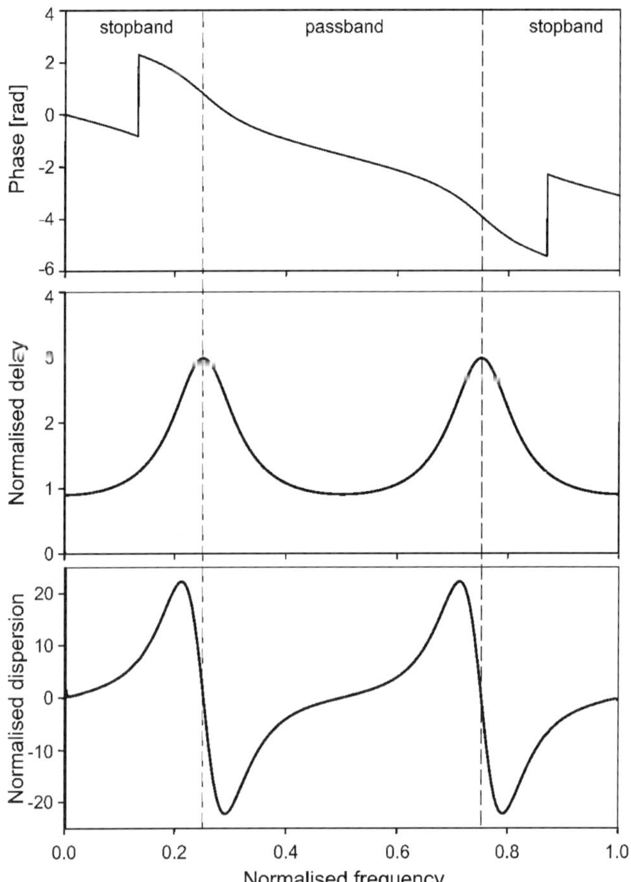

Fig. 9.23. Phase, normalised delay, and normalised dispersion of the cross transfer of MZI + ring filter

9.5 Cascaded Interleavers

As indicated in the introduction, interleavers can be used as building blocks to form more complicated optical functions. One way of cascading interleavers has been extensively discussed in Sect. 9.3.1. In this lattice architecture, the overall cascade had the same basic functionality as a single interleaver, but its transfer function was strongly improved.

One way of extending the functionality is to build a cascade based on a binary tree architecture as shown in Fig. 9.4. Full demultiplexing of N channels requires N-1 slicers. Therefore, optical integration technology is essential for economic fabrication of such composite devices, as pioneered by Verbeek et al. [4] and further developed by NTT, e. g. [39]. Compared to other multiplexers, for example the arrayed waveguide grating (AWG, see Chap. 4), this architecture does not seem to scale very well with the number of multiplexed optical channels. However, it offers a large degree of flexibility for engineering the transfer function, e. g. [23, 24], and especially for (fine-)tuning to individual channels and reconfiguring the assignment of optical signal channels to the output ports. The signals at the two output ports of an MZI interleaver are swapped by introducing a π phase shift in one of the MZI branches. Since this can be done for each of the slicers in the binary tree, each wavelength channel can be routed to each of the output ports, and many – but by no means all – channel permutations can thus be obtained. In this way, a binary tree slicer arrangement combines the functionality of optical frequency (de)multiplexing with switching which may make it an efficient architecture for a number of wavelength routing applications.

If full demultiplexing is not required, for example for dropping a single channel from a multiplexed signal, a partial binary tree may be used. Only $\log_2 N$ slicers are required if the unwanted channels can be discarded. If the other channels should be kept, the partially demultiplexed signal must be remultiplexed using another identical partial binary tree. An example of a 1 from 8 add-drop multiplexer using this principle is shown in Fig. 9.5. The add-function comes "for free" with the drop function if a 2×2 interleaver is applied for the deepest stage, and a total of $2 \log_2 N - 1$ slicers is needed.

It is worth noting that the "quality" of the transfer function is not necessarily equal for each of the slicers in a cascade. The most critical stage is the first one which has the smallest free spectral range since its function is to separate adjacent channels. Its properties mainly determine the flatness and width of the passband and the adjacent channel isolation of the composite device. A lower order lattice filter can be used for the following stages since the channel width will be a smaller fraction of the FSR of those stages. The principle is shown in Fig. 9.24.

The isolation of the drop channel with respect to the add channel deserves special attention because these channels – separated by only a single large-FSR slicer in the configuration of Fig. 9.5 – have the same nominal frequency, and the level of the locally generated add signal may be relatively high. This may cause strong interference products within the bandwidth of the detection system (in-band crosstalk). The potential problem arising from this is elegantly solved by adopting an add-after-drop configuration at the expense of a single additional slicer, one of the "D" blocks, as illustrated in Fig. 9.25.

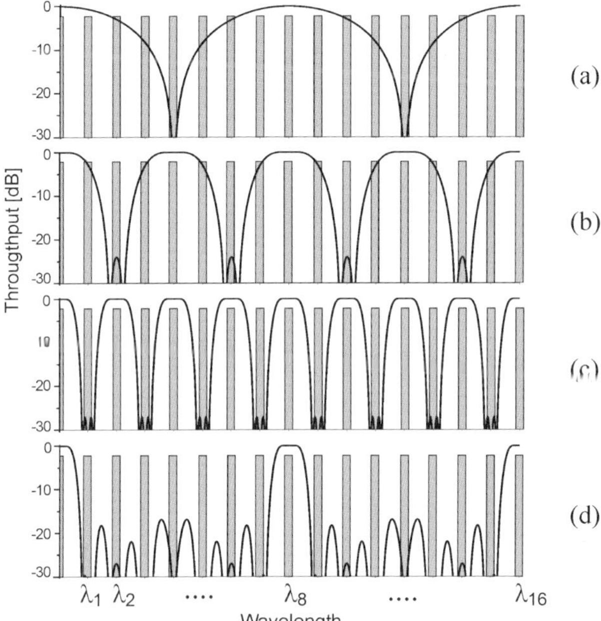

Fig. 9.24. Transfer functions of different lattice filters. (**a**) Single-stage MZI. (**b**) Two-stage (third order). (**c**) Three-stage (fifth order). The ratios of the free spectral ranges of the slicers are 4:2:1 for slicers a, b, and c, respectively. (**d**) Transfer function from the input to one of the outputs of the binary tree cascade of these three interleavers

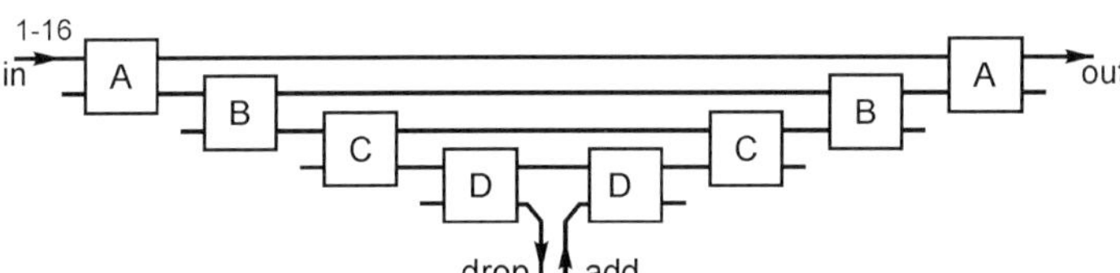

Fig. 9.25. Schematic drawing of 1 from 16 add-drop multiplexer with add-after-drop functionality for a 200 GHz channel spacing. FSR ratios are 1:2:4:8 for A:B:C:D-type slicers, respectively. For obtaining a 25 dB isolation over > 85 GHz stopband, A-type slicers were chosen to have 3 stages, B-type 2 stages, and C and D-types 1 stage [5]

An example of the lay-out of a thermo-optically tuneable 1 from 16 add-after-drop multiplexer, based on partial binary trees of different order lattice filters, and designed to select channels on a 200 GHz ITU grid in the C-band, is given in Fig. 9.26. The device was fabricated using silicon oxynitride technology [40, 41]. Its in-to-drop and in-to-out characteristics when tuned to a given channel are shown in Fig. 9.27.

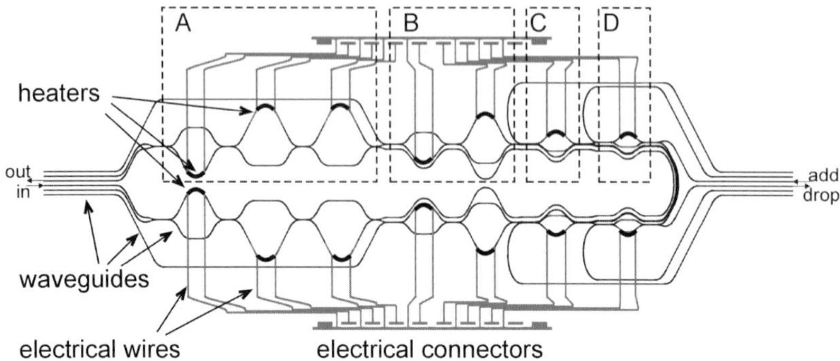

Fig. 9.26. Lay-out of a tuneable 1 from 16 add-drop multiplexer using the architecture shown in Fig. 9.25 [5]. The optical circuit has been folded about the vertical centre line. The locations of A, B, C, and D type slicers are indicated. Metal connections of the heaters used for thermo-optically tuning the individual delay sections are also shown

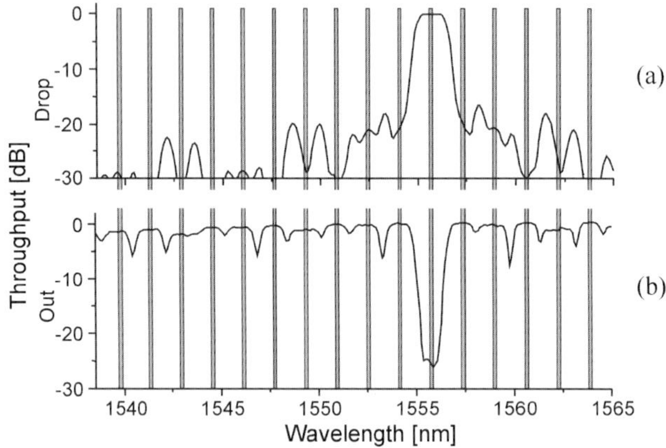

Fig. 9.27. Measured transfer functions: in-to-drop (**a**) and in-to-out (**b**) of the multiplexer shown in Fig. 9.26. The grey vertical bars indicate channels of the 200 GHz ITU grid for which the filter was designed. Measurements were done with unpolarised light

In Sect. 9.3.2 the combination of an MZI with a ring resonator has been discussed. A different kind of such a combination, placing the ring outside the MZI (in fact just cascading ring and MZI), has been proposed by Vázquez et al. [42]. They introduce gain into the ring, leading to a strongly peaked transmission spectrum instead of the allpass characteristic described in Sect. 9.3.2. This might make the applicability of their configuration as an interleaver or demultiplexer in WDM systems questionable.

9.6 Realisations of Interleavers in various Fabrication Technologies

As stated in the introduction, interleavers can be built from any type of wavelength filter having a periodic frequency response. This section aims at giving a flavour of recent realisations of interleavers based on different operating principles and using different technologies.

9.6.1 Bulk and Fibre Optics

Although this chapter focuses on integrated optical realisations of interleavers, most commercially available devices at the time of writing are based on bulk optics.

Lattice Filters

The implementations, based on birefringent crystals, that are equivalent to MZI lattice filters have been known for a long time as Lyot [43] or Solc [44] filters. Many different variations of the basic principle have been realised [1]. In such filters (see Fig. 9.28), which found their early applications mainly in astronomic instruments, the two fundamental polarisations in the crystals, the ordinary (o) and extraordinary (e) waves, experience a different refractive index ($n_e - n_o = \Delta n$), and hence a frequency-dependent phase difference $\Delta\varphi$ given by

$$\Delta\varphi = \frac{2\pi}{c} f \Delta n\, L \qquad (9.50)$$

where L is the length of the crystal. Equation (9.50) is very similar to (9.2). The angular offset between the optic axes of consecutive crystals causes a mixing of the o- and e-waves from one crystal to the next, just as the directional couplers do between consecutive delay sections in an MZI lattice filter. Instead of rotating the crystals, half-waveplates may be located in between the aligned crystals. The device of Fig. 9.28 works only with a single input polarisation, but, exploiting the complementary mixing properties of orthogonal polarisations in a configuration as shown in Fig. 9.29, polarisation-independent operation can be obtained [45].

Instead of birefringent crystals, also artificial birefringent units can be used, consisting of polarisation beam splitters providing two different propagation paths (through different isotropic materials, e. g. glass blocks) for the two orthogonal polarisations, as illustrated in Fig. 9.30 [46].

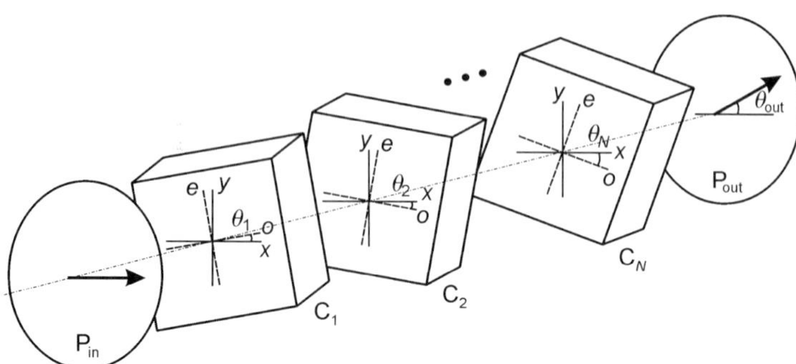

Fig. 9.28. Principle of a lattice filter based on birefringent crystals. Due to the refractive index difference for the ordinary (o) and extraordinary (e) waves, each crystal produces a differential delay of the fundamental polarisation components. The angular offsets between consecutive crystal's optic axes determine the coupling between the o- and e-waves from one crystal to the next. P: polarizer, C: crystal. After [47]

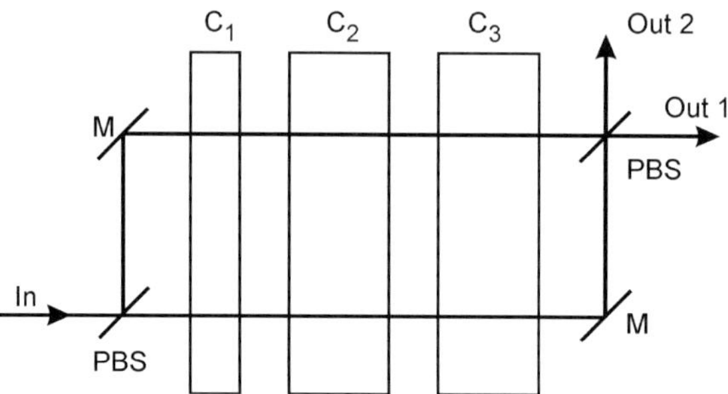

Fig. 9.29. Polarisation-independent birefringent crystal lattice filter using polarisation diversity. M: mirror; PBS: polarisation beam splitter; C_i: crystal i. After [45]

The same authors designed an electro-optically tuned interleaver as a variation of the structure of Fig. 9.30 where one of the blocks has been replaced by an electro-optic lithium niobate crystal. It requires about 4 kV for tuning over its free spectral range [48].

Since the crystals or the glass blocks will usually need to be many thousands of wavelengths long, the effects of thermal expansion can be severe. Compensation of thermal drift can be obtained by applying carefully matched combinations of different crystals. Although lattice-type filters can be designed to have zero dispersion (see Sect. 9.4), it has also been proposed to use matched interleavers as multiplexers and demultiplexers, respectively, one having positive and the other negative group delay. An interleaver that can be changed to have positive or negative group delay by moving a half-waveplate to a different position is shown in Fig. 9.31 [49].

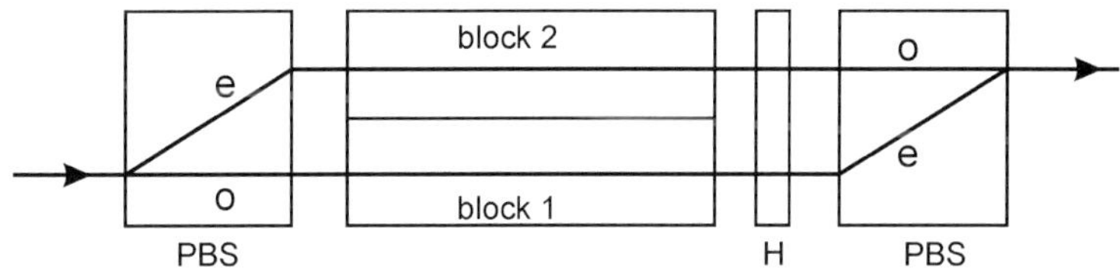

Fig. 9.30. Differential delay unit using "artificial" birefringence by providing separate propagation paths for the orthogonal polarisations through two blocks of different materials, after [46]. PBS: birefringent walk-off crystal serving as polarising beam-splitter; H: half-waveplate rotating the polarisation by 90 degrees

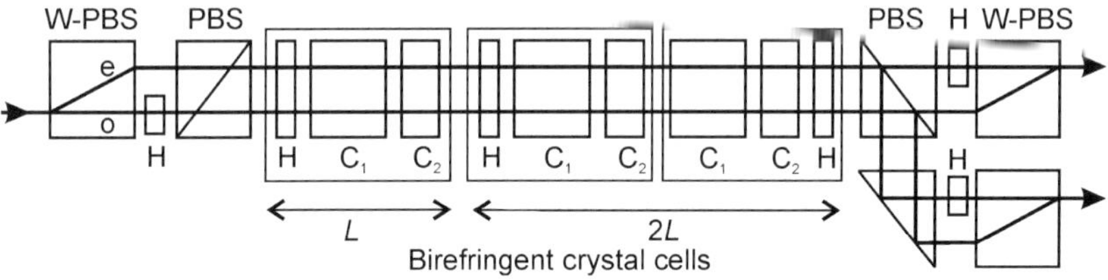

Fig. 9.31. Interleaver with selectable positive or negative group delay, after [49]. Selection is done by moving the input half-waveplate from the o-beam to the e-beam. W-PBS: walk-off polarising beamsplitter; PBS: polarising beamsplitter; H: half-waveplate; C_1, C_2 temperature-compensating birefringent crystals

Michelson–Gires–Tournois Interferometers

The equivalent of a ring resonator is the Fabry–Perot resonator in bulk optics. A Fabry–Perot resonator with one 100% mirror, which can only be used in reflection, is known as a Gires–Tournois resonator (GTR) [50], see also Chap. 6, Sect. 6.3.3. It acts as an allpass reflection filter having a periodic and strongly nonlinear phase response, just as the ring resonator coupled to a single waveguide does in transmission (Fig. 9.19).

Then, the bulk equivalent of the Mach–Zehnder interferometer with a ring resonator coupled to one of its branches is the Michelson interferometer (MI), where one or two of the mirrors of the MI is/are replaced by a GTR, thus forming a so-called Michelson–Gires–Tournois interferometer, see Fig. 9.32 and [51, 52]. If both MI mirrors are replaced by GTR's, a more rectangular-shaped passband can be obtained [53]. Although one of the interleaver output signals will be back reflected, it can be separated from the input signal by tilting some of the mirrors. More complicated designs and several references to the patent literature can be found in [1].

A fibre analogue of a Michelson–Gires–Tournois interferometer has been demonstrated in [54].

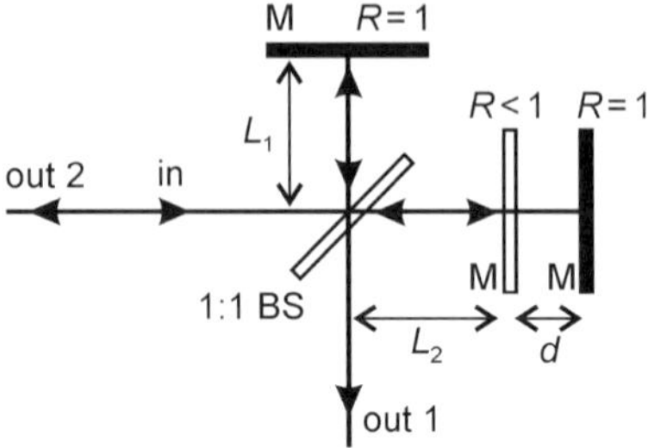

Fig. 9.32. Michelson–Gires–Tournois interferometer filter, after [52]. The back-reflected light has a frequency response that is complementary to that of the transmitted light, and the input signal will appear frequency interleaved at the reflection and transmission port of the device

Fabry–Perot Resonator Arrays

Fabry–Perot (FP) resonators are discussed in Chap. 6. By arranging a number of such resonators in series, while carefully choosing the mirror reflectances, the rectangular-shaped transfer function that is desirable for interleaver operation can be well approximated, see the configuration of Fig. 9.33 [55].

This technique is not limited to conventional bulk FP etalons, but it can also be applied to the design of thin-film filters [56].

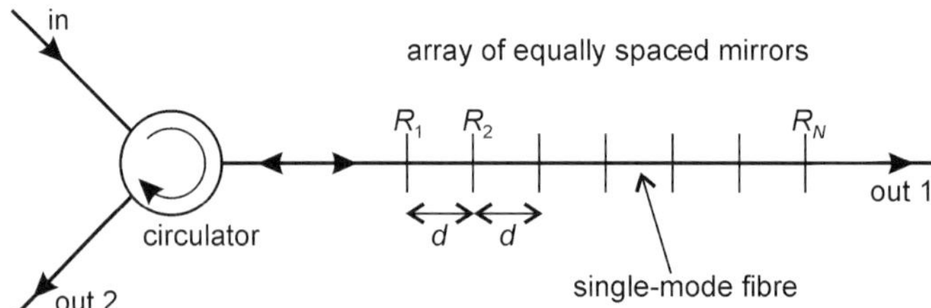

Fig. 9.33. Interleaver based on Fabry–Perot resonator arrays using a circulator for separating input and reflected output, after [55]. Outputs 1 and 2 are complementary, producing frequency-interleaved signals

Fibre-based Interleavers

Fibre Bragg gratings (FBG's) are very suitable for fabricating high-quality filter elements, see Chap. 5. Interleavers can be built by combining several equidistantly tuned FBG's, see e. g. [57]. The principle is shown in Fig. 9.34.

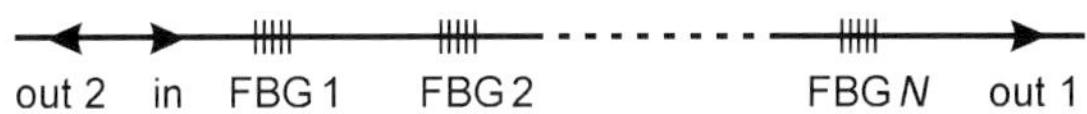

Fig. 9.34. Array of fibre Bragg gratings (FBG) [57]

As in Fig. 9.33, the reflected output can be separated from the input by using a circulator. A related technology is engineering a fibre Bragg grating in such a way, by sampling and chirping, that it reflects multiple channels [25, 58] which can be selected to provide interleaver functionality. This technique has been taken further by combining the fibre Bragg grating-based FP resonators with a fibre MZI interleaver [59].

Besides the fibre Bragg-grating technology mentioned above, lattice-type fibre MZI interleavers have been realised. Due to the relatively long fibre length of the branches – almost inevitable because of handling constraints –, inaccuracies of cutting fibre lengths, and thermal drift tuning facilities and active stabilisation circuits are generally needed. An interesting approach using a reference light source with a wavelength outside the regular transmission band is shown in Fig. 9.35 [60, 61]. Other variations include a reflective fibre MZI using a loop mirror or a fibre Bragg grating [62] and a configuration using highly birefringent fibres [63]. Li et al. [64] exploited the interference between two modes in an 8 cm section of two-moded fibre for obtaining a compact thermally tuneable MZI-like filter with 1.72 nm FSR.

A 3×3 (or 1×3) interleaver based on three-arm fibre MZI filters, using symmetric 3×3 fused fibre directional couplers, was investigated by Wang et al. [65]. Fewer of such filters would need to be cascaded for separating a given number of channels, compared to 2×2 interleavers. A 100/300 GHz interleaver with 40 GHz passband width at the −0.5 dB level has been demonstrated. The same authors propose a three-armed MZI with allpass ring-resonators coupled to each of the arms [66]. They show the design of

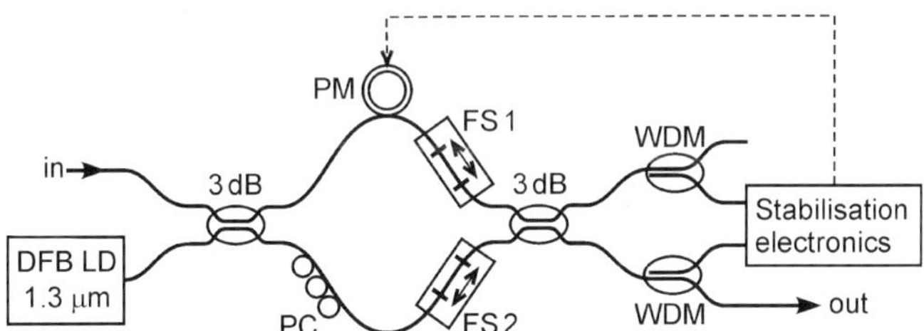

Fig. 9.35. Fibre MZI, stabilized using a reference light source. The filter can be tuned by changing the reference wavelength. DFB LD: distributed feedback laser diode, 3 dB: 3 dB-directional coupler, PM: phase modulator (piezoelectric microstretcher), PC: polarisation controller, FS: fibre stretcher (mm range), WDM: coarse wavelength division multiplexing fibre coupler [61]

a 100/300 GHz 3×3 interleaver with a 85 GHz passband width. However, the transmission curves of adjacent channels overlap somewhat in the transition region.

A binary tree cascade of fibre MZI's resulting in a 16-channel demultiplexer with 100 GHz channel spacing has also been demonstrated [67].

9.6.2 Planar Optical Waveguide Interleavers

Most work on interleaver filters in planar technology has been done in silica-based waveguide systems, especially by AT&T/Lucent [4, 33, 15, 21, 68, 69] and NTT [70, 7, 3, 71, 2], but also others [72]. Temperature dependence in such filters can be compensated by introducing short grooves filled with a material having a thermo-optic coefficient of the opposite sign as that of silica [73]. A clear trend in this technology is towards higher refractive index contrast in order to allow for smaller bend radii and hence more compact designs, as illustrated by the interleaver described in [74].

Stronger refractive index contrast can be obtained by using different materials, e. g. silicon oxynitride (SiON) [40, 41]. This technology has been used by IBM and the University of Twente for fabricating resonant-coupler-based filters and binary tree cascaded interleavers [75–77, 5].

Still higher contrasts are possible with semiconductor materials like GaAs [78] or silicon [79]. Although these materials certainly allow for more compact designs, they also put high demands on etching process quality because roughness will cause strong scattering loss due to the high contrast. Also, fibre-chip coupling becomes increasingly difficult as contrast increases.

Polymers can be attractive for these applications [80, 81] because of their simple processing compared to the inorganic materials mentioned before and hence possibly lower cost. Also, they generally have an order of magnitude larger thermo-optic coefficient so that tuning power requirements can be lower. These materials may, however, have problems with long-term stability.

Lithium niobate is an attractive material because of its electro-optic properties and mature technology. Filters fabricated in this technology are often based on frequency-selective mode conversion, e. g. [82]. Electro-optically tuneable interleavers based on this principle are described in [83], Fig. 9.36.

Cusmai et al. [84] proposed a device based on a three-arm MZI, using three-way directional couplers instead of the usual 2×2 devices. Different from the device reported in [65], the power coupling ratio of the directional couplers is not 1:1:1 here. The device and its equivalent built from

2×2 MZI's are shown in Fig. 9.37. The 2×2 cascade may have the advantage that a better isolation can be obtained for a given fabrication accuracy.

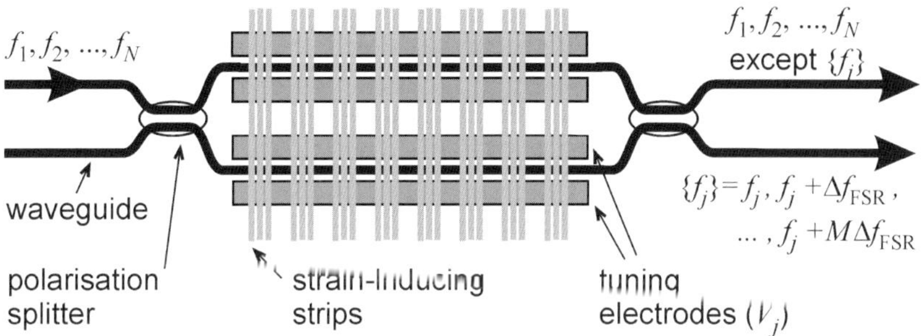

Fig. 9.36. Electro-optically tuneable lithium niobate interleaver [83]. The strain-inducing strips produce an off-diagonal element in the refractive index tensor leading to TE-TM mode conversion. The frequencies at which the phase matching condition for efficient mode conversion occurs is tuned by a voltage on the electrodes, changing the birefringence of the lithium niobate crystal

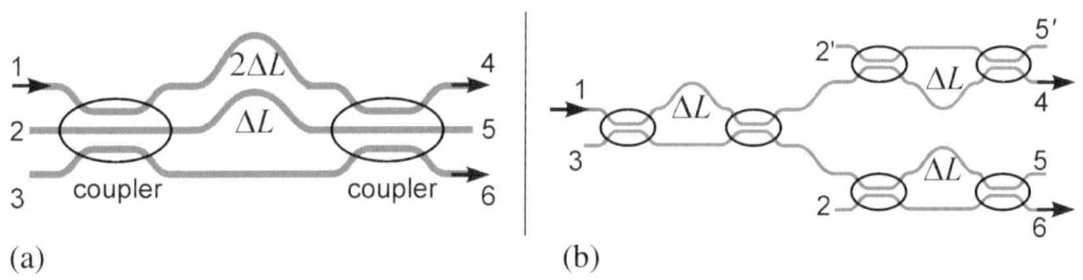

Fig. 9.37. (**a**) Three-arm MZI, and (**b**) its equivalent built from conventional two-arm MZI's (after [84]). The 3-way directional coupler is designed to have a 50:0:50 power splitting ratio from the central input channel to the respective output channels and a 25:50:25 ratio when excited at one of the outside channels. In normal operation, the 3-arm MZI is applied as a 2×2 device, using only the outside input (1 and 3) and output ports (4 and 6)

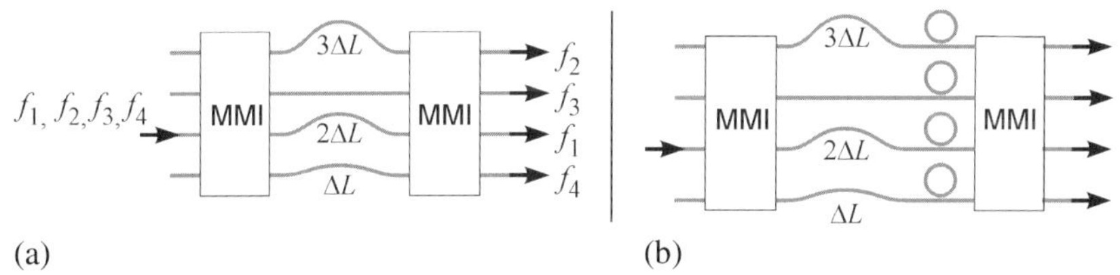

Fig. 9.38. Multiple-branch MZI interleavers (MMI: multi-mode interferometer). (**a**) Basic periodic 4×4 demultiplexer; for (1×4) interleaver functionality only a single input is used. (**b**) Improved transfer function by adding nonlinear phase shifters (ring resonators) to each branch (after [87])

As can be inferred from the equivalent circuit, the 3-arm MZI has the squared response of a single 2-arm MZI, thus providing a limited degree of passband flattening and stopband broadening. Compared to the 2-arm MZI, it also has reduced sensitivity to fabrication errors affecting the splitting ratio of the couplers. It should be noted that another approach by Oguma et al. [71] provides a method for obtaining highly accurate splitting ratios using a composition of inaccurate couplers and delay-lines.

Generalised multiple-branch MZI's have been introduced before where multi-mode interferometers (MMI's) act as multi-port couplers (Fig. 9.38a), e. g. van Dam et al. [85] and Lierstuen and Sudbø [86]. This type of device, shown in Fig. 9.38 for 4 input $\times$ 4 output ports, can be generalised to $N \times N$ ports and bears a strong similarity with the arrayed waveguide gratings (AWG's) that have been discussed in Chap. 4. MMI-based devices may be designed to have smaller insertion loss than AWG's, but they have fewer degrees of freedom for optimisation of the transfer functions. The concept was further optimised for passband flattening by Madsen [87] by introducing nonlinear phase shifters in each of the MZI branches (Fig. 9.38b), similar to the MZI + ring discussed in Sect. 9.3.2. Xiao and He recently discussed crosstalk reduction of MMI-based demultiplexers by cascading each of the output channels with a 1×1 multiple-branch MZI interferometer (consisting of a cascade of a $1 \times M$ and an $M \times 1$ MMI, interconnected by M waveguides providing the appropriate phase shifts) acting as a band-pass filter. They also modelled a 1×16 demultiplexer built from a binary-tree-type cascade of 1×4 interleavers [88].

In Sect. 9.1 the application of an interleaver for reducing the channel spacing and improving the passband shape of an AWG router by Oguma et al. [3] has already been mentioned. Although it is not an interleaver application in the strict sense, it is worth mentioning that Doerr et al. [68, 69] demonstrated significant passband flattening of an AWG by directly connecting the two output channels of the 3 dB output coupler of a single-stage MZI periodic filter to the AWG input coupler. The field distribution at the combined MZI output channels shifts periodically with frequency.

If the FSR of the MZI is equal to the AWG channel spacing, the combination of the Gaussian imaging properties of a basic AWG with the periodic shift of its input field leads to a passband-flattened overall response.

Photonic crystals [89] are an emerging technology for integrated optics. They promise extremely compact optical circuits, and their strong dispersive properties make them interesting for designing optical filters. As yet, the technology is still immature, the main problems being the optical loss and the fabrication accuracy of the sub-micron features of these structures.

Fig. 9.39 shows a proposed interleaver which operates like an extremely compact two-mode interference device [90, 91].

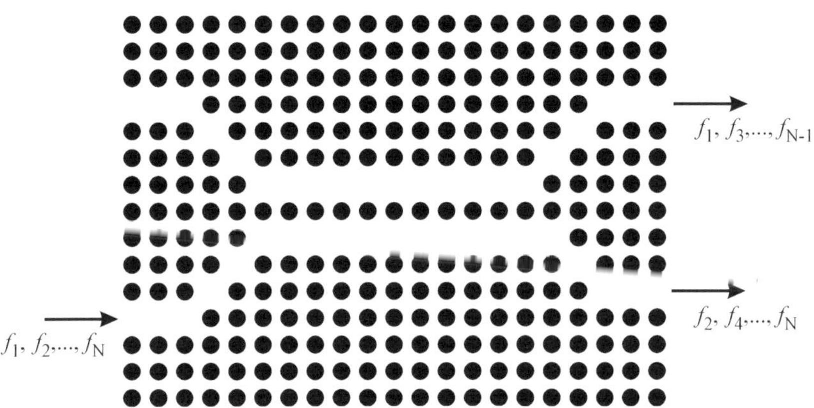

Fig. 9.39. Two coupled waveguides in a photonic crystal form a two-mode interferometer acting as an interleaver [90]

9.6.3 Some Commercially Available Interleavers

As can be seen from the references in the preceding sections, several large companies such as NTT, Lucent, and IBM have investigated interleaver technologies. However, although these results may have been applied in proprietary systems, these companies do not seem to have marketed such interleavers as separately available products. Also, several interleaver patents have been assigned to other companies that do not seem to be active in this market; see e. g. the references in [1]. Most of the commercially available interleavers that were visible on the Internet in the period of June 2004 to May 2005 are represented in Table 9.1. More details about the suppliers and model designations are given in Table 9.2.

Most manufacturers do not supply detailed information on the technology that is used; no data was available on the specific designs. A distinguishing feature of these interleavers is the width of the passband that ranges from 25 to 68 % of the free spectral range.

Table 9.1. Specifications of commercially available interleavers

Ref. Code [a]	AOC	AVA	FNR	GLD	HIT	ISO	JDS	KCT	OEL	OLK	OPX	TLE	TXN	WSR
Technology [b]	X(B?)	X	X(A?)	B	E	A	X	F	X(B?)	X	X	X	D	C
Ports [in×out]	1×2	1×2	1×2	1×2	1×2	1×2	1×2	1×2	1×2	1×2	1×2	1×2	1×2	1×2
Channel spacing [GHz]	50	50	50	50	50	100	50	50	50	100	50	100	50	50
Wavelength range [c]	C/L	C/L	C/L	C/L	C/L	C/L	C/L	C/L	C/L	C/L	C/L/cs	C	C/L	C/L
Passband (min) [GHz] [d]	21	20	20	15	20	68	16 B	20 C	21	30 B	30	25 C	20	22
Insertion loss (max) [dB]	2	2	1.8	1.5	2.5	0.5	2	2	2.2	1.8	2.2	1.2	0.3	2
Passband ripple (max) [dB]	0.5		0.5			0.5	0.4	0.2	0.5	0.4	0.2	0.2	0.2	0.5
Isolation, adj. ch. (min) [dB] [e]	25	27	25	35	28 A	25	25	22	25	25	25	25	23	22 A
Return loss (min) [dB]	45		45	60	40	50	45	50	45	45	45	50		45
PDL (max) [dB]	0.3		0.2	0.2		0.05	0.4	0.15	0.3	0.4	0.2	0.2	0.1	0.3
PMD (max) [ps]	0.2		0.2				0.2	0.1	0.2	0.3	0.1	0.2	0.2	
Chrom. disp. (max) [ps/nm]	±15	±35	±20		±15		±20	±35	±15	±20	±35			15
Operating temperature [°C]	−5/+65	−5/+65	−5/+70	0/+70	0/+65	0/+70	0/+65	0/+70	0/+65	0/+65	0/+65	0/+45	−5/+70	0/+65
Telcordia 1209/1221	yes	yes	yes						yes		yes		yes	
Features [f]		B	A,B	A		A,B	A	B	B	A	A			
Remarks [g]	B	B,C	B	A	B,C		B	B	B		B		B	

A representative interleaver has been selected from each brand, if possible with 50 GHz channel spacing.

[a] See Table 9.2 for an explanation of codes.

[b] A = crystal; B = fibre; C = fibre MZI; D = fibre Bragg grating; E = planar; F = MZI; X = unknown.

[c] Definitions vary among manufacturers. C = C-band (~1530 – 1560 nm); L = L-band (~1569 – 1604 nm); C/L = C or L; cs = custom.

[d] Passband defined at −0.5 dB level, except: A = at 0.6 dB level; B = "clear bandwidth"; C = condition unspecified.

[e] A = crosstalk.

[f] Advertised features: A = athermal; B = passive.

[g] A = This device has possibly been withdrawn; B = other channel spacings available; C = other port counts available.

Table 9.2. Reference codes and suppliers

Ref. Code	Brand	Type	Web site	Headquarters location
AOC	AOC Technologies	AOC 50/100 Opt. Interleaver	www.aoctech.com	Dublin, CA, USA
AVA	Avanex	PowerMux	www.avanex.com	Fremont, CA, USA
FNR	Finisar	50 G Interleaver	www.finisar.com	Sunnyvale, CA, USA
GLD	Gould	50 GHz Interleaver	www.gouldfo.com	Millersville, MD, USA
HIT	Hitachi	1×2 25 – 50 GHz Wavelength Splitter	www.hitachi-cable.co.jp	Tokyo, Japan
ISO	Isomet	3rd Generation Interleaver	www.isomet.com	Springfield, VA, USA
JDS	JDS Uniphase	IBC 50/100 (de)mux Interleaver	www.jdsu.com	San Jose, CA, USA
KCT	Koncent	100 GHz Interleaver	www.photoptech.com/koncent	Fuzhou, China
OEL	O/E Land Inc.	WINT 50/100 GHz Interleaver	www.o-eland.com	Saint Laurent, Quebec, Canada
OLK	Oplink	CF01 100	www.oplink.com	San Jose, CA, USA
OPX	Optoplex	Nova Interleaver	www.optoplex.com	Fremont, CA, USA
TLE	Teleste	DVO553	www.teleste.com	Turku, Finland
TXN	Teraxion	TF-WDM	www.teraxion.com	Sainte-Foy, Canada
WSR	Wavesplitter	Waveprocessor F3T	www.wavesplitter.com	Fremont, CA, USA

9.7 Outlook

Interleaver technology provides a route towards a cost-effective increase of bandwidth utilisation of optical fibre communication links. They can be used as add-on components for upgrading existing systems as well as for building blocks in innovative wavelength routers and multiplexers with a high channel count.

Although much effort has been devoted to optimising interleavers in bulk-type technologies which has given rise to highly sophisticated devices, most progress is to be expected from planar waveguide technologies with their inherent potential for integration. This integration will enable the realisation of large cascades of interleavers, each of which may consist of many stages. Such composite devices allow the elimination of many fibre connections, thus promising increasing functionality, improved reliability, and reduced insertion loss in a small package. In order to accommodate the desired functionality on a limited chip area, there is a clear trend towards increasing the integration density by using stronger refractive index contrast. This high contrast enables emerging VLSI photonics technologies like micro ring resonator arrays and photonic crystals. Planar technology also provides the potential for low-voltage and low-power electrical tuneability of the interleavers, using thermo-optic or electro-optic effects.

It should be emphasized that the resonant coupler approach, possibly in combination with ring-resonator based allpass filters, which has been shown in Sect. 9.3 for interleaver design, has a much wider applicability to filter design. For example, (gain) equalisers, dispersion compensators, and deconvolution filters may be designed using a very similar approach and may be realised in the same technology, thus leading to a cross-fertilisation among these fields.

If hitherto largely unused wavelength ranges in optical fibre communications (e. g. the range between the 1300 and 1500 nm windows) will be opened up, this will have a strong effect on the requirements for future interleaver technology. Depending on available optical amplifiers, a strong need may arise for banded interleavers (see Fig. 9.2c) having a very sharp roll-off characteristic that can separate adjacent groups of wavelength channels. Alternatively, interleavers may be desired that operate in an extremely wide wavelength range of 1250 – 1650 nm, separating more than 1000 channels. Such extreme requirements will really stress both fabrication technology and device design.

References

1. S. Cao, J. Chen, J. N. Damask, C. R. Doerr, L. Guiziou, G. Harvey, Y. Hibino, H. Li, S. Suzuki, K.-Y. Wu, and P. Xie: "Interleaver technology: comparisons and applications requirements," J. Lightwave Technol. **22**, 281–289 (2004)
2. T. Mizuno, T. Kitoh, M. Oguma, Y. Inoue, T. Shibata, and H. Takahashi: "Mach–Zehnder interferometer with a uniform wavelength period," Opt. Lett. **29**, 454–456 (2004)
3. M. Oguma, T. Kitoh, K. Jinguji, T. Shibata, A. Himeno, and Y. Hibino: "Passband-width broadening design for WDM filter with lattice-form interleave filter and arrayed-waveguide gratings," IEEE Photon. Technol. Lett. **14**, 328–330 (2002)
4. D. H. Verbeek, C. H. Henry, N. A. Olsson, K. J. Orlowsky, R. F. Kazarinov, and B. H. Johnson: "Integrated four-channel Mach Zehnder multi/demultiplexer fabricated with phosphorous doped SiO$_2$ waveguides on Si," J. Lightwave Technol. **6**, 1011–1015 (1988)
5. C. G. H. Roeloffzen, F. Horst, B. J. Offrein, R. Germann, G. L. Bona, H. W. M. Salemink, and R. M. de Ridder: "Tunable passband flattened 1-from-16 binary-tree structured add-after-drop multiplexer using SiON waveguide technology," IEEE Photon. Technol. Lett. **12**, 1201–1203 (2000)
6. C. Kostrzewa and K. Petermann: "Bandwidth optimization of optical add/drop multiplexers using cascaded couplers and Mach–Zehnder sections," IEEE Photon. Technol. Lett. **7**, 902–904 (1995)
7. K. Jinguji, N. Takato, Y. Hida, T. Kitoh, and M. Kawachi: "Two-port optical wavelength circuits composed of cascaded Mach–Zehnder interferometers with point-symmetrical configurations," J. Lightwave Technol. **14**, 2301–2310 (1996)
8. M. Sharma, H. Ibe, and T. Ozeki: "Optical lattice-type add-drop multiplexing filters and their use in WDM networks," Opt. Fiber Technol. **4**, 117–134 (1998)
9. T. Liu, Y. C. Soh, Y. Zhang, and Z. Fang: "Parameter optimization of an all-fiber Fourier filter flat-top interleaver," Opt. Eng. **41**, 3217–3220 (2002)
10. G. Cincotti: "Fiber wavelet filters," IEEE J. Quantum Electron. **38**, 1420–1427 (2002)
11. Q. Wang and S. He: "Optimal design of a flat-top interleaver based on cascaded M–Z interferometers by using a genetic algorithm," Opt. Commun. **224**, 229–236 (2003)
12. Q. J. Wang, T. Liu, Y. C. Soh, and Y. Zhang: "All-fiber Fourier filter flat-top interleaver design with specified performance parameters," Opt. Eng. **42**, 3172–3178 (2003)
13. S. W. Kok, Y. Zhang, C. Wen, and Y. C. Soh: "Design of all-fiber optical interleavers with a given specification on passband ripples," Opt. Commun. **226**, 241–248 (2003)
14. T. Zhang, K. Chen, and Q. Sheng: "A novel interleaver based on dual-pass Mach–Zehnder interferometer," Microw. Opt. Technol. Lett. **42**, 253–255 (2004)
15. C. K. Madsen and J. H. Zhao: "A general planar waveguide autoregressive optical filter," J. Lightwave Technol. **14**, 437–447 (1996)
16. O. Schwelb: "Characteristics of lattice networks and spectral filters built with 2×2 couplers," J. Lightwave Technol. **17**, 1470–1480 (1999)

17. O. Schwelb: "Transmission, group delay, and dispersion in single-ring optical resonators and add/drop filters – A tutorial overview," J. Lightwave Technol. **22**, 1380–1394 (2004)
18. K. Jinguji and M. Oguma: "Optical half-band filters," J. Lightwave Technol. **18**, 252–259 (2000)
19. C. J. Kaalund and G. D. Peng: "Pole-zero diagram approach to the design of ring resonator-based filters for photonic applications," J. Lightwave Technol. **22**, 1548–1559 (2004)
20. M. M. Spühler and D. Erni: "Towards structural optimization of planar integrated lightwave circuits," Opt. Quantum Electron. **32**, 701–718 (2000)
21. C. K. Madsen and J. H. Zhao: "Postfabrication optimization of an autoregressive planar waveguide lattice filter," Appl. Opt. **36**, 642–647 (1997)
22. C. K. Madsen: "Efficient architectures for exactly realizing optical filters with optimum bandpass designs," IEEE Photon. Technol. Lett. **10**, 1136–1138 (1998)
23. G. Cincotti and A. Neri: "Logarithmic wavelength demultiplexers," J. Lightwave Technol. **21**, 1576–1583 (2003)
24. G. Cincotti and A. Neri: "Design and performances of logarithmic wavelength demultiplexers," IEEE Photon. Technol. Lett. **16**, 1325–1327 (2004)
25. R. Slavik and S. LaRochelle: "Large-band periodic filters for DWDM using multiple-superimposed fiber Bragg gratings," IEEE Photon. Technol. Lett. **14**, 1704–1706 (2002)
26. G. Lenz, B. J. Eggleton, C. R. Giles, C. K. Madsen, and R. E. Slusher: "Dispersive properties of optical filters for WDM systems," IEEE J. Quantum Electron. **34**, 1390–1402 (1998)
27. C. K. Madsen and J. H. Zhao: *Optical Filter Design and Analysis, a signal processing approach* (John Wiley & Sons, New York, 1999)
28. A. V. Oppenheim and R. W. Schafer: *Digital Signal Processing* (Prentice-Hall, Englewood Cliffs, NJ, 1975)
29. S. K. Mitra: *Digital signal processing: a computer-based approach*, 2nd edition (McGraw-Hill, New Delhi, 2001)
30. J. G. Proakis and D. G. Manolakis: *Digital signal processing: principles, algorithms, and applications*, 2nd edition (Macmillan Publishing, New York, 1992)
31. B. Moslehi, J. W. Goodman, M. Tur, and H. J. Shaw: "Fiber-optic lattice signal processing," *Proc. IEEE* **72**, 909–930 (1984)
32. M. Kuznetsov: "Cascaded coupler Mach–Zehnder channel dropping filters for wavelength-division-multiplexed optical systems," J. Lightwave Technol. **12**, 226–230 (1994)
33. H. H. Yaffe, C. H. Henry, M. H. Serbin, and L. G. Cohen: "Resonant couplers acting as add-drop filters made with silica-on-silicon waveguide technology," J. Lightwave Technol. **12**, 1010–1014 (1994)
34. K. Jinguji and M. Kawachi: "Synthesis of coherent two-port lattice-form optical delay-line circuit," J. Lightwave Technol. **13**, 73–82 (1995)
35. K. Oda, N. Takato, H. Toba, and K. Nosu: "A wide-band guided-wave periodic multi/demultiplexer with a ring resonator for optical FDM transmission systems," J. Lightwave Technol. **6**, 1016–1023 (1988)
36. S. Suzuki, M. Yanagisawa, Y. Hibino, and K. Oda: "High-density integrated planar lightwave circuits using SiO_2-GeO_2 waveguides with a high refractive index difference," J. Lightwave Technol. **12**, 790–796 (1994)

37. M. Kohtoku, S. Oku, Y. Kadota, Y. Shibata, and Y. Yoshikuni: "200-GHz FSR periodic multi/demultiplexer with flattened transmission and rejection band by using a Mach–Zehnder interferometer with a ring resonator," IEEE Photon. Technol. Lett. **12**, 1174–1176 (2000)

38. C. K. Madsen: "General IIR optical filter design for WDM applications using all-pass filters," J. Lightwave Technol. **18**, 860–868 (2000)

39. N. Takato, A. Sugita, K. Onose, H. Okazaki, M. Okuno, M. Kawachi, and K. Oda: "128-Channel polarization-insensitive frequency-selection-switch using high-silica waveguides on Si," IEEE Photon. Technol. Lett. **2**, 441–443 (1990)

40. K. Wörhoff, P. V. Lambeck, and A. Driessen: "Design, tolerance analysis and fabrication of silicon oxynitride based planar optical waveguides for communication devices," J. Lightwave Technol. **17**, 1401–1407 (1999)

41. R. Germann, H. W. M. Salemink, R. Beyeler, G. L. Bona, F. Horst, I. Massarek, and B. J. Offrein: "Silicon oxynitride layers for optical waveguide applications," J. Electrochem. Soc. **147**, 2237–2241 (2000)

42. C. Vázquez, S. E. Vargas, J. M. S. Pena, and A. B. Gozalo: "Demultiplexers for ultranarrow channel spacing based on Mach–Zehnders and ring resonators," Opt. Eng. **43**, 2080–2086 (2004)

43. B. Lyot: "Un monochromateur à grand champ utilisant les interférences en lumière polarisée," C. R. Acad. Sci. (Paris) **197**, 1593–1595 (1933)

44. I. Šolc: "Birefringent chain filters," J. Opt. Soc. Am. **55**, 621–625, (1965)

45. W. J. Carlsen and C. F. Buhrer: "Flat passband birefringent wavelength-division multiplexers," Electron. Lett. **23**, 106–107 (1987)

46. J. Zhang, L. Liu, and Y. Zhou: "A tunable interleaver filter based on analog birefringent units," Opt. Commun. **227**, 283–294 (2003)

47. J. Zhang, L. Liu, and Y. Zhou: "Novel and simple approach for designing lattice-form interleaver filter," Opt. Express **11**, 2217–2224 (2003)

48. J. Zhang, L. Liu, and Y. Zhou: "Optimum design of a novel electro-optically tunable birefringent interleaver filter," J. Opt. A: Pure Appl. Opt. **6**, 1052–1057 (2004)

49. J. Chen: "Dispersion-compensating optical digital filters for 40-Gb/s metro add-drop applications," IEEE Photon. Technol. Lett. **16**, 1310–1312 (2004)

50. F. Gires and P. Tournois: "Interféromètre utilisable pour la compression d'impulsions lumineuses modulées en fréquence," C. R. Acad. Sci. **258**, 6112–6115, (1964)

51. B. B. Dingel and M. Izutsu: "Multifunction optical filter with a Michelson–Gires–Tournois interferometer for wavelength-division-multiplexed network system applications," Opt. Lett. **23**, 1099–1101 (1998)

52. B. B. Dingel and T. Aruga: "Properties of a novel noncascaded type, easy-to-design, ripple-free optical bandpass filter," J. Lightwave Technol. **17**, 1461–1469 (1999)

53. C. H. Hsieh, R. Wang, Z. J. Wen, I. McMichael, P. Yeh, C. W. Lee, and W. H. Cheng: "Flat-top interleavers using two Gires–Tournois etalons as phase-dispersive mirrors in a Michelson interferometer," IEEE Photon. Technol. Lett. **15**, 242–244 (2003)

54. Q. J. Wang, Y. Zhang, and Y. C. Soh: "Efficient structure for optical interleavers using superimposed chirped fiber Bragg gratings," IEEE Photon. Technol. Lett. **17**, 387–389 (2005)

55. S. Yim and H. F. Taylor: "Spectral slicing optical waveguide filters for dense wavelength division multiplexing," Opt. Commun. **233**, 113–117 (2004)

56. H. Chen, P. Gu, Y. Zhang, M. Ai, W. Lv, and B. Jin: "Analysis on the match of the reflectivity of the multi-cavity thin film interleaver," Opt. Commun. **236**, 335–341 (2004)

57. J. Kim, J. Park, S. Chung, N. Park, B. Lee, and K. Jeong: "Bidirectional wavelength add/drop multiplexer using two separate MUX and DEMUX pairs and reflection-type comb filters," Opt. Commun. **205**, 321–327 (2002)

58. H. Lee and G. P. Agrawal: "Add-drop multiplexers and interleavers with broadband chromatic dispersion compensation based on purely phase-sampled fiber gratings," IEEE Photon. Technol. Lett. **16**, 635–637 (2004)

59. R. Slavík and S. LaRochelle: "All-fiber periodic filters for DWDM using a cascade of FIR and IIR lattice filters," IEEE Photon. Technol. Lett. **16**, 497–499 (2004)

60. J. T. Ahn, H. K. Lee, K. H. Kim, M. Y. Jeon, D. S. Lim, and E. H. Lee: "A stabilised fibre-optic Mach–Zehnder interferometer filter using an independent stabilisation light source," Opt. Commun. **157**, 62–66 (1998)

61. J. T. Ahn, H. K. Lee, M. Y. Jeon, D. S. Lim, and K. H. Kim: "Continuously tunable multi-wavelength transmission filter based on a stabilised fibre-optic interferometer," Opt. Commun. **165**, 33–37 (1999)

62. H. Yonglin, L. Jie, M. Xiurong, K. Guiyun, Y. Shuzhong, and D. Xiaoyi: "High extinction ratio Mach–Zehnder interferometer filter and implementation of single-channel optical switch," Opt. Commun. **222**, 191–195 (2003)

63. Y. Lai, W. Zhang, J. A. R. Williams, and I. Bennion: "Bidirectional nonreciprocal wavelength-interleaving coherent fiber transversal filter," IEEE Photon. Technol. Lett. **16**, 500–502 (2004)

64. Q. Li, C.-H. Lin, and H. P. Lee: "A novel electrically tuned all-fiber comb filter," *Opt. Fiber Commun. Conf.* (OFC'04) Techn. Digest (Los Angeles, CA, USA, 2004) Vol. **1**, 23–27 (2004)

65. Q. J. Wang, Y. Zhang, and Y. C. Soh: "All-fiber 3x3 interleaver design with flat-top passband," IEEE Photon. Technol. Lett. **16**, 168–170 (2004)

66. Q. J. Wang, Y. Zhang, and Y. C. Soh: "Design of 100/300 GHz optical interleaver with IIR architectures," Opt. Express **13**, 2643–2652 (2005)

67. C. H. Huang, H. Luo, S. Xu, and P. Chen: "Ultra-low loss, temperature-insensitive 16-channel 100-GHz dense wavelength division multiplexers based on cascaded all-fiber unbalanced Mach–Zehnder structure," *Opt. Fiber Commun. Conf.* (OFC'99), Techn. Digest (San Diego, CA, USA, 1999) Vol. **1**, 79–81 (1999)

68. C. R. Doerr, L. W. Stulz, R. Pafchek, and S. Shunk: "Compact and low-loss manner of waveguide grating router passband flattening and demonstration in a 64-channel blocker/multiplexer," IEEE Photon. Technol. Lett. **14**, 56–58 (2002)

69. C. R. Doerr, L. W. Stulz, and R. Pafchek: "Compact and low-loss integrated box-like passband multiplexer," IEEE Photon. Technol. Lett. **15**, 918–920 (2003)

70. N. Takato, K. Jinguji, M. Yasu, H. Toba, and M. Kawachi: "Silica-based single-mode waveguides on silicon and their application to guided-wave optical interferometers," J. Lightwave Technol. **6**, 1003–1010 (1988)

71. M. Oguma, T. Kitoh, Y. Inoue, T. Mizuno, T. Shibata, M. Kohtoku, and Y. Hibino: "Compact and low-loss interleave filter employing lattice-form structure and silica-based waveguide," J. Lightwave Technol. **22**, 895–902 (2004)

72. D. Di Mola, G. Sanvito, M. Lenzi, and E. Fioravanti: "Flat-band add-drop FIR lattice filter design," IEEE J. Select. Topics Quantum Electron. **5**, 1366–1372 (1999)
73. S. Kamei, M. Oguma, M. Kohtoku, T. Shibata, and Y. Inoue: "Low-loss athermal silica-based lattice-form interleave filter with silicone-filled grooves," IEEE Photon. Technol. Lett. **17**, 798–800 (2005)
74. T. Mizuno, Y. Hida, T. Kitoh, M. Kohtoku, M. Oguma, Y. Inoue, and Y. Hibino: "12.5-GHz spacing compact and low-loss interleave filter using 1.5% Δ silica-based waveguide," IEEE Photon. Technol. Lett. **16**, 2484–2486 (2004)
75. B. J. Offrein, G. L. Bona, F. Horst, H. W. M. Salemink, R. Beyeler, and R. Germann: "Wavelength tunable optical add-after-drop filter with flat passband for WDM networks," IEEE Photon. Technol. Lett. **11**, 239–241 (1999)
76. B. J. Offrein, R. Germann, F. Horst, H. W. M. Salemink, R. Beyeler, and G. L. Bona: "Resonant coupler-based tunable add-after-drop filter in silicon-oxynitride technology for WDM networks," IEEE J. Select. Topics Quantum Electron. **5**, 1400–1406 (1999)
77. B. J. Offrein, F. Horst, G. L. Bona, H. W. M. Salemink, R. Germann, and R. Beyeler: "Wavelength tunable 1-from-16 and flat passband 1-from-8 add-drop filters," IEEE Photon. Technol. Lett. **11**, 1440–1442 (1999)
78. M. H. Hu, Z. Huang, K. L. Hall, R. Scarmozzino, and R. M. Osgood: "An integrated two-stage cascaded Mach–Zehnder device in GaAs," J. Lightwave Technol. **16**, 1447–1455 (1998)
79. Y. J. Lin, S. L. Lee, and C. L. Yao: "Four-channel coarse-wavelength division multiplexing demultiplexer with a modified Mach–Zehnder interferometer configuration on a silicon-on-insulator waveguide," Appl. Opt. **42**, 2689–2694 (2003)
80. C. Kostrzewa, R. Moosburger, G. Fischbeck, B. Schüppert, and K. Petermann: "Tunable polymer optical add/drop filter for multiwavelength networks," IEEE Photon. Technol. Lett. **9**, 1487–1489 (1997)
81. B. Chen, H. Jia, J. Zhou, D. Zhao, H. Lu, Y. Yuan, and M. Iso: "Optimized design of fluorinated polyimide based interleaver," Appl. Opt. **42**, 4202–4207 (2003)
82. P. Tang, O. Eknoyan, and H. F. Taylor: "Rapidly tunable polarisation independent optical add-drop multiplexer in Ti:LiNbO$_3$," Electron. Lett. **38**, 242–244 (2002)
83. H. F. Taylor: "Tunable spectral slicing filters for dense wavelength-division multiplexing," J. Lightwave Technol. **21**, 837–847 (2003)
84. G. Cusmai, F. Morichetti, R. Costa, A. Melloni, and M. Martinelli: "An integrated optical interleaver based on a three-arms Mach–Zehnder interferometer," *Proc. 12th Europ. Conf. Integr. Optics* (ECIO'05), Grenoble, France, pp. 418–421 (2005)
85. C. van Dam, M. R. Amersfoort, G. M. ten Kate, F. P. G. M. van Ham, M. K. Smit, P. A. Besse, M. Bachmann, and H. Melchior: "Novel InP-based phased-array wavelength demultiplexer using a generalized MMI-MZI configuration," *Proc. 7th Europ. Conf. Integr. Optics* (ECIO'95), Delft, The Netherlands, pp. 275–278 (1995)
86. L. O. Lierstuen and A. Sudbø: "8-Channel wavelength division multiplexer based on multimode interference couplers," IEEE Photon. Technol. Lett. **7**, 1034–1036 (1995)
87. C. K. Madsen: "A multiport frequency band selector with inherently low loss, flat passbands, and low crosstalk," IEEE Photon. Technol. Lett. **10**, 1766–1768 (1998)

88. Y. Xiao and S. He: "An MMI-based demultiplexer with reduced cross-talk," Opt. Commun. **247,** 335–339 (2005)
89. J. D. Joannopoulos, R. D. Meade, and J. N. Winn: *Photonic crystals: molding the flow of light* (Princeton University Press, Princeton, NJ, USA, 1995)
90. S. Boscolo, M. Midrio, and C. G. Someda: "Coupling and decoupling of electromagnetic waves in parallel 2-D photonic crystal waveguides," IEEE J. Quantum Electron. **38**, 47–53 (2002)
91. J. Zimmermann, M. Kamp, A. Forchel, and R. März: "Photonic crystal waveguide directional couplers as wavelength selective optical filters," Opt. Commun. **230,** 387–392 (2004)

Appendix

A.1 Wavelength Grids

A.1.1 DWDM

The worldwide interoperability of wavelength division multiplex (WDM) fibre optic systems requires standardization of key parameters which includes recommended operation wavelengths.

One of the international bodies focusing on standardisation in fibre optic communication is the International Telecommunication Union (ITU) and in particular its Telecommunication Standardization Sector (ITU-T), see e. g. www.itu.int.

ITU-T (created on 1 March 1993) replaces the former International Telegraph and Telephone Consultative Committee (CCITT), whose origins go back to 1865. ITU-T's mission is "to ensure an efficient and on-time production of high quality standards (recommendations) covering all fields of telecommunications", and for that purpose ITU-T closely cooperates with the public and the private sectors worldwide.

ITU-T Recommendation G.694.1 specifies frequencies/wavelengths for Dense WDM (DWDM) applications. The frequency grid covering the C-band is anchored at 193.1 THz and extends from 196.1 THz (1528.77 nm) to 192.1 THz (1560.61 nm) with channel separations of 100 GHz and multiples and submultiples thereof. An extension to the L-band covers the frequency range between 191.4 THz (1566.31 nm) and 185.9 THz (1612.65 nm).

Conversion between frequencies ν and wavelengths λ can be made according to

$$\lambda = c/\nu$$

using $c = 2.997\ 924\ 58 \times 10^8$ m/s as the speed of light in vacuum.

A.1.2 CWDM

ITU-T Recommendation G.694.2 provides a wavelength grid for coarse wavelength division multiplexing (CWDM) applications. This grid supports a channel spacing of 20 nm and comprises 18 channels with wavelengths 1270 nm, 1290 nm, ... 1610 nm, as illustrated in Fig. A.1.

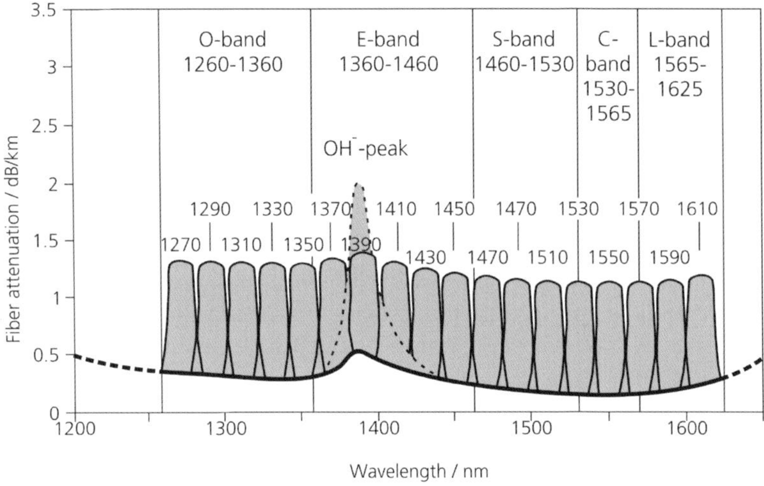

Fig. A.1. METRO CWDM wavelength grid as specified by ITU-T G.694.2

A.2 Passive Optical Networks (PONs)

Wavelength standardization for Passive Optical Networks (PONs) is developed and promoted by several standards bodies with two major groups predominating: the FSAN/ITU-T and the EFMA/IEEE.

FSAN/ITU-T: During the 1980s many of the largest carriers around the world started to develop optical access solutions for their networks. However, most of these never exceeded trial application status due to the high cost and relatively lower demand for such services at the time. With the introduction of the Internet, however, and the subsequent demand in the 1990s for more bandwidth, the need for efficient broadband access became more prevalent. Seven major telecommunications service providers established the Full Service Access Networks (FSAN) organization in 1995 to derive a common set of requirements among providers to give to equipment vendors in an attempt to accelerate the commercial deployment of optical access systems. (Currently 21 service providers are represented within FSAN.)

The most relevant corresponding standardizations are:

APON (ATM-based PON), ITU G983.1
upstream: 1260–1360 nm ("1310 nm"), (155 Mbit/s, digital);
downstream: 1480–1580 nm ("1550 nm"), (622 Mbit/s)

BPON (broadband PON), ITU G983.3
upstream: 1310 nm (622 Mbit/s, digital);
downstream: 1490 nm (622 Mbit/s, digital) and 1550 nm
                     (622 Mbit/s, analogue)

The characteristics of an enhanced BPON are compiled in Table A.1.

Table A.1. BPON WDM enhancement G.983.3 (after K. Saito and H. Ueda, "Optical access technology using FSAN B-PON technology", OPTIMIST workshop, Amsterdam, The Netherlands, 30.09.2001)

		Application examples
1.3 µm wavelength band		For use in ATM-PON upstream
lower limit	1260 nm	
upper limit	1360 nm	
Intermediate wavelength band		for future use
lower limit	1360 nm	(for allocation by ITU-T)
upper limit	1480 nm	
Basic band		For use in ATM-PON downstream
lower limit	1480 nm	
upper limit	1500 nm	
Enhancement band (option 1)		For additional digital service use
lower limit	1539 nm	
upper limit	1565 nm	
Enhancement band (option 2)		For video distribution service
lower limit	1550 nm	
upper limit	1560 nm	
Future L-band		For future use, reserved for allocation
lower limit	>1580 nm (?)	by ITU-T
upper limit	for further study	

GPON (Gigabit PON), ITU G983.4
upstream: 1260–1360 ("1310") nm, (up to 2.5 Gbit/s, digital);
downstream: 1480–1500 ("1490") nm, (up to 2.5 Gbit/s, digital)
            and 1550–1560 ("1550") nm, (622 Mbit/s, analogue)
            1539–1565 nm for digital use

A.3 Ethernet-related Standardization

The EFMA/IEEE (Ethernet in the First Mile Alliance by the Institute of Electrical and Electronics Engineers, Inc.) promotes standards-based Ethernet in the First Mile technology as a key networking technology for local subscriber access networks.

A corresponding standard, designated Wide WDM (WWDM), uses 24.5 nm channel spacing instead of 20 nm (see above). The standard originated from the attempt to allow existing cabling in buildings and campuses to be used for 10 Gigabit Ethernet (10 GbE). It is intended for both multimode and single-mode fibres, for which excessive dispersion at 10 Gbit/s single channel bit rate would be a problem. As a consequence four lanes are defined, each to be operated at a rate of 3.125 Gbit/s. These lanes are

    Lane 0   1269.0 nm – 1282.4 nm   (1275.7 nm nominal)
    Lane 1   1293.5 nm – 1306.9 nm   (1300.2 nm nominal)
    Lane 2   1318.0 nm – 1331.4 nm   (1324.7 nm nominal)
    Lane 3   1342.5 nm – 1355.9 nm   (1349.2 nm nominal)

Another standard, comparable to APON (see above) is

EPON (Ethernet-based PON), IEEE 802.11ah
symmetric 1.25 Gbit/s up/down; upstream: 1310 nm, downstream: 1550 nm

References

1.   www.rbni.com (redfern broadband networks inc.)

Glossary

Adjacent channel crosstalk (see 'crosstalk')

Attenuation (also called 'loss')
Decrease in power from one point to another. In optical fibres: measured in dB/km at a specified wavelength.

Back reflection
Ratio α_{back} of sum of light power reflected back into coupling waveguide to total input power P_{in}. P_r is the reflected power from source r.

$$\alpha_{back} = -10\log(\sum_r \frac{P_r}{P_{in}})dB \ .$$

Back scattering
The return of a portion of scattered light to the input end of a fibre with propagation direction of the back scattered light opposite to its original direction (due to Rayleigh-scattering e. g.), in dB.

Band pass
transmits light within a carefully defined wavelength range, while light outside this range is blocked.

Bandwidth

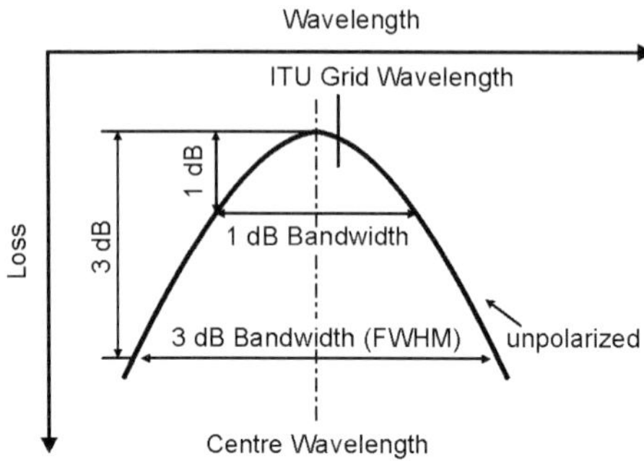

Fig. G.1. Definition of bandwidth

1 dB bandwidth

The difference between the longer and shorter wavelengths at a level 1 dB below the peak. This parameter is measured with unpolarised light input and defined as the narrowest value of all separated ports.

3 dB bandwidth (FWHM)

The difference between the longer and shorter wavelengths at a level 3 dB below the peak. This parameter is measured with unpolarised light input and defined as the narrowest value of all separated ports.

Coupling loss

The loss of optical power α_c when light is coupled from one optical device j to another device k:

$$\alpha_c = -10\log(\frac{P_k}{P_j})dB$$

with P_j, P_k being the power in device j and k, respectively.

Coupling ratio

The ratio of optical power from one output port to the total output power.

Crosstalk

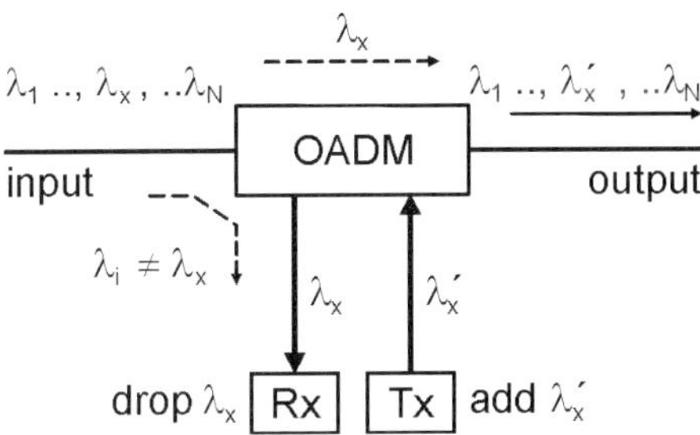

Fig. G.2. Crosstalk in an OADM where a channel of wavelength λ_x is dropped and a channel of wavelength λ_x' is added (the suffix denotes the wavelength). The dashed arrows denote the unwanted crosstalk: i.e. the interband crosstalk from channel $\lambda_i (\lambda_i \neq \lambda_x)$ into the drop port and the intraband crosstalk where the remaining field of the λ_x channel interferes with the added λ_x' channel at the output

Intrachannel crosstalk
Crosstalk due to unwanted signals with identical wavelength or wavelength separation smaller than the channel width of the filter defining the desired signal (e. g. due to cascaded demux/mux or at the output of add/drop multiplexers). Sometimes also called in-band crosstalk or coherent crosstalk.

Interchannel crosstalk
Crosstalk due to unwanted signals with wavelength separation exceeding the channel width of the filter defining the desired signal. Sometimes also called out-of-band crosstalk.

Far-end crosstalk (also called 'crosstalk')
Optical crosstalk C_{ik} of wavelength channel i on channel k

$$C_{ik} = -10\log(\frac{P_{ik}}{P_{kk}})$$

where P_{ik} is the residual optical power of wavelength i in channel k and P_{kk} is the power at wavelength k in channel k. If i and k are adjacent channels, the crosstalk is called **adjacent channel optical crosstalk** and P_{ik} is taken as the highest output power of the longer or shorter wavelength channel. For the **non-adjacent channel optical crosstalk**, P_{ik} is given by the channel with highest optical output power in the nonadjacent channels.

The **total optical crosstalk** C_i^t in channel i in a multichannel configuration is

$$C_i^t = -10\log(\sum_{k \neq i} \frac{P_{ik}}{P_{kk}})dB\,.$$

Near-end crosstalk or directivity
Ratio of light power emitted from a non-excited input waveguide to the power guided in (an) excited waveguide(s) of a multiple-input device such as a mux/demux or an optical isolator

$$\alpha_{dir} = -10\log(\frac{P_{in,nonex}}{P_{in,exc}})dB$$

where $P_{in,exc}$ is the power launched into one input waveguide of a multiport device and $P_{in,nonex}$ is the power measured at another input waveguide.

Excess loss
Ratio of sum of light power at all output waveguides to light power in the input waveguide

$$\alpha_{excess} = -10\log(\sum_{i} \frac{P_{out,i}}{P_{in}})dB$$

where P_{in} is the power in the excited input waveguide, and $P_{out,i}$ is the output power in waveguide i.

Extinction ratio of a modulator
Ratio of output power for the mark P_1 to output power for the space P_0

$$f_{ext} = 10\log\left(\frac{P_1}{P_0}\right)dB\,.$$

Inband crosstalk see 'intrachannel crosstalk'

Insertion loss
The loss of power α_{ins} in dB that results from inserting a component, such as a filter, mux/demux, connector, or splice, into a previously continuous path:

$$\alpha_{ins} = -10\log(\frac{P_2}{P_1})dB$$

with P_1 (P_2) being the optical power before (after) inserting the component.

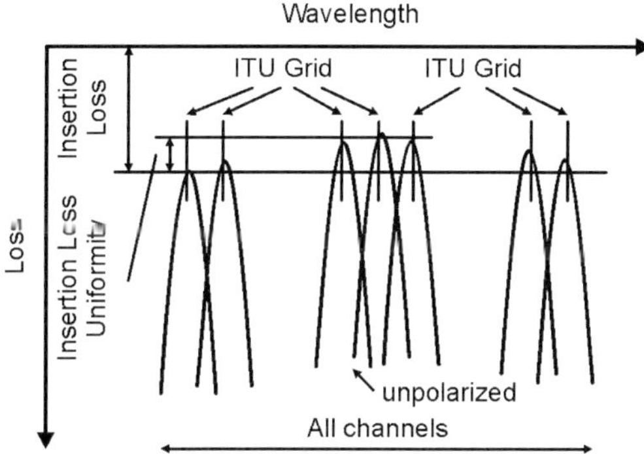

Fig. G.3. Insertion loss and insertion loss uniformity

Insertion loss uniformity
The difference between the minimum and maximum optical losses at the wavelengths of a multiport device.

ITU grid wavelength
The wavelength per ITU grid reference ITU-T G.692 (cf. Appendix, Sect. A.1.1).

Long-wave pass, (long-pass filter)
transmits longer wavelengths and reflects shorter wavelengths (opposite: short-wave pass).

Loss (see 'attenuation').

Non-adjacent channel crosstalk (see 'crosstalk').

Notch filter
reflects a narrow wavelength region with high transmission outside that region.

Optical return loss
The ratio of optical power reflected by a component or an assembly to the optical power incident on a component port when that component or assembly is introduced into a link or system.

Out-of-band crosstalk see 'interchannel crosstalk'

Polarisation

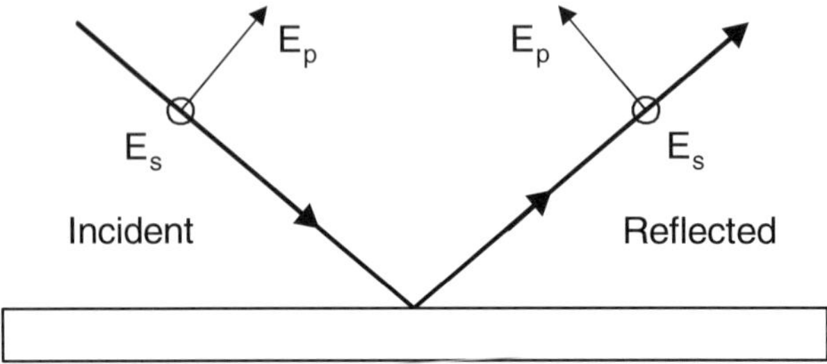

Fig. G.4. Definition of s- and p-polarisation (E_s perpendicular to the plane of the paper)

s-polarisation or TE-polarisation: electric field vector perpendicular to plane of incidence,
p-polarisation or TM-polarisation: electric field vector parallel to plane of incidence.

Polarisation-dependent loss

Ratio α_{PDL} of lowest power P_{low} transmitted at any state of polarisation to highest power P_{high} transmitted at any state of polarisation:

$$\alpha_{PDL} = -10\log(\frac{P_{low}}{P_{high}})dB .$$

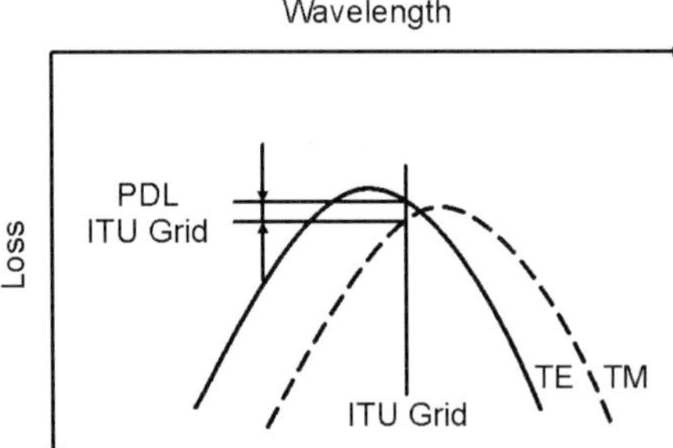

Fig. G.5. Polarisation-dependent loss

Polarisation-dependent shift

Shift of centre wavelength due to polarisation.

Polarisation mode dispersion

Pulse spreading caused by locally varying birefringence in a device or fibre.

Short-wave pass (short-pass filter)

transmits shorter wavelengths and reflects longer wavelengths (opposite: long-wave pass, long-pass filter).

Uniformity

Ratio α_{uni} of lowest power P_{min} emitted from an output waveguide to highest power P_{max} emitted from an output waveguide of a multiple-output device or optical element:

$$\alpha_{uni} = -10\log(\frac{P_{min}}{P_{max}})dB \ .$$

Wavelength isolation

Isolation of light signal in the desired optical channel from the unwanted channels. Synonym: far-end crosstalk.

Biographical Notes

Alfred Driessen (*1949) obtained his M.Sc. degree in Physics and his Ph.D. in experimental physics on Quantum Solids from the University of Amsterdam. After a period as postdoc at the Free University Amsterdam on Metal Hydrides he joined the Lightwave Device (now Integrated Optical MicroSystems) Group at the University of Twente as associate professor in 1988. His working field from then is integrated optics for optical communication. Since 2003 he is full professor of IOMS and his interest is increasingly focused on nanophotonic structures, like micro resonators and photonic wires that eventually could lead to VLSI photonics. He is (co-)author of more than 180 refereed journal and conference papers and holds 3 patents.

Douwe H. Geuzebroek received an MSc degree in Electrical Engineering at the University Twente, the Netherlands in 2001 on the field of optical telecommunication networks. After that he started his Ph.D. research at the Integrated Optical MicroSystems group of the same university and received a Ph.D. degree in 2005 with a thesis focusing on flexible optical network components based on densely integrated microring resonators. He is currently involved in the start-up phase of a new company called Xio-Photonics, that will commercialize the microring resonator technology.

Charles A. Hulse received his Bachelor's and Master's Degrees in electrical engineering from the University of California, Davis. Charles joined the Optical Coating Laboratory, Inc. (OCLI), in 1995, and has held a variety of technical and leadership roles in coating development, optics, and laser development. Charles is a member of the Laser and Optics development group in the Science and Technology department of JDSU.

Kyriacos Kalli, C.Phys., M.Inst.P., received the B.Sc. (Hons) in Theoretical Physics (1988) and Ph.D. in Physics (1992) from the University of Kent at Canterbury. His Ph.D. thesis investigated linear and non-linear optical phenomena using high finesse ring resonators. In 1993 he was a visiting research scholar with the Fiber and Electro-Optics Research Center, Virginia Tech, investigating optical fibre-based intruder detection schemes and temperature sensors for on-board-ship fire monitoring. From 1994 - 1996 he undertook research at the University of Kent into the use of fibre Bragg gratings in multiplexed sensor arrays and Raman spectroscopy for pollution monitoring of ground water. He joined the University of Cyprus in 1996, engaged in research into integrated gas flow and gas sensors based on porous silicon micromachining, fluorescence spectroscopy for environmental pollution studies, and non-destructive evaluation of semiconductors using photoreflectance and photothermal measurements. Since 2001 he is a Permanent Lecturer of Physics at the Higher Technical Institute in Cyprus, where he heads the Photonics Research Laboratory. His research interests are in Bragg grating and optical fiber sensors, photonic switching devices, laser material interactions and photosensitivity in glass and polymer materials. He has more than 80 journal and conference publications and several book contributions. Dr. Kalli is a member of the Institute of Physics, Optical Society of America and Institute of Electrical and Electronics Engineers.

Ton Koonen is a full professor in the Electro-Optical Communication Systems group, a member of the COBRA Institute, at Eindhoven University of Technology in the Netherlands since 2001. Prior to that, he spent more than 20 years at Bell Labs in Lucent Technologies as a technical manager of applied research. He also is a Bell Labs Fellow (the first in Europe) since 1999. Next to his industrial position, he has been a part-time professor in Photonic Networks at Twente University from 1991 to 2000. His main interests are currently in broadband fibre access networks and optical packet-switched networks. He has initiated and led several European and national R&D projects in this area, a.o. in dynamically reconfigurable hybrid fibre access networks for fibre-to-the-curb and fibre-to-the-building, and on label-controlled optical packet routing. He also is involved in radio over fibre techniques, and in techniques for short-range polymer optical fibre networks. He (co-)authored more than 200 conference and journal papers.

Berndt Kuhlow received the diploma and Ph.D. degree from the Technical University of Berlin (TUB) in 1969 and 1974, respectively. Working for the Federal Institute for Materials Research and Testing (BAM) from 1974-76 and as an assistant professor in physics at the TUB from 1976-1982, he joined HHI in 1983, where he first investigated optical light valve systems for HDTV projection followed by research on optical signal processing systems. His current work includes free space optics and planar waveguide optics for optical communication systems. He is (co)author of over 80 scientific publications. Since 1996 he holds a professorship in

Physics at the TUB. He was awarded the Karl-Scheel Preis of the German Physical Society in 1976.

Jean-Pierre Laude received an engineering degree from the Ecole Supérieure d'Optique de Paris in 1963 and a doctorate in spectrometry from Laboratoire A. Cotton, Orsay University in 1966. From 1963 to 1966 he was a lecturer and researcher at the Centre National de la Recherche Scientifique, director of the grating department laboratory of Jobin Yvon from 1966 to 1974, and from 1974 to 2000 research director of Jobin Yvon that became part of the Horiba group in 1997. Since 2000 he is an independent researcher and scientific consultant. J. P. Laude is Fellow of SPIE and the current President of the SPIE Technology Achievement Award Committee. He holds 50 patents and has published extensively in optics and spectroscopy (e. g. books on WDM: Masson 1991, Prentice Hall 1992, Artech 2002). He received several technology awards : 1972 Paris, Centre National d'Etudes Spatiales Medal (Gratings for space missions). 1997 San Diego, SPIE Technology Achievement Award (WDM and gratings). 1995 Baltimore, Photonics Circle of Excellence Award with K. Liddane (DWDM). 1996 Baltimore, Lasers and Optronics Award with S. Slutter (wavelength routers). 2001 Nice, Information Society Technology Prize (Ultra Dense WDM components).

Xaveer J.M. Leijtens (1962) studied physics at the University of Amsterdam. He received his master´s degree with honours in 1988 and his Ph.D. degree in 1993 on research in the field of high energy physics. In 1993 he joined the Photonic Integrated Circuits group at the Delft University of Technology. At present he works at the COBRA research institute in the Opto-Electronic Devices group at the Eindhoven University of Technology. His work on Photonic Integrated Circuits includes research on Arrayed Waveguide Grating (de)multiplexers, crossconnects, multiwavelength receivers and transmitters and the development of a CAD-tool for photonic ICs. Xaveer Leijtens is chairman of the board of the IEEE-LEOS Benelux chapter and he (co-)authored over 100 scientific publications and conference contributions.

Alain Mugnier Alain Mugnier was born on April 8, 1972 in Lons-le-Saunier, France. He graduated from the Ecole Supérieure d'Optique (French engineering school in optics) in 1996. In 2000 he joined Highwave Optical Technologies, where he developed fibre Bragg grating-based devices for advanced optical communications systems (gain flattening filters, tuneable dispersion compensators). He was later involved in the development of high-power fibre lasers for scientific and industrial applications.

Currently he is senior R&D engineer in the fibre laser department of Quantel.

Andreas Othonos received his B.Sc. in theoretical physics (1984) and M.Sc. in high energy physics (1986) from the University of Toronto. Between 1986-1990 he conducted research in the area of Ultrafast Laser Semiconductor Interactions at the the National Research Council of Canada (NRC) and the University of Toronto where he was awarded a Ph.D. degree. From 1990 to 1996 he was with the Ontario Laser and Lightwave Research Center at the University of Toronto, working on ultrafast dynamics in semiconductors and fiber photosensitivity. Since 1996 he has been with the Department of Physics at the University of Cyprus and a principle investigator of the Research Center of Ultrafast Science. His current research interests involve ultrafast laser induced dynamics, laser-matter interactions, nonlinear optics, laser physics, fiber optic photosensitivity, and telecommunications. He has published more than 70 journal papers along with several invited reviews and chapters in books.

Christophe Peucheret received the Dipl.-Ing. degree from Ecole Nationale Supérieure des Télécommunications de Bretagne, Brest, France in 1994, the M.Sc. in Microwaves and Optoelectronics from University College London, also in 1994, and the Ph.D. degree from the Technical University of Denmark in 2003. He has been involved in research and education in the field of optical communication systems at the Department of Electromagnetic Systems, then Research Center COM (now COM•DTU) at the Technical University of Denmark since 1997. His research interests are in the area of signal generation and transmission. In this context he has been involved in studies on filtering effects in WDM networks, with special emphasis on the influence of filter dispersion.

David Pureur was born in 1970 in the north of France. He received his Ph.D. in Physics from the University of Lille in 1997 and then worked at Stanford University (USA) as an invited post-doc. After that he joined Highwave Optical Technologies in 1998 as an R&D engineer. He was promoted to director of R&D in 2000 and led all product development of the company, and became Chief Technology officer of the company in 2003. In 2006 David joined the Quantel group as R&D manager. During his tenure at Highwave. David led research and development on specific fibres, fibre Bragg gratings, non linear effects, Erbium-doped fibres and fibre lasers, and he has the same focus with Quantel. David has more than 12 years of experience in the telecom, lasers, and sensors industries. He has written more than 80 international papers, conference contributions, and patents. He is also Technical Committee member of international conferences. David is married and has two daughters.

René M. de Ridder was born 1950 in Amsterdam, the Netherlands. He studied Electrical Engineering at the University of Twente, Enschede, The Netherlands, where he obtained a masters degree in 1978 and a Ph.D. in 1988. His thesis subject was on sensitive thin-film magnetic-field sensors. After that he turned to integrated optical devices for optical communication. A basis for his work in this field was laid during a half year of sabbatical leave, spent in 1989 at (then) AT&T Bell Labs, Holmdel, NJ, USA. Since 1981 René de Ridder is employed as a lecturer and research scientist at the University of Twente, where he is now a member of the Integrated Optical MicroSystems group in the MESA+ Research Institute. His current research interests include photonic crystal structures and wavelength-selective devices.

Chris Roeloffzen was born in Almelo, The Netherlands, in 1973. He received the M.Sc. degree in applied physics and a Ph.D. degree in electrical engineering from the University of Twente, Enschede, The Netherlands, in 1998 and 2002, respectively. From 1998 to 2002 he was engaged in research on integrated optical add-drop demultiplexers in Silicon Oxinitride waveguide technology, in the Integrated Optical MicroSystems Group at the University of Twente. In 2002 he became an Assistant Professor in the Telecommunication Engineering Group at the University of Twente. He is presently involved in research and education on optical fiber communications systems. His current research interests are optical signal processing, optical beam forming networks, optical wavelength filtering, and optical dispersion compensation.

Robert B. Sargent received his Bachelor's Degree in Physics from the University of California, Berkeley, and his Ph.D. in Optical Sciences from the University of Arizona, Tucson. Robert has been with OCLI and JDSU for 16 years, serving in a variety of engineering and leadership roles. His work has concentrated on the development of energetic thin film deposition processes for manufacturing filters used in applications such as optical instrumentation and wavelength division multiplexing (WDM) for fibre optic telecommunications. Robert is currently a program manager in JDSU's Optics and Display Products group, and is working on the introduction of a new fast-cycle filter deposition platform.

Meint K. Smit was born in Vlissingen, the Netherlands, in 1951. He graduated in Electrical Engineering at the Delft University of Technology in 1974, and received his Ph.D. in 1991, both with honours. In 1974 he started work on radar and radar remote sensing and joined the Delft University of Technology in 1976. He switched to optical communications in 1981, focusing on the research field: Integrated Optics. In 1994 he became leader of the Photonic Integrated Circuits group of the Delft University. He is the inventor of the Arrayed Waveguide Grating, for which he received a LEOS Technical Achievement award in 1997. He was appointed professor in 1998, and in that year his group, together with the groups of prof. Wolter and prof. Khoe (COBRA, TU Eindhoven) received a research grant of 40 M€ for establishing a National Research Center on Photonics. In 2002 he moved with his group to the Technical University of Eindhoven, where he is presently leader of the Opto-Electronic Devices group at the COBRA Research Institute. In 2002 he was appointed LEOS Fellow for contributions in the field of Opto-Electronic Integration.

Markus K. Tilsch received his Ph.D. in physics from the University of Technology, Darmstadt, Germany in 1997. His thesis was concerned with precision optical coatings produced through ion beam sputtering. He joined Optical Coating Laboratory, Inc. (OCLI), Santa Rosa, CA, USA in 1998, which was acquired by JDSU in 2000. He held research and development positions in thin-film filter developments for telecommunications, visual, and infrared coatings. Currently he is managing the coating process development group in the Science and Technology department of JDSU.

Carl-Michael Weinert received his physics diploma in 1975 at the Ludwigs-Maximilian-Universität München and his Dr. rer. nat. degree in 1980 from the Technische Universität Berlin where he was engaged in theoretical work on deep impurities in semiconductors. After his work as postdoc at Freie Universität Berlin and the Physikalisch Technische Bundesanstalt in Braunschweig he joined the Integrated-Optics Division at Heinrich-Hertz-Institute in 1987. He is engaged in work on modelling and numerical simulations of photonic devices and photonic crystal components. His current work also includes investigations of optical signal processing and optical communication systems.

Index

Springer Series in
OPTICAL SCIENCES